Environmental Mutagens and Carcinogens

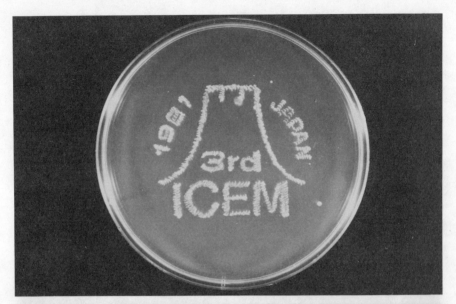

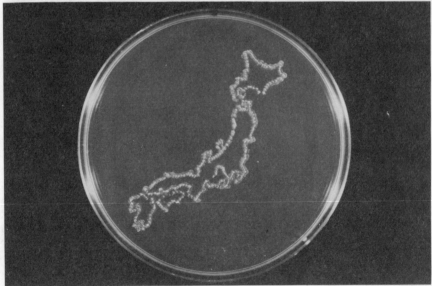

These "mutagenic drawings" were obtained by suitably streaking sensitive bacteria (*S. typhimurium*, strain TA98) on the surface of top agar incorporating a promutagen (2-aminofluorene), without S-9 mix. 2-Aminofluorene had been previously exposed either to ultraviolet light (upper plate [ICEM]) or to sunlight (lower plate [JAPAN]). For further details on this subject see the article by De Flora in this volume (pp. 527-542).

Environmental Mutagens and Carcinogens

Proceedings of the 3rd International Conference on Environmental Mutagens, Tokyo, Mishima and Kyoto, September 21-27, 1981

Editors
TAKASHI SUGIMURA
National Cancer Center, Tokyo

SOHEI KONDO
Osaka University, Osaka

HIRAKU TAKEBE
Kyoto Univesity, Kyoto

RC 268.6
I 58
1981

Alan R. Liss, Inc.

Published in North, Central, and South America, Europe, including The United Kingdom, The Middle East, and Africa by Alan R. Liss, Inc., 150 Fifth Avenue, New York, N.Y.10011, U.S.A.

Library of Congress Cataloging in Publication Data

Main entry under title:

Environmental mutagens and carcinogens.

   International Conference on Environmental Mutagens and Carcinogens (3rd: 1981: Tokyo, Japan, etc.)
   Includes index.
   1. Carcinogens—Congresses. 2. Chemical mutagenesis—Congresses. 3. Environmentally induced diseases—Congresses.
I. Sugimura, Takashi. II. Kondo, Sohei, 1922- III. Takebe, Hiraku, 1934- IV. Title.
RC268.6.I58   1981   616'.042   82-15231
ISBN 0-8451-3007-2   AACR2

© UNIVERSITY OF TOKYO PRESS, 1982
UTP 3047-68672-5149

All rights reserved. No part of this publication may be reproduced or transmitted in any form or by any means, electronic or mechanical, including photocopy, recording, or any information storage and retrieval system, without permission in writing from the publisher.

Jointly published by
University of Tokyo Press, Tokyo

and

Alan R. Liss, Inc., New York

Printed in Japan

Contents

Preface

IAEMS President's Address

PLENARY AND SPECIAL LECTURES

A View of a Cancer Researcher on Environmental Mutagens
  Takashi Sugimura 3
A History of Attempts to Quantify Environmental Mutagenesis
  Alexander Hollaender 21
A View from the Ivory Tower
  Charlotte Auerbach 37
Some DNA Repair-Deficient Human Syndromes and Their Implications for Human Health
  B. A. Bridges 47
Factors Affecting Mutagenicity of Ethylnitrosourea in the Mouse Specific-Locus Test and Their Bearing on Risk Estimation
  W. L. Russell 59
Role of DNA Damage in Radiation and Chemical Carcinogenesis
  Arthur C. Upton 71
The First Five Years of ICPEMC; The International Commission for Protection against Environmental Mutagens and Carcinogens
  F. H. Sobels and J. Delehanty 81
A Brief Sketch of Environmental Mutagen Studies in Japan
  Yataro Tazima 91
Twenty Years Later! (Comments on Tazima's lecture)
  Alexander Hollaender 101

MECHANISMS OF MUTAGENESIS

Mechanisms of Mutagenesis in Bacteria
  D. W. Mount, J. W. Little, B. Markham, H. Ginsburg, C. Yanisch, and S. Edmiston 105

Plasmid Mediated Enhancement of Chemical Mutagenesis
   Karen L. Perry, Stephen J. Elledge, Marilyn R. Lichtman, and Graham C. Walker 113

Non-DNA Primary Targets for the Induction of Genetic Change
   R. H. Haynes, J. G. Little, B. A. Kunz, and B. J. Barclay 121

Mechanisms of Induced Mutagenesis in Yeast
   Christopher Lawrence 129

The Importance of Mutant Yield Data for the Quantification of Induced Mutagenesis
   R. H. Haynes, F. Eckardt, B. A. Kunz, and W. Göggelmann 137

Examining the Mechanism of Mutagenesis in DNA Repair-Deficient Strains of *Drosophila melanogaster*
   P. Dennis Smith, Ruth L. Dusenbery, Sheldon F. Cooper, and Clifford F. Baumen 147

Mechanisms of Mutagenesis in Cultured Mammalian Cells
   Roger Cox 157

MUTATION IN GERM CELLS

Germ-Cell Sensitivity in the Mouse: A Comparison of Radiation and Chemical Mutagens
   Antony G. Searle 169

Dependence of Mutagenesis on Germ Cell Stage and Sex
   F. H. Sobels 179

Dependence of Mutagenesis in *Drosophila* Males on Metabolism and Germ Cell Stage
   Ekkehart W. Vogel 183

MUTAGENESIS AND CARCINOGENESIS

Comparison of Mutagenesis and *in vitro* Transformation Models in Detecting Potential Carcinogens
   Verne A. Ray 197

Correlations between Mutagenesis and Neoplastic Transformation in *in vitro* Systems
   Leonard M. Schechtman and Richard E. Kouri 209

Free Oxygen Radicals— Modulator of Carcinogens
   Walter Troll 217

Role of DNA Damage and Repair in Carcinogenesis
   Taisei Nomura 223

Increased Cancer Incidence in the Progeny of Male Rats Exposed to Ethylnitrosourea before Mating
    L. Tomatis, J. R. P. Cabral, A. J. Likhachev, and V. Ponomarkov... 231

TOXICOLOGY OF ENVIRONMENTAL MUTAGENS

Introduction to the Symposium on Toxicology of Environmental Mutagens
    René Truhaut 241

Place of Mutagenicity Tests in Toxicological Evaluation of Chemicals Likely to Be Present in the Environment
    Ivan Chouroulinkov... 249

Mutagenic Hazard and Genetic Risk Evaluation on Environmental Chemical Substances
    Nicola Loprieno.. 259

Classification of Carcinogens as Genotoxic and Epigenetic as Basis for Improved Toxicologic Bioassay Methods
    John H. Weisburger and Gary M. Williams 283

What Is the Predictive Value of Short-Term Mutagenicity Testing to Determine Carcinogenic Potential of Compounds?
    Dietrich Schmähl and Beatrice Luise Pool 295

MUTAGENS AND CARCINOGENS IN THE ENVIRONMENT

Overview of Mutagens and Carcinogens in the Environment
    Lawrence Fishbein... 307

Styrene—A Widespread Mutagen: Conclusions from the Results of Testing
    Gösta Zetterberg... 315

Detection of Worker Exposure to Mutagens in the Rubber Industry by Use of the Urinary Mutagenicity Assay
    Marja Sorsa, Kai Falck, and Harri Vainio 323

Mutagenicity Screening Studies on Pesticides
    Yasuhiko Shirasu, Masaaki Moriya, Hideo Tezuka, Shoji Teramoto, Toshihiro Ohta, and Tatsuo Inoue 331

Knowledge Gained from the Testing of Large Numbers of Chemicals in a Multi-Laboratory, Multi-System Mutagenicity Testing Program
    Errol Zeiger 337

MODIFICATION OF MUTAGENESIS

Enhancement and Suppression of Genotoxicity of Food by Naturally Occurring Components in These Products

Hans F. Stich, Chiu Wu, and William Powrie 347

Mechanisms and Genetic Implications of Environmental Antimutagens
Tsuneo Kada 355

Antimutagens and the Problem of Controlling the Action of Environmental Mutagens
Urkhan Alekperov 361

CHEMICAL STRUCTURE OF MUTAGENS

An Overview of the Structural Features of Mutagenic S-Chlorallylthio and Dithiocarbamate Pesticides and Trichlorfon, Dichlorvos and Their Metabolites
Lawrence Fishbein 371

The Dependence of Mutagenic Activity on Chemical Structure in Two Series of Related Aromatic Amines
Majdi M. Shahin 379

Arylamine-DNA Adduct Formation in Relation to Urinary Bladder Carcinogenesis and *Salmonella typhimurium* Mutagenesis
F. F. Kadlubar, F. A. Beland, D. T. Beranek, K. L. Dooley, R. H. Heflich, and F. E. Evans 385

MUTAGENESIS IN HIGHER PLANTS

Mutagenicity Testing of Chemical Mutagens in Higher Plants
Taro Fujii 399

Maize as a Monitor for Environmental Mutagens
Michael J. Plewa 411

CYTOGENETICS

The Role of Cytogenetics in Evaluation of Human Genetic Risk
Nikolay P. Bochkov 423

Basis and Biological Significance of Sister Chromatid Exchange
S. A. Latt, R. R. Schreck, E. Sahar, I. J. Paika, and C. Kittrel 431

Virus Induced Cellular Genetic Changes
Warren W. Nichols 443

Environmental Mutagens and Karyotype Evolution in Mammals
Tosihide H. Yosida 455

The Induction of Chromosome Aberrations in Germ Cells: A Discussion of Factors That Can Influence Sensitivity to Chemical Mutagens

R. Julian Preston 463

Genetically Determined Chromosomal Instability in Man
   Masao S. Sasaki 475

Formation of Chromatid-Type Aberrations from Chemically Induced DNA Lesions: Information Obtained with Hydroxyurea and Caffeine
   B. A. Kihlman, H. C. Andersson, K. Hansson, B. Hartley-Asp, and F. Palitti 483

METABOLISM OF MUTAGENS AND CARCINOGENS

Metabolic Rationale for the Weak Mutagenic and Carcinogenic Activity of Benzo(e)pyrene
   James K. Selkirk, Michael C. MacLeod, Betty K. Mansfield, and Anne Nikbakht... 497

N-Hydroxylation of Mutagenic Heterocyclic Aromatic Amines in Protein Pyrolysates by Cytochrome P-450
   Ryuichi Kato, Yasushi Yamazoe, Kenji Ishii, and Tetsuya Kamataki.. 505

Microsomal Metabolic Activation, Mutagenicity and Antimutagenicity
   Marcel Roberfroid, Chehab Razzouk, François Hervers, and Heinz Günter Viehe 511

Modulation of Mutagenesis by Biological Substances
   Hikoya Hayatsu... 521

Biotransformation and Interaction of Chemicals as Modulators of Mutagenicity and Carcinogenicity
   Silvio De Flora.. 527

MUTAGENS AND CARCINOGENS IN THE DIGESTIVE TRACT

Polynuclear Aromatic Hydrocarbons and Catechol Derivatives as Potential Factors in Digestive Tract Carcinogenesis
   Stephen S. Hecht, Steven Carmella, Keizo Furuya, and Edmond J. La Voie... 545

Approaches to the Identification of Human Colon Carcinogens
   Michael C. Archer, Mark T. Goldberg, and W. Robert Bruce.... 557

Endogenous Formation of N-Nitroso Compounds and Gastric Cancer
   S. R. Tannenbaum, P. Correa, P. M. Newberne, and J. G. Fox ... 565

Ingestion and Excretion of Nitrate in Humans and Experimental Animals
   Hajimu Ishiwata and Akio Tanimura 571

Quantitative Estimation of Endogenous Nitrosation in Humans by Measuring Excretion of N-Nitrosoproline in the Urine

Hiroshi Ohshima and Helmut Bartsch 577

GENETIC FACTORS AND EPIDEMIOLOGY

Genetic Factors in the Response to Environmental Mutagens
    H. J. Evans.. 589

Diseases of Environmental-Genetic Interaction: Preliminary Report on a Retrospective Survey of Neoplasia in 268 Xeroderma Pigmentosum Patients
    Kenneth H. Kraemer, Myung M. Lee, and Joseph Scotto 603

Surveillance of Human Populations for Germinal Cytogenetic Mutations
    Ernest B. Hook and Philip K. Cross........................... 613

Towards Documenting Human Germinal Mutagens: Epidemiologic Aspects of Ecogenetics in Human Mutagenesis
    John J. Mulvihill 625

Epidemiological Follow-Up Study on Mutagenic Effects in Self-Poisoning Persons
    Andrew Czeizel 639

The Role of the Epidemiologist in Evaluating Human Genetic Risk
    James R. Miller 647

The Clinician's Role in Monitoring for Diseases from Chemicals in the Environment
    Robert W. Miller 655

RISK ESTIMATES AND ASSESSMENT

Carcinogenic Potency
    Bruce N. Ames, Lois S. Gold, Charles B. Sawyer, and William Havender 663

The Alkylation of Macromolecules as a Basis for the Evaluation of Human Genetic Risk
    Carl Johan Calleman 671

Radiation as a Mutagen/Carcinogen Model and a Regulatory Norm
    Per Oftedal 679

Hiroshima and Nagasaki: Three and a Half Decades of Genetic Screening
    William J. Schull, James V. Neel, Masanori Otake, Akio Awa, Chiyoko Satoh, and Howard B. Hamilton...................... 687

A Model for Low Dose Risk Estimates
    Carl Johan Calleman 701

Risk Estimations Based on Germ-Cell Mutations in Mice

U. H. Ehling 709

The Mutational Specificity of Gamma-Rays Is Influenced by Dose: Implications for Thresholds and Risk Estimation
   Barry W. Glickman 721

Apparent Threshold and Its Significance in the Assessment of Risks Due to Chemical Mutagens
   Yataro Tazima. 729

Thresholds and Negative Slopes in Dose-Mutation Curves
   R. C. von Borstel 737

INTERNATIONAL COOPERATION AND PROBLEMS IN DEVELOPING COUNTRIES

Problems of Environmental Mutagens and Carcinogens in Developing Countries: Introduction to Panel Discussion
   Per Oftedal 745

Evaluation of Short-Term Tests for Mutagenicity in the International Program for Chemical Safety
   Frederick J. de Serres 747

Problems of Environmental Mutagens and Carcinogens in Developing Countries: Synthesis of the Problems
   C. Ramel. 753

Southeast Asian Conference on the Detection and Regulation of Environmental Mutagens, Carcinogens and Teratogens
   Clara Y. Lim-Sylianco 759

A Report of the Southeast Asian Workshop on the Methods for Detecting Environmental Mutagens and Carcinogens, February 4-17, 1979
   Sumin Smutkupt 765

Environmental Mutagenesis and Carcinogenesis in Latin America
   Cristina Cortinas de Nava 771

Author Index 777

Preface*

Scientists working on environmental mutagens gathered together on the occasion of the 3rd International Conference on Environmental Mutagens which was held on September 21-27, 1981, in Japan. It was a wonderful opportunity to discuss in depth the issue of environmental mutagens. We hope that everyone made the most of this splendid opportunity to exchange valuable information and perhaps even make some breakthroughs in this field of research.

Our scientists can take pride in the remarkable accomplishments already achieved in the short history of this field. Studies on environmental mutagens cover various areas. Environmental mutagens affect germ cells as well as somatic cells of humans. Not only humans, but also bacteria, plants and animals, are continuously exposed to environmental mutagens, and biological evolution has occurred in the presence of enormous numbers of mutagens. What role have environmental mutagens played in this evolution? Have mutagens always had adverse effects on creatures? Studies on environmental mutagens are intimately involved with the central issues of life science.

In addition, studies on environmental mutagens are very important for devising ways to protect humans from hazardous effects that may produce genetic diseases, birth defects, cancer and aging. Thus scientists in the field of environmental mutagens have the great responsibility of providing the general public and their governments with precise information for procedures to ensure that the health of all humanity is protected from potentially harmful agents.

We on the Organizing Committee in Japan were very glad to have been asked by the International Association of Environmental Mutagen Societies to hold this 3rd International Conference on Environmental Mutagens in Japan. The Conference was held in Tokyo and Mishima, so that the conference members had a chance to see a big city like Tokyo and a small city like Mishima, where traditional Japanese life still remains. We also held an official Satellite Meeting in Kyoto, a historical city where the traditional culture of Japan survives to this day.

I am greatly indebted to all the members of the Organizing Committee. I also wish to thank the various societies, industrial enterprises and many individual persons for the generous financial support that made this conference possible.

*This preface is mainly based on a speech given at the opening ceremony of the Conference.

This book contains plenary and special lectures, symposium papers and selected papers presented at panel discussions and contributor sessions of the Conference and has been edited according to the contents of the papers.

I think that this volume of Proceedings reflects the fruitful scientific outcome of the 3rd International Conference on Environmental Mutagens.

October 1, 1981

Takashi Sugimura
President
Third International Conference
on Environmental Mutagens

IAEMS President's Address

Your excellency Mr. Murayama, your honor Dr. Kurokawa, President Dr. Sugimura, Dr. Auerbach, Dr. Hollaender, Friends and Colleagues. It is a great and profound pleasure to participate in the opening of the Third International Conference on Environmental Mutagens.

It seems like a very short time ago that we met in Edinburgh— and even in Asilomar—where it all started. And yet the development of our science during these years has been tremendous. The International Association now has six regional societies, with a total of close to two thousand members. We have seen a proliferation — and some elimination — of test systems, and an ever growing stream of test results. The mapping of our total environment for mutagens has been going at a tremendous rate. At the same time, a great deal has become known of the mutagenic process, but admittedly our understanding and identification of underlying general principles is still spotty and incomplete.

Our warnings against the threat from mutagens has coincided with the general awakening to the reality of our planet's limits to growth. Our test methods and our concepts of carcinogenic and mutagenic effects have become everyday words in newspapers as well as in regulatory statements.

Concern today is largely about the carcinogenic effects—which mainly affect older people. Genetic effects may be catastrophic but are given less attention. They need attention and more knowledge.

Conferences like this are like fish eye photographic cameras: the whole field of our science is brought together in one picture. Accents, contrasts, profiles and similarities are brought into focus and made recognizable and negotiable. Ideas are created, propagandaed for — or punctured.

As I said, our little corner of the biological sciences has become a fashionable bit of society's establishment. Conferences like this serve — among other things — to keep the scientific wheels turning, and to prevent a premature petrification of rules and regulations on unsatisfactory bases of facts and understanding.

It is my firm belief that our Japanese hosts will provide all the facilities — as well as much of the science — to make this conference a successful and memorable event which I am happy to declare opened.

Per Oftedal
President
International Association of Environmental Mutagen Societies

Plenary and Special Lectures

A View of a Cancer Researcher on Environmental Mutagens

Takashi Sugimura

National Cancer Center Research Institute, Tsukiji, Tokyo, Japan

ABSTRACT Environmental mutagens can be classified into three categories according to their chronological appearance; naturally occurring substances, pyrolysis products and chemicals made by modern industries. Based on overlap of mutagens and carcinogens, search for mutagens/carcinogens in pyrolysates and charred food are in progress. New heterocyclic amines with very high specific mutagenic activities were identified and some of them were proved to be carcinogenic. Mutagens including flavonoids were isolated from many edible plants, although their carcinogenicities are still controversial. In addition to environmental mutagens/carcinogens, environmental tumor promoters are important. Indole alkaloids and polyacetates, which are both new types of tumor promoters, were isolated from fungi and seaweeds. Elimination of environmental mutagens/carcinogens and promoters should be important in cancer prevention.

Where and what are the problems?

Numerous mutagens have been found in the human environment. Some of them, such as alkaloids and flavonoids in plants and mycotoxins, are naturally occurring. This class of mutagens has existed from prehistoric times, and evolution of animals has proceeded under continuous exposure to these mutagens. When humans first used fire for cooking and warming themselves and for light, a second class of mutagens, produced by pyrolysis, was introduced into the environment. This occurred about 10^5 years ago. A third class of mutagens has been introduced into the environment continuously since about 150 years ago when synthesis of organic substances first started. Mutagens formed by the consumption of fuel have increased tremendously in this 50 years, and the impact of man-made mutagens became sizeable when dye industries using coal tar developed at the beginning of this century. Many mutagens are also produced in large amounts by petrochemical industries developed after the Second World War. The problem of man-made mutagens must be solved by progress in modern science.

Recent research has indicated the importance of mutagenic oxygen radicals produced at the expense of life under air. Furthermore, many mutagens have been found to be formed in our bodies from non-mutagenic nitrosable substances and nitrite. Nitrite is easily produced in our body from nitrate, and the ingestion of nitrate is unavoidable. Ultraviolet and ionizing radiations cause mutations and exposure of humans to cosmic rays and internal emission of β-rays from the naturally occurring radioactive potassium isotope is also unavoidable. These factors and naturally occurring mutagens and mutagens formed during cooking indicate the necessity of establishing the concept of background exposure to environmental mutagens. This concept will make it easy to consider the risk-benefit issue of environmental mutagens.

In our environment, there are many different mutagens with various extents of mutagenic impact on humans. The question arises of "Which is more important, mutation in germ cells or mutation in somatic cells?" Even after the tragic exposure of human beings to the atomic bomb, however, it was difficult to demonstrate an increased incidence of genetic diseases in the offspring of survivors. In spite of the presence of numerous mutagens in our environment, it is hard to prove that any single chemical produces a heritable genetic human disease. A newly produced mutation in a single germ cell could hardly be recognized because the population of humans on the earth is only 10^9. On the other hand, a quarter of all humans suffer from cancer. Cancer cells are produced from normal cells by initiation, namely somatic cell mutation. The human body consists of 10^{13-14} cells. Some cells are fully differentiated and so no longer have no chance to develop cancer, but single initiated cells among a tremendous number of normal cells can produce a visible cancer mass as a result of clonal growth and it is certain that somatic mutations produced by environmental mutagens are very important in production of cancer.

There are many reports of typical carcinogens that are mutagens (1,2,3,4). Many of these typical mutagens are carcinogenic. With these compounds the term "environmental mutagen" is a synonym of "environmental carcinogen". Furthermore, as described later, the carcinogenic process is composed of at least two steps: first step is initiation, which is related with mutagenesis, and their promotion, which is thought not to be related to mutagenesis.

If the causes of diseases are known, elimination of these causes should result in prevention of the diseases. This should also be true for cancer. Thus, studies on environmental mutagens, including their detection, elucidation of their biological mechanisms, and their elimination, are crucial for prevention of cancer.

Why do mutagens in food warrant urgent study?

Studies on immigrants indicated that the pattern of cancers in various organs is changed by change in the life style. A typical example is Japanese immigrants to Hawaii and California. They showed a decreased incidence of stomach cancer but an increased incidence of colon cancer. Food is also a major source of environmental mutagen/carcinogen uptake, since a person consumes an average of 50 tons of food in a life time.

Environmental mutagens/carcinogens in food include mycotoxins, nitrosamines, aromatic hydrocarbons, residual pesticides, plant products such as quercetin and rutin and inappropriate food additives. However, there are also many unidentified mutagens in ordinary cooked foods.

Cooking and storage of food yield many mutagens/carcinogens. Pyrolysis at above 200°C produces various types of mutagens through successive radical reactions. The browning reaction (Maillard reaction), involving Amadori rearrangement, also yield many mutagens that have not yet been well identified. Studies on the formation of nitrosamines and nitrosamides in the stomach from precursors in food are also important.

Recent results on mutagens/carcinogens produced by pyrolysis of amino acids, proteins and proteinaceous foods has been reported (5,6,7,8,9). Most of these mutagens are heterocyclic amines. Their chemical names, abbreviations, structures, sources on first isolation, activities and locations are shown in Table 1. The specific mutagenic activities of some of them are higher than those of aflatoxin B_1, AF-2 and 4-nitroquinoline 1-oxide. Their specific mutagenic activities toward *Salmonella typhimurium* TA98 and TA100 are given in Fig. 1, with those of typical environmental mutagens/carcinogens. These environmental mutagens/carcinogens are activated by microsomal enzymes, mainly cytochrome P-450 with an absorption maximum at 448 nm, with a CO-difference spectrum that can be induced in rat liver by PCB or 3-methylcholanthrene. Activated forms of heterocyclic amines are hydroxyamino compounds and adducts with guanine base were elucidated in case of Trp-P-2 and Glu-P-1, as shown in Fig. 2. Some of them are mutagenic and also clastogenic to cultured mammalian cells.

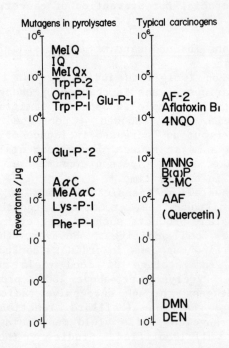

Fig. 1. Comparison of specific mutagenicities of pyrolysate mutagens with those of typical mutagens/carcinogens (*Salmonella typhimurium* TA98 or TA100)

Table 1. Mutagens isolated from pyrolysates

Full name (Abbreviation)	Structure	Source of isolation	Spec. Mut. Act. Revertants/μg,TA98,+S9 mix	Present in
2-Amino-3,4-dimethylimidazo-[4,5-f]quinoline (MeIQ)		Broiled sardine	661,000	
2-Amino-3-methylimidazo-[4,5-f]quinoline (IQ)		Broiled sardine	433,000	Fried beef, heated beef extract
2-Amino-3,8-dimethylimidazo-[4,5-f]quinoxaline (MeIQx)		Fried beef	145,000	
3-Amino-1-methyl-5H-pyrido[4,3-b]indole (Trp-P-2)		Tryptophan pyrolysate	104,000	Broiled sardine
4-Amino-6-methyl-1H-2,5,10,10b-tetraazafluoranthene (Orn-P-1)		Ornithine pyrolysate	56,800	
2-Amino-6-methyldipyrido[1,2-a:3',2'-d]imidazole (Glu-P-1)		Glutamic acid pyrolysate	49,000	
3-Amino-1,4-dimethyl-5H-pyrido[4,3-b]indole (Trp-P-1)		Tryptophan pyrolysate	39,000	Broiled sardine, broiled beef
2-Aminodipyrido[1,2-a:3',2'-d]imidazole (Glu-P-2)		Glutamic acid pyrolysate	1,900	Broiled dried-cuttelfish
2-Amino-9H-pyrido[2,3-b]indole (AαC)		Soybean globulin pyrolysate	300	Cigarette smoke
2-Amino-3-methyl-9H-pyrido[2,3-b]indole (MeAαC)		Soybean globulin pyrolysate	200	Cigarette smoke
3,4-Cyclopentenopyrido[3,2-a]-carbazole (Lys-P-1)		Lysine pyrolysate	86	
2-Amino-5-phenylpyridine (Phe-P-1)		Phenylalanine pyrolysate	41	

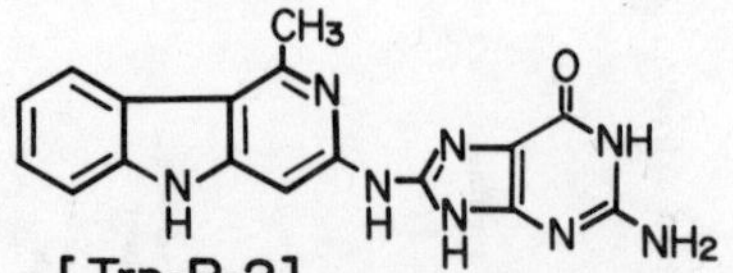

[Trp-P-2]
3-(C⁸-guanyl)amino-1-methyl-5H-pyrido[4,3-*b*]indole

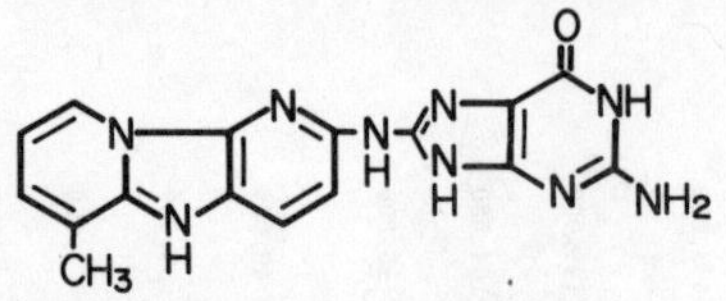

[Glu-P-1]
2-(C⁸-guanyl)amino-6-methyldipyrido[1,2-*a*:3',2'-*d*]imidazole

Fig. 2.   Adducts of Trp-P-2 and Glu-P-1 with guanine base (10,11)

The carcinogenicities of Trp-P-1, Trp-P-2, Glu-P-1, Glu-P-2, AαC and MeAαC have already been proved by *in vivo* long term animal tests. High incidences of hepatomas were found in mice treated with Trp-P-1, Trp-P-2, Glu-P-1 and Glu-P-2. Female mice were more sensitive than males.

As previously reported, charred parts of broiled fish and meat showed marked mutagenicity. The mutagenic potency was too strong to be accounted for by the content of benzo[*a*]pyrene and other aromatic hydrocarbons and nitrosamines and this provided a clue to the existence of new heterocyclic amines. The presence of some of the latter in cooked food has been proved and about half the mutagenic activity in charred parts of food can be explained by their presence. A method for quantitative chemical analysis of these compounds is being developed.

Although Trp-P-1, Trp-P-2 and Glu-P-1 are highly mutagenic, being more mutagenic than aflatoxin B_1, their carcinogenicities seem to be roughly equal to that of dimethylbenzanthracene, as shown in Fig. 3. This may be the result of their high initiating activities but low promoting activities. In support of this idea, administration of phenobarbital to rats given previously treated with Trp-P-1 markedly enhanced the number of enzyme altered foci in the liver, indicating that Trp-P-1 produced

Environmental Mutagens and Cancer Prevention   9

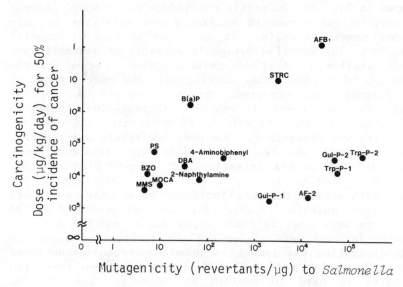

Fig. 3.   Relation of mutagenic and carcinogenic potencies (Some points have been rearranged by Kawachi from reference (12))

many initiated cells that could progressed to cancer when a tumor-promoter was applied later. Pyrolysis products in food are examples of new mutagens/carcinogens. Many other classes of mutagens/carcinogens have been found already and investigations on the structures and carcinogenicities of these compounds and their quantitative determinations are very important.

Is mutagenic quercetin, a ubiquitous flavonoid in plants carcinogenic?

Quercetin is a flavonoid found in many edible plants. Quercetin and kaempferol and their glycosides, rutin and astragalin are widely distributed and their contents in some foods are fairly high. For instance, a can of fruit juice may contain these flavonoids at a level of milligrams.

Quercetin and kaempferol have been reported to be mutagenic toward *Salmonella typhimurium*, and their glycosides, rutin and astragalin, were mutagenic after hydrolysis by hesperidinase or fecalase (13,14,15,16), as shown in Fig. 4. Quercetin can induce chromosomal aberrations, sister chromatid exchanges and mutations in cultured mammalian cells. It can also produce micronuclei *in vivo*. The specific mutagenic activity of quercetin was very similar to that of benzo[a]pyrene. Based on the overlap of mutagens and carcinogens, the carcinogenicity of quercetin was expected. However, two previous reports indicated that quercetin was not carcinogenic. After the discovery of the mutagenicity of quercetin, its carcinogenicity was reexamined. One positive result was reported: when quercetin was fed to Norwegian rats, many carcinomas were induced in the intestine and urinary bladder (17). However, studies in Japan indicated that animals fed quercetin showed the same incidence of spontaneous tumors as control animals (18,19). ACI rats, ddY mice and golden hamsters were used in these experiments, but the doses of quercetin used were higher and the duration of quercetin administration was longer in these experiments than those using Norwegian rats. The discrepancy between these two results is hard to explain at present. However, the carcinogenicity of quercetin does not seem to be strong, even though its moderate mutagenicity has been confirmed by many investigators.

In plant foods such as fruit juice, pickles, spices and tea, quercetin and kaempferol and their glycosides account for most of the mutagenicity, indicating that human beings are continously exposed to flavonoids. Thus the problem of whether flavonoids are carcinogenic is really crucial from two view points: If flavonoids are carcinogenic, as expected from mutation data and from ex-

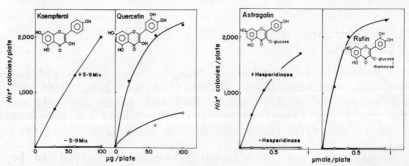

Fig. 4. Mutagenicities of flavonoids and their glycosides (*Salmonella typhimurium* TA98)

periments on Norwegian rats, what is the best suggestion to make to the general public? Is it necessary to reduce intake of foods containing flavonoids? If flavonoids are not carcinogenic, as deduced from data on ACI rats, ddY mice and golden hamsters, what does their positive mutagenicity mean? The specific mutagenic activity of quercetin is not very low. The problems of many substances with much lower specific mutagenic activity than quercetin and having much less chance to be consumed by humans are being debated seriously by consumers, factory workers, scientists and officers in regulatory agencies.

Flavonoids inhibit metabolic activation of many promutagens (20). In addition, it was recently found that quercetin inhibits the biological activities of a typical tumor promoter, 12-O-tetradecanoyl phorbol-13-acetate (TPA) such as its induction of ornithine decarboxylase (ODC) in mouse skin and enhancement of phospholipid metabolism in cultured cells. Flavonoids are mutagenic, but they may suppress the step of promotion in carcinogenesis. Anyway conclusions must be reached on whether quercetin is carcinogenic and if it is carcinogenic on how it can be eliminated. The carcinogenicity of flavonoids is an unavoidable issue for scientists studying environmental mutagens.

How is tumor promotion related to environmental mutagens?

Carcinogenesis is understood to be consist of at least two steps: initiation and promotion (21). The former is directly related to mutagenesis, but the latter is not. Substances catalysing promotion are called "promoters". This two step concept of carcinogenesis is supported by *in vivo* experiments, *in vitro* experiments and epidemiological data. As shown in Fig. 5, initiation (mutation) is not itself sufficient to convert normal cells to cancer cells. A problem is whether initiation is a bottleneck or a promotion step in human carcinogenesis. Recently, various lines of evidence have suggested that the promotion step is the more important. Most typical carcinogens may have some promoter activity together with strong initiating activity. Tobacco tar contains mutagens and also tumor promoters. Epidemiology studies showed that when heavy smokers stopped smoking their probability of onset of lung cancer decreased immediately.

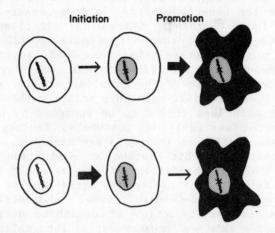

Fig. 5. Scheme of tumor initiation and promotion

The terminology of tumor promoters is ambiguous, because the mechanism of tumor promotion is not well known. I shall divide tumor promoters into two classes; well-defined tumor promoters and ill-defined tumor promoters. TPA is of the former type. Its effective dose is very small, and its effective chemical structure for biological activity has been well investigated. Various characteristic *in vitro* and *in vivo* biological activities of TPA have been demonstrated, including the inductions of ODC and adhesion of cultured HL-60 cells, inhibition of differentiation of Friend leukemic cells, binding with specific receptor, release of arachidonic acid and prostaglandin and activation of expression of the EB virus genome. Ill-defined tumor promoters include, in addition to cigarette tar, bile acids for colon carcinogenesis, saccharin for bladder carcinogenesis, phenobarbital for hepatocarcinogenesis and estrogenic hormones for breast carcinogenesis. There may also be unidentified tumor promoters of the second class with a variety of other chemical structures.

We launched a project to look for new tumor promoters besides TPA, since TPA has been the only well-defined tumor promoter for the past 15 years. We adopted four screening steps. Among 263 compounds tested, 54 gave a positive reaction in the first screening method, reddening of mouse ear on application of the compounds. Of these 54 compounds, 15 induced ODC *in vivo* in the second screening test. Of these, 5 induced adhesion of cultured HL-60 cells *in vitro* in the third test. Finally in the fourth test, a long term test on animals, 3 were identified as true tumor promoters *in vivo*. Including 1 other compound, we now know 4 newly discovered tumor promoters, teleocidin B, dihydroteleocidin B, lyngbyatoxin A (probably identical with teleocidin A) and aplysiatoxin (22,23,24,25,). These compounds are all effective at very low concentrations equivalent to the effective concentration of TPA. Teleocidin was isolated from *Streptomyces* and lyngbyatoxin and aplysiatoxin were isolated from seaweed and blue green algae. The structures of these new tumor promoters are given in Fig. 6. A typical experiment indicated that on treatment with 100 μg of dimethylbenzanthracene, only 2 of 49 mice had tumors, but on treatment with the same amount of dimethylbenzanthracene followed by painting with 2.5 μg of dihydroteleocidin B twice a week for 20 weeks 48 of 50 mice had tumors. Dihydroteleocidin B alone did not produce any tumors in 50 mice. It should be emphasized that TPA and teleocidin share the same receptor on the cell surface although teleocidin is not similar in structure to TPA. Furthermore it is very interesting that debromoaplysiatoxin, which lacks the bromine atom present in aplysiatoxin, seems to have similar ODC induction activity but

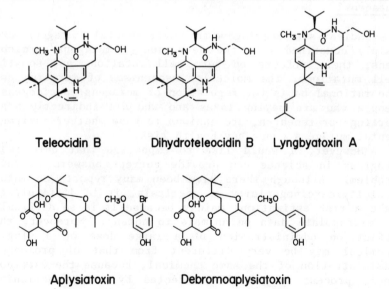

Fig. 6. Structures of TPA and newly found tumor promoters

quite low ability to cause adhesion of HL-60 cells or *in vivo* tumor promotion.

In tests on less than 300 compounds, we were fortunate to discover several compounds with similar tumor promoting potencies to that of TPA. We suppose that our environment may contain many more tumor promoters of both classes. If environmental mutagens have more effect in somatic cell carcinogenesis than in germ cell mutation and if the carcinogenesis process is completed by the promoting processes, detection and quantitation of promoters, elucidation of their mechanism of action and their elimination from the environment should be as important as similar studies on environmental mutagens. Should the International Commission for Protection Against Environmental Mutagens and Carcinogens, ICPEMC, added "and Promoters", ICPEMCP, to its name?

Recent investigations indicate that typical tumor promoters, including TPA and teleocidin, enhance chemically induced mutations, induce sister chromatid exchanges and cause gene amplification. Further study may shed a light for understanding the mechanism of tumor promotion.

What is the civic duty of scientists studying environmental mutagens?

Scientists dealing with environmental mutagens are mainly concerned with the detection of mutagens/carcinogens, their effects on germ cell mutations and somatic cell mutations, the molecular mechanisms of mutation and the rational basis for regulation of mutagens/carcinogens. People who are paying taxes and who are indirectly supporting our research, are anxious to know whether environmental mutagens cause any health risk.

However, I would like to emphasize that only real progress in science can provide correct answers to this problem. Although there have been many reports on mutagenicity/carcinogenicity of chemicals, it is difficult to make a risk estimation, mainly because it is so difficult to extrapolate data on animals to humans. Moreover, the effect on administration of a large dose of a single chemical may be very different from that of prolonged administration of the same chemical, because the carcinogenic process is probably affected by many functioning modulators.

Here I may cite an interesting example of two compounds. AF-2 is a notorious case of a food additive once used in Japan, that shows both mutagenicity and carcinogenicity. People were very much concerned about the risk of this man-made food additive and consumers reacted strongly on hearing it was carcinogenic. On the other hand bracken, an edible plant, is also known to contain a carcinogen(s). The carcinogenic potencies of AF-2 and bracken were both calculated from animal experiments and expressed in TD_{50} values; i.e. the doses inducing tumors in 50% of the test animals when fed continuously. The total annual production of AF-2 and consumption of bracken fern were obtained from Governmental statistics and the average daily intakes of these compounds per head of population were calculated. The value obtained by dividing the average intake by the TD_{50} reflects the size of the risk. Ironically the values for AF-2 and bracken fern were 0.002 and 0.003, respectively (26). Bracken fern has long been eaten in Japan and people readily think that this plant that grows wild in the mountains is free from carcinogens. It is also very strange that no regulatory measure against bracken has even been considered by regulatory agencies in Japan and there has been no propaganda about the risk of bracken fern.

The factors that are considered important by society are not necessarily important scientifically, and vice versa. Mutagens in coffee were first reported in 1978 from our laboratory as direct acting mutagens toward *Salmonella typhimurium* TA100 (27). Their positive activity in the inductest was also demonstrated. Brewed coffee, instant coffee and caffeine-free instant coffee showed about the same potencies. Moreover, an epidemiological correlation between coffee drinking and pancreatic cancer has recently been reported (28). It is still uncertain whether the mutagenic substance(s) in coffee is really carcinogenic. In everyday conversation people discuss whether they should stop drinking coffee. Scientists are aware of the fact that some human populations, such as the Seventh Day Adventists who do not drink coffee have an abnormally low incidence of cancer. So more information on mutagens in coffee should be obtained in the laboratory, and more epidemiological data should be collected. Otherwise the general population will come to distrust scientists, and this will be a great loss to our society. Incidentally the mutagenicity of coffee can be abolished by addition of a low concentration of sulfite, which is approved as a food additive and is added as a preservative to some wines. Not only coffee but also non-volatile residues of whiskey and brandy contain directly acting mutagens toward *Salmonella typhimurium* TA100. We do not have sufficient information about whether mutagens existing in certain beverages are carcinogenic or not, but it is our duty to give right answers to people: What should they do or not do?

Why is cancer prevention important?

The National Cancer Center Hospital, Japan, was established in 1962. Since then about 6,000 patients have been cured in terms of five-year survival at Hospital of the Center. This means about 300 lives are saved annually, while about 500 patients die in the hospital. The annual budget for the hospital is ¥6×10^9 ($$3 \times 10^7$). Thus the cost to save a life is ¥2×10^7 ($$10^5$). In Japan, about 2×10^5 cancer patients are seen annually. A simple calculation shows that to save 2×10^5 lives at of ¥2×10^7 ($$10^5$) a life costs ¥$4 \times 10^{12}$ ($$2 \times 10^{10}$). This is a tremendous amount of money. If we add an indirect cost related to the burdens on the family and place of work, the cost would be astronomical. Neither money, nor human resources, trained physicians, surgeons and nurses can be provided for this purpose. On the contrary, if the effort of scientists could reduce only 10% of the new cancer patient annually, we could readily save ¥4×10^{11} ($$2 \times 10^9$). Cancer prevention is clearly a very pragmatic goal which is in the hands of scientists working on environmental mutagens.

What can be recommended to people? What research should have priority?

By integrating the present knowledge on environmental mutagens/carcinogens and tumor promoters, such as their existence, potency, quantity, size of exposure, and modulation, we have proposed 12 tentative points to the general population. Since none of these points have been fully elucidated experimentally, clinically or epidemiologically, these proposals are necessarily tentative (29). These are our 12 points.

1. Take a well balanced diet, in terms of both taste and nutrition.
2. Do not eat the same foods repeatedly or exclusively, and exercise caution in taking the same medicines over long periods.
3. Avoid over-eating.
4. Avoid drinking too much alcohol.
5. Refrain from excessive smoking.
6. Take optimal daily doses of vitamins A, C and E. Include a moderate amount of fibrous food ("roughage") in your diet.
7. Avoid excessive intake of salty food and do not drink very hot water, tea or coffee.
8. Avoid eating too much of burnt parts of food, such as burnt parts of charcoal grilled meat and fish.
9. Avoid moldy food other than foods like cheese that are meant to be moldy.
10. Avoid excessive exposure to the sun.
11. Avoid overwork, which lowers resistance to disease.
12. Take a bath or shower frequently.

Beside the subjects described in these articles, there are many points which might be considered urgent by scientists or which might be encouraged by funding agencies. Examples are as follows:

1. Endogenous formation of mutagens/carcinogens and promoters
2. Mutagens/carcinogens and promoters in air pollutants, especially nitroarenes
3. Mutagens/carcinogens and promoters in fermented products and beverages
4. Factors modulating formation and metabolic activation of mutagens/carcinogens, the mutagenesis process and tumor-promotion steps
5. DNA damage responsible for carcinogenesis: Are big structural changes in chromosomes such as rearrangements, insertions, transpositions and translocations more important than simple gene mutation?
6. Activation of oncogenes by mutagens/carcinogens
7. What is the true nature of initiation mutation in carcinogenesis?
8. Genetic and nongenetic host factors modifying initiation and promotion
9. Mutagens/carcinogens and tumor promoters in pesti-

cides, insecticides and intermediates of industrial chemicals
10. Mutagenic/carcinogenic and tumor promotion potencies of drugs to be used for long periods
11. The theoretical approach to risk estimation
12. Are infections with viruses that are not oncogenic in a strict sense and chronic inflammation related to carcinogenesis?

Of course these are only a few examples and do not exclude many other important subjects. Integrated efforts by scientists in different disciplines will be a royal way to reach the goal of cancer prevention.

Acknowledgements

The author wishes to thank many collaborators in Japan and other countries. He is also grateful for the financial support given by the Japanese Govenment and Various Societies. This work was greatly encouraged by the US-Japan Cooperative Cancer Research Program and the US-Japan Medical Cooperative Program.

References

Our review type articles and a minimum of key references are cited owing to limitation of space.

(1) McCann, J., and Ames, B.N. *Proc. Natl. Acad. Sci. USA* 73, 950-954 (1976)
(2) Sugimura, T., Sato, S., Nagao, M., Yahagi, T., Matsushima, T., Seino, Y., Takeuchi, M., and Kawachi, T. In ; P.N. Magee, S. Takayama, T. Sugimura, and T. Matsushima (eds.), Fundamental in Cancer Prevention, pp. 191-215, Univ. of Tokyo Press, Tokyo/Univ. Park Press, Baltimore (1976)
(3) Purchase, I.F.H., Longstaff, E., Ashby, J., Styles, J.A., Anderson, D., Lefevre, P.A., and Westwood, F.R. *Nature 264*, 624-627 (1976)
(4) Nagao, M., and Sugimura, T. *Ann. Rev. Genet. 12*, 117-159 (1978)
(5) Sugimura, T., Nagao, M., Kawachi, T., Honda, M., Yahagi, T., Seino, Y., Sato, S., Matsukura, N., Matsushima, T., Shirai, A., Sawamura, M., and

Matsumoto, H. *In*; H.H. Hiatt, J.D. Watson, and J.A. Winsten (eds.), Origins of Human Cancer, Book C, pp. 1561-1576, Cold Spring Harbor Laboratory, New York (1977)

(6) Nagao, M., Yahagi, T., Kawachi, T., Seino, Y., Honda, M. Matsukura, N., Sugimura, T., Wakabayashi, K., Tsuji, K., and Kosuge, T. *In*; D. Scott, B.A. Bridges, and F.H. Sobels (eds.), Progress in Genetic Toxicology, pp. 259-264, Elsevier/North-Holland, Amsterdam (1977)

(7) Sugimura, T., Kawachi, T., Nagao, M., and Yahagi, T. *In*; G.R. Newell, and N.M. Ellison (eds.), Nutrition and Cancer: Etiology and Treatment, pp. 59-71, Raven Press, New York (1981)

(8) Matsukura, N., Kawachi, T., Morino, K., Ohgaki, H., Sugimura, T., and Takayama, S. *Science 213*, 346-347 (1981)

(9) Sugimura, T. *Mutation Res. 55*, 149-152 (1978).

(10) Hashimoto, Y., Shudo, K., and Okamoto, T. *Chem. Pharm. Bull. 27*, 1058-1060 (1979)

(11) Hashimoto, Y., Shudo, K., and Okamoto, T. *Biochem. Biophys. Res. Commun. 92*, 971-976 (1980)

(12) Meselson, M., and Russell, K. *In*; H.H. Hiatt, J.D. Watson, and J.A. Winsten (eds.), Origins of Human Cancer, Book C, pp. 1473-1481, Cold Spring Harbor Laboratory, New York (1977)

(13) Brown J.P. *Mutation Res. 75*, 243-277 (1980)

(14) Nagao, M., Morita, N., Yahagi, T., Shimizu, M., Kuroyanagi, M., Fukuoka, M., Yoshihira, K., Natori, S., Fujino, T., and Sugimura, T. *Environmental Mutagenesis 3*, 401-419 (1981)

(15) Tamura, G., Gold, C., Ferro-Luzzi, A., and Ames, B.N. *Proc. Natl. Acad. Sci. USA 77*, 4961-4965 (1980)

(16) McGregor, J.T., and Jurd, L. *Mutation Res. 54*, 297-309 (1978)

(17) Pamukcu, A.M., Yalciner, S., Hatcher, J.F., and Bryan, G.T. *Cancer Res. 40*, 3468-3472 (1980)

(18) Saito, D., Shirai, A., Matsushima, T., Sugimura, T., and Hirono, I. *Teratog. Carcinog. Mutagen 1*, 213-221 (1980)

(19) Hirono, I., Ueno, I., Hosaka, S., Takanashi, H., Matsushima, T., Sugimura, T., and Natori, S. *Cancer Lett. 13*, 15-21 (1981)

(20) Gelboin, H.V., Donna, W., Gozukara, E., Natori, S., Nagao, M., and Sugimura, T. *Nature 291*, 659-661 (1981)

(21) Berenblum, I. *Cancer Res. 1*, 807-814 (1941)
(22) Fujiki, H., Mori, M., Nakayasu, M., Terada, M., and Sugimura, T. *Biochem. Biophys. Res. Commun. 90*, 976-983 (1979)
(23) Fujiki, H., Mori, M., Nakayasu, M., Terada, M., Sugimura, T., and Moore, R.E. *Proc. Natl. Acad. Sci. USA 78*, 3872-3876 (1980)
(24) Sugimura, T., Fujiki, H., Mori, M., Nakayasu, M., Terada, M., Umezawa, K., and Moore, R.E. *In*; E. Hecker, N. Fusenig, F. Marks, and W. Kunz (eds.), Carcinogenesis, Vol. 7, pp. 69-73, Raven Press, New York (1982)
(25) Umezawa, K., Weinstein, I.B., Horowitz, A., Fujiki, H., Matsushima, T., and Sugimura, T. *Nature 290*, 411-412 (1981)
(26) Kinebuchi, M., Kawachi, T., Matsukura, N., and Sugimura, T. *Fd Cosmet. Toxicol. 17*, 339-341 (1979)
(27) Nagao, M., Takahashi, Y., Yamanaka, H., and Sugimura, T. *Mutation Res. 68*, 101-106 (1979)
(28) McMahon, B., Yen, S., Trichopoulos, D., Warren, K., and Nardi, G. *N. Eng. J. Med. 304*, 630-633 (1981)
(29) Sugimura, T. ; M.D. Anderson (ed.), Molecular Interrelations of Nutrition and Cancer, 34th Annual Symposium on Fundamental Cancer Research (1981, in press)

A History of Attempts to Quantify Environmental Mutagenesis

Alexander Hollaender

Associated Universities, Inc., The Council for Research Planning in Biological Sciences, Inc., Washington, D.C., U.S.A.

ABSTRACT It became obvious in the early 1960's that the ready recognition of mutations produced by chemicals could have a profound influence on the refinement of methods to detect environmental mutagens. The experience derived over the previous 30 years in characterizing the effects of ionizing and ultraviolet radiation on the genetic mechanism came to serve us in good stead. Although the effects of chemicals are considerably more complicated and often require the analysis of individual substances, nonetheless, the area has developed rapidly in recent decades. The establishment and historical background of the International Association of Environmental Mutagen Societies (IAEMS) will be discussed. An attempt at the quantitation of chemical effects has been developed in comparison with radiation mutagenesis. As a first step, a definition of the "Mutagen Burden" or unavoidable exposure to chemicals will be discussed. A mathematical approach (Haynes/Eckhardt) will be considered and finally an outline for the comprehensive investigation of detailed interscience study will be made of less than six chemicals.

INTRODUCTION

The study of induced mutagenesis had its beginning more than 54 years ago, as all of us are well aware, with the work of Herman Muller (1) who developed a quantitative method for the recognition of mutations in <u>Drosophila</u> and who was able to establish the mutagenic effect of ionizing radiation. During the 1930's, the field of ultraviolet induction of mutagenesis developed very well, especially toward the end of the 1930's with the recognition that the most effective wavelength is the one absorbed by the nucleic acids. I had many interesting discussions with Dr. Muller in regard to the importance of these areas and it became obvious to us that the next field to be studied would be the effects of chemicals that result in mutation in exposed tissue. The recognition of chemical mutagenesis should be credited to Auerbach and Robson (2), to Rapoport (3), and to Oehlkers (4) whose work during the second World war was published in 1946.

The detection and evaluation of environmental mutagens was a very logical development but it took almost 20 years before practical methods became available for their routine measurement. When Milislav Demerec retired from the Directorship of the Cold Spring Harbor Laboratory he concentrated all of his research on developing Salmonella as organisms for the detection of chemical mutagens (5). Later, his elegant mutation techniques were further developed by Philip Hartman (6) of Johns Hopkins University. Many chemicals were missed at that time which later turned out to be mutagens since it was found that liver extract is needed for the activation of numerous mutagens. This requirement was first identified by Heinrich Malling (7). Liver microsomes are now used routinely in chemical mutation work.

The most significant development came when Bruce Ames of Berkeley developed Salmonella as a most ready tool for the detection of many mutagens (8). Ames developed many new strains of Salmonella which have specific ability to detect certain types of mutagens. The extensive work coming from Ames' laboratory has resulted in many important findings including the recognition of the broad overlap between mutagenesis and carcinogenesis. This is not to say that Salmonella hasn't any limitations, despite its use in very many laboratories all over the world. Salmonella is most useful for very preliminary screening and as a part of a well chosen battery of tests.

In the early 1960's it became obvious to some of us that the time had come to do something systematic so as to apply the knowledge derived through improving methods of mutagen detection to the problems of environmental pollutants. The first question was, where could the work be done most effectively? The Ford Foundation gave me a small travel grant in the mid-sixties to explore the possibilities at several universities where research centers could be developed for chemical mutagen testing. I was helped very much in this exploratory period by Marvin Legator (then at the Food and Drug Administration) and Samuel Epstein (then at the Children's Cancer Hospital of the Harvard University Medical School). Workshops were initiated at Brown University[a] and at the University of Zurich[b]. Thereafter, our appeal became a little easier and others became more and more convinced that something should be done in an organized way.

Through my friend, Joe Slater, I applied to the newly established Anderson Foundation for support of my ideas and the urgent need for giving this field an immediate boost. I received a grant of $25,000 with no real strings attached to use as I saw fit to promote the field of environmental mutagenesis. In 1969, I turned over these funds to the newly organized Environmental Mutagen Society[c] (EMS). While it gave the necessary boost to get the EMS underway, the Society itself developed rather slowly, hitting its stride in the last several years.

After 1969, as all of you know, societies were organized in Europe, Japan, and India. By 1973 we had four major societies for environmental mutagenesis. At the EMS meeting at Asilomar which followed the International Congress of Genetics, the International Association of Environmental Mutagen Societies (IAEMS)[d] was, logically, instituted. This was followed by the international conference in Edinburgh in 1977, and now our third conference here in Japan. The Australian-New Zealand EMS has just joined the other four societies of the IAEMS. A number of additional societies have been organized and are in the process of applying to join the IAEMS.

---

[a]Providence, Rhode Island (1971) "Workshop on Mutagenicity"
[b]University of Zurich, Switzerland (1973) "Workshop on Mutagenicity"
[c]Environmental Mutagen Society, 4720 Montgomery Lane, 506, Bethesda, Maryland 20014 (Verne Ray, President)
[d]International Association of Environmental Mutagen Societies, c/o Institute of General Genetics, Box 1031, Blindern, Oslo, Norway (Per Oftedal, President)

A number of very important things had to be done in connection with the development of the field of environmental mutagenesis. Many members of the Society were very much interested to promote the field as an essential part of the study of the environment. We then established an information collective, the Environmental Mutagen Information Center (EMIC)[e], which is now one of the most successful information sources in the area of environmental studies. EMIC was organized in 1969 under the direction of John Wassom with the support of Heinrich Malling.

In the meantime, many new methods for the detection and characterization of environmental mutagens were developed. As a matter of fact, the number increased so that we have today a great abundance of methods including some very excellent ones. The training of scientists so that they can use these methods for the intelligent detection of mutations has become increasingly important to this area. In earlier days, not many researchers saw the importance of this field. However, Galveston[f] instituted such a curriculum and has since become an important center for training in this field. Today we have many good methods for the detection of mutagens developed in many laboratories. However, in the long run, it is up to industry to use these methods. They have been most cooperative in this area, becoming an essential cog in the development of environmental mutagenesis research.

A number of workshops and training courses were organized to help individuals in developing countries by demonstrating techniques for the recognition of potentially dangerous chemicals used extensively in the environment (i.e., pesticides). We now have three important centers in the developing countries of Mexico, the Philippines, and Egypt.

The field of Environmental Mutagenesis had gone through the usual chain of development of any relatively new field, first, with the recognition of the phenomenon, then the

---

[e]Environmental Mutagen Information Center (EMIC) Oak Ridge National Laboratory, P. O. Box Y, Oak Ridge, Tennessee 37830
[f]University of Texas Medical Branch, Galveston, Texas, Annual Genetic Toxicology Workshops

development of methods for its detection, followed by the qualitative observations, and finally a quantitative evaluation which contributes to an understanding of the underlying basic mechanisms. We had gone through these same steps to understanding radiation mutagenesis with many interesting results. Even after more than 50 years of intensive study, there are still quite a number of unsolved problems related to radiation mutagenesis, just as there are very many problems facing us with environmental mutagenesis.

There is really a great difference between radiation and chemical mutagenesis. One has to consider that most of the background radiation is attributed to sources outside of the human organism, whereas chemical exposures can occur from both outside influences or, since chemicals represent the building blocks of which we are made, the different internal pathways by which these compounds are absorbed and become an integral part of the body. By comparison, radiation is relatively simple with only a few types and chemicals are very numerous, with many specific metabolic pathways. Some chemicals are readily eliminated, others go through a number of stages of metabolism before they break down and then combine with other compounds or tissues. Unfortunately, these complications are not always realized as one becomes involved in working with chemicals. Occasionally, the effect of some compounds may be transitory and it has been found that some are even found in the gonads, despite their identity as somatic mutagens without germinal effect.

When it became necessary to define the basis for radiation mutagenesis some 30 years ago, it was thought essential to determine the unavoidable radiation to which man is exposed (i.e., cosmic radiation, radiation from potassium in the body, and essential medical X-rays). "Permissible limits" were based on multiples of unavoidable background radiation exposure. As a matter of fact, this served as an elegant way to proceed in determining the idea of a "permissible _dose_", or multiples of unavoidable background radiation.

Some of us thought that a similar approach could be applied to defining chemical exposure and effects. We know that there are certain chemicals in our bodies and many others which might be regularly ingested, inhaled, or absorbed through the skin. A few are well established mutagens. Task Force 4 of the International Commission for Protection against Environmental Mutagens (ICPEMC)[g], consisting of Takashi Sugimura, Verne Ray, Lorenzo Tomatis, and myself (as Chairman), has been working to define a few of the basic chemical mutagens which all of us carry within us and also those which we cannot avoid from external sources or which are produced internally. Of the multitude of chemicals with which man is in daily contact, we decided to begin with three common compounds (which are also established mutagens), Aflatoxin, Benzo(a)pyrene, and Nitrosamine.

AFLATOXIN is a fungus which grows on agricultural products and, therefore, is easily ingested in foods. This mutagen is of particular importance to developing countries where fungal infection can spread through crops that are stored, often without refrigeration. Aflatoxin is a very serious mutagen and also a very strong carcinogen for both somatic and germinal cells.

BENZO(A)PYRENE is a by-product of combustion and is present in smoke and in automobile exhaust and is therefore most prevelant. This compound has been detected in body tissues and is a well-established somatic mutocarcinogen, although it has not been established as being mutagenic to germinal tissue where it is found in small quantities. It probably doesn't reach the chromosomes in the form which is detected by chemical analysis in the gonads. It is easily detectable and data are available from a variety of sources, especially from laboratories in Germany (9). Some figures are given in Table I.

NITROSAMINES are ingested and are possibly also produced by the nitrites in saliva or by intestinal bacteria. However, nitrosamines are also found in germ-free animals. There must, then, be some enzyme system which is responsible for internally producing nitrosamine. The enzymes responsible for this are not yet clear. There is considerable literature in this area and much controversy. One problem, of course, is that nitrosamines pass through the body very

---

[g]ICPEMC Task Force 4, studying the "Mutagen Burden". Three meetings held in 1979-80 to establish preliminary measurements of chemical intake and effect.

Table I
An Estimate of Man's Exposure to Mutagenic Chemicals
Carried Within the Human System and Unavoidably Ingested,
Absorbed, or Inhaled from the Environment

Chemical	Measurement
AFLATOXIN[a] (B1 = 312)	250 ng/day/person - 3000 ng/day/person ($= 8 \times 10^{-7}$ mmoles)
Intake:	10 ng/hr/person - 120 ng/hr/person ($= 3 \times 10^{-8}$ mmoles)
	C_1 (plasma) 10 ng/kg - 120 ng/kg ($= 3 \times 10^{-8}$ mmoles)
	C_2 (tissue) 29 ng/kg - 480 ng/kg ($= 9 \times 10^{-8}$ mmoles)
BENZO(A)PYRENE[b]	0.2 g - 15.6 ng/100 g dry tissue ($= 8 \times 10^{-7}$ mmoles - 6×10^{-5} mmoles)
NITRITE (NO_2)[c]	0.012 - 0.88 mmoles/person/day
NITRATE (NO_3)[c]	0.3 - 9.3 mmoles/person/day
NITROSAMINE[d]	0.001 - 0.05 mmoles/person/day

The exposure data supplied above were compiled by:
[a]Dennis Hsieh, Univ. California, Davis; [b]James Selkirk, Oak Ridge National Laboratory, Tennessee; [c]Philip Hartman, Johns Hopkins University, Baltimore; [d]Anthony Pegg, Pennsylvania State University.

---

readily and are probably present not only in the intestine but also in the fecal matter. As a matter of fact, a special symposium following this conference[h] will consider the different problems of these compounds.

If we consider the three obvious mutagens as a whole, it is remarkable how few detrimental effects have been recognized with the relatively significant quantities of

---

[h]Seventh International Meeting on Analysis and Formation of N-Nitroso Compounds, Tokyo, September 28 - October 1, Helmut Bartsch, IARC, Lyon, France (Chairman)

mutagens or carcinogens we carry within the body while still maintaining our present integrity of genetic mechanism.

This is an interesting problem which requires further analysis. There is also the possibility that we carry sufficient "anti-mutagens" to stop the activity of the mutagens within. There is good indication that such compounds exist, but whether these are adequate to take care of the stability of our genetic make-up is not certain (i.e., ascorbic acid, uric acid).

As a basis for consideration, these three chemical compounds represent common quantities and exposures. Those found in smaller quantities ought not to be overlooked as potentially effective mutagens as, for example, William L. Russell found of Ethylnitrosourea (ENU) (10), the strongest mutagen ever found in the mouse yet formerly considered rather harmless. There may be other substances that are just as effective but which exist in lesser quantities and have not yet been recognized to contribute to the Mutagen Burden. This is another important aspect which deserves further investigation.

I would like now to discuss some of the quantitative aspects of Environmental Mutagenesis as I see it from my narrow point of view. An interesting approach has been used by Committee 17 of the Environmental Mutagen Society ("Environmental Mutagenic Hazard" In Sciences (1975) 187: 503-514) in developing the "REC" or rem-equivalent-chemical on the basis of the very extensive data which are available in radiation mutagenesis. Although this attempt is a very promising one, it is not adequate. A cautionary note is attached stating that the calculation implicitly assumes the response of the organism, while modulated by numerous pharmocological factors, remains constant in time. This, of course, is not entirely correct because the physiology of the human is a very dynamic system.

Therefore, reliance solely upon data derived from means determined for radiation mutagenesis are not adequate for an accurate evaluation of the more complex steps in chemical mutagenesis. At least this 1975 study was a good attempt and should be encouraged with some modification. The systems we are discussing are based primarily on animal work. Extrapolating from the mouse to man is a somewhat dangerous transition. However, at the present time, this is the best that we can do if no other data are available on which to proceed. This is the reason I thought that the "Mutagen Burden" concept would offer some kind of base for human evaluation and I will come back to this point later.

The highly significant mutation data obtained on the specific locus method by Russell and Lyon and Morris on somewhat different loci in the mouse gave, until lately, the best data on the quantitative genetic effects of ionizing radiation on certain hybrid strains of mice. The genetic effects of chemicals on mammals _in vivo_ is considerably more difficult for the reasons which we have already discussed. There is a shortage of quantitative data on the _in vivo_ effects of chemicals in mammals. Studies show that the chemical effects on the mouse produce a very low number of gene mutations measured by the specific locus method of Russell. This was changed somewhat with Russell's finding that ENU is the most effective mutagen ever found.

When Ehling reported the effect of ionizing radiation on the skeleton as a ready way to study dominant mutations in the mouse, the picture changed. These studies have been considerably extended by Selby at Oak Ridge and some of the results are shown on Table II.

Table II
Frequency of Presumed Dominant Mutations
Affecting the Skeleton of the Mouse

Dose (R)	Interval between dose fractions	Stage	No. of F_1 skeletons	Presumed mutations	
				(No.)	(%)
0			1739	1	0.06
600	0	Post-spermatogonial	569	10	1.8
600	0	Spermatogonial	754	5	0.7
100 + 500	24 hours	Spermatogonial	277	5	1.8
500 + 500	10 weeks	Spermatogonial	131	2	1.5

Reproduced from "Radiobiological Equivalents of Chemical Pollutants", Int. Atomic Energy Agency, Vienna, p 78 (1980).

Table III
Specific Enzyme Activities (Expressed in % Activity of
Normal Lenses) in the Lenses of Mice with Nzc-Cataract

Enzyme	Cataract Mutants	
	Heterozygote	Homozygote
Glucose-6-phosphatdehydrogenase	96	102
Malatdehydrogenase	111	132
Isocitratdehydrogenase	97	119
Hexokinase	161	243
Lactatdehydrogenase	130	159
Glyceraldehydphosphatdehydrogenase	116	121
Pyruvatkinase	89	111
Enolase	126	143
Phosphoglyceratkinase	102	124

Reproduced from "Elektrophorese Forum '80", 2. Diskussionstagung Technische Universitat, Munchen, Munchen (1980).

Still newer developments include the observation of the inherited cataract method by Ehling's group who recognized that these can be observed easily with a slit lamp on a large number of animals especially since these again are dominant effects. It was first developed with the evaluation of ionizing radiation but has proven most promising with chemicals. I believe that we can look forward to many interesting data available with in vivo genetic effects of chemicals in the mouse.

One of the most interesting discussions in regard to quantitative effects has been given by Haynes and Eckhardt who developed mathematical equations which take into account not only the mutation rate, but the survival rate and repair mechanisms as well. The set of curves thus developed show that even with the present rather primitive data, it is possible to get significant results. Some of the first statistics obtained are given in Figure 1. This approach would, of course, be more successful if more publications will include, besides mutation data, also exact survival curves. Joyce McCann, in a Chapel Hill discussion, recently pointed out very well the difficulties in obtaining such data by the plate method.

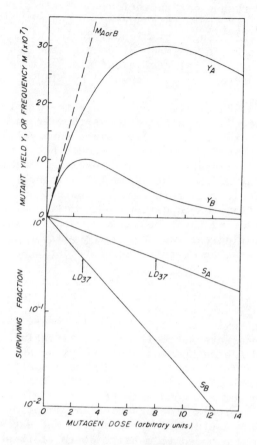

Fig. 1. Mutant yields Y(x), frequencies M(x) and survivals S(x) plotted over dose in arbitrary units for the purely linear kinetic response pattern. For both case A and B, $m_1 = 10^{-6}$ (units)$^{-1}$. In A, $k_1 = 1.23 \times 10^{-1}$ units)$^{-1}$; in B, $k_1 = 3.7 \times 10^{-1}$ (units)$^{-1}$; that is, the cross-section for killing in B is 3-fold greater than in A. Note that the frequency curve (dashed line) is the same for both A and B whilst the areas under the yield curves are very different. In each case Y_{max} is at the dose for 1 lethal hit ($\simeq$LD37). Reproduced from "Quantitative Measures of Mutagenicity and Mutability based on Mutant Yield Data" Friederike Eckhardt and Robert H. Haynes. Mut. Res. 74: 439-458 (1980).

At a workshop on Statistical Analysis of In Vitro Tests for Mutagenicity held this Spring at Chapel Hill, chaired by David Hoel and managed by Michael Shelby with Fred de Serres, a careful analysis was made of some of the best data available especially on Salmonella but also of a few tissue culture tests. (Proceedings sponsored by NIEHS & NTP, in press).

This workshop illustrated again the difficulty of interpretation of mutation data as they are now appearing in the literature, especially since survival data are usually not given. However, it should be pointed out that more and more investigators are becoming conscious of the complications of the plate test for Salmonella. It is especially difficult with this test to control and determine the survival ratio, the condition that individual organisms experience during the test and many other points discussed by McCann during this workshop. We will be looking forward to the publication of this very interesting workshop.

Ehrenberg's group in Stockholm, in a very interesting paper, recommended the use of the appearances of alkylating amino acids in hemoglobin to be used as a measure of exposure to genotoxic agents, alkylating agents per se, and compounds that are metabolized to alkaylating agents. They illustrated the methods on ethylene oxide and dimethyl nitrosoamine. The latter compound respires metabolic oxygenation which gives rise to alkylating compounds which are responsible for the genotoxic effects. This method is of considerable promise for practical application in mice and could be very useful for the detection of exposure of people inadvertantly exposed to massive doses of these compounds (11).

Per Oftedal, in a review of Problems in the Reevaluation of Genetic Risk from Radiation and Other Environmental Hazards, gave a very important discussion of the shape of the mutation curve produced by radiation in Drosophila but also discussed the effects of chemicals and their relation to radiation effects. He also compared the risk estimated by the UNSCEAR and the BIER committees. Even after the many investigations to date, it was not possible for these two very important committees to agree on the doubling dose at low level exposure. So it is not surprising that we have little agreement in regard to chemical exposure.

In all of these discussions, I have tried to avoid the

problems of the recovery phenomena. Of course, the problem
of repair is one of the most fundamental ones first observed
after exposure of bacteria and yeast to ultraviolet radiation. It was first recognized by carefully controlling
conditions of exposure to radiation at temperatures where
practically no metabolism takes place and exposing the culture after to conditions of a broad variety which makes
possible the recognition of the recovery phenomenon. All
such techniques which make quantitative evaluation possible
are somewhat difficult to control with chemicals.

Another area which I have not touched on is the work
in cytogenetics - not that I think that it is not important,
rather, it has been discussed so very well in symposia and
special papers at this conference. The cytogenetic approach
lends itself readily to quantitative work as it was first
developed by Carl Sax almost 50 years ago and carried on so
well by his co-workers. The leadership of H.E. Evans and
his group, the new developments initiated by J. H. Taylor and
later by S. Latt and then further developed and extended
into new lines by S. Wolff are most impressive.

Most of the attempts in regard to evaluation of the
quantitative aspects of the effects of chemical mutagens
are based on experience on radiation effects. There the
success for quantitative estimation was developed by a systematic planning approach that tackled the problem in a
carefully planned way which took advantage of all of the
different methods that radiation biology offered. This
approach to radiation effects utilized all possible
angles - working with viruses, microorganisms, plants, <u>Dro-
sophila</u>, and mouse - from nucleic acid to proteins, using
all possible tools available 35 years ago, taking advantage
of the developments which were new at that time, the use
of tracers included. Thus, it was possible to develop a
fairly comprehensive picture of the effects of ionizing
and ultraviolet radiation. Since this approach was done
from a very fundamental point of view, it had also considerable impact on other biological sciences. Something of
this type will have to be done with chemical mutagenesis.

Where do we go from here? I would like to see an
agreement, possibly sponsored by the IAEMS, on the following
points:
    1. Let a group of us, who are interested in the
       basic aspects of chemical mutagenesis, agree on a

very short list of compounds to be investigated as thoroughly as possible, including all traditional as well as modern biological methods. If we concentrate on a few compounds, we will make more progress than if we spread our efforts on the wide variety of compounds we are presently studying.

2. What are the compounds on which one should concentrate? A committee of the IAEMS should prepare the recommendation. I would like to see us start with the three compounds which form the basis for the Mutagen Burden, i.e., Aflatoxin, Benzo(a)pyrene, and Nitrosamine. One or two additional compounds should be recommended.

3. The IAEMS should take the leadership and coordinate the investigators, collect all the data, see if they are adequate for risk assessment possibly for use in setting tolerance limits. H. Bartsch will have a workshop on Nitrosamines following this conference in Tokyo. However, this type of workshop is very useful, but if too scattered, may fail to emphasize the basic genetic approaches. The approach by the Lyon group is extremely useful.

4. If the investigation of these few compounds is done well enough, I visualize that it could serve as a model to other studies. I believe it is time to get away from reports on one or two Salmonella strains as a means of evaluation of such a wide variety of compounds. Each of these types of reports should make at least an attempt to study the underlying mechanisms.

This should not distract us from all of the important genetic and toxicological work going on. I feel the completion of the Gene-Tox work of the U.S. Environmental Protection Agency (EPA), the different approaches of the ICPEMC , the efforts of the European Economic Community, and the U.S. National Academy of Sciences Committee (NAS) should be encouraged in every way possible. All of these efforts, I believe,will contribute important data to the larger picture which is forming on the complex problems of environmental mutagenesis. I should also compliment the many individual investigators who have done so much to promote this field and who are helping to bring order into this complex field. Indeed, we have in front of us some very promising developments in this relatively new field which hardly existed twenty years ago. I want to pay my special compliments to our Japanese colleagues who have done such a magnificent job in making the study of environmental mutagenesis a very prominent field.

Acknowledgment

This work was supported in part by U.S. Department of Energy Contract EY-76-C-02-0016 with Associated Universities, Inc., Brookhaven National Laboratory.

REFERENCES

1. Muller, H. (1928) The Production of Mutations by X-rays. Proc. Natl. Acad. of Sci., U.S., 14:714-716.
2. Auerbach, C. and J.M. Robson (1946) Chemical Production of Mutations. Nature, 157:302.
3. Rapoport, J.A. (1946) Carbonyl Compounds and the Chemical Mechanism of Mutations. Compt. Rend. Acad. Sci., U.S.S.R., 54:65-67.
4. Oehlkers, F. (1943) Die Auslosung von Chromosomen-mutationen in der Meiosis Durch Einwirkung von Chemikalien 2. Indukt. Abstamm. -u. Vererbungslehre, 81: 313-341.
5. Demerec, M. (1956) Genetic Studies with Bacteria. Carnegie Institute of Washington Publication #612.
6. Hartman, P.E., Z. Hartman, R.C. Stahl, and B.N. Ames (1971) Classificiation and Mapping of Spontaneous and Induced Mutations in the Histidine Operon of <u>Salmonella</u>. Adv. in Gen., 16:1-34.
7. Malling, H.V. (1971) Dimethylnitrosamine: Formation of Mutagenic Compounds by Interaction with Mouse Liver Microsomes. Mut. Res., 13:425-429.
8. Ames, B., J. McCann, E. Yamasabi (1975) Methods for Detecting Carcinogens and Mutagens with the <u>Salmonella</u> Mammalian Microsome Mutagenicity Test. Mut. Res., 31: 347-364.
9. Graf, W. (1975) Carcinogenic Polycyclic Aromatic Hydrocarbons in Humans and Animal Tissues. Infektions Krankheiten und Hygiene, 161:85-103.

10. Russell, W.L., E.M. Kelly, P.R. Hunsicker, K.W. Bangham, S.C. Maddux, and E.L. Phipps (1979) Specific-Locus Test Shows Ethylnitrosourea to be the Most Potent Mutagen in the Mouse. Proc. Natl. Acad. of Sci., U.S., 76: 5818-5819.
11. Osterman-Golkar, S., L. Ehrenberg, D. Segerback, and I. Haeestrom (1976) Evaluation of Genetic Risks of Alkylating Agents II Haemoglobin as a Dose Monitor. Mut. Res., 34:1-10.
12. Oftedal, P. (1976) Problems in the reevaluation of genetic risks from radiation and other environmental hazards. In: Radiation Research, O. F. Nygaard, H. I. Adler, and W. K. Sinclair, eds. Academic Press, N. Y. pp.169-181.

A View from the Ivory Tower

Charlotte Auerbach

Department of Genetics, University of Edinburgh, Edinburgh, Scotland

Let me start by thanking the organisers of this congress for making me one of the honorary presidents. It is a great pleasure for me to preside at a meeting that witnesses progress in a field that I have helped to open up and that has developed so quickly in the forty years of its existence. It took me some time to decide on a suitable topic for my address. In finally choosing it, I reasoned somewhat like this.

(1) My claim to be worth listening to rests nowadays mainly on my age and long experience in chemical mutagenesis. Therefore I shall talk about things past, as I have done on other occasions.

(2) I shall be addressing workers in a field that is expanding very rapidly, with new problems turning up all the time. Therefore I must choose a topic which, although reaching back into the past, has a bearing on present and future research.

(3) And finally, since old people are usually allowed a certain amount of reminiscing, I shall not talk so much about the historical development of chemical mutagenesis as about my reactions to it. So, in the end I was led to the topic that you find in your programme: "View from the Ivory Tower".

The ivory tower is, of course, that in which the researcher tends to immure himself. From its height the bustle of the market place is not easily conceived. It is in such a tower that I started mutation work, and I have never quite left it. As I see it, there are two main aspects to research in an ivory tower: one good, one bad. Let me take the good one first. It consists of the demand for a very high standard of scientific honesty and thoroughness, and for severe criticism of one's own results as well as those published by others. This attitude does not exclude speculation or even quite wild flights of fancy as long as these are presented as such and not as facts. It has remained in full force for fundamental research, although it has been somewhat modified by the present organization of science. By this I mean the fact that nowadays the same problem is often studied in several different laboratories, and that the competition for positions and grants favours those who publish much and quickly. As a result, the replications, controls and ancillary tests that used to form parts of one long paper now are often distributed over several shorter ones, and the truth of a finding is established not so much by the thoroughness of one investigator as by the concordance or mutual confirmation of several.

Now let me turn to the bad aspect of ivory-tower psychology. This is that it tends to make the investigator neglectful and even, sometimes, contemptuous of applied work. This is a very harmful effect of such high-perched isolation. Fortunately, not everybody in the ivory tower is affected by it. Among our group of EMS-workers I may point to Bruce Ames, who often climbs down from the ivory tower into the market place, to the great benefit of his fellow men. Muller, who initiated fundamental mutation research, was strongly motivated by concern for human welfare. Almost from the start of his X-ray work he was concerned about the possible damaging effect of radiations on human heredity. Indeed he was the perfect example of a scientist who did his research in the ivory tower, but broadcast its results in the market place.

Let me tell you a reminiscence which I share with several among you. In the early 1950's, the Russells reported that, in spermatogonia and oogonia of mice, a given dose of X-rays produced more mutations when applied as acute exposure than when given as chronic treatment. Obviously, this finding was of great importance for the assessment of radiation damage, and Muller decided to test for a dose rate effect in oogonia of Drosophila. He reported on this work in an evening lecture at a symposium that Dr. Sobels had organized in Leiden. Muller's first results had agreed with the findings in mice. But at this point his ivory tower attitude made him play devil's advocate and look for possible flaws in his physical and biological set-up. In long-lasting and painstaking experiments he ruled out some possible flaws, e.g. an effect of dose rate on germ-cell maturation, but discovered some others, mainly concerned with discrepancies between dose as measured by physical dosimetry or biologically by the frequencies of sex-linked lethals in spermatozoa. Thus, in the end, the corrected data from three years of intensive experimentation could not be taken as proving a dose rate effect, only as strongly suggesting it. One of the participants at the conference found this "heart-breaking", but I do not think I was the only one who, on the contrary, found it heartening that even a scientist who took human welfare so much to heart felt he could not compromise his scientific honesty and thoroughness for its sake. Moreover, as Muller himself pointed out, the results of his tests were of value for future radiation experiments by ruling out certain sources of error and pointing to the existence of others.

The advent of chemical mutagens did not at first change much in the ivory tower attitude of the investigators. It is true that, very early on, Dr. Robson and I wrote a Letter to Nature, warning against an increased mutational risk in the progeny of cancer patients who had been successfully treated with nitrogen mustard. But for quite a few years, treatment of cancer with a few alkylating agents was the only suspected genetic risk to mankind, and mutation workers on pro- and eukaryotes felt free to deal with research on chemical mutagenesis in true ivory tower fashion. It was in those years that the fundamental facts about the mechanism of chemical mutagenesis at the molecular and supramolecular level were discovered.

Even Rapoport's finding that formaldehyde - a chemical so extensively used in civilized society - is strongly mutagenic for $D_r$osophila larvae did not affect this attitude. We did, indeed, get letters from some pig breeders who were worried that formalin-sterilized milk might produce mutations in the pigs, or

even in the consumers of pork. All we could and did do
at that time was to advise against feeding such milk to
animals used for breeding. Instead we spent several years
studying the highly peculiar action of formaldehyde on
Drosophila larvae: its strict limitation to one sex and
one germ cell stage, its marked dependence on both amount
and quality of nutrition, the negative correlation between
damage to the treated larvae and mutation frequency in
their progeny, and the tendency of lethals to occur in
clusters at closely neighbouring loci, suggesting action
at the replication point. Apart from this last conclusion, which recently has been confirmed for bacteria,
one may query the value of these years of laborious
experimentation. My answer is that, similar to the dose
rate work of Muller, these experiments revealed factors
of development and metabolism which may profoundly modify
the mutagenic effects of treated larval food. In the
case of formaldehyde, one and the same concentration fed
to larvae of one and the same strain may produce anything
from 0 to over 10% sex-linked lethals, depending on the
conditions of treatment. I was reminded of this when it
was found recently that a carcinogen, 7,12-DMBA, is
highly mutagenic when fed to larvae of one strain of
Drosophila but nearly non-mutagenic for another strain.
It seems at least possible that this surprising result,
too, may be due to developmental or metabolic differences
between these strains. If so, then analysis at this
level may bring us a step nearer to an understanding of
the mechanism of mutagenesis by polycyclic hydrocarbons.

Gradually, as more and more environmental mutagens
became known, the need for monitoring suspected genetic
toxins became so urgent that the ivory tower attitude is
now no longer tenable, and we have to base our assessment
of genetic risk on less stringently secured conclusions.
I am sure that most of us are worried about this,
especially those who, while not yet having reached my
age, have known work in the ivory tower. One of these
obviously is Per Oftedal who, at the Budapest meeting,
said "It might be a subject for thought to try to decide
where science ends and social responsibility takes over".
I do not know the answer to this question but I should
like to give you some of my ideas on it.

The major uncertainty arises, of course, in attempts
to extrapolate from lower organisms to man. This forms

the subject of many discussions, and I do not feel that I can contribute anything new to it. Actually, it seems to me that the difficulty of extrapolation may be exaggerated, at least where a substance is clearly mutagenic for a lower eukaryote. It is, of course, possible that such a substance may not penetrate to the human germ cells or may be detoxicated before it does so or may not be transformed into the ultimate mutagen. However, since we have to err on the side of caution where human welfare is concerned, I feel we are justified in warning against a substance that is clearly mutagenic in another eukaryote. But how do we decide whether or not a substance is "clearly mutagenic" in any one system? It is here where I have misgivings, and I want to share them with you.

(1) Let me deal first with the *collection of data*. The need for data is so urgent and speed of publication is so important that I think *all* well established data are worth publishing even if they should be based on only one carefully and critically performed experiment. But I also think that the one and only purpose of such limited data is brief information for other EMS workers; discussion and conclusions at this stage are at best a waste of journal space; at worst they create prejudices in the mind of the reader or even - worst of all - in that of the public. Conclusions can be drawn and hypotheses can be discussed only when the first data have been repeated, confirmed and extended to modified kinds of treatment, and when doubtful points have been cleared up by the necessary ancillary tests.

(2) Next let me turn to the question of **statistics**. I am afraid that here my ideas are somewhat heterodox. It is, of course, indispensable that conclusions must be drawn from statistically significant data. Much attention is given to the refinement of statistical techniques in mutagen testing, with the aim of lending significance to even small differences in mutation frequencies. But we should not forget that the control cells are never completely comparable to the treated ones because of treatment effects on cellular components other than DNA, which may promote or hinder the origin or expression of mutation. These effects are superimposed on those uncontrolled factors that cause differences between spontaneous mutation frequencies in the same system, the

same laboratory, and sometimes published in the same paper in different tables. A weakly positive treatment effect may be statistically significant because it happens to be paired with a very low negative control value, while it does not differ significantly from higher control values to other experiments. I think it is this kind of situation which the Vice President of the National Academy of Sciences, USA, had in mind when he said in the last Annual Report "Sometimes precision is not readily attainable. In these cases, the use of numbers, measurements, statistics and other quantitative indicators might serve to give an air of accuracy to conclusions that are intrinsically uncertain".

I am not a statistician and am better at criticising than at making positive suggestions; but I feel quite sure that the correct statistical treatment of mutation experiments should take more care of the homo- or heterogeneity between replicate tests than of the significance of individual results. I know from experience that a very significant freak result may be produced by some unknown, uncontrolled and unrepeatable difference between control and experimental series. I would never feel justified in drawing conclusions from a single experiment, on whatever astronomical scale and with whatever statistical significance. It is sometimes assumed that dose effect curves can replace replications, but this is not necessarily true. Such experiments are usually carried out on the same population of cells with the same treatment given for different periods of time or at different dilutions. Any uncontrolled peculiarities of the biological or chemical material will affect all series, and the vagaries of biological experiments are such that uncontrolled factors can never be excluded

(3) Thirdly, let me say a few words about *references to literature*. I admit that it is now next to impossible for an experimental worker to keep up with literature even in his own limited field. All the same I feel it is mandatory that every author should carefully read and evaluate those papers whose results he quotes as confirming, contradicting or otherwise illuminating his own results. If he refers to work with a system with which he is not familiar, he should consult a colleague about the validity of the published conclusions. All too

often references to what appears to be established fact are based on cautiously worded suggestions by a critical author or on the less cautiously worded claims by one of those optimistic authors who believe that their shaky data are as well founded as the leaning tower of Pisa. Such uncritical literature references not only can be grossly misleading; they also bar the way to a more fruitful analysis.

(4) This danger is especially great for reviews, and this brings me to my final misgiving: the *advantages and disadvantages of reviews* on mutagenic chemicals. The advantages are obvious. The literature on chemical mutagens increases so rapidly that nobody who starts research on a substance can be familiar with all previous work in this area. He can, of course, obtain very complete reference lists from the Environmental Mutagen Information Centre; but it will usually not be possible for him to read all or even most of the listed papers. So he feels lucky when he finds a review on his chosen substance. But here a snag arises. No reviewer unless he is lucky enough to deal with a newly discovered mutagen, is able to read all relevant publications. Likewise, no reviewer can deal critically with data obtained in systems with which he is not familiar. I am sure that all of you will have come across discrepancies and misunderstandings in certain sections of otherwise good reviews. Obviously a reviewer has to be critical in order to be useful. I agree with Professor Melchers, who recently wrote in an article for the International Review of Cytology. "Secondary literature like this volume is only justifiable when its authors sort out the primary literature (into) 1) genuinely reproducible facts, 2) findings that are single but uncertain ---- , 3) unconfirmed assertions. These (last-named) may stimulate new experiments."

Not only do I agree with this statement. I would go farther and say that an uncritical review is not only useless but harmful because the reader tends to accept statements in a review on trust, so that very soon a doubtful or even wrong conclusion becomes accepted as established fact. I can think of two ways to overcome this difficulty: one is to have reviews written by a group of authors, each familiar with one or a few systems, and between them covering all systems. Alternatively, one might use such a multidisciplinary group for refereeing reviews by one or two authors. My method of choice would be the first or a combination of them both. In any case, the problem of reliable reviewing is of the utmost importance for our work and will have to be solved if we do not want to confuse the newcomer and perpetuate wrong statements.

After having spent most of my time in throwing bricks from the ivory tower into the market place, I want to use the rest of it for encouraging traffic in the opposite direction, that is, the throwing of bricks - building bricks - from the market place into the ivory tower. Tests of mutagens often reveal peculiarities which cannot be dealt with adequately in applied research but may open up new and promising avenues for investigations on the basic mechanisms of mutagenesis. Let me mention a few of them. In plants and animals, storage increases the frequency of chromosome breaks, but not of gene mutations. In Drosophila as well as fungi, certain loci are refractory to mutation by a particular mutagen, while responding to others. In yeast and bacteria, reverse mutations and supersuppressor mutations respond differently to plating medium modifications. In bacteria, the plating medium affects mutation fixation and division delay to different degrees for different loci and different mutagens. Changes in the genetic background may profoundly affect the mutational spectrum with or without mutagenic treatment. The base chain of DNA contains hot spots for different mutagens, and the amino acid chain of an enzyme contains hot spots for mutational response in general. Some of these phenomena have already been investigated in the ivory tower and have yielded interesting results and speculations on DNA-replication and repair; on the interaction between neighbouring bases in DNA; on the relation of mutational hot spots to the tertiary structure of an enzyme and the secondary structure of DNA. These studies are likely to be continued and to offer further insight into the mechanism of mutagenesis.

On the contrary, those peculiarities that obviously occur at the expression level have been neglected because - rightly - their explanation will be found not in the general mechanism of mutagenesis but in peculiarities of cellular metabolism under the combined

stresses of mutagenic treatment and the opening-up of a
new metabolic pathway or the closing-down of an old one.
Yet it seems to me that analysis of these phenomena is
likely to make important contributions to the elucidation
of biochemical pathways and their connections with each
other, perhaps also to the still mysterious cross-pathway
regulation.

   In conclusion, then, I should like to impress on
this audience that our work, in order to retain its
vitality, requires a constant two-way traffic between
ivory tower and market place, with benefit to research
in both places.

Some DNA Repair-Deficient Human Syndromes and Their Implications for Human Health*

B. A. Bridges

MRC Cell Mutation Unit, University of Sussex, Falmer, Brighton, England
*This paper is based on a review lecture given in London to the Royal Society on February 12th, 1981 and published in full in *Proc. Roy. Soc. B*, **212**, 263-278 (1981).

ABSTRACT Several human syndromes are described with which has been associated a deficiency in the ability to repair damage to cellular DNA. This deficiency is generally manifested as a sensitivity to DNA damaging agents. In xeroderma pigmentosum a high frequency of light-induced skin cancers is correlated with hypermutability of fibroblasts following UV irradiation in vitro, providing telling support for the somatic mutation theory of cancer. Closer inspection, however, reveals that a pseudopromoting action of DNA damage may be equally important in skin carcinogenesis in both XP and normal individuals. That more than an enhanced frequency of somatic mutation may be necessary for early skin neoplasms is illustrated by Cockayne syndrome where XP-like changes in the skin are not observed despite an apparently enhanced UV-mutability of cultured fibroblasts.

Xeroderma pigmentosum, ataxia-telangiectasia and Cockayne syndrome all show progressive neurological disease, suggesting that common factors are involved in DNA repair and the normal development and function of the nervous system.

Patients with ataxia-telangiectasia also show a severely depressed immune response and a search for other individuals with impaired immunocompetence revealed patient 46BR whose cells are sensitive to a wide range of DNA damaging agents. It is suggested that common factors may be involved in DNA repair and the proper development and functioning of the immune system.

INTRODUCTION

As a glance through the proceedings of this conference will show, DNA damaging agents are of concern from a health point of view because of their potential for causing genetic disease and cancer. Clearly, repair of DNA damage is a matter of fundamental importance in determining human response to mutagens and one of the chief approaches in man, as in bacteria and yeast, involves the study of variants believed to be deficient in the ability to carry out such repair. These variants tend to be hypersensitive to mutagens such as ultraviolet light and ionizing radiation and to be prone to develop malignant disease. As I shall show, however, they have revealed unexpected associations between DNA repair and neurological and immunological effects which may have implications for human health much wider than mutagenesis and carcinogenesis. In addition, even the carcinogenic action of mutagens on the skin cannot be considered solely in terms of somatic mutation but must take account of the immunosuppressive effect of such agents.

Xeroderma pigmentosum

DNA repair deficiency

The hereditary disease xeroderma pigmentosum (XP) is generally, and rightly, regarded as the classic example of a DNA repair-deficient syndrome. The disease is an autosomal recessive disorder and afflicted individuals are extremely sun-sensitive, the exposed skin exhibiting at an early age many of the features seen when fair-skinned Celts are exposed to a lifetime of sunshine: freckles, hyperkeratoses and eventually malignant neoplasms.

Fibroblasts from most XP individuals do not possess normal ability to excise pyrimidine dimers from their DNA (Cleaver, 1968; Setlow et al., 1969). As a result the cells are hypersensitive to the lethal effect of ultraviolet (UV) light and gene mutations are induced by very low doses of UV (Maher et al., 1976; Arlett, 1980b). This hypersensitivity extends to many chemical mutagens and carcinogens and to DNA adducts formed by the photosensitized reaction with 8-methoxypsoralen and 360 nm UV (PUVA). Genetic control of excision is complex; we recently described the seventh complementation group of XP (Keijzer et al., 1979) cells from one member of which have proved also to be sensitive to ionizing radiation (Arlett et al., 1981). There is no evidence yet that any of the XP genes codes for an enzyme involved in excision and some evidence that other repair enzyme activities may also be deficient (Kuhnlein et al., 1978; Sutherland et al., 1975). Nevertheless the excision repair deficiency in most XP individuals is real enough and the correlation between hypersensitivity of the skin to sunlight tumorigenesis and the hypersensitivity of somatic cells to UV mutagenesis is perhaps the strongest argument in favour of the somatic mutation theory of cancer. As I shall hope to show now, this is not quite as clear-cut as it may seem.

Immune effects and carcinogenesis

For some years I have been worried by an unfortunate observation made by Reed and his colleagues more than six years ago (Reed, 1974; Reed et al., 1977). They treated four XP patients with PUVA, hoping to induce a tan that might protect them against sunlight. As was to be expected from the known DNA damaging and mutagenic effect of PUVA, however, the result was the appearance of skin neoplasms. What was unexpected was the extreme rapidity with which these appeared, in as little as one or two months after the beginning of PUVA treatment. In view of the long latent periods generally associated with skin cancer such an effect seemed inconsistent with an initiating action of PUVA and suggested that this treatment was more likely allowing already transformed foci to express themselves as visible tumours, a phenomenon that might be termed pseudopromotion. More recently Al-Saleem et al., (1980) have reported that inoperable skin tumours in seven out of eight XP patients were made to regress by treatments designed to block immune suppressor cell production and function. There is also the convincing evidence of Kinlen et al., (1979) that patients on immunosuppressive drugs show a markedly enhanced incidence of skin tumours. Accordingly, we proposed (Bridges and Strauss, 1980) that DNA damaging treatments to the skin may cause the appearance of neoplasms from pre-existing foci by a pseudopromoting effect, perhaps by blocking immune control processes.

There is now a formidable body of evidence that profound disturbance of the immune system, both local and systemic, can occur when animals are exposed to ultraviolet light. More recent data show similar effects with

PUVA (for references see Bridges et al., 1981).

In man we recently reported that around 50 per cent of psoriasis patients undergoing PUVA treatment showed a reduced or absent delayed cellular hypersensitivity response to dinitrochlorobenzene (DNCB) (Strauss et al., 1980). We have also observed subsequently that a lesser but detectable degree of impairment of delayed cellular hypersensitivity response in sensitized individuals can be seen in areas masked during PUVA treatment (Bridges et al., 1981). Thus immune impairment in man also has both a local and systemic component. We now know that suppression of delayed cellular hypersensitivity response to DNCB can also be suppressed by an erythemogenic exposure to sunlight (Bridges, unpublished data).

The existence of immune impairment in the skin after treatment with DNA damaging agents such as PUVA and UV, although not yet well understood mechanistically, may be regarded as established (Bridges, 1981d; Kripke, 1981). The evidence that such agents may act as pseudopromoters in skin carcinogenesis is less secure, although suggestive. If true, however, where does this leave the classical correlation of sunsensitivity for skin tumorigenesis and hypersensitivity of fibroblasts for UV mutagenesis. Does it still support the somatic mutation theory of cancer? It probably does, but in a less simple way than is usually assumed. If sunlight does indeed suppress cell-mediated immunity to a much greater extent in XPs than in normal individuals then any initiated neoplastic cell may have a much greater chance of developing into a tumour than a similar cell in a normal individual. The enhanced tumour yield in XPs might then represent a considerable exaggeration of the actual initiation (mutation) rate compared with normal individuals (Bridges, 1981b). It should not be forgotten that the skin may well be a special case, both on account of the large amount of DNA damage it frequently sustains and on account of the immune control mechanisms it clearly possesses which may be not entirely unconnected with a need to control the consequences of the DNA damage.

Neurological effects

Apart from the extensive changes in the skin, the most interesting characteristic of XP is the prevalence of neurological abnormalities. Although not all patients show such effects, when these occur they can be very severe and are commonly progressive. Robbins et al. (1974) list many neurological abnormalities and have suggested that neurons with unrepaired DNA damage might be unable to function or survive and that the progressive neurological dysfunction may be due to the failure of neurons in XP patients to repair damage due to an unknown chemical mutagen. If an endogenous mutagen is involved, it would have implications for all of us since the same hypothetical mutagen is presumably damaging neurons in individuals without a DNA repair deficiency. Since DNA repair is unlikely to be 100% efficient or accurate, then just as with the action of sunlight on normal human skin, one might expect the hypothetical chemical mutagen to cause progressive and random neuron loss in normal individuals. Although this would occur at a much slower rate than in XP individuals it might nevertheless conceivably contribute to the neurological problems of old age such as senile dementia.

Failure to repair chemical mutagen damage to neurons is not, however, the only interpretation of the neurological defect in XP. Some findings are perhaps more consistent with a primary genetic lesion in cell differentiation or in some factor which may affect many processes involving DNA. In this connection it may be mentioned that some XP patients show immature sexual development and a severe reduction in overall growth (de Sanctis and Cacchione, 1932).

Cockayne syndrome

UV sensitivity

Patients with Cockayne syndrome (CS) are extremely sensitive to sunlight and, as with XP, their fibroblasts are hypersensitive to UV (Wade and Chu, 1979). Although no defect in DNA repair has yet been observed, the cells show an abnormal pattern of nucleic acid synthesis after UV. Whereas in normal cells synthesis of both RNA and DNA is depressed after irradiation but reovers within a few hours in CS cells depression but no recovery is seen, as with XP cells (Lehmann et al., 1979).

CS patients do not show any XP-like freckling, hyperkeratoses, or skin neoplasms, nor is there any evidence for any cancer proneness in CS patients (Guzzetta, 1972). Nevertheless, the fibroblasts on one CS individual proved to be hypermutable by UV to a degree comparable with that of an XP variant (Maher et al., 1975; Arlett, 1980a).

This may be regarded as surprising if XP is taken to show that somatic mutations initiate skin cancer. It becomes understandable, however, if CS patients differ from XP patients in not showing a hypersensitivity to the pseudo-promoting effect of UV, as discussed above. Their skin may contain initiated cells but their outgrowth is checked by immune control processes. It could be that CS individuals would show an enhanced incidence of skin cancer following a normal lengthy latent period if they were to live long enough. At any rate, they seem to indicate that any enhanced rate of somatic mutation alone is not enough to generate early skin lesions.

Neurological defects

CS is also characterized by progressive neurological disease. Although apparently normal for the first few months, or even years, affected individuals appear to undergo a crisis after which both physical and mental development is severely retarded.

Ataxia-telangiectasis

DNA repair deficiency

Patients with the hereditary disease ataxia-telangiectasia (A-T) are very prone to develop cancer particularly of the lymphoid tissue. There are also in the radiobiological literature several reports of adverse reaction to radiotherapy. These two properties stimulated David Harnden and I to initiate a study of the radiosensitivity of fibroblasts from A-T patients. We found that they showed considerably hypersensitivity to ionizing radiation (Taylor et al., 1975) just as A-T lymphocytes showed more chromosome damage after irradiation than normal cells (Higurashi and Conen, 1973; Rary et al., 1974). These observations suggested that A-T cells might be deficient in repair of radiation damage. Although the nature of the deficiency is still controversial there is fairly general agreement that this interpretation is correct (see e.g. Bridges and Harnden, 1981).

In one respect A-T cells behave superficially like recA bacteria, and are distinguished clearly from XP cells as to the type of repair-deficiency. A-T cells are not hypermutable by ionizing radiation and may well be hypo- or non-mutable (Arlett, 1980a). It is currently thought that ionizing radiation does not induce base-pair substitution mutations in mammalian cells and that the mutations induced in normal humam fibroblasts, and which are either induced normally or at a reduced level in A-T cells, are small deletions. Putting together the cellular phenomenology of A-T cells one may plausibly suggest that there exists a specific type of radiation damage that in normal cells is repaired or dealt with in a manner that results in small deletions. In A-T cells, however, the pathway is incomplete and leads to chromosomal breakage and cell death.

Although A-T is a single gene mutation, present data do not allow us to conclude that it leads to a deficiency of a single repair enzyme although this is still a possibility. There are some signs that the basic defect may be one of cell differentiation so that the expression of genes is abnormal Some A-T patients, for example, have no thymus, some have hypoplastic gonads and some have indications of embryonal cells in the liver.

Neurological effects

A-T patients manifest a progressive cerebellar degeneration due to neuron loss. As with XP, it is not immediately obvious why DNA repair should be so important to cells which are not required to divide and which are in a relatively well-protected environment. It seems more likely that the abnormality in neurons may have arisen during development. Once a number of cells exist with a built-in self-destruct capability, the eventual loss of these may, through the many feed-back systems in the cerebellum result in greater stress upon other abnormal cells, so hastening their own demise and resulting in progressive disease, as suggested by Kidson and Dambergs (1981).

Immune effects

A particularly interesting feature of A-T is the immunodeficiency. Although A-T patients have T-lymphocytes in almost normal numbers, they are deficient in B cells and in helper T cells. Moreover, they frequently make no IgA or IgE, despite the fact that the α gene (at least) is present (T. Waldmann, pers. comm.). It is possible that the explanation of the immunodeficiency may be found in the hypothesis that A-T is a disease of cell differentiation. A more specific hypothesis (although it may be only a special case of the above) is suggested by the analogy with recA bacteria. The recA protein, besides having important roles to play in DNA repair, is also required for the cutting and splicing of bacterial DNA (genetic recombination). In particular, it catalyses the

the invasion of a superhelical DNA duplex by a free single-stranded DNA end, the first step in genetic recombination. Immunoglobulin genes in mammals appear to exist as several separate regions which have to be cut out and spliced before transcription occurs (Brack et al., 1978; Coleclough et al., 1980); Kataoka et al., 1980). A similar process may well occur in the generation of the specific surface receptors on lymphocytes. A deficiency in an enzyme or cofactor needed for gene splicing and DNA repair could well account for both the immuno-deficiency and the radiosensitivity and chromosome instability in A-T patients.

The cancer-proneness of A-T patients cannot be accounted for on the basis of the somatic mutation theory in an analogous manner to XP. The spontaneous rate of gene mutation in A-T fibroblasts is not outside the normal range and there is no hypermutability after exposure to ionizing radiation (even if ionizing radiation were present naturally in significant doses). The enhanced cancer incidence is more likely a consequence of the spontaneous chromosomal instability. Although chromosomal breaks are probably not likely to lead to malignancy, rarer stable chromosomal aberrations may well be able to do so and may arise rather more frequently than in normal individuals. Nor should one ignore the facts that A-T patients have a profoundly disturbed immune system and that the majority of their tumours are lymphomas and leukemias.

Heterozygotes

Although A-T is a rare disease (perhaps 1 in 40,000 births), the frequency of heterozygous carriers is high. Swift et al. (1976) have calculated that, assuming there is one allele, 1 per cent of the population should carry the gene if the 1 in 40,000 A-T frequency is correct. If more than one allele is involved, the carrier frequency should be even higher. This is of some relevance since Swift et al. (1976; 1981) have produced evidence that blood relatives of A-T patients are also more likely to develop malignant disease than the general population, in particular early onset lymphomas and leukemias and cancers of the ovary, stomach, and bile duct. In culture, cells from heterozygotes are difficult, if not impossible, to distinguish from wild-type cells by currently available techniques. If the cancer-proneness of heterozygotes is real (and a more comprehensive epidemiological study is desirable) it reveals the existence of a genetic predisposition to malignancy by one (or perha.ps a few) DNA repair genes in

the heterozygous state. Given that we know of only a few
DNA repair genes in man, and that others are likely to
exist in the heterozygous state even if non-viable as
homozygotes, the effect of DNA repair genes on current
cancer incidence may be more profound than is generally
supposed.

A more detailed treatment of all aspects of A-T may be
found in Bridges and Harnden (1981).

An immune-depressed patient - 46BR

The association of immune-deficiency with a DNA repair
deficiency in A-T suggested that other immune deficiency
syndromes might show a similar association. My colleagues
have recently reported on an 18-year old girl whose fibroblasts we have designated 46BR, who may represent such a
condition (Webster et al. (1981). The patient is severely
immuno-compromised; in particular she has no IgA and her
lymphocytes fail to transform with mitogens. She has
indications of dwarfism and no secondary sexual characteristics. Her fibroblasts are hypersensitive to UV, gamma
radiation, dimethylsulphate, methyl nitrosourea and nitrosoguanidine, they also show hypersensitive induction of
sister chromatid exchanges after UV. Although 46BR
fibroblasts are still being characterized, the overall
picture so far clearly points to the existence of a DNA
repair deficiency.

The absence of neurological abnormalities and the
sensitivity of the fibroblasts to a wide range of chemical
mutagens distinguish 46BR from A-T, but both have deficiencies in sexual differentiation that may be significant.

Concluding remarks

Our initial interest in identifying human syndromes showing
a deficiency in DNA repair was to build a collection of
mutant cell strains blocked in specific pathways analogous
to those existing in bacteria, that would enable us to
probe the molecular basis for DNA repair in man. It must
be conceded that the strains so far found have not yet
taken us far along this road, even if one includes the
many others not discussed here.

Notwithstanding, the mutant strains and syndromes
so far identified have shed valuable light on several
aspects of human health and disease. In addition to the
role of DNA repair in protecting us against malignant
disease and heritable mutations, there is now a strong
suggestion either that DNA repair is essential for the

proper development and function of the nervous and immune systems, or that all three are dependent upon some common genes.

ACKNOWLEDGEMENTS
Much of the work discussed that has originated in the author's laboratory has been the responsibility of my colleagues Colin Arlett, Alan Lehmann, Ian Teo, Lynne Mayne, Sue Harcourt, Stevie Stevens and Gary Strauss and it is a pleasure to acknowledge their enthusiastic labours. The work on human DNA repair mutants has been jointly supported by the MRC and the European Community under contract 166-76-1-BIOUK.

REFERENCES
Al-Saleem, T., Ali, Z.S. and Qassah, M. *Lancet 2*, 264-265 (1980).
Arlett,C.F. in *Progress in Environmental Mutagenesis* (ed. M. Alacevic) (161-174) Amsterdam:Elsevier/North Holland (1980).
Arlett, C.F., Harcourt, S.A., Lehmann, A.R., Stevens, S., Ferguson-Smith, M.A. and Morley, W.N. *Carcinogenesis 1*, 745-751 (1981).
Brack, C., Hirama, M., Lenhard-Schuller, R. and Tonegawa, S. *Cell 15*, 1-14 (1978).
Bridges, B.A. *Proc.Roy.Soc.B. 212*, 263-278 (1981a).
Bridges, B.A. *Carcinogenesis 2*, 471-472 (1981b).
Bridges, B.A. and Strauss, G.H. *Nature 283*, 523-524 (1980).
Bridges B.A., Strauss, G.H., Hall-Smith, P. and Price, M. in *Psoralens in Cosmetics and Dermatology*, New York/Plenum Press, in press.
Bridges, B.A. and Harnden, D.G. (eds.). *Ataxia-telangiectasia: a link between cancer, immunology and neuropathology*. London:John Wiley and Sons (1981).
Cleaver, J.E. *Nature 218*, 652-656 (1968).
Coleclough, C., Cooper, D. and Perry, R.P. *Proc.Nat.Acad. Sci. US 77*, 1422-1426 (1980).
de Sanctis, C. and Cacchione, A. *Sper.Freniatr. 56*, 269-292 (1932).
Guzzetta, F. in *Handbook of Clinical Neurology* (ed. P.J. Vinken, E.W. Bruyn) (431-440) Amsterdam:North Holland. (1972).
Higurashi, M. and Conen P.E. *Cancer 32*, 380-383 (1973)
Kataoka, T., Kawakami, T., Takahashi, N. and Honjo, T. *Proc.Nat.Acad.Sci. US 77*, 919-923 (1980).

Keijzer, E., Jaspers, N.G.J., Abrahams, P.J., Taylor, A.M.R., Arlett, C.F., Zelle, B., Takebe, H., Kinmont, P.D.S. and Bootsma, D. *Mutation Res. 62*, 183-190 (1979).
Kidson, C. and Dambergs,R in *Ataxia-telangiectasia: a link between cancer, immunology and neuropathology* (ed. B.A. Bridges, D.G. Harnden) London:John Wiley and Sons (1981).
Kinlen, L. et al. *Brit.Med.J. 2*, 1461-1466 (1979).
Kripke, M.L. *Photochem.Photobiol. 32*, 837-839 (1980).
Kuhnlein, U., Lee, B., Penhoet, E.E. and Linn, S. *Nucl.Acid Res. 5*, 951-958 (1978).
Lehmann, A.R., Kirk-Bell, S. and Mayne, L. *Cancer Res. 39*, 4237-4241 (1979).
Maher, V.M., Ouellette, L.M., Curren, R.D. and McCormick, J.J. *Nature 261*, 593-595 (1975).
Maher, V.M., Curren, R.D., Ouellette, L.M. and McCormick, J.J in *Fundamentals in Cancer Prevention* (ed. P.N. Magee, S. Takayama, T. Sugimura, T. Matsushima) (363-379) Tokyo: Univ. of Tokyo Press/Baltimore:Univ. Park Press (1976).
Rary, J.M., Bender, M.A. and Kelly, T.E. *Amer.J.Hum.Genet. 26*, 70A (1974).
Reed, W.B. *Acta Dermatovener 56*, 315-318 (1976).
Reed, W.B., Sugarman, G.I. and Mathis, R.A. *Arch.Dermatol. 113*, 1561-1563 (1977).
Robbins, J.H., Kraemer, K.H., Lutzner, M.A., Festoff, B.W. and Coon, H.G. *Annals Int.Med. 80*, 221-248 (1974).
Setlow, R.B., Regan, J.D., German, J. and Carrier, W.L. *Proc.Nat.Acad.Sci. US 64*, 1035-1041 (1969).
Strauss, G.H., Bridges, B.A., Greaves, M., Hall-Smith, P., Price, M. and Vella-Briffa, D. *Lancet 2*, 556-559 (1980).
Sutherland, B.M., Rice, M. and Wagner, E.K. *Proc.Nat.Acad.Sci. US 72*, 103-107 (1975).
Swift, M., Sholman, L., Perry, M. and Chase, C. *Cancer Res. 36*, 209-215 (1976).
Swift, M.in*Ataxia telangiectasia: a link between cancer, immunology and neuropathology* (ed. B.A. Bridges, D.G. Harnden) London:John Wiley and Sons (1981).
Taylor, A.M.R., Harnden, D.G., Arlett, C.F., Harcourt, S.A., Lehmann, A.R., Stevens, S. and Bridges, B.A. *Nature 258*, 427-429 (1975).
Wade, M.H. and Chu, E.H.Y. *Mutation Res. 59*, 49-60 (1979).
Webster, D., Arlett, C.F., Harcourt, S.A., Teo, I. and Henderson,L. in *Ataxia-telangiectasia: a link between cancer, immunology and neuropathology* (ed. B.A. Bridges, D.G. Harnden) London:John Wiley and Sons (1981).

Factors Affecting Mutagenicity of Ethylnitrosourea in the Mouse Specific-Locus Test and Their Bearing on Risk Estimation

W. L. Russell

Biology Division, Oak Ridge National Laboratory, Oak Ridge, Tennessee, U.S.A.

ABSTRACT The high mutagenic effectiveness of ethylnitrosourea (ENU) in the mouse specific-locus test has permitted rapid progress in investigating many of the complexities of mutagenic action in the mouse. These include effect of sex and cell stage, dose-response curve, dose fractionation and nature of the mutations. The mutagenic effect of related chemicals is markedly different. In view of the high mutational response to ENU, the weak or zero response to several chemicals that are potent mutagens in other systems cannot be attributed to insensitivity of the specific-locus test. It is probably due to an efficient repair mechanism in mouse spermatogonia. Dose fractionation experiments indicate that there is extensive repair even against ENU-induced mutation when the dose is low enough. Many of these results raise questions about the reliability of some approaches to risk estimation.

INTRODUCTION

The induction of mutations by radiation in mice has long been known to be a complex process affected by many factors, including some important ones that were not predicted from work on non-mammalian organisms (7). It is reasonable to suppose that chemical mutagenesis in mammals will be even more complex and largely unpredictable. Therefore, in order to evaluate probable risks to man, it seems desirable to conduct comprehensive, in-depth studies on transmitted mutations induced in mammals by a variety of model mutagens. The questions to be investigated should include the effect of sex and cell stage, dose response curve, dose fractionation, nature of the mutations, and many others.

Until recently, work of this nature has been hampered by the fact that most chemical mutagens tested for the induction of transmitted mutations in mice have proved to give only weak or zero effect in the spermatogonial stem cells, the cells which are of prime concern in human risk estimation. Even procarbazine, which was the most effective mutagen for these cells, produced, at its most potent dose, only about one third as many mutations as could be obtained with 600 R, the most effective single dose of X radiation (3). So, even with this compound, it has taken considerable research effort to make any progress in the analysis of the variables that might affect its mutagenic action. The recent discovery (16) that ethylnitrosourea (ENU) is a 'supermutagen' in mouse stem-cell spermatogonia has identified a chemical with which rapid progress can be made in the investigation of the complexities of mutagenic action in the mouse. The major findings obtained so far are reported here.

METHOD

The results presented all come from the use of the specific-locus method for visible mutations (10). Treated wild-type mice are mated to individuals from a test stock carrying seven visible markers in homozygous condition. F_1 offspring are scored for mutations at any of the seven gene loci. The method detects gene mutations and deficiencies. The phenotypes of animals homozygous for the mutations obtained range from death in early embryonic stages to near normality, i.e., to a phenotype intermediate between wild type and the expression of the viable test-stock allele.

MUTAGENIC EFFECTIVENESS OF ENU

The most startling property of ENU is its effectiveness compared with that of other mutagens so far tested in the mouse. In one experiment in which a dose of 250 mg/kg of ENU was injected into 9-12 week old males, 72 mutations at the specific loci were obtained in the 10,146 offspring derived from cells that were stem-cell spermatogonia at the time of treatment. This is 8 times greater than the mutation frequency obtained with 600 R, the most effective acute dose of X rays. It is 23 times as high as the mutation rate from 600 R of chronic gamma rays. Compared to the peak mutation rate produced by procarbazine, heretofore the most mutagenically effective chemical known for mouse spermatogonia, the ENU rate proves to be 24 times higher. The above ENU mutation rate is 134 times that of the spontaneous mutation rate. In terms of hazard, it is disturbing to realize that this mutation frequency from ENU is more than 80,000 times greater than that considered as a maximum permissible level of risk from a whole year of exposure to radiation. Clearly, the mutagenic effectiveness of ENU not only facilitates the exploration of the factors affecting its mutagenic action, it also indicates some urgency in doing this, in case man is exposed to any chemicals with similar potency.

EFFECT OF SEX AND CELL STAGE

Current experiments on the male are showing that, whereas doses of 100 or 250 mg/kg of ENU yield high mutation rates in stem-cell spermatogonia, the same doses to postspermatogonial stages are giving mutation rates approximately one order of magnitude lower. This difference is in the same direction as that obtained with mitomycin C (8), but does not parallel the results obtained with, for example, procarbazine and ethylmethanesulfonate (EMS). Procarbazine appears to have roughly similar mutagenic effect in spermatogonial and postspermatogonial stages, while EMS, which is mutagenic in postspermatogonial stages, has, so far, produced no mutations in spermatogonia (8). None of these results resemble those from acute X irradiation, in which postspermatogonial stages give about twice the mutation frequency found in spermatogonia. These contrasting findings provide one illustration of how biological factors in the mammal introduce much complexity into chemical mutagenesis.

The mouse results with ENU apparently differ from those obtained in Drosophila, where exposure of spermatogonial stages is reported to yield mutation rates similar to, or less than, those obtained in postspermatogonial stages (5,7). However, the actual comparison desired is not yet possible, because, to my knowledge, no mutation frequencies have been determined in treated spermatogonial stem cells in Drosophila.

Mutation frequencies in the offspring so far collected from females injected with ENU (mostly at 100 mg/kg) are also low. In conceptions occurring within six weeks after treatment, which involve germ cells that were mature or maturing oocytes at the time of exposure, the mutation rate, like that in postspermatogonial stages, is currently running at about one order of magnitude below the rate in spermatogonia. From exposure of arrested oocytes, scored in conceptions occurring more than six weeks after treatment, only one mutation has been observed in over 10,000 offspring, a figure not significantly different from the control.

DOSE-RESPONSE CURVE

Preliminary results are available on the shape of the dose-response curve for exposure of spermatogonia over a range of doses from 25 to 250 mg/kg, with extensive data only at 100 and 250 mg/kg. The curve now appears to be S-shaped with the highest response per unit of exposure at 100 mg/kg. The 250 mg/kg results fall below an extrapolation of a linear fit through the 0 and 100 mg/kg points. The absolute mutation frequency at 250 mg/kg is, however, considerably higher than that at 100: it does not fall below it to produce a humped dose-response curve like that which occurs with X rays (11), and is claimed for procarbazine (3), for comparable levels of spermatogonial killing. The cell killing, as judged by the length of the sterile period, at 100 mg/kg of ENU is closely similar to that obtained at 600 R of acute X rays or with 600 mg/kg of procarbazine; and it is above these levels that the dose-response curves for X rays and procarbazine take on negative slopes.

At each of the three doses of 25, 50, and 75 mg/kg, the mutation frequency per unit dose is currently lower than that for 100 mg/kg, but more data are needed before statistically valid conclusions can be drawn regarding these single-dose exposures. However, the importance, for risk evaluation, of a possible drop below a linear response led to the studies reported in the next section.

EFFECT OF DOSE FRACTIONATION

Doses of 10 mg/kg of ENU were injected into males at 10 weekly intervals, and the resultant mutation frequency in spermatogonia from this total exposure of 100 mg/kg is now being scored for a comparison with the result from a 100 mg/kg dose given in a single injection. To date, only 4 mutations have been obtained in 12,664 offspring from the fractionated exposure. This is less than one tenth of what would be expected from our total of 63 mutations in 21,253 offspring for the single exposure. The actual ratio of the induced mutation frequency (experimental minus control) with the fractionated exposure to that with the single exposure is 0.09. If one-tenth of the mutation frequency obtained with the fractionated exposure can be considered an indication of what would result from a single exposure of 10 mg/kg, then we can conclude that reducing the dose from 100 to 10 mg/kg would reduce the mutation frequency by a factor of more than 100.

This result is obviously of vital importance for risk estimation, especially when considered in conjunction with an important related finding in our laboratory by Carricarte and Sega (2). Earlier, Sega et al (17) had demonstrated that the amount of unscheduled DNA synthesis occurring in mouse spermatids following injection of mutagenic chemicals is often directly proportional to the amount injected. The new work by Carricarte and Sega shows that this is true for ENU over the range of 100 to at least 10 mg/kg, and possibly lower. Assuming that the concentration of ENU reaching spermatogonia is similar to that indicated for the spermatids, the above results suggest that the much lower than proportional mutational response obtained when the injected amount of ENU is reduced from 100 to 10 mg/kg is due, not to failure of the chemical to reach the gonad, but to something that happens in the spermatogonia. This could be either an ability to detoxify the chemical before it attacks the DNA, or, perhaps more likely, a capacity to repair genetic damage when the repair process is not swamped by a high dose.

NATURE OF THE MUTATIONS

The frequency distribution among the seven loci of the mutations obtained from ENU is strikingly similar to that observed for X and gamma irradiation of spermatogonia, with one exception. With the chemical or with irradiation, the mutation rate is relatively low at the a and se loci, intermediate at the c locus and high at the b, d and p loci. With radiation, however, the frequency is highest of all at the s locus, whereas with ENU the frequency at s is low, not much higher than that at the a and se loci. In view of the evidence, presented below, that the ENU-induced mutations may be primarily gene mutations, the general correspondence of the distribution among the loci with that for irradiation of spermatogonia provides some support for the view that the radiation-induced mutations may also be primarily gene mutations. This argument is based on the observation that when deficiencies are known to constitute a larger proportion of the mutations scored, as is the case with X irradiation of postspermatogonial stages or neutron irradiation of spermatogonia, then the distribution among the loci is noticeably changed (12).

More than 25% of the mutations investigated to date have a phenotypic expression intermediate between those of the viable test-stock allele and wild-type. This suggests that perhaps most of the events are gene mutations. This view is further strengthened by the observation that not a single one of the 267 independent specific-locus mutations obtained, to date, in spermatogonia in our ENU experiments is a d-se deficiency, a type which is not uncommon when the conditions of a radiation experiment are known to result in deletions.

Information from tests on the viability of the ENU-induced mutations in homozygous condition is far from complete, but some interesting results have already emerged. At all loci, except d and s, the proportion of mutations that are lethal in homozygous condition is very low, certainly less than 5%. At the d locus, most spontaneous mutations in homozygous condition are juvenile lethal, the animals dying at about weaning age. More than 80% of the ENU-induced d locus mutations now tested are similarly juvenile lethal as homozygotes. At the s locus, only 13 ENU-induced mutations have been tested, to date. Of these, 6 are viable and 6 are postnatally lethal in homozygous condition. The remaining one may be a prenatal lethal. Similar results have been obtained on a smaller sample of mutations induced by procarbazine (3). The conclusion that emerges from these findings is that, whether a mutation is going to be viable or lethal in homozygous condition depends, not only on the nature of the mutagen, but also largely on the locus. Disturbances at some loci are apparently much more likely, than those at other loci, to be fatal to the organism.

Results from other specific-locus-locus experiments, such as that being conducted by Johnson and Lewis (6) are likely to provide additional information on the nature of the ENU-induced mutational events.

EFFECT OF ETHYLNITROSOURETHANE

In preliminary testing of compounds related to ENU, we have observed only 1 specific-locus mutation in 3,877 offspring of males injected with 312 mg/kg of ethylnitrosourethane, and 1 mutation in 636 offspring of males that received a dose of 400 mg/kg. All these data involved exposure of the spermatogonial stage. Clearly, this compound is much less mutagenically effective than ENU in mouse spermatogonia.

MUTATION FREQUENCIES WITH METHYLNITROSOUREA

The mutation results obtained with methylnitrosourea (MNU) are most surprising. This compound is much more toxic than ENU, and we have, therefore, had to limit our injected amounts to 80 mg/kg and below. To date, the mutation frequency in spermatogonial stem cells for doses in the range of 70 to 80 mg/kg is only 1 in 15,757 offspring. This does not approach being significantly above the control rate. A comparable ENU exposure would have given about 30 mutations in an equivalent number of offspring. Thus, MNU, which is a powerful mutagen in other biological systems (in some cases more potent than ENU), is having little or no effect in mouse spermatogonia.

Adding to this surprise is another unexpected result. In the past few months, we have been exploring the effect of MNU on post-stem-cell stages (i.e. including differentiating spermatogonia and postspermatogonial cells). In a total of only 4,631 offspring from males exposed to doses of 50 or 75 mg/kg, we have observed 7 specific-locus mutations induced in these stages. This is almost as high a frequency as that which ENU would produce from comparable exposures to spermatogonia. It also appears to be considerably higher than the ENU-induced mutation rate in post-stem-cell stages, although exact comparison is currently hampered by different proportions of offspring from the various post-stem-cell stages. Thus, in the mouse specific-locus test, ENU is a supermutagen in spermatogonia, where MNU does little or nothing; and MNU is a powerful mutagen in post-stem-cell stages, where ENU appears to be weak. It is worth noting, in connection with these results, that MNU is much more effective than ENU in inducing dominant lethals in postspermatogonial stages (4).

RISK ESTIMATION

Several conclusions concerning risk estimation can be drawn from the information presented in this paper. The extreme sensitivity of mouse spermatogonia to the induction of specific-locus mutations by ENU shows that the specific-locus test is not, as has been suggested, an insensitive measure of chemical mutagenesis. Therefore, the long list of chemicals that are highly mutagenic in other systems, but not mutagenic, or only weakly so, in the specific-locus test on mouse spermatogonia, demonstrates, not that the test is insensitive, but that mouse spermatogonia are insensitive or have a powerful repair capacity. This conclusion certainly seems valid for compounds like EMS and MNU, which are known to reach other germ-cell stages in active form, and, therefore, almost certainly reach the spermatogonia as well.

The fact that a single injection of 6 mg of ENU per mouse can induce a mutation frequency more than 80,000 times greater than that considered as a maximum permissible level of risk from a year of exposure to radiation raises much concern that there may be chemicals to which man is exposed that have a similar mutagenic effectiveness. The worry is somewhat reduced, however, by the finding of the fractionation effect. If the explanation of the effect lies in repair, then it indicates that, even for the most effective mutagen known in the mouse, the spermatogonia can, at low doses, repair at least 90% of the genetic damage induced. This raises the question of whether the percentage of damage repaired might be even greater at still lower doses.

In calling ENU the most effective chemical mutagen known for mouse spermatogonia, the word "effective" is used in the sense of producing the most mutations at subtoxic levels of administration. On a molar basis, calculated from the doses that give peak effects, two other chemicals, triethylenemelamine (TEM) and mitomycin C are more mutagenic than ENU in mouse spermatogonia, but these compounds are so toxic that only low doses can be given, and, therefore, only relatively low mutation rates can be obtained. TEM is so toxic that the maximum single dose given in mutation experiments was 2 mg/kg (9). An equimolar dose of ENU would be 1.1 mg/kg. At those doses it is likely, on the basis of the fractionation results with ENU, that the TEM/ENU potency ratio would be even greater than the potency ratio computed by equi-molarity extrapolations from the experimental results that were obtained at a high dose of ENU and a low dose of TEM. Thus, if man were being exposed to these chemicals at the 1 to 2 mg/kg dose levels, TEM would constitute by far the greater hazard, assuming, of course, that the human response is like that of the mouse. This example illustrates one of the considerations that must obviously be kept in mind when estimating relative mutagenic risks of chemicals.

It has been proposed that mutation frequency induced by a chemical in mice can be effectively estimated by doing mutation and dosimetry studies in Drosophila, and then simply following this with dosimetry in mouse germ cells. The fractionation effect with ENU presents difficulties for this proposed extrapolation. The concept might have held if the reduced effect with fractionation had been due to a smaller proportion of the total dose reaching the germ cells. However, the results on unscheduled DNA synthesis mentioned earlier strongly indicate that this is not the case.

In other publications (14,15) I have emphasized the danger of oversimplifying the problem of risk estimation. The data on ENU and related compounds provide further illustrations of this danger. One example is the problem of extrapolating from Drosophila to mice. We tested ENU in the mouse because it is a powerful inducer of gene mutations in Drosophila. However, its effect in Drosophila appears to be greatest in postspermatogonial stages, and in these stages in the mouse it is relatively ineffective. Before testing ENU, we had done an extensive test on diethylnitrosamine which is also a strong mutagen in Drosophila. In the mouse specific-locus test on spermatogonia, it gave only three mutations in 60,179 offspring (16), a rate almost identical with that in the control. MNU, another strong mutagen in Drosophila postspermatogonial stages, may be said to have a parallel effect in the mouse, but there was no prediction that it would be ineffective in spermatogonia, the cell stage of primary importance in risk estimation.

I would be horrified if any of these present difficulties led to a lessening of research on Drosophila. I would urge that, on the contrary, investigation of the differences in response between mice and Drosophila should be seized upon as a valuable approach to an understanding of the mechanisms of mutagenesis. A similarly useful approach is offered by the variation in response between different germ-cell stages. In this connection, I would urge a major expansion of Drosophila research on spermatogonia, and particularly on the stem-cell spermatogonia, the mutagenic response of which, as far as I know, has not been investigated.

Drosophila was picked only as an example to illustrate the difficulties of extrapolating chemical mutagenesis results from non-mammalian systems to mammalian germ cells. Many other examples could be cited. Even results on mammalian cells in vitro are not proving to be reliably predictive of effects in spermatogonia. One relatively short-term test that appears to give results showing good correlation with those from the mouse specific-locus test in spermatogonia, is the mammalian spot test (9).

As was pointed out in the Drosophila-mouse comparison, differences in response between different test systems should provide useful insights into the nature of mutagenesis. A fuller understanding of this may, in turn, eventually eliminate some of the difficulties in arriving at simple test systems that reliably predict what we really want to know for genetic risk evaluation.

The findings reported here will, I hope, stimulate research at the molecular level. For example, in addition to finding out what methylation and ethylation do to DNA in vitro, or in simple organisms, we need to know what biological factors make the relative mutational responses of stem-cell and post-stem-cell stages to MNU and ENU so diametrically opposite.

In view of the fact that ENU apparently produces predominantly gene mutations, rather than deletions or other chromosomal aberrations, it was important for risk evaluation to find out whether it would produce, in significant numbers, the kinds of dominant skeletal mutations that occur with ionizing radiation and which have proved so valuable in estimating the risk of serious disorders in first-generation offspring (1). Selby (18) has already obtained 7 probable skeletal mutations in 309 offspring from treated spermatogonial stages of males injected with 150 mg/kg of ENU. Three of these 7 would probably be considered serious clinical problems if they occurred in humans.

In the discussion on the nature of the mutations, it was concluded that some loci (two out of the seven used in the specific-locus test) are more prone than the others to have mutations that are lethal in homozygous condition. Selby's results suggest that there are a sizeable number of loci throughout the genome that are prone to have mutations with serious dominant effects. If so, and if ENU can be considered as a model of chemical mutagens that induce primarily gene mutations, it would appear that even this class of mutagens could have serious effects in first-generation offspring.

Research jointly sponsored by the National Institute for Environmental Health Sciences under IAG #222 Y01-ES-10067 and the Office of Health and Environmental Research, U.S. Dept. of Energy, under contract W-7405-eng-26 with the Union Carbide Corporation.

(1) BEIR III *Report of the Advisory Committee on the Biological Effects of Ionizing Radiation* National Academy Press, Washington, D.C. (1980)
(2) Carricarte, V. A. and Sega, G. A. Personal communication.
(3) Ehling, U. H. and Neuhäuser, A. *Mutat. Res. 59*, 245-256 (1979)
(4) Generoso, W. M. Personal communication.
(5) Grell, E. H., Grell, R. F. and Generoso, E. E. *Environ. Mutagenesis 3*, 381 (1981)
(6) Johnson, F. M. and Lewis, S. E. *Proc. Natl. Acad. Sci. USA 78*, 3138-3141 (1981)
(7) Ondrej, M. *Folia Biol. (Prague) 16*, 230-236 (1970)
(8) Russell, L. B., Selby, P. B., von Halle, E., Sheridan, W. and Valcovic, L. *Mutat. Res. 86*, 329-354 (1981)
(9) Russell, L. B., Selby, P. B., von Halle, E., Sheridan, W. and Valcovic, L. *Mutat. Res. 86*, 355-379 (1981)
(10) Russell, W. L. *Cold Spring Harbor Symp. Quant. Biol. 16*, 327-335 (1951)
(11) Russell, W. L. *Genetics 41*, 658 (1956)
(12) Russell, W. L. *Jpn. J. Genet. Suppl. 40*, 128-140 (1965)
(13) Russell, W. L. *An. da Acad. Brasileira de Ciencias 39*, 65-75 (1967)
(14) Russell, W. L. *Arch. Toxicol. 38*, 141-147 (1977)
(15) Russell, W. L. *Genetics 92*, s187-s194 (1979)
(16) Russell, W. L., Kelly, E. M., Hunsicker, P. R., Bangham, J. W., Maddux, S. C. and Phipps, E. L. *Proc. Natl. Acad. Sci. USA 76*, 5818-5819 (1979)

(17) Sega, G. A., Owens, J. G. and Cumming, R. B. *Mutat. Res. 36*, 193-212 (1976)
(18) Selby, P. B. in *Mutagenicity: New Horizons in Genetic Toxicology* ed. Heddle, J. A. Academic Press, New York (in press)

Role of DNA Damage in Radiation and Chemical Carcinogenesis

Arthur C. Upton

Institute of Environmental Medicine, New York University, New York, New York, U.S.A.

ABSTRACT Inherited differences in susceptibility to cancer, the frequent occurrence of chromosomal abnormalities in cancer cells, and the correlation between genotoxicity and carcinogenicity, implicate genetic determinants in the pathogenesis of the disease. The kinetics of carcinogenesis *in vivo* and cell transformation *in vitro* imply that a series of genetic and epigenetic changes are involved. Analysis of the nature of such changes, with identification of the transforming genes in some cases, suggests that recombinational changes in DNA, in addition to germinal and somatic mutations, may constitute a mechanism of viral, physical, and chemical carcinogenesis.

INTRODUCTION

The importance of genetic factors in the etiology of cancer has long been recognized. Familial differences in susceptibility to the disease have been known since the 17th century (36), and the concept that carcinogenesis might result from newly acquired damage to the genetic apparatus of a somatic cell--the so-called somatic mutation hypothesis--dates back at least to the beginning of the present century (6). Within the past 20 years, advances in pharmacokinetics, molecular genetics, cytogenetics, gene sequencing, and somatic cell genetics have enabled the role of DNA damage in carcinogenesis to be studied in increasing detail.

THE MULTISTAGE NATURE OF CARCINOGENESIS

Attempts to correlate the development of cancer with alterations in DNA must take into account evidence that carcinogenesis is a multi-step process. In adult populations, the incidence of neoplasms increases with age as a power function, which varies from one neoplasm to another, approximating 4 for non-melanomatous cancers of the skin, 5-6 for cancers of the gastrointestinal tract and 11 for cancers of the prostate, in a given human population (13). In a population exposed chronically to a carcinogenic stimulus, the resulting excess of cancers also typically increases as a power function of the duration of exposure, with rates for different dose levels which form a family of curves paralleling the curve for controls (1). A single exposure to a carcinogen, such as whole-body ionizing radiation, may also cause a dose-dependent increase in the death rate from neoplasms without appreciably altering the slope of the power relationship between mortality rate at a given dose and advancing age (58). These power law relationships have been interpreted to signify that the process of carcinogenesis involves two or more successive causal events, some of which may be mutational changes, and that the age-distribution of "spontaneous" as well as "induced" cancers denotes the cumulative effects of exposure to successive carcinogenic stimuli (13, 14).

Although some carcinogenic effects--such as those of radiation--appear only after a long latent period, others may appear more rapidly; for example, the risk of lung cancer in ex-smokers stops rising almost immediately after cessation of cigarette smoking (13), implying that smoking can exert relatively prompt effects on late stages in carcinogenesis. Distinction between agents that act at early stages in carcinogenesis and those that act at later stages has become a major focus of contemporary cancer research. "Promoting" agents as a class--in contrast to "initiating" agents--are inferred to affect relatively late stages of carcinogenesis and to act through epigenetic mechanisms rather than through damage to DNA. Although initiating agents, which are genotoxic as a class, are generally thought to act through genetic mechanisms, this conclusion remains to be proven (49).

Because carcinogens are, in general, cytotoxic, their effects are complicated *in vivo* by cell killing and resulting alterations in cell population kinetics, organ function, endocrine balance, immunological reactivity, etc. Dose-response relationships for carcinogenic effects *in vivo* may thus reflect a diversity of mechanisms, depending on the conditions of exposure. Although analysis of dose-incidence data cannot, therefore, be expected to identify the mechanisms of carcinogenesis, the dose-incidence curve for certain carcinogen-induced cancers is consistent with a non-threshold function, at least in the low-to-intermediate dose range (46, 57). Hence, such a curve is compatible with the hypothesis that cancer may result from the transformation of a single, suitably susceptible cell through damage to its genes or chromosomes (46, 57).

INFLUENCE OF GENETIC BACKGROUND ON SUSCEPTIBILITY

The correlation between susceptibility to cancer--spontaneous or induced--and various hereditary traits provides further evidence in support of the somatic mutation hypothesis, since mutations affecting susceptibility to cancer which are acquired via germinal cells (31, 43, 56) should in principle also affect susceptibility when acquired via somatic cells, at least in certain instances (31). Inherited traits associated with increased susceptibility to cancer include chromosomal disorders, Mendelian recessive diseases, dominant tumor syndromes, and various other conditions (31, 39). In many of these traits susceptibility to specific types of cancer is increased by orders of magnitude. The particular genes responsible for the observed changes in susceptibility remain to be identified, but rapid advances in cytogenetics, gene cloning, and gene sequencing have made the identification and characterization of such genes increasingly feasible (31).

Many of the disorders with heightened susceptibility to cancer are also characterized by chromosomal abnormalities, chromosomal instability, and defects in DNA repair (39, 45). In a high proportion of human cancers, moreover, the cells show abnormal clonal karyotypes (50), and in some such instances cells with the chromosomal abnormalities have been detected in advance of the clinical onset of malignancy (50). These findings, considered in the light of the genotoxicity of initiating agents and complete carcinogens (2, 40, 64), strongly point to alteration of genes and chromosomes as putative mechanisms of carcinogenesis.

DNA DAMAGE AND CARCINOGENICITY

The development of assay methods containing suitable metabolic activation systems has disclosed an increasing proportion of complete carcinogens to be capable of causing DNA damage (63, 64) and mutations (2, 40, 48, 53). Although the ability of such agents to damage DNA may not necessarily account for their carcinogenic action, certain DNA lesions may be inferred to be carcinogenic on the basis of more-direct evidence. For example, O^6-alkylguanine adducts in DNA have been implicated in the preferential tumorigenic effects of nitrosourea on the rat brain, the brain being the only tissue of exposed animals in which repair systems fail to prevent such adducts from accumulating (19, 37). Similarly, pyrimidine dimers have been implicated in the tumorigenic effects of ultraviolet irradiation on fish cells, since the effects are preventable by photoreversal with visible light (20).

The relationship between DNA damage and carcinogenicity has been elucidated further through studies using *in vitro* systems, in which damage to genes and chromosomes can be compared concomitantly with neoplastic transformation in cells of one and the same population. Studies with such systems have revealed that the frequency of mutations at a given genetic locus is orders of magnitude lower than the frequency of chromosome aberrations (16) or neoplastic transformations (3, 7, 24). They have also indicated that transformation is a multi-step process, usually becoming manifest only after 20 or more cell doubling divisions (4, 17). Post-treatment with caffeine can either enhance or inhibit *in vitro* transformation, depending on the dose and the cells in question, which has been interpreted to implicate misrepair of DNA damage as a possible mechanism of carcinogenesis (15, 32), in keeping with observations on the effects of caffeine *in vivo* (44, 65). In general, however, the kinetics of induction and repair of transformational damage resemble more closely the kinetics for chromosome aberrations than the kinetics for mutations (18, 29, 33) or cell killing (25, 34).

While cell culture systems provide an increasingly powerful analytical tool, interpretation of the kinetics of *in vitro* transformation is complicated by the fact that the ultimate rate of transformation in cells which have been exposed to radiation or a chemical carcinogen can be subsequently modified by a diversity of factors, including many agents that promote or inhibit carcinogenesis *in vivo* (5, 26, 27, 38, 47, 62). In fact, it would appear that the number of cells which ordinarily express transformational changes is but a small percentage of those which are potentially able to do so, for reasons which remain to be determined. It is noteworthy that essentially 100 per cent of cells are apparently transformable under certain conditions (22, 27).

GENETIC RECOMBINATION AND TRANSPOSITION

Error-prone recombinational changes in the genome, which have come to be recognized as an important mechanism of normal differentiation (11, 23), have been implicated increasingly as a mechanism of carcinogenesis since Temin (54) first proposed such a possibility to account for the oncogenic effects of oncorna viruses. Various types of recombinational changes can be envisaged to result directly or indirectly from damage to DNA. The activation or amplification of normal genes by such shuffling of genetic sequences, as has been observed in cells transformed by tumor viruses (21, 30, 33, 42), constitutes an increasingly plausible mechanism through which the carcinogenic effects of physical and chemical agents also might be mediated (8, 30, 42, 60).

It is conceivable, furthermore, that tumor promotion, which is itself a multi-step process (52), may reflect successive recombinational changes, and that these may be produced through epigenetic mechanisms. Evidence that various chemically transformed cells may contain only a single transforming gene (51) can be reconciled with the multi-stage nature of carcinogenesis, if the activation of such a gene is postulated to entail more than one recombinational event.

Although mutagens vary in the extent to which they increase the relative frequency of transpositions, as compared with mutations (8, 9), an enhanced probability of recombinational changes is to be expected in view of the large numbers of polynucleotide strand breaks that are likely to be produced when DNA is damaged and/or repaired. For example, the amount of whole-body radiation that is required to double the lifetime risk of fatal cancer--namely, 3-5 Sv (59)--is orders of magnitude larger than the dose (0.02 - 0.05 Sv) that is required to double the frequency of chromosome aberrations in lymphocytes (35). Furthermore, this dose can be calculated to cause hundreds of DNA strand breaks in every cell of the body (10). The fact that carcinogen-treated cells typically require many cell divisions to express their transformational damage implies that the transforming lesion evolves step-wise. Hence, a succession of alterations in the genome, including recombinational changes initiated by destablilizing effects of DNA damage (12) and evolving under the influence of continuing selection pressure (28, 55, 60) can be conceived to constitute a possible unifying mechanism of physical, chemical, and viral carcinogenesis (8). As has been emphasized above, however, alternative possibilities cannot be excluded in our present state of knowledge. Thus without further understanding of cell regulation and differentation, inferences about the role of DNA damage in carcinogenesis must remain speculative.

ACKNOWLEDGEMENTS:

Support for the preparation of this manuscript was provided by Grant #ES00260 from the National Institute of Environmental Health Sciences and #CA13343 from the National Cancer Institute. The author is grateful to K. Griffin, J. Smith and L. Witte for their assistance in the preparation of this manuscript.

REFERENCES

(1) Albert, R. E. and Altshuler, B. In: *Radionuclide Carcinogenesis*, C. L. Sanders, R. H. Busch, J. E. Ballou, and D. D. Mahlum, Eds. (U. S. Atomic Energy Commission, Oak Ridge, Tennessee, 1973), pp. 233-235.
(2) Ames, B. N., McCann, J., and Yamasaki, E. *Mutation Res. 31*, 347-363 (1975).
(3) Barrett, C. J. and Ts'o, P. O. P. *Proc. Natl. Acad. Sci. U.S.A. 75*, 3297-3301 (1978).
(4) Barrett, C. J. and Ts'o, P. O. P. *Proc. Natl. Acad. Sci. U.S.A. 75*, 3761-3765 (1978).
(5) Borek, C., Miller, R., Pain, C., and Troll, C. *Proc. Natl. Acad. Sci. U.S.A. 76*, 1800-1803 (1979).
(6) Boveri, T. Zur Frage der Entstehung maligner Tumoren (Fischer, Jena, East Germany, 1914).
(7) Brookes, P. *Mutation Res. 86*, 233-242 (1981).
(8) Cairns, J. *Nature 289*, 353-357 (1981).

(9) Carrano, A. V., Thompson, L. H., Lindl, P. A., and Minkler, J. L. *Nature 271*, 551-553 (1978).
(10) Cole, A., Mayn, R. E., Chen, R., Corry, P. M., and Hittelman, W. In: *Radiation Biology in Cancer Research*, R. E. Meyn and H. R. Withers, Eds. (Raven Press, New York, 1980), pp. 33-58.
(11) Coleclough, C., Perry, R. P., Karjalainen, K., and Weigert, M. *Nature 290*, 372-378 (1981).
(12) Cornelis, J. J., Su, Z. Z., Ward, D. C., and Rommelaere, J. *Proc. Natl. Acad. Sci. U.S.A. 78*, 4480-4484 (1981).
(13) Doll, R. In: *Proceedings of the 10th International Cancer Congress, Vol. 5* (Year Book Medical Publishers, Chicago, 1971), pp. 1-28.
(14) Doll, R. *Cancer Res. 38*, 3573-3583 (1978).
(15) Donovan, P. J. and DiPaolo, J. A. *Cancer Res. 34*, 2720-2727 (1974).
(16) Evans, H.J. and Vijayalaxmi. *Nature 292*, 601-605 (1981).
(17) Fernandez, A., Mondal, S. and Heidelberger, C. *Proc. Natl. Acad. Sci. U.S.A. 77*, 7272-7276 (1980).
(18) Gehly, E. G., Landolph, J., Nagasawa, H., Little, J. B., and Heidelberger, C. *Cancer Res.* (in press).
(19) Goth, R. and Rajewsky, M. F. *Proc. Natl. Acad. Sci. U.S.A. 71*, 639-643 (1974).
(20) Hart, W.R., Setlow, R. B., and Woodhead, A. D. *Proc. Natl. Acad. Sci. U.S.A. 74*, 5574-5578 (1977).
(21) Haward, W. S., Neel, B. G., and Astrin, S. M. *Nature 290*, 475-480 (1981).
(22) Heidelberger, C. *Advan. Cancer Res. 18*, 317-366 (1973).
(23) Hieter, P. A., Korsmeyer, S. J., Waldmann, T. A., and Leder, P. *Nature 290*, 368-372 (1981).
(24) Huberman, E., Mager, R. and Sachs, L. *Nature (London) 264*, 360-361 (1976).
(25) Kakunaga, T. and Crow, J. D. *Science 209*, 505-507 (1980).
(26) Kennedy, A. R., Mondal, S., Heidelberger, C., and Little, J. B. *Cancer Res. 38*, 439-443 (1978).
(27) Kennedy, A. R., Fox, M., Murphy, G., Little, J. B. *Proc. Natl. Acad. Sci. U.S.A. 77*, 7262-7266 (1980).
(28) Kinsella, A. R. and Radman, M. *Proc. Natl. Acad. Sci. U.S.A. 75*, 6149-6153 (1978).
(29) Kinsella, A. R. and Radman, M. *Proc. Natl. Acad. Sci. U.S.A. 77*, 3544-3547 (1980).

(30) Klein, G. In: *Cancer - Achievements, Challenges and Prospects for the 1980s, Vol 1*, J. H. Burchenal and H. F. Oettgen, Eds. (Grune & Stratton, New York, 1981), pp. 81-100.
(31) Knudson, A. G., Jr. In: *Cancer - Achievements, Challenges, and Prospects for the 1980s, Vol. 1*, J. H. Burchenal and H. F. Oettgen, Eds. (Grune & Stratton, New York, 1981), pp. 381-396.
(32) Kondo, S. In: *Fundamentals in Cancer Prevention*, P. N. Magee, S. Takayama, T. Sugimura, and T. Matsushima, Eds. (Univer. of Tokyo Press, Tokyo, 1976), pp. 417-429.
(33) Lavi, S. and Etkin, S. *Carcinogenesis 2*, 417-423 (1981).
(34) Little, J.B., Hatsumi, N., and Kennedy, A. R. *Radiation Res. 79*, 241-255 (1979).
(35) Lloyd, D. C. and Purrott, R. J. *Radiation Protection Dosimetry 1*, 19-28 (1981).
(36) Lynch, H. T. *Recent Results in Cancer Research, Vol. 12* (Springer-Verlag, New York 1967).
(37) Margison, G. P. and Leihues, P. *Biochem J. 148*, 521-525 (1975).
(38) Mondal, S. and Heidelberger, C. *Nature 260*, 710-711 (1976).
(39) Mulvihill, J. J., Miller, R. W., and Fraumeni, J. F., Eds. *Genetics of Human Cancer, Progress in Cancer Research and Therapy, Vol. 3*, (Raven Press, New York 1977).
(40) Nagao, M., Sugimura, T. and Matsushima, T. *Ann. Rev. Genet. 12*, 117-159 (1978).
(41) National Academy of Sciences. *The Effects on Populations of Exposure to Low Levels of Ionizing Radiation*. Advisory Committee on the Biological Effects of Ionizing Radiation (BEIR) (National Academy of Sciences-National Research Council, Washington, D. C. 1980).
(42) Neel, B. G., Haward, W. S., Robinson, H. L., Fang, J., and Astrin, S. M. *Cell 23*, 323-334 (1981).
(43) Nomura, T. In: *Tumors in Early Life in Man and Animals*, L. Severi, Eds. (Perguia, Italy, 1978), pp. 873-891.
(44) Nomura, T. *Cancer Res. 40*, 1332-1340 (1980).
(45) Paterson, M. C. In: *Carcinogens: Identification and Mechanisms of Action*, A. C. Griffin and C. R. Shaw, Eds. (Raven Press, New York, 1979), pp. 251-276.

(46) Peto, R. In: *Origins of Human Cancer,* H. H. Hiatt, J. D. Watson, and J. A. Winsten, Eds. (Cold Spring Harbor Laboratory, Cold Spring Harbor, New York, 1977), pp. 1403-1428.
(47) Poiley, J. A., Raineri, R., and Pienta, R. J. *Br. J. Cancer 39,* 8-14 (1979).
(48) Rinkus, S. J. and Legator, M. S. *Cancer Res. 39,* 3289-3318 (1979).
(49) Rubin, H. *J. Natl. Cancer Inst. 64,* 995-1000 (1980).
(50) Sandberg, A. A. *The Chromosomes in Human Cancer and Leukemia* (Elsevier North-Holland, New York 1980).
(51) Shilo, B. and Weinberg, R. A. *Nature 289,* 607-609 (1981).
(52) Slaga, T. J., Fischer, S. M., Nelson, D., and Gleason, G. L. *Proc. Natl. Acad. Sci. U.S.A. 77,* 3659-3663 (1980).
(53) Sugimura, T., Sato, S., Nagao, M., Yahagi, T., Matsushima, T., Seino, Y., Takeuchi, M., and Kawachi, T. In: *Fundamentals in Cancer Prevention,* P. N. Magee, S. Takayama, T. Sugimura, T. Matsushima, Eds. (University of Tokyo Press, Tokyo, 1976), pp. 191-215.
(54) Temin, H. M. *Perspectives Biol. Med. 14,* 11-26 (1970).
(55) Toman, Z., Dambly, C., and Radman, M. In: *Molecular and Cellular Aspects of Carcinogen Screening Tests,* R. Montesano et al (Eds.), IARC Publication No. 27, International Agency for Research on Cancer, Lyon, pp. 243-255.
(56) Tomatis, L. and Goodall, C. M. *Int. J. Cancer 4,* 219-225 (1969).
(57) Upton, A. C. *Radiat. Res. 71,* 51-74 (1977).
(58) Upton, A. C., Kimball, A. W., Furth, J., Christenberry, K. W., and Benedict, W. H. *Cancer Res. 20,* 1-62 (1960).
(59) Upton, A. C. In: *Cancer - Achievements, Challenges and Prospects for the 1980s,* Vol 1, J. H. Burchenal and H. F. Oettgen, Eds. (Grune & Stratton, New York, 1981), pp, 185-198.
(60) Varshavsky, A. *Proc. Natl. Acad. Sci. U.S.A. 78,* 3673-3677 (1981).
(61) Weinstein, I. B. *Bull. N.Y. Acad. Med. 54,* 366-383 (1978).

(62) Weinstein, I. B., Yamasaki, H., Wigler, M., Lee, L. S., Fisher, P. B., Jeffrey, A., and Grunberger, D. In: *Carcinogenesis: Identification and Mechanisms of Action*, A. C. Griffin and C. R. Shaw, Eds. (Raven Press, New York, 1979), pp. 399-418.
(63) Williams, G. M. *J. Assoc. Off. Anal. Chem. 62*, 857-863 (1979).
(64) Williams, G. M., Kroes, R., Waaigers, H. W., and van de Poll, K. W., Eds. *The Predictive Value of Short-Term Screening Tests in Carcinogenecity Evaluation* (Elsevier North-Holland, Amsterdam 1980).
(65) Zajdela, F. and Latarjet, R. *C. R. Acad. Sci., Paris, t.277*, 1073-1076 (1973).

The First Five Years of ICPEMC; The International Commission for Protection against Environmental Mutagens and Carcinogens

F. H. Sobels and J. Delehanty*

Department of Radiation Genetics and Chemical Mutagenesis, University of Leiden, Leiden, The Netherlands

Secretariat, ICPEMC, Rijswijk, The Netherlands

In scientific conferences like the one we are attending, where the excitement of new data and fresh ideas will soon be upon us, one may ask what is the purpose of allocating precious time to an organization such as ICPEMC. After all, ICPEMC possesses no laboratories, has no research budget, and cannot present novel experimental results. Why am I here then? All those present here have witnessed during the past ten years the rapid development of the field of chemical mutagenesis. This growth is a direct consequence of the concern for human ill-health resulting from damage to the genetic material. Initiated in the 1960's and continuing at a seemingly exponential rate, research has revealed that mutagenic chemicals occur in the food we eat, the water we drink and the air we breathe. Moreover, there is an ominous parallel between the mutagenic and carcinogenic potential of a large number of these chemicals.

The problems of adequately protecting the human population against mutagenic and carcinogenic hazards are enormous and far exceed national boundaries. Intensive collaboration on a world wide scale was urgently needed. Thus, ICPEMC, the International Commission for Protection against Environmental Mutagens and Carcinogens, came into existence in January 1977 to expand the role of the International Association of Environmental Mutagen Societies; ICPEMC is mainly sponsored by the Institut de la Vie. It is dedicated to identifying and evaluating genotoxic chemicals in the environment and to suggesting approaches to regulatory and legal decision-making. It is an independent organization of scientists from universities, research institutes, government agencies, and industry.

ICPEMC developed rapidly. A founding Commission was organized to identify the most appropriate problems in the vast field of environmental mutagenesis and carcinogenesis, and special study groups called Committees were formed.

The charge and expertise in Committee 1, chaired by Dr. B.J. Kilbey and Dr. D. Brusick, is directed specifically towards assessing chemical mutagenicity. It attempts to link results of mutagenicity screening tests to potential human mutagenicity by comparing data from whole-animal mammalian tests to other tests for mutagenicity ranging from microbial to mammalian cell culture. The whole-mammal tests that the Committee are using as standards are the specific locus mutation test, the heritable translocation test, and the dominant lethal mutation test.

ICPEMC believes that this approach is most valuable. In chosing mammalian test data as standards, Committee 1 concentrates on a biological system that most resembles the human. As will become more apparent during my discussion of other ICPEMC activities, the concern for factors such as whole-body chemical metabolism, chemical access to protected mammalian germ cells, and chemical excretion and circulation is crucial, but impossible to gauge in short-term assays.

Committee 2, chaired by Dr. D.B. Clayson, deals with the relationship between mutagenicity and carcinogenicity. The question is debated intensely as one of fundamental biology and also is of immediate concern in assessing mutagenicity tests for human carcinogenicity prediction.

The key to understanding this relationship is the definition of carcinogenesis. Operationally, Committee 2 has distinguished between two categories of carcinogens:

"A genotoxic carcinogen is an agent that significantly increases the number of tumors in a population and has the capability of inducing alterations in genetic information in test systems and presumably in the cells from which the cancer arises".

"A nongenotoxic carcinogen significantly enhances the incidence of tumors in a population but does not have the ability to induce genetic alterations in test systems (4)".

Confirmation that a carcinogen may be nongenotoxic will depend on identifying its mechanism of action. However, the recognition of the existence of nongenotoxic substances will strengthen the predictive power of present prescreening tests for genotoxic carcinogens.

Both genotoxic and nongenotoxic carcinogens have been evaluated in several test systems. For example Paul Howard-Flanders (3) examined mutagenesis in mammalian cells in culture. His study is particularly thorough, focusing not only on biological problems such as selection of cell lines, mutant selection schemes, and toxicity, but also on chemical problems such as metabolic activation and the potential for chemically modified bases to become premutational lesions. From this we may conclude that mutagens can be monitored accurately by mammalian cell culture methods and that results from liver microsome-activated bacterial mutagen test systems and mammalian cell systems reasonably agree.

The causal relationship between mutagenicity and carcinogenicity testing in mammalian systems is not defended so strongly. The method of choice in this context is the mammalian cell transformation assay, which is the subject of another ICPEMC review by Dr. Peter Brookes (1).

Although initially received enthusiastically, cell transformation assays have gained the reputation for being difficult to perform reproduceably, especially among different laboratories. This, of course, is cause for anxiety. On theoretical grounds, it is suspect because the biological relationship between in vitro transformation and in vivo malignancy is itself unclear. Uncertainty also exists as to the criteria which define transformation.

On the positive side, much transformation data agree with that of whole-animal carcinogenicity tests. This, coupled with the still popular hypothesis that in vitro cell transformation is related directly to carcinogenesis - a contention supported largely by the fact that transformed cells often produce tumors when injected into syngenic test animals - encourages further research to determine if such similarities are superficial or represent common biological characteristics.

Nongenotoxic carcinogens have been discussed by David Clayson (2) and Andrew Sivak (6). They agree that much of the discrepancy between animal carcinogenicity tests and short-term prescreening tests that rely on mutational endpoints could be explained by the existence of carcinogenesis enhancers. Because carcinogenesis appears to be a multistage process, experiments must be designed to distinguish between the initiation of carcinogenesis and tumor development.

Microbial tests, at least in the literature, dominate the field of mutagenicity and carcinogenicity testing. Georges Mohn has published an ICPEMC report on this subject (5). In general, confidence in these tests to detect genotoxic carcinogens has been expressed, since practically all chemicals that are mutagenic in mammals are also mutagenic in microbes. Of course, this correspondence is achieved only when rigorous calibration with appropriate reference mutagens and carcinogens is performed and mammalian metabolic activation is included.

Here too, confidence and suspicion coexist. Chemicals that cause specific, rare mutations or that may be activated by nonmicrosomal enzymes, often go undetected. And often, of course, nongenotoxic carcinogens are impossible to spot. Causing even more anxiety are chemicals whose mutagenic activity are detected in microbial tests but are not satisfactorily found in animal carcinogenicity tests. A great deal of flexibility must be exercised in dealing with chemical testing for genotoxicity.

More than 100 predictive tests have been developed thusfar. In Committee 2's opinion, the only sensible approach to validate predictive tests must employ calibration techniques, such as I have mentioned.

Based upon limited validation, predictive tests may be described as developing, developed, or established. Only nine developed tests have been designated by Committee 2 and only one, the Salmonella/microsome test, is said to be established.

The tasks of Committees 1 and 2 that I described just now, largely pertain to purely scientific issues. However, now ICPEMC reaches a critical junction. We are led from the existence and chemistry of mutations to their consequences. We examined the nature of these consequences and found most to be destructive or unhealthy. At this point, the societal importance of chemical mutagenesis becomes more evident. Science has been compelled to react, to advise, to inform.

Aware of this growing concern, the founding Commission created ICPEMC Committee 3, chaired by Drs. A.C. Kolbye and E. Poulsen. Although the members of Committee 3 are trained scientists, their professions encompass the interface of science and society. They are regulators (Kolbye, Poulsen, Miller, Munro), lawyers (Kolbye, Damme), and specialists in international scientific law (Fluss) and policy (Ott).

Committee 3 has taken a scientific approach to examining the state of regulation of mutagens and carcinogens. Yet, it has done so from the point of view of those who are confronted with the pressure of deciding the rules of control.

As I have said, the Committee's approach has been analytical even if it cannot be definitive. Its first task has been to examine existing legal instruments of all types of formality. From this evaluation a clear pattern emerged. In existing legislation, many more actions specify testing requirements rather than responses to test results. The regulatory process must be expanded. Hazards must be identified qualitatively; risks must be quantified; and once risks are estimated, decisions must be made to accept or reject them.

This has been the area of acute concern to Committee 3. It is a decision not for scientists to make exclusively, though their advice should clearly be sought.

Nevertheless, regulators are required to estimate risk acceptability. According to the Committee, four components contribute to the decision: (1) the magnitude of the risk above background, (2) comparison of the chemical risk versus the accepted risk from other activities (e.g. driving an automobile or coal mining), (3) the balance of risk between absence or presence of a chemical in the environment, and (4) the socio-economic impact the rejection of a chemical might have.

The heart of the argument on legal controls for genotoxic chemicals lies in the answers science can provide. If in attempting to clarify things an agency requires additional data, the implication arises immediately - it is necessary to respond to those data. It is imperative that once submitted, data are evaluated. The fear exists that an enormous amount of expensive data may be generated, filed, and forgotten.

Thus, in the view of the Committee, regulatory strategies should include the following: (1) the degree of rigor with which a regulation requires, requests, recommends, or suggests that data be collected, (2) the nature of the data, (3) the criteria for assessing test results, and (4) the stringency of an agency's response.

The regulatory experience of the Committee recognized that control must include a flexible array of countermeasures ranging from outright prohibition to public education and product labelling.

Lastly, data pertaining to mutagenicity and carcinogenicity should be only one component of a toxicological profile, especially in view of the uncertainty of test performance and data interpretation.

There is an obvious gap in our knowledge of chemical mutagenesis and carcinogenesis prediction. It is the same gap those of us in radiation biology faced 30 years ago. How does one estimate genetic risk from chemicals in the environment? How do we gauge exposure of short duration and low dose?

Committee 4 has been formed to face these questions. Quantitative extrapolation has serious scientific and social consequences. The genetic damage caused at low doses by chemicals that have significant medical or economic value must be determined precisely. Only on this basis can regulators hope to make unequivocal risk-benefit decisions.

As a means to exploring these problems, Dr. Mary Lyon's group has examined several topics. Knowledge of differential sensitivity of various germ cell stages is relevant in several ways. It may point to needs to protect special sections of the population or to avoid procreation within short time of exposure.

Dose-response relationships are of course the crux of the Committee's analysis. Here one is confronted with an obvious discrepancy between the high doses of strong mutagens required to demonstrate genotoxic effects in mammalian bioassays and the low concentrations most often encountered in the environment. Dose response is discussed in terms of the chain of events from exposure to response; of metabolic processing and repair; and of methods for measuring dose at the molecular level of DNA adducts.

The best quantitative extrapolation at present can be made from mice, using the specific locus mutations and dominant mutations carrying skeletal abnormalities or cataracts of the eye.

When the endpoints cannot be measured directly, an indirect approach can be used which I have termed the parallelogram (7, 8). By comparing endpoints that can be determined experimentally, for example DNA adducts in mammalian cells in vitro and in mouse germ cells, and relating this to induced mutation frequencies in the in vitro system, an impression can be gained of the damage to be expected in the germ cells at comparable level of alkylation. The right hand side of Fig. 1 shows another extension of the parallelogram to estimate the mutation

Fig. 1

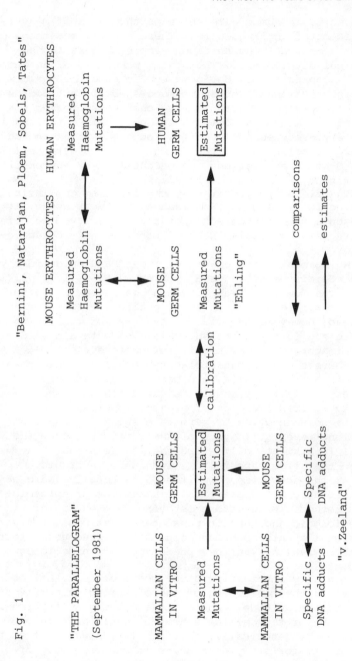

frequencies in human germ cells on the basis of measured somatic mutation frequencies. These somatic mutations can now be detected in human blood samples thanks to new technical developments. The system makes use of an automatic scanning system designed by Professor J.S. Ploem and his group and fluorescent mono-specific antibodies against hemoglobin variants developed by Professor L.F. Bernini. By measuring comparable hemoglobin mutations in mouse erythrocytes and specific locus mutations in the mouse germ cells, estimates of mutations in germ cells of exposed human populations can be obtained by appropriate comparisons, as shown in Fig. 1.

The fifth and last issue, the founding Commission selected initially was, in a sense, the most human. It represents the only direct approach to observe the effects of mutagenic chemicals in human populations - epidemiology.

My comments on the work of the ICPEMC epidemiology group, Committee 5, chaired by Prof. J.R. Miller, will be brief. The Committee was formed a year after the other ICPEMC committees, and thus its evaluations are not as advanced. However, the most blatant reason for having less to say is the astonishing lack of data and epidemiological approaches for use in mutagenic surveys.

Whereas cancer registries and other sources of information of human data have been pursuing the occurrence of cancer, any relationship to mutagenesis has been largely ignored. Perhaps of greater importance is the lack of documentation of heritable human disease and mutational events. In view of the contention that the actual impairment to life from heritable disease may exceed that from cancer, this ignorance is tragic.

At the same time, Committee 5 is equally aware of the theoretical and practical limitations of epidemiologic methods. Much effort is being devoted to defining epidemiologic and genetic parameters that would reflect mutational events in human populations. One such effort is the revival of sentinel phenotypes as epidemiologic endpoints.

A highly interesting study contributed by Dr. Andrew Czeizel is the monitoring of the offspring of survivors of attempted suicide, especially involving self-poisoning. Since many of the chemicals are mutagenic and can be assessed cytogenetically, this population would offer the possibility for studying changes in the frequency of genetic disease. The same tack is being explored by

Dr. John Mulvihill in his study of reproduction of cancer patients. To supplement new directions, Dr. Ei Matsunaga has undertaken the arduous task of collecting and evaluating the existing data on the prevalence of genetic disease.

The Commission of ICPEMC has done more than oversee scientific direction and make policy. Within the Commission itself, several small groups have been formed to examine specific urgent problems. Such groups are termed Task Groups.

Certainly, the most successful Task Group has been that chaired by Prof. Bryn Bridges, Task Group 1. The charge of this group is to make preliminary analysis of chemicals or situations of possible mutagenic risk. Within the past year studies have been published on the epidemiologic evidence for hair dye carcinogenicity and on the genotoxicity of epichlorohydrin, dichlorvos, isoniazid, and on compounds and techniques used in psoriasis treatment.

Two Task Groups recently have been set up at the request of the European Chemical Industry Ecology and Toxicology Centre (ECETOC). The first, Task Group 5, has been asked to develop criteria for distinguishing between genotoxic and epigenetic carcinogens. Chaired by Dr. Art Upton, Task Group 5 is analyzing a data base of about 400 chemicals in 16 tests to discover a minimum battery of tests for nongenotoxicity. The group has selected specific examples to demonstrate nongenotoxic processes. These include action in the liver and bladder as well as the participation of antioxidants, hormones, and nutrition.

The second group is Task Group 6, chaired by Dr. Udo Ehling. They are reviewing the experimental and human evidence of thresholds for genotoxic chemicals with regard to the induction of genetic effects. A preliminary draft has been circulated. Among the topics studied in this report are the nature of dose-response and the criteria for thresholds, the metabolism of mutagens, DNA repair mechanisms, and the effects of genotoxic chemicals on mammalian cells and in animals. Germ cell mutations have been receiving particular attention.

ICPEMC is now completing its first five-year term. During this period, eight Commission Meetings and 21 Committee Meetings have been held. As a fruit of these activities, 25 ICPEMC papers have been generated sofar on a great variety of topics in the field of ICPEMC's interests. At present we are at the stage of delivering our major products, the reports of the various ICPEMC Committees. We expect that at least three of these will appear next year, i.e., Committee 1's report on short term testing, Committee 2's on "mutagenesis testing as an approach to carcinogenesis" and Committee 3's on "regulatory policies". Those on risk estimates and epidemiology will follow a little later. At present, ICPEMC is engaged in reviewing and evaluating these major Committee reports and those of the ongoing studies of the various Task Groups. Obviously, our work is not finished after these five years, and many of the problems sofar discussed need further elaboration and surveillance. At the same time, considerable effort is spent to define new problem areas for future ICPEMC evaluation. Polygenes, their mutation frequencies, DNA repair, recombinant DNA methodology and transposition mediated mutation are being considered as topics.

Let me end, by gratefully acknowledging the support that has enabled ICPEMC to develop: Professor Marois, President of the Institut de la Vie and Madame Guisan, his secretary general, Professor Sugimura and his Japanese colleagues, who established a Japanese ICPEMC group that also published the ICPEMC brochure for this meeting, the Gesellschaft für Strahlen- und Umweltforschung in Munich, the European Community and ECETOC. All Commission and Committee activities have been characterized by great dedication and willingness to carry out the often arduous tasks. If I single out one individual who deserves special acknowledgement, it is our secretary Dr. Paul Lohman; without his unfailing enthusiasm and organisational talents, ICPEMC would simply not have developed to what it is today.

References
(1) Brookes, P. Mutation Res. 86, 233-242 (1981)
(2) Clayson, D.B. Mutation Res. 86, 217-229 (1981)
(3) Howard-Flanders, P. Mutation Res. 86, 307-327 (1981)
(4) ICPEMC-Committee 2, final report, Mutation Res. (in press)
(5) Mohn, G.R. Mutation Res. (in press)
(6) Sivak, A. Mutation Res. (in press)
(7) Sobels, F.H. Mutation Res. 46, 245-260 (1977)
(8) Sobels, F.H. Arch. Toxicol. 46, 21-30 (1980)

A Brief Sketch of Environmental Mutagen Studies in Japan

Yataro Tazima

National Institute of Genetics, Mishima, Japan

ABSTRACT Soon after the establishment of Environmental Mutagen Society of Japan in 1972, a potent mutagenicity was discovered for AF-2, a food preservative then used in Japan. The incident exerted a significant impact not only on the improvement of screening procedure of carcinogenic compounds but also on legislative rationale for mutagenicity. Use of convenient bacteria systems demonstrated a close correlation between mutagens and carcinogens which attracted the attention of oncologists to the availability of these systems. Potent muta-carcinogenic components were discovered in a charred surface of broiled fish and meat. Further a wide variety of foodstuffs was found to contain mutagenic agents. Several modulating factors of mutagenicity, comutagens, antimutagens and desmutagens, were demonstrated which added the complexity in the evaluation of human risks. Exploitation of efficient monitoring system(s) for overall effects is urgently needed.

INTRODUCTION

It was not until the establishment of EMS in USA that we recognized seriously the need for environmental mutagen research. Being informed by Dr. Alexander Hollaender that EMS was established in USA in 1969, we organized a research group under a grant in aid of the Ministry of Education, Science and Culture. One year later we established a small society, the inauguration of which was honored by the attendance of Dr. E. B. Freese, then the president of EMS,

Drs. F. J. de Serres and W. W. Nichols. The society has grown up far more rapidly than we expected and is going to hold 10th annual meeting this fall. It may be of worthy to review the progress of the study on this occasion.

Mutagenicity of AF-2   The first research project we took up was the development of sensitive test system(s). Each member, using his most familiar experimental organism(s), devoted himself to the development of test systems sensitive to weak mutagens. In order to compare the sensitivity among different test systems, we chose, as standard test chemicals, nitrofuran derivatives, including AF-2. The compound was then widely used in Japan as a food preservative and mutagenicity and carcinogenicity had already been demonstrated for some nitrofuran derivatives.

Soon after the start of the research in 1972, Tonomura and Sasaki (37) discovered a strong clastogenicity of AF-2 using human lymphocytes. They immediately informed this to the group members. Thereafter, the mutagenicity of this compound was demonstrated by many members of the group with various test systems; *E. coli, B. subtilis*, yeast and silkworm germ line (see the reference 36). Independently of this group Sugimura and his collaborators (40) confirmed strong mutagenicity of AF-2 in bacteria using both *Salmonella* repair test and *E. coli* reversion test. Although mutagenic activity had not yet been detected by *in vivo* experiments with mammals, it was conjectured that the compound might also be mutagenic to man.

We, therefore, warned our government of the possible genetic danger of this compound and requested them to perform safety tests on mammals. But they did not listen to us. Since then, long and disagreeable disputes arose between government authorities and ourselves. I have no intention to go into further detail of the dispute. In September 1974, one and half years after our warning, the government took action to ban AF-2. After that, numerous evidences for carcinogenicity of this compound were obtained one after another. This can be mentioned as a very good example that the mutagenicity was first taken into consideration for safety control (36).

Short term tests   In those days, Sugimura and his group were in search of rapid and reliable prescreening method(s) for carcinogenic compounds. They compared several microbial assay methods using, as critelia, mutagenicity and/or DNA attacking effects. They found that *Salmonella* test system

(1) could be used as a quick convenient test method for screening mutagens and perhaps carcinogens. Cases that show a good correlation between the carcinogenicity and mutagenicity accumulated gradually (2,3). However, when they tested the mutagenicity of 27 nitrofuran derivatives with *Salmonella* system, they found that it failed to screen up all nitrofuran derivatives. Since then, improvement of the screening capability of the *Salmonella* reversion system was energetically pushed forward by Ames' group. A year later, McCann *et al.* (18) succeeded in the synthesis of strains TA98 and TA100 which do respond to compounds of wider range. Further, improvement of this method was energetically made by Sugimura and his group particularly with regard to metabolic activation (26).

Kada developed another test system named "Rec-assay" (9,11). The principle is to utilize differential repairability of chemically produced DNA-lesions between rec^- and rec^+ strains. The method does not detect the mutagenicity but detect DNA-damaging capacities of test chemicals.

By employing those systems, innumerable compounds have so far been tested of their mutagenicity. Those tested include medical and agricultural drugs, industrial chemicals, food additives, foodstuffs, drinks, cosmetics, and air and water pollutants (26). According to Matsushima (20) of total 105 papers presented before the annual meeting of JEMS in 1980 51% used *Salmonella* reversion system, which contrasted markedly to the similar statistics of EMS and EEMS showing 27% and 28% respectively.

However, Ames' *Salmonella* system is not omnipotent. The study on mycotoxins represents a very good example on this point. A fairly large number of mutagenicity and carcinogenicity test data has so far been accumulated concerning these toxins. Ueno (39) classified the test data on mycotoxins into 8 groups according to the chemical structure of toxins and pointed out that detectability of those toxins differed remarkably depending on the test system used.

Bisfurans showed, without exception, positive results on Ames' assay, all requiring S-9 activation. The rec-assay also gave mostly positive results, showing a fairly good accord with the known carcinogenicity. On the contrary, strong carcinogenic lactones were negative on the Ames' assay even when tested with S-9, whereas they were positive on the rec-assay.

Anthraquinoid hepatocarcinogens,(-)luteoskyrin and (+)ruglosin, and chlorinated carcinogens such as chloropeptids, ochratoxin A and griseofulvin were negative on

Ames' assay. In rec-assay, the former two were positive but the latter three were negative.

Pyrolysates Cigarette smoke is known to enhance significantly the incidence of lung cancer. The most suspicious agent(s) had been considered to be cyclic hydrocarbons, such as benzo(α)pyrene, benzanthracen, *etc.*, produced by burning of the cigarette paper. However, strong mutagenic agent(s) were discovered in tars by Kier *et al.* (17) and Hutton *et al.* (8). According to Kawachi *et al.* (16) the mutagenic activity was much too high to be accounted for by the amounts of benzo(α)pyrene present in the smoke condensate, which suggested that cigarette smoke condensates might involve mutagenic compounds other than benzo(α)pyrene. Matsumoto *et al.* (19), using *Salmonella* TA98 strain, found that pyrolysates of most amino acids exhibited significant mutagenic activity and the strongest was L-tryptophan.

Extending their study on tobacco smoke (23), Sugimura and his group demonstrated that both smoke condensates and extracts of charred surfaces of fish and beef contained mutagenic substance toward *Salmonella* TA100 and TA98 (24,25). Smoke condensates of some species of fish required metabolic activation to induce mutagenicity. They also demonstrated that pyrolysates of some amino acids, such as tryptophane serine, and glutamic acid had strong mutagenicities to *Salmonella*. Further, they obtained two mutagenic principles in crystalline form. Those were named Trp-P-1 and Trp-P-2 and later determined of their chemical structure (32).

Both Trp-P-1 and Trp-P-2 were β-carborin derivatives and strongly mutagenic to *Salmonella* in the presence of metabolic activation. Their mutagenic activities were far stronger than that of aflatoxin B_1.

Nagao *et al.*(25) investigated the mutagenicity of pyrolisate of sugars. Glucose, arabinose, fructose and sorbitol showed remarkable mutagenicity on TA100 without metabolic activation. Glucosamine, an amino-sugar, was also strongly mutagenic.

Cooked foods Several foodstuffs have been examined of their pyrolysates with regard to mutagenicity to *Salmonella* TA98 and TA100. Ueda *et al.*(38) performed mutagenicity testing of 50 foodstuffs including rice, flour, soybean curd, egg, pork, chicken, cabbage, egg plant, mashed potato, sugar, *etc*. Of those, 36 items were found to contain mutagenic agents when tested with metabolic activation.

The mutagenic activity was hardly observed at the

pyrolysis condition of 200°C for 10 min. But the activity became observable at the treatment of 250°C for 10 min in dried bonito and soybean flour. The activity reached maximum at 300°C but fell down at more higher temperatures.

It may be worthy to mention here that Commoner et al. (5) in U.S. discovered that strong mutagenic substance to Salmonella was produced at temperatures which do not exceed 105°C when he cooked beef extracts. This condition occurs in common including the preparation of humbergers on electrically heated hot plates. Recently, Sugimura et al. (15) isolated IQ (2-amino-3-methyl imidazo[4,5-f]quinoline) and its methyl derivatives from sardines grilled under ordinary cooking conditions.

Mutagenicity of flavonoids Flavon and its derivatives are contained in leaves of many plant species. It is highly likely that those substances have a chance to be taken into human body by ingestion. Positive mutagenicity of those substances has been discovered with Salmonella. Sugimura et al.(33) observed that a flavon derivative, kaempherol, was mutagenic on TA100. Further, they tested other flavon derivatives and found a fairly strong mutagenicity to TA98 for quercetin and galangin and weak mutagenicity for fistein. All those compounds were mutagenic to TA100 too.

Recently, Nagao et al.(27) examined the mutagenicity of several kinds of spices, tea, coffee, pickles, wine and liquor. Mutagenicity was tested on extracts with DMSO or methanol. About 50% tested were found mutagenic to Salmonella. Mutagenic agents thus detected are not yet chemically known but some have been identified as kaempherol.

Comutagens In the tar of tryptophane pyrolysate comutagenic agents, nonharman and harman, were found plenty co-existing with mutagenic compounds, Trp-P-1 and Trp-P-2. Both groups of compounds resembled each other: the former group having β-carboline moiety and the latter group γ-carboline moiety.

During the process of isolation of former mutagenic agents Nagao et al.(22) noticed that the mutagenic activity was markedly reduced at the step of the separation of their compounds from norharman and harman. From this finding they discovered strong enhancing effect of norharman and harman on the mutagenic activity of Trp-P-1 and Trp-P-2. Norharman and harman per se show no mutagenicity.

When norharman was added to Trp-P-1 or Trp-P-2, mutagenicity to TA98 increased approximately ten fold, requiring metabolic activation by S-9 mix. Harman showed similar degree of the activation.

The comutagenic activity of norharman and harman were also observed for various mutagens, which require metabolic activation by S-9 mix, *eg.* 4-dimethyl-aminoazobenzen, benzo(α)pyrene, N-2-fluorenylacetamide, 4-dimethylaminostilben, but not observed for AF-2 and 4-NQO which did not require metabolic activation.

The mechanism of the comutagenic action of norharman has not been elucidated yet, but there is an indication that the intercalation of this compound to DNA molecule seems to play an important role. Inhibition of metabolism by norharman is also known as the cause of enhancement of mutagenicity of benzo(α)pyrene (26).

Another type of comutagenicity is the production of strong mutagenic substance(s) by reaction between non- or weak mutagenic compounds. By mixing sorbic acid and sodium nitrite a new potent DNA-damaging as well as mutagenic agent was produced (10). Sorbic acid itself is negative and sodium nitrite is positive only at high concentrations, especially at acidic pH. This reaction occurs even at room temperature but is enhanced by heating the mixture at 100°C.

Hayatsu and his collaborators (6,35) analyzed the above reaction products and obtained more than three different mutagenic substances. One of them was crystallized and determined of its structure. Namiki and his collaborators (28) also purified one of the reaction products and they identified ethylnitrolic acid as one of the main products, which was highly positive to rec-assay.

<u>Antimutagens and desmutagens</u> So far at least two types of antimutagens have been found. One acts antagonistically to already existing mutagens, by inactivating them *in vitro*. Whereas, the other inhibits the production of mutations in the cell. Kada named the former desmutagen and the latter antimutagen.

The first example, desmutagen, is vegetable juice. Kada *et al.*(12,21) noticed that number of his^+ revertants induced by Trp-P (a crude extract of tryptophan pyrolysate) was markedly reduced by treatment with cabbage or radish juice and this effect was abolished by boiling of the juice. Then, they examined many other vegetables of their mutagenicity. The samples were homogenized and centrifuged at 9000 x g for 30 min. Supernatants were assayed of their mutagenicity on *Salmonella* TA98 with S-9 mix.

Of 59 vegetables and fruits examined those possessed

strong antimutagenicity were: cabbage, broccoli, green pepper, egg plant, apple, burdock, shallot, ginger, pineapple and mint leaf. Those moderately effective were radish, sweet potato, grape, Japanese ginger, cauliflower, and enokidake mushroom and shimeji mashroom.

Studies on the characterization and purification of the antimutagenic factor from cabbage juice indicated that it is not precipitable by centrifugation at 200,000 x g and is resistant to treatment with DNAse and RNAse, but sensitive to treatment with pronase.

The second type is the factor that inactivates the mutagenicity of known mutagens or formation of mutagens in the cell. Kada and Kanematsu (13) observed that MNNG-induced mutations are remarkably reduced in $E.\ coli$ by the presence of cobalt chloride. The mutation assay system they used was reversion to prototrophy from tryptophane requiring mutant. By adding $CoCl_2$ at different concentration to agar media they first tested the survival and mutability. The presence of cobalt chloride, up to 20 µg/ml, did not modify significantly the survival of bacteria and number of try^+ colonies per plate was very low. Second they treated bacteria first with MNNG and plated on agar. Mutations were induced markedly at 0 metal concentration but reduced drastically in plates at higher metal concentrations.

Third, they treated bacteria with MNNG either in the presence or absence of cobalt chloride, washed the cells and plated them on cobalt free medium. Revertants appeared plenty, irrespective of post treatments with or without cobalt.

Thus the presence of cobalt was known to abolish the mutagenicity of MNNG. Recently, vitamine C was found to play the latter type of antimutagenic action by suppressing the formation of potent mutagenic substance between sorbic acid and nitrite (30).

CONCLUDING REMARKS

So far it has been known that mutagenic agents, either strong or weak, are distributed in foodstuffs of surprisingly wide variety. For instance, charred surface of broiled or roasted fish and meat contain strong mutagenic compounds. Furthermore, even uncooked foods have been known to contain mutagenic agents, since several flavonoids have been demonstrated to have potential mutagenicity.

Spices, tea, coffee, wine, soybean paste and soybean sauce were all found to contain potent mutagens to *Salmonella* (14,27,34).

Evidences are known that primitive man *Sinanthropus pekinensis* used the fire. Since then man must have taken several mutagenic agents into his body for more than 500,000 years. This raises an important question "Why man could have survived on so strong mutagenic compounds until today?" So far, the mutagenic activity has been tested mainly in bacteria but scarcely confirmed in mammals. Therefore, following possibilities can be considered.

First: these compounds are mutagenic to bacteria but not mutagenic to man *in vivo*. This is not valid because some mutagenic agents to bacteria are also known to be carcinogenic in experimental animals or even in man (26).

Second: those compounds are mutagenic (and carcinogenic) to man but unable to reach to germ cells *in situ* contained in the gonad.

Third: those agents are decomposed to non-mutagenic substances before they enter into human body. As mentioned above many desmutagenic agents exist in nature. So the third possibility clearly exists.

The answer I would like to discuss here in particular is on the second points. When xenobiotic compounds are taken into human body, several mechanisms may act to remove or destroy those substances. For instance, Nishioka and his group (29) demonstrated that saliva contains substances which can inhibit the mutagenicity of several known mutagenic compounds. Shimodaira et al.(31) discovered that mutagenic activity of MNNG was inactivated in mammalian blood. Based on this finding Arimoto et al.(4) discovered hemin as one of the effective inhibitors. It inhibited strongly the mutagenicity of polycyclic aromatic hydrocarbons: as benzo(α)pyrene, 3-methylcholanthorene, *etc.*, but not reacted upon such compounds as AF-2, 4 NQO or MNNG. They also revealed that biliverdin, bilirubin and chlorophyllin were effective inhibitors. Further they found that unsaturated fatty acids were also antimutagenic (7). In this way, several kinds of metabolic activities may operate to remove the mutagenic substances and protect human genomes from chemical danger.

Although we do not have comprehensive knowledge yet, I believe that man must be equipped with several builtin safeguard systems that tolerate or destroy the mutagenic potentiality of those compounds. As aforementioned, several substances existing in nature act mutagenic, comutagenic and/or antimutagenic in each other. Mechanisms functioning between those agents appear to be very complicated. The overall effects may sometimes be enhanced but in other times depressed. Accordingly, even the epidemiological approaches might not be easy to ascribe the cause of specific ailment to a single chemical substances. At present we can assess the overall damage on DNA either at chromosomal or molecular level. It is the only way that we can assess the genetic effects inflicted on man. However, we do not know how to translate such DNA damage(s) into realistic ailment at an individual level. Since incidence of such ailment is a product of the interaction between the mutagens and reactivity of the individuals, I think that the exploitation of efficient monitoring system(s) for overall effects are the most important research subjects for risk assessment. In this connection, investigation of inidividual sensitivity in relation to the genetic constitution may also be pertinent.

REFERENCES

(1) Ames, B. N. et al. in *Chemical Mutagenesis*, A. Hollaender ed. Plenum Press *1*, 267-282 (1971)
(2) Ames, B. N. et al. *Proc. Nat. Acad. Sci. U.S. 70*, 2281-2285 (1973)
(3) Ames, B. N. et al. *Mutation Res. 31*, 347-364 (1975)
(4) Arimoto, S. et al. *Cancer Letters 11*, 29-33 (1980)
(5) Commoner, B. et al. *Science 201*, 913-916 (1978)
(6) Hayatsu, H. et al. *Mutation Res. 30*, 417-419 (1975)
(7) Hayatsu, H. *Env. Mutagen Res. Comm. 3* , 29-32 (1981)
(8) Hutton, J. J. et al. *Cancer Res. 35*, 2461-2468 (1975)
(9) Kada, T. et al. *Mutation Res. 16*, 165-174 (1972)
(10) Kada, T. *Ann. Rept. Nat. Inst. Genet.* (24), 43 (1974)
(11) Kada, T. *IARC Scientific Publication* (12), 105-115 (1975)
(12) Kada, T. et al. *Mutation Res. 53*, 351-354 (1978)
(13) Kada, T. et al. *Proc. Japan Acad. 54B*, 234-237 (1978)
(14) Kaneda, T. et al. *Abst. 7th Ann. Meeting EMS Japan,* p.17 (1978)
(15) Kasai, H. et al. *Proc. Japan Acad. 56*, 278-283 (1980)
(16) Kawachi, T. et al. *Mutation Res. 54*, 217 (1978)
(17) Kier, L. D. et al. *Proc. Nat. acad. Sci. U.S. 71*, 4159-4163 (1974)
(18) McCann, J. et al. *Proc. Nat. Acad. Sci. U.S. 72*, 979-983 (1975)

(19) Matsumoto, T. et al. *Mutation Res. 48,* 279-286 (1977)
(20) Matsushima, T. *Env. Mutagen Res. Comm. 3,* 1-2 (1981)
(21) Morita, T. et al. *Agric. Biol. Chem. 42,* 1235-1238 (1978)
(22) Nagao, M. et al. *Proc. Japan Acad. 53,* 95-98 (1974)
(23) Nagao, M. et al. *Cancer Letters 2,* 221-226 (1977)
(24) Nagao, M. et al. *Cancer Letters 2,* 335-340 (1977)
(25) Nagao, M. et al. in *Progress in Genetic Toxicology,* Scott, Bridges and Sobels, eds. pp. 259-264 (1977)
(26) Nagao, M. et al. *Ann. Rev. Genet. 12,* 117-159 (1978)
(27) Nagao, M. et al. *Abst. 7th Ann. Meeting EMS Japan* p.15 (1978)
(28) Namiki,M. et al. *Agric. Biol. Chem. 39,* 1335-1336 (1975)
(29) Nishioka, H. *Env. Mutagen Res. Comm. 3,* 33-39 (1981)
(30) Osawa, T. et al. *Biochem. Biophys. Comm. 95,* 835-841 (1980)
(31) Shimodaira, T. *Abst. 7th Ann. Meeting EMS Japan,* p.30 (1977)
(32) Sugimura, T. et al. *Proc. Japan Acad. 53,* 59-61 (1977)
(33) Sugimura, T. et al. *Proc. Japan Acad. 53B,* 194-197 (1977)
(34) Takahashi, Y. et al. *Abst. 7th Ann. Meeting EMS Japan,* p.16 (1978)
(35) Tanaka, K. et al. *Fd. Cosmet. Toxicol. 16,* 209-215 (1978)
(36) Tazima, Y. *Env. Health Perspectives 29,* 183-187 (1979)
(37) Tonomura, A. et al. *Japan. J. Genet. 48,* 291-294 (1973)
(38) Ueda, M. et al. *Env. Mutagen Res. Comm. 1,* 17-18 (1978)
(39) Ueno,Y. *Iden 34,* 12-21 (1980)
(40) Yahagi, T. et al. *Cancer Res. 34,* 2266-2273 (1974)

Twenty Years Later!*

Comments by Alexander Hollaender

Council for Research Planning in Biological Sciences, Inc., Washington, D.C., U.S.A.

When I visited Japan twenty years ago, I was asked many questions about the development of biology at the Oak Ridge National Laboratory. How was it possible to build a strong biology group in an isolated place such as Oak Ridge, Tennessee? Now, twenty years later, we have witnessed a magnificent development of this relatively new field here in Japan. Of course, you must remember the basic sciences were already prominent here twenty years ago. I would just like to remind you of the very important genetic group well known all over the world. Also, biochemistry, biophysics, and cell biology were already in the forefront of world science. Where you have excellent basic sciences, it is logical that applied sciences will flourish as well. It is not surprising with the outstanding oncological work and with microbiology - including fermentation technology - that the environmental mutagenesis studies, closely related to oncology, have reached a high stage of development and have contributed to the welfare of man.

Careful planning, concentrating on all efforts of what one visualizes as being most important, giving everything else second or third place, seeing science as a whole, guiding it to the area where it will contribute most to our understanding of basic problems - all of these will help to increase our knowledge of the relation of basic research to human studies.

* These comments followed the Special Lecture by Dr. Yataro Tazime on September 26, 1981 at the Mishima, Japan meeting of the 3rd International Conference on Environmental Mutagens.

Taking the long point of view, first one has to have a solid foundation, then quantitative approaches will in the long run contribute substantially to closing the gap between basic and applied work. Both of these have to be considered simultaneously because one will fertilize the other. Close association between what some of us consider to be fundamental and the application of this to practical problems will also continue to encourage our colleagues in industry and agriculture who have contributed so very much in this area.

Mechanisms of Mutagenesis

Mechanisms of Mutagenesis in Bacteria

D. W. Mount, J. W. Little, B. Markham, H. Ginsburg, C. Yanisch, and S. Edmiston

Department of Molecular and Medical Microbiology, College of Medicine, University of Arizona, Tucson, Arizona, U.S.A.

ABSTRACT Cellular functions which are induced by DNA damage play a major role in the response of E. coli to many mutagens. Agents such as ultraviolet light and ionizing radiation induce a set of functions, called SOS functions, which increase DNA repair and mutagenesis. Induction involves specific proteolytic cleavage of a repressor, thus derepressing the genes which code for these functions. Only some of these inducible repair functions are associated with mutations (i.e. are error-prone). Mutagenesis by alkylating agents is influenced by an adaptation mechanism. Prolonged treatment of cells with low levels of alkylating agents induces a protein which specifically removes potentially mutagenic 0^6-methyl groups from damaged guanine bases.

INTRODUCTION

In recent years, mutagenesis has been intensively studied in E. coli and its bacteriophages. We shall restrict discussion to work done with this organism in the past two years. There are some excellent reviews of earlier work (1,2).
E. coli responds to many mutagenic agents by expressing new functions, of which some increase and others decrease the number of subsequent mutagenic events. These new functions, along with genes that control their expression, can be mutated themselves to forms that enhance or block the mutagenic response of E.

coli (2). The goal of work in this area has been to understand the mechanisms of regulation and the functions of the genes which are under their control.

There have been two systems described which are inducible by DNA damage and which influence the mutagenic response of E. coli. The first is the so-called SOS system which responds to many treatments which inhibit DNA synthesis such as irradiation by ultraviolet light or ionizing radiation. The SOS system leads to increased DNA repair and mutagenesis but is complex in that some of the repair functions remove the mutagenic lesion without causing a mutation, whereas others always have mutagenesis associated with them and are therefore assumed to be error-prone (2). The second system is an adaptive response to DNA damage caused by many alkylating agents and decreases mutagenesis by direct removal of the presumed mutagenic lesion before it influences mutagenesis (for review, ref. 3). We shall first discuss the SOS response and then the adaptive response.

THE SOS RESPONSE OF E. COLI

The SOS functions of E. coli include increased DNA repair, inhibition of cell division, induced mutagenesis and phage induction, and are called SOS functions because their coordinate expression is thought to lead to increased survival. This system is regulated by a repressor, the product of the lexA gene, and a protease, the recA gene product, that cleaves this repressor.

In a regulatory model now based on considerable genetic and biochemical data (4-6), the lexA protein is a repressor of a number of unlinked genes, which are believed to play a role in the SOS response. In undamaged cells during normal growth, the lexA protein represses these target genes. When the DNA is damaged, a signal molecule is generated which reversibly activated a specific protease activity of the recA protein, and this protease cleaves the lexA repressor, inactivating its function. The products of the SOS target genes are then made at much higher levels, and the increased levels of these gene products causes the various SOS functions, including mutagenesis, to be expressed. Finally, when DNA damage is repaired, the level of the signal molecule drops, the protease activity of the recA protein disappears, and the lexA repressor again represses the SOS genes.

Recent work has revealed that at least 12 unlinked genes (lexA, recA, uvrA, uvrB, umuC, sfiA, himA, and several din genes defined by in vivo fusions of damage-inducible promoters with lacZ) are controlled by lexA repressor (4,5,7-12). A 20 base-pair binding site (4/5 consensus -ACTGTAT-T-CA--CAG--) for lexA repressor is located at the beginning of the four genes which have been analyzed (4,5,13,14). The lexA gene is unique in having two such sites (4,5). Operator-like mutations in recA have been isolated by us and by others (7,13) and several have been sequenced (7).

In damage-induced cells and in vitro, lexA repressor is cleaved into two fragments of roughly equal size (4,6) in a reaction requiring single-stranded DNA, a nucleoside triphosphate cofactor and recA protein in a reaction very similar to that described for phage lambda repressor by Roberts and colleagues (15,16). The prevailing view is that, in the damage-induced cell, single stranded DNA or one or more nucleotides are the inducing signal for the SOS system.

The lexA gene has been sequenced (17,18) to reveal an ala-gly peptide near the middle of the repressor. Phage lambda, 434 and P22 repressors have a similar peptide and in these cases this site is the site of cleavage by recA protease (19). The mutant lexA3 gene which encodes a repressor very resistant to cleavage, and is totally deficient in SOS mutagenesis, bears a G to A mutation at the position corresponding to the ala-gly bond and which would alter the repressor sequence to ala-asp at the presumed cleavage site (18). These observations are completely consistent with cleavage of the wildtype lexA repressor at the ala-gly site as a prerequisite for expression of the SOS mutagenesis functions.

Two groups of SOS genes influence mutagenesis. Mutations in the uvrA and uvrB genes increase sensitivity to agents such as UV light, but they also increase mutation yields (1). Our explanation for this observation is that the DNA lesions remain unrepaired and keep the SOS genes induced for longer times, and are more likely themselves to become sites of mutation. Mutation yields also decrease if UV-damaged, excision-proficient cells are placed in a buffer to prevent growth (1). In this case, the time for the excision mechanism to remove the DNA lesions should be increased, and incubation in buffer should inhibit protein synthesis necessary for SOS gene expression. The second group of genes necessary for inducible mutagenesis is represented by umuC (14). Mutations in this gene greatly decrease mutagenesis following DNA damage but influence survival to a lesser extent. The role of umuC in inducible mutagenesis is not known at the present time. However, it is likely that this protein plays a direct role in the mutagenic response of E. coli.

We have analyzed further the genetic requirement of the recA and lexA functions in inducible mutagenesis. First, operator mutations in recA do not alleviate the mutagenesis defect in lexA⁻ cells (12, and unpublished observations), suggesting that one or more genes under lexA repression, such as umuC (14), must be induced for mutagenesis to occur. Second, in addition to its role in cleaving lexA repressor, presumably leading to induction of specific mutagenesis genes, recA function is also required for mutagenesis in cells lacking lexA repressor, i.e. mutagenesis is not observed in spr⁻recA⁻ cells. Although we do not understand the nature of this secondary role of recA in mutagenesis, we have considered the following possibilities: (a) a direct structural role at sites of DNA damage, (b) proteolytic modification of proteins involved in DNA replication, and (c) cleavage of other regulatory proteins which influence mutagenesis.

ADAPTIVE RESPONSE TO ALKYLATING AGENTS

Following short periods of growth in medium containing low concentrations of alkylating agents, E. coli becomes resistant to both the killing and mutagenic effects of those agents, as well as to other alkylating agents (3). This response, named the adaptive response, does not give resistance to all DNA damaging agents. For instance, adapted cells remain as sensitive and as prone to mutagenesis by UV light as normal cells. Therefore, the mechanisms appear to be different than those of the SOS functions although, as described in the next section, they can interact.

Agents which methylate DNA, such as N-methyl-N'-nitro-N nitrosoguanidine (MNNG), methyl methane sulfonate (MMS), and methyl nitrosourea all induce the response, whereas ethylating agents do not, although the methylating agents can confer resistance to ethylating agents (3). Therefore, it appears that methylated bases induce the system, but once induced, the system can repair both methylated and ethylated bases. Much less 0^6 methyl-guanine, which appears to be the mutagenic lesion in MNNG treated cells, is found in adapted cells than in normal cells (20). Similarly, there is very little enzymatic activity in extracts of normal cells which removes the 0^6 methyl-guanine (21). However, extracts of adapted cells contain substantial activity of this enzyme (22). Both physiological and biochemical studies have suggested that the 0^6 methyl group is transferred to the reactive molecule, and that each enzyme molecule is able to act only once (20,22-24). The system can therefore be saturated.

Studies with mutants have indicated that the mutagenic and lethal DNA lesions produced by alkylating agents are separate and are repaired by at least two separate pathways, both of which are inducible. Mutants which can adapt to the mutagenic effects of alkylating agents but are deficient in adapting to their killing effects have been described (25,26). Others are deficient in both processes (26), while a third class is constitutive (27).

INTERACTION OF THE SOS AND ADAPTATION RESPONSES

In adapted cells, the ability of alkylating agents to induce SOS functions is considerably reduced, presumably because the lesions removed by the adaptive mechanism are also required to induce the SOS system. There is also a reduced ability to respond to other inducing treatments, perhaps for a similar reason (28).

In unadapted cells carrying lexA$^-$ (resistant to recA cleavage) or umuC$^-$ mutations, MMS induced mutagenesis is considerably reduced (29). These results indicate that induction of the SOS system, including umuC, is necessary for MMS induced mutagenesis. In contrast, mutagenesis by EMS and MNNG is reduced in lexA$^-$ cells, but not in umuC$^-$ cells (29,30). Therefore, the mutagenic action of these agents depends upon induction of SOS functions, but not specifically upon umuC. Moreover, the significant levels of induced mutagenesis observed in lexA$^-$ strains indicates that there are at least two pathways for mutagenesis by these agents, one dependent upon the SOS system and a second SOS-independent pathway. Since most mutations induced by EMS and MNNG are GC to AT transitions (31), these different mechanisms have the same end effect.

One likely role of the SOS functions on alkylating agent-induced mutagenesis is that the uvr excision repair system can remove bulky DNA lesions (21). When the SOS system is induced, these lesions may be more rapidly removed as a result of increased expression of the uvrA and uvrB genes.

Acknowledgements:

Our research is supported by grants GM24496 and GM24178 from the National Institutes of Health and by grant number PCM 79-12059 from the National Science Foundation.

References:

(1) Witkin, E.M. Ann. Rev. Microbiol. 23, 487-514 (1969).
(2) Witkin, E.M. Bacteriol. Rev. 40, 869-907 (1976).
(3) Cairns, J., Robins, P., Sedgwick, B., Talmud, P. Prog. Nucl. Acids Res. Molec. Biol. 26, 237-244 (1981).
(4) Little, J., Mount, D.W., Yanisch, C. Proc. Natl. Acad. Sci. (USA) 78, 4199-4203 (1981).
(5) Brent, R., Ptashne, M. Proc. Natl. Acad. Sci. (USA) 78, 4204-4208 (1981).
(6) Little, J.W., Edmiston, S., Pacelli, L., Mount, D.W. Proc. Natl. Acad. Sci. (USA) 77, 3225-3229 (1980).
(7) Sancar, A., Clark, A.J., Rupp, D., Volkert, M. Personal communication.
(8) Kenyon, C., Walker, G. Proc. Natl. Acad. Sci. (USA) 77, 2819-2823 (1980).
(9) Kenyon, C., Walker, G. Nature 289, 808-810 (1981).
(10) Fogliano, M., Schendel, P.F. Nature 289, 196-198 (1981).
(11) Huisman, O., D'Ari, R. Nature 290, 797-799 (1981).
(12) Miller, H., Echols, H. Personal communication.
(13) Volkert, M., Margossian, L.J., Clark. A.J. Proc.

Natl. Acad. Sci. (USA) 78, 1786-1780 (1981).
(14) Kenyon, C., Walker, G. Personal communication.
(15) Craig, N.L., Roberts, J. Nature 283, 26-29 (1980).
(16) Phizicky, E., Roberts, J. J. Mol. Biol. 139, 319-328 (1980).
(17) Horii, T., Ogawa, T., Oqawa, H. Cell 23, 689-697 (1981).
(18) Markham, B., Little, J., Mount, D.W. Nucl. Acids Res. (in press).
(19) Yocum, R., Ptashne, M., and also Sauer, R. Personal communications.
(20) Schendel, P.F., Robins, P.E. Proc. Natl. Acad. Sci. (USA) 75, 6017-6020 (1978).
(21) Warren, W., Lawley, P.O. Carcinogenesis 1, 67-78 (1980).
(22) Olsson, M., Lindahl, T. J. Biol. Chem. 255, 10569-10571 (1980).
(23) Robins, P.E., Cairns, J. Nature 280, 74-76 (1979).
(24) Karran, P., Lindahl, T., Griffin, B. Nature 280, 76-77 (1979).
(25) Jeggo, P. J. Bacteriol. 139, 783-791 (1979).
(26) Jeggo, P., Defais, M., Samson, L., Schendel, P. Molec. gen. Genet. 162, 299-305 (1978).
(27) Sedgwick, B., Robins, P. Molec. gen. Genet. 180, 85-90 (1980).
(28) Defais, M., Jeggo, P., Samson, L., Schendel, P.F. Molec. gen. Genet. 177, 653-659 (1980).
(29) Schendel, P., Defais, M. Molec. gen. Genet. 177, 661-665 (1980).
(30) Schendel, P., Defais, M., Jeggo, P., Samson, L., Cairns, J. J. Bacteriol. 135, 466-475 (1978).
(31) Coulondre, C., Miller, J.H. J. Mol. Biol. 117, 577-606 (1977).

Plasmid Mediated Enhancement of Chemical Mutagenesis

Karen L. Perry, Stephen J. Elledge, Marilyn R. Lichtman, and Graham C. Walker

Biology Department, Massachusetts Institute of Technology, Cambridge, Massachussetts, U.S.A.

ABSTRACT The plasmid pKM101 is one of a number of naturally occurring plasmids which both increase the susceptibility of bacterial cells to mutagenesis and also increase their resistance to UV-killing. These effects are $recA^+lexA^+$-dependent and pKM101 can suppress the mutagenesis and repair deficiencies of umuC mutants suggesting that pKM101 carries an analog of the umuC gene. A region of pKM101 termed muc is responsible for these effects. The muc region is approximately 2000 bp long and codes for two polypeptides. The expression of at least one of these gene products is induced by UV and preliminary genetic analysis of its regulation suggests that the lexA protein is the repressor for this gene. DNA homologous to the pKM101 muc region has been found on other naturally occurring plasmids.

INTRODUCTION

A considerable number of naturally occurring plasmids have been reported which confer upon their host both increased resistance to killing by UV irradiation and increased susceptibility to UV and chemical mutagenesis (1,8,24,33). R46 and its derivative, pKM101, (26,27) and R205 (20,33) are representative examples in Escherichia coli and Salmonella typhimurium and pMG2 is an example in Pseudomonas aeruginosa (17).

One plasmid in particular, pKM101, has become the object of intensive scrutiny. pKM101 is a 35.4 kb N-incompatabilty group plasmid (16) that was derived from its clinically isolated parent R46 (27) by a series of in vivo manipulations (27) that resulted in the deletion of a 13.8 kb region of DNA containing several of the drug resistance of R46 (16). Because of its ability to increase the susceptibility of cells to UV and chemical mutagenesis pKM101 was introduced into the Ames tester strains (21) and it has played a major role in the success of the system (22,23).

E. coli or S. typhimurium cells containing pKM101 exhibit an increased susceptibility to both base-substitution and frameshift mutagenesis with a variety of agents (21,26,27) and an increased resistance to killing by UV (25,27,38). In addition pKM101 increases the survival of UV-irradiated phage in both unirradiated cells (38) and irradiated cells (38,40) and causes a modest increase in the spontaneous reversion rate of certain point mutations (9,26,38). In this paper we will summarize our current understanding of the mechanisms by which pKM101 exerts its effects and the relationship of these plasmid-mediated processes to the cellular processes involved in chemical mutagenesis.

RELATIONSHIP OF pKM101-MEDIATED PROCESSES TO CELLULAR GENE FUNCTIONS

pKM101 has been shown to increase both susceptibility to mutagenesis and resistance to killing in uvrA, uvrB, uvrC, polA, recB recC, recF and uvrD strains (10,26,37,38) suggesting that the products of these genes are not required for the pKM101-mediated processes. In contrast, the ability of pKM101 to influence mutagenesis and repair is dependent on the $recA^+ lexA^+$ genotype in E. coli (10,25,26,37,38) as is the bacterial "error-prone repair" system (28,45). However pKM101 does not exert its effects by causing a general induction of the $recA^+ lexA^+$ dependent phenomena since the synthesis of the recA protein (19,41) and the stability of λ lysogens (10) are unaffected by pKM101. Instead pKM101 seems to increase specifically the capacity of a bacterial cell to carry out those processes usually referred to as "error-prone repair" (28,45). This hypothesis is most strongly supported by the observation that pKM101 is able to suppress the deficiencies of umuC (UV-nonmutable) mutants (12,35) in mutagenesis and Weigle reactivation (36,41). The simplest interpretation of these results is that pKM101 supplies the cell with an analog of the umuC protein, thereby providing the cell with an increased amount of a biochemical function needed for error-prone repair. We have discussed alternative interpretations of this result elsewhere (32).

IDENTIFICATION OF THE muc REGION OF pKM101

We have been able to isolate mutants of pKM101 which were unable to increase the susceptibility of cells to base substitution mutagenesis (32,39). These same plasmid mutants were also defective in their ability to increase frameshift mutagenesis with the appropriate mutagen implying a much closer relationship between these two processes than is often considered. This result does not necessarily mean that base substitution and frameshift mutagenesis occur by the same biochemial mechanism but does suggest that the two processes share a common step at either the mechanistic or regulatory level. Similarly pKM101 mutants which had lost the ability to increase mutagenesis had also lost their ability to increase UV resistance and their ability to reactivate UV-irradiated phage implying a similarly close connection between these processes (32,39).

These pKM101 mutants were derived either by the insertion of the transposable element Tn5 (32) or by mutagenesis with N-methyl-N'-nitro-N-nitrosoguanidine (39). By restriction endonuclease cleavage we were able to map the position of twenty, independent Tn5 insertions that eliminated the plasmid's ability to increase mutagenesis and resistance to UV-killing. All of these insertions mapped within an approximately 1900 bp region which we termed the muc (mutagenesis: UV and chemical) region which is necessary for the plasmid-mediated effects (32). By analyzing random Tn5 insertions we obtained two insertions about 2250 bp apart that flank the muc region and have no effect on mutagenesis or UV-resistance; these placed an upper limit on the size of the muc region (32).

The muc region has recently been cloned into a derivative of the vector pBR322 and we have found that the hybrid plasmid, which contains a 2200 bp fragment of pKM101 DNA, is capable of increasing both the susceptibility of cells to mutagenesis and their resistance to UV killing. Thus the muc region of pKM101 is both necessary and sufficient for pKM101-mediated mutagenesis and UV-resistance. If pKM101 does in fact work by providing an analog of the umuC gene product, then the muc region codes for that analog.

INDUCIBILITY OF THE muc GENE PRODUCT

Our observation that pKM101 suppressed the deficiency of umuC mutants in W-reactivation raised the possibility that the expression of a muc function of pKM101 was induced by DNA damage. In order to examine this directly we constructed a muc-lac gene fusion. A restriction fragment containing the portion of the muc region counterclockwise from the Bgl site was enzymatically joined to the carboxy terminal portion of the β-galactosidase gene (6). When a plasmid containing the resulting muc-lac fusion was introduced into a wild-type strain low level β-galactosidase activity was observed indicating that a hybrid protein was being produced. If the cells were UV-irradiated an increase in the level of β-galactosidase was observed. Thus it appears that at least one gene product from the muc region is inducible by UV. In addition, this muc gene is read in a clockwise direction on the pKM101 map (15,16).

The availability of the plasmid carrying the muc-lac fusion provided us with an easy way of genetically analyzing the regulation of this muc gene. If the plasmid was introduced into $recA^-$ or $lexA^-$ cells, β-galactosidase activity was present at low levels and was not induced by UV-irradiation. In contrast, if the muc-lac plasmid was introduced into either spr or spr recA strains β-galactosidase activity was present constitutively at higher levels and was not induced further by DNA damage. The genetic dependence of expression of this muc-lac fusion is the same as that of the din-lac fusions (13) which we have examined previously (2,13,14). The results are consistent with a model (4,14,19,42) in which lexA is the repressor of this muc gene and induction requires cleavage of the lexA protein by the activated recA protease. The results suggest a further similarity between the pKM101 muc gene(s) and the chromsomal umuC gene(s) - both are induced by DNA damage and both are repressed by the lexA protein. In addition the $recA^+ lexA^+$-dependent inducibility of a pKM101 gene product could help explain several previous observations - the synergistic effect of pKM101 and tif mutations on mutagenesis (38) and the weak pKM101-mediated mutagenesis seen in strains carrying an allele of recA (lexB) that is deficient in the recA protease activity (3,29,37).

THE muc REGION OF pKM101 CODES FOR TWO POLYPEPTIDES

In order to identify the protein products of the muc region the small hybrid plasmid described above was introduced into maxicells (30) derived from a spr recA strain. Besides the proteins coded for by the vector, two additional polypeptides were observed, one of approximately 43,000 daltons and one of approximately 16,000 daltons. By $\gamma\delta$ (30) or Tn5 (32) insertion mutagenesis we obtained muc⁻ derivatives of the hybrid plasmid which could no longer increase the susceptibility of cells to mutagenesis or increase their resistance to UV-killing. The positions of these insertions within the muc region were determined by restriction endonuclease cleavage. The protein products coded for by these plasmids were then determined using spr recA maxicells. By this type of analysis we have determined that the larger polypeptide is encoded in the clockwise end (32) of the muc region and the smaller polypeptide is encoded in the counterclockwise end.

Both polypeptides seem to be required for the plasmid-mediated effects on mutagenesis since in preliminary experiments these insertion mutants of the hybrid plasmid appear to complement, at least partially, the deficiencies of the appropriate pKM101 insertion mutants in mutagenesis enhancement and resistance to UV-killing. Furthermore it seems likely that the genes for the two polypeptides are within the same operon since a Tn5 insertion within the gene coding for the small polypeptide resulted in the loss of both polypeptides.

We have recently cloned the umuC gene(s) from the E. coli chromosome and we are attempting to compare its gene product(s) with those of the muc region of pKM101. We hope that our work on the pKM101 muc genes and the E. coli gene will allow a systematic approach to be taken towards determining the molecular mechanism of error-prone repair.

OTHER NATURALLY-OCCURRING PLASMIDS CARRYING muc GENES

In the course of analyzing the functional organization of pKM101 we discovered that the muc region was surrounded by inverted repeats raising the possibility that the genes either are, or once were, part of a transposable genetic element (15). Because of this we were interested in establishing whether other mutagenesis-enhancing plasmids found in nature carried sequences homologous to the muc genes of pKM101. Thus we examined the DNA of a number of such plasmids by Southern blot analysis (34) using a probe consisting of an internal restriction fragment from the pKM101 muc region. To date we have established that the plasmids N3, R45, R48 (33) have homology to the muc region of pKM101.

It is not yet clear why genes capable of suppressing umuC mutations would be found on plasmids. However a number of bacteria such as Haemophilus spp., Proteus spp., Methylococcus sp., and methylobacter sp. (5,11,31) are nonmutable with UV; perhaps they naturally lack umuC function. We have previously suggested (7,43,44) that S. typhimurium LT2 may be deficient in umuC function. The acquisition of a plasmid carrying umuC-like genes would provide such cells with both the increased resistance to killing which is due to $umuC^+$ or muc^+-dependent processes and also susceptibility to mutagenesis by UV and a variety of chemical agents.

ACKNOWLEDGMENTS

We thank J. Geiger, C. Kenyon, J. Krueger, B. Mitchell, K. Paek, P. Pang, S. Rabinovitch, W. Shanabruch, and S. Winans for their support, encouragement, and useful discussions. We especially thank L. Withers for her help in preparing the manuscript.

This work was supported by Public Health service grant 5-R01-CA21615-05 from the National Cancer Institute. G.C.W. was a Rita Allen Scholar during this research. K.L.P. and S.J.E. were supported by an NIH Predoctoral Training Grant Number GM07287. M.R.L. was a participant in the Undergraduate Research Opportunities Training Program at the Massachusetts Institute of Technology and was supported in part by a Wellesley/UROP Summer Grant.

REFERENCES

(1) Arai, T. and Ando, T. Keio J. Med. 29, 47-54 (1980)
(2) Bagg, A., Kenyon, C.J. and Walker, G.C. Proc. Natl. Acad. Sci. USA, in press (1981)

(3) Blanco, M. and Rebollo, J. Mutat. Res. 81, 265-275 (1981)
(4) Brent, R. and Ptashne, M., Proc. Natl. Acad. Sci. USA 78, 4204-4208 (1981)
(5) Bridges, B.A. In "Second International Symposium on the Genetics of Industrial Microorganisms" p7-14, Ed. MacDonald, K.D., Academic Press, New York (1976)
(6) Casadaban, M.J., Chou, J. and Cohen, S.N. J. Bacteriol. 143, 971-980 (1980)
(7) Dobson, P.P. and Walker, G.C. Mutat. Res. 71, 25-41 (1980)
(8) Drabble, W.T. and Stocker, B.A.D. J. Gen. Microbiol. 53, 109-123 (1968)
(9) Fowler, R.G., McGinty, L., Mortelmans, K.E. J. Bacteriol. 140, 929-937 (1979)
(10) Goze, A. and Devoret, R. Mutat. Res. 61, 163-179 (1979)
(11) Hofmeister, J., Kohler, H., and Filppanov, V.D. Molec. Gen. Genet. 176, 265-273 (1979)
(12) Kato, T. and Shinoura Y. Mol. Gen. Genet. 156, 121-131 (1977)
(13) Kenyon, C.J., and Walker, G.C. Proc. Natl. Acad. Sci. USA, 77, 2819-2823 (1980)
(14) Kenyon, C.J. and Walker, G.C. Nature, 289, 808-810 (1981)
(15) Langer, P.J., Shanabruch, W.G. and Walker, G.C. J. Bacteriol 145, 1310-1316 (1981)
(16) Langer, P.J. and Walker, G.C. Molec. Gen. Genet 182, 268-272 (1981)
(17) Lehrbach, P., Kung, A.H.C., and Lee, B.T.O. J. Gen. Microbiol. 98, 167-176 (1976)
(18) Little, J. and Hanawalt, P.C. Mol. Gen. Genet 150, 237-248 (1977)
(19) Little, J.W., Mount, D.W. and Yanisch-Perron Proc. Natl. Acad. Sci. USA 78, 4199-4203 (1981)
(20) MacPhee, D.G. Mutat. Res. 19, 356-359 (1973)
(21) McCann, J., Spingarn, N.E., Kobori, J. and Ames, B.N. Proc. Natl. Acad. Sci. USA 72, 979-983 (1975)
(22) McCann, J., Choi, E., Yamasaki, E. and Ames, B.N. Proc. Natl. Acad. Sci. USA 72, 5135-5139
(23) McCann, J. and Ames, B.N. Proc. Natl. Acad. Sci. USA 73, 950-954 (1976)
(24) Molina, A.M., Babudri, N., Tamaro, M., Venturini, S. Monti-Bragadin, C. FEMS Microbiol Letters 5, 33-37 (1979)
(25) Monti-Bragadin, C., Babudri, N. and Samer, L. Mol.

Gen. Genet. 145, 303-306 (1976)
(26) Mortelmans, K.E. and Stocker, B.A.D. J. Bacteriol. 128, 271-282 (1976)
(27) Mortelmans, K. and Stocker, B.A.D. Mol. Gen. Genet 167, 317-328 (1979)
(28) Radman, M. In "Molecular mechanisms for repair of DNA" p 283-291, Eds. Hanawalt, P.C. and Setlow, R.B., Plenum Press, New York (1975)
(29) Roberts, J.W. and Roberts, C.W. Nature 290, 422-424 (1981)
(30) Sancar, A., Wharton, R.P., Seltzer, S., Kacinski, B.M., Clarke, N., and Rupp, W.D. J. Molec Biol 148, 45-62 (1981)
(31) Notani, N.K. and Setlow, J.E. J. Bacteriol. 143, 516-519 (1980)
(32) Shanabruch, W.G. and Walker, G.C. Mol. Gen. Genet. 179, 289-297 (1980)
(33) Shapiro, J.A. In "DNA Insertion Elements, Plasmids and Episomes" p 15-22, Eds. Bukhari, A.I., Shapiro, J.A., and Adhya, S.L., Cold Spring Harbor Laboratory, Cold Spring Harbor, New York (1977).
(34) Southern, E.M. J. Molec. Biol. 98, 503-517 (1975)
(35) Steinborn, G. Molec Gen. Genet. 165, 87-93 (1978)
(36) Steinborn, G. Molec. Gen. Genet. 175, 203-208 (1979)
(37) Waleh, N.S. and Stocker, B.A.D. J. Bacteriol 137, 830-838 (1979)
(38) Walker, G.C. Mol. Gen. Genet 152, 93-103 (1977)
(39) Walker, G.C. J. Bacteriol 133, 1203-1211 (1978)
(40) Walker, G.C. J. Bacteriol 135, 415-421 (1978)
(41) Walker, G.C. and Dobson, P.P. Mol. Gen. Genet 172, 17-24 (1979)
(42) Walker, G.C., Kenyon, C.J., Bagg, A., Langer, P.J. and Shanabruch, W.G. J. Natl. Cancer Instit. Monograph, in press (1981)
(43) Walker, G.C., Elledge, S.J., Perry, K.L., Bagg, A., and Kenyon, C.J. In "Induced Mutagenesis: Molecular Mechanisms and Their Implications for Environmental Protection" Eds. Lawrence, C.W., Prakash, L. and Sherman, F., Plenum Press, New York, in press (1981)
(44) Walker, G.C. In "In vitro Toxicity Testing of Environmental Agents" Eds. Kolber, A.R. and Wang, T.K., Plenum Press, New York, in press (1981)
(45) Witkin, E.M. Bacteriol. Rev. 40, 869-907 (1976)

Non-DNA Primary Targets for the Induction of Genetic Change

R. H. Haynes, J. G. Little, B. A. Kunz and B. J. Barclay

York University, Department of Biology, Toronto, Canada

ABSTRACT Thymidylate deprivation induced in yeast by FdUMP or antifolate drugs is lethal and highly recombinagenic, though not mutagenic, for nuclear genes. On the other hand, excess thymidylate is a potent mutagen, and also lethal, though apparently not a strong recombinagen. Since imbalances in DNA nucleotide precursor pools can be produced by drugs and metabolic disturbances which do not involve initial chemical attack on DNA, it is clear that cells must be regarded as containing non-DNA primary targets for the induction of genetic change.

INTRODUCTION

For the past twenty years speculation on the mechanisms of induced genetic change has been based largely on the DNA damage-repair hypothesis (18). It has been established that radiations and many chemical mutagens are capable of producing directly, or subsequently to metabolic activation, a variety of structural defects in DNA, and that such lesions are subject to various modes of DNA repair (17). These processes of repair can be divided operationally into two categories, error-free repair (EFR) and error-prone repair (EPR). Lesions repaired by error-free mechanisms have no genetic consequences, whereas mutations, especially in higher cells, are considered to be fixed largely as a result of the action of error-prone repair processes (10, 33). In addition, repair enzymes also may be involved in genetic recombination, and the induction of recombinants by radiations and chemicals may reflect the action of 'recombinagenic' repair mechanisms which, in yeast, can be distinguished genetically from EFR and EPR (18, 25). Much has been learned about the lesions produced by mutagens which attack DNA directly and the macromolecular basis of repair. However, less attention has been paid to the genetic consequences of disturbances in DNA synthesis which can be produced by agents or conditions which affect DNA precursor metabolism (1).

Experiments with various *in vitro* DNA synthesis systems have shown that replicational fidelity is strongly dependent on correctly balanced concentrations of nucleotide substrates in the reaction mixture (13, 16, 19, 22, 30). In bacteria it has been known for some years that thymidylate deprivation is mutagenic (9). In addition, it has been shown in mammalian cells that the mutagenicity of the base analogs bromodeoxyuridine and 2-aminopurine can arise not only from base-pairing ambiguities but also from perturbations of normal deoxyribonucleoside triphosphate pools in cells growing in the presence of these analogs (20, 21). Finally, it is well established in a number of mammalian systems that excessive doses of thymidine can provoke imbalances in DNA precursor pools, and that the toxic and mutagenic effects of these imbalances arise, at least in part, from the depletion of endogenous deoxycytidine triphosphate (dCTP) pools caused by the allosteric effect of dTTP on ribonucleotide reductase (6, 14, 29). Thus it is reasonable to think that the intracellular concentrations of the nucleotide precursors required for DNA synthesis may be regulated so as to minimize the frequency of genetic change, and that agents which alter nucleotide pool sizes can have profound genetic consequences (1).

GENETIC EFFECTS OF THYMIDYLATE STRESS IN YEAST *(SACCHAROMYCES)*

Among the four nucleotides which are necessary substrates for DNA synthesis, deoxythymidine triphosphate (dTTP) occupies a special position (Figure 1). First, its biosynthetic pathway is longer, more complex and requires a greater expenditure of energy than that of the other nucleotides. Second, its synthesis involves the production of another potential DNA precursor, deoxyuridine triphosphate (dUTP) which means there are in fact *three* pyrimidine triphosphate precursors potentially available for DNA synthesis. And third, the thymidylate pathway is coupled in an important way with folate metabolism: N^5, N^{10}-

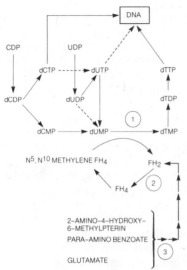

Figure 1

methylenetetrahydrofolate serves as the methyl donor in the conversion of dUMP to dTMP by thymidylate synthetase, and this reaction constitutes a significant drain on intracellular folate pools. Agents which inhibit thymidylate synthetase, or limit the supply of reduced folates, can reduce or eliminate dTMP biosynthesis (2, 12, 23, 27, 32). In yeast, 5-fluorodeoxyuridine monophosphate (FdUMP) inhibits thymidylate synthetase (Fig. 1, reaction 1) (5), antifolates, such as aminopterin or methotrexate, inhibit dihydrofolate reductase (Fig. 1, reaction 2) and the sulfa drugs, as analogs of para-amino benzoate, impede *de novo* folate synthesis (Fig. 1, reaction 3) (12, 27, 32). The combination of aminopterin (or methotrexate) plus sulfanilamide provides a very tight block on dTMP biosynthesis. We have found that all of these agents are capable of inducing 'thymineless death' in yeast and are potent recombinagens for nuclear genes; these lethal and genetic effects can be abolished by concurrent provision of dTMP along with the drugs. Thus, the effects of FdUMP and the various antifolates can be attributed specifically to depletion of intracellular dTMP pools (1, 24). On the other hand, dTMP deprivation in yeast is not mutagenic for nuclear genes though it does cause mitochondrial point mutations and cytoplasmic petites (3). The fact that nuclear genes of yeast, a simple eukaryote, are refractory to mutation induction by thymidylate stress, whereas several prokaryotes are readily mutated, may reflect some fundamental, but hitherto unrecognized, difference in the organization and control of DNA synthesis between these two groups of organisms.

For our studies on the genetic effects of drugs which produce thymidylate stress in yeast we constructed a derivative of the diploid strain D7 of Zimmermann *et al* (34) which is capable of taking up and utilizing exogenous dTMP for DNA synthesis (*tup-* character). This strain, designated

D7-T2, also can take up FdUMP, and has the following genotype.

$$\frac{\alpha \quad tup- \quad ade2-40 \quad cyh2 \quad trp5-12 \quad ilv1-92}{a \quad tup- \quad ade2-119 \quad + \quad trp5-27 \quad ilv1-92}$$

In such strains it is possible to measure mitotic recombination at ADE2 and CYH2, gene conversion at TRP5 and reverse mutation at ILV1; our previous papers can be consulted for further details on how these various genetic effects are identified and measured (23, 24).

As examples of the aforementioned phenomena we show here the recombinagenicity of FdUMP (Fig. 2) and sulfathiazole (Fig. 3). After six hours of growth at $34°C$ in YPDP broth (27, 34) containing FdUMP (500 µg/ml) cultures of D7-T2 consist almost entirely of mother-daughter doublets having the 'dumbbell' morphology characteristic of cells in which DNA synthesis has been inhibited (7, 27). Over this same time period, cell growth is inhibited, viability is reduced below 10% with kinetics characteristic of 'thymineless death', and the frequency of mitotic recombinants, measured either at ADE2 or as gene conversion at TRP5, increases dramatically (Fig. 2, open circles). In parallel cultures which also contain dTMP (300 µg/ml) the lethal and recombinagenic effects of FdUMP are abolished and the cells grow normally (Fig. 2, closed circles). We have observed similar recombinagenic effects in D7-T2 with aminopterin, methotrexate, sulfanilamide, sulfathiazole, sulfactamide,

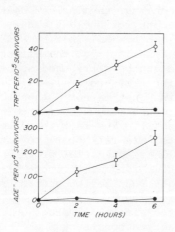

Figure 2

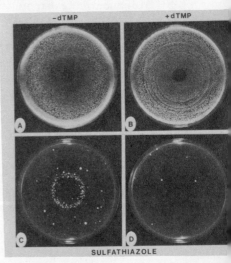

Figure 3

sulfamethoxazole and various combination of these drugs; in all cases the recombinagenic effects can be abolished specifically by provision of dTMP (1). Figure 3 illustrates these effects for sulfathiazole in our well-test for mitotic recombination which is based on the segregation of cycloheximide resistance in D7-T2 (24). In this test a lawn of cells was spread on YPDP agar both without (Fig. 3A) and with (Fig. 3B) dTMP (100 µg/ml). Sulfathiazole (5 mg in aqueous solution) was added to the center well in each plate. After 24 hours incubation at 34°C the masterplates were replicated onto YPD medium containing cycloheximide (5 µg/ml) (Fig. 3C) and cycloheximide plus dTMP (Fig. 3D). The ring of cycloheximide resistant colonies shown in Fig. 3C indicates that sulfathiazole induces mitotic recombination; the absence of such a ring in Fig. 3D shows that this effect is abolished by concomitant provision of dTMP. The lethal effect of sulfathiazole caused by thymidylate depletion is shown by the lack of cell growth in the annular ring surrounding the center well in Fig. 3A; the abolition of this effect by dTMP is indicated by the occurrence of cell growth in the equivalent annular zone in Fig. 3B.

It should be noted that cells treated with antifolates require not only thymidylate, but also adenine, histidine and methionine for growth; however it is only thymidylate deprivation which induces mitotic recombination. In other experiments we have shown that purine starvation is lethal but that this killing is not accompanied by any observable genetic effects. We also carried out reconstruction experiments to show that the effects illustrated in Fig. 2 did not arise by any possible growth advantage possessed by spontaneous recombinants in the presence of FdUMP. Finally, although cycloheximide resistance in D7 also can arise by induced monosomy, genetic tests revealed that the *maximum* frequency of such events present in Fig. 3C would be about 25% (24). The recombinagenic effects of thymidylate deprivation can be elicited without the aid of drugs either by growing strains homozygous for the temperature sensitive mutation *cdc21* (which occurs in the structural gene for thymidylate synthetase) at the restrictive temperature (23), or by direct starvation of a D7 derivative which is auxotrophic for dTMP (F. Eckardt, personal communication).

In none of the experiments described here was there any indication of FdUMP- or antifolate-induced reversion of *ilv1*-91, nor were we able to detect induced forward mutation at any one of several genes in the biosynthetic pathway of adenine by use of the *ade2 adex* system (11) in a haploid derivative of D7-T2 (23). On the other hand, Barclay and Little (4) have shown that *excess* dTMP is a potent mutagen, (it also is lethal (31)) though apparently not a strong recombinagen. Very high concentrations of dTMP will produce cycloheximide-resistant colonies in a well-test but this apparent recombinagenicity could arise from mutation at the heterozygous marker. Thus, it would appear that in yeast the frequencies of reverse mutation and mitotic recombination can be modulated independently by variations in thymidylate pool sizes.

At present we can only speculate on the mechanism of mutagenesis by excess dTMP (1, 4). However, we have obtained macromolecular evidence for the occurrence of DNA strand breaks in yeast during thymidylate deprivation, and it is reasonable to think that these breaks are responsible for the concomitant elevation of recombination frequencies (1). Elsewhere we have argued that such breaks could arise by a process of reiterative uracil incorporation and excision in DNA associated with a depressed ratio of thymine to uracil nucleotides (1), a process which also occurs in mammalian cells (15).

CONCLUDING REMARKS

The dramatically different effects produced by thymidylate deprivation and excess on recombination and mutation, respectively, emphasizes the importance of distinguishing between these two end-points in genetic toxicology. This distinction is especially important if genetic exchange pheonomena, rather than point mutations, should prove to be the critical early events in carcinogenesis (8, 26, 28). In this connection it is interesting to note that ionizing radiation is a more efficient inducer of neoplastic transformation than 254 nm ultraviolet light in C3H 10T½ cells (M. M. Elkind, personal communication). And in yeast, ionizing radiation is a more efficient recombinagen than UV, whereas UV is the more efficient mutagen (11a). Recombinagens also could induce recessive homozygosis or hemizygosis in the developing embryo. Such genetic exchange events could lead to the expression of recessive lethal mutations and thereby account, at least in part, for the teratogenicity of certain antifolates (27a).

This work has shown that there exists at least one

class of agents, those which impede thymidylate biosynthesis, which are capable of producing genetic effects neither by direct attack on DNA, nor by metabolic activation into DNA-damaging products. Rather, their primary targets are enzymes involved in DNA precursor synthesis and the alterations which do occur in DNA appear to be secondary to disturbances in DNA synthesis. It is possible that other classes of antimetabolites also could produce genetic changes in such a manner. Thus, the classical DNA damage-repair hypothesis will have to be broadened, and made more explicit biochemically, if the results described here are to be accomodated within it.

REFERENCES

(1) Barclay, B.J., Kunz, B.A., Little, J.G. and Haynes, R.H. Can. J. Biochem. March 1982 (in press).
(2) Barclay, B.J., and Little, J.G. J. Bacteriol. 132, 1036-1037 (1977).
(3) Barclay, B.J., and Little, J.G. Molec. Gen. Genetics 160, 33-40 (1978).
(4) Barclay, B.J., and Little, J.B. Molec. Gen. Genetics 181, 279-281 (1981).
(5) Bisson, L., and Thorner, J. J. Bacteriol 132, 44-50 (1977).
(6) Bradley, M.O., and Sharkey, N.A. Nature 274, 607-608 (1978).
(7) Brendel, M., and Langjahr, U.G. Molec. Gen. Genetics 131, 351-358 (1974).
(8) Cairns, J. Nature 289, 353-357 (1981).
(9) Coughlin, C.A., Adelberg, E. A. Nature 178, 531-532 (1956).
(10) Drake, J.W. In: Progress in Genetic Toxicology (D. Scott, B.A. Bridges & F.H. Sobels, eds.), Elsevier/North Holland, Amsterdam (1977), pp. 43-55.
(11) Eckardt, F., and Haynes, R.H. Mutation Res. 43, 327-338 (1977).
(11a) Eckardt, F., Kunz, B.A., and Haynes, R.H. Environ. Mutagen. 3, 318 (1981).
(12) Fäth, W.W., Brendel, M., Laskowski, W., and Lehmann-Brauns, E. Molec. Gen. Genetics 132, 335-345 (1974).
(13) Fersht, A.R. Proc. Natl. Acad. Sci. U.S.A. 76, 4946-4950 (1979).
(14) Fox, R.M., Tripp, E.H., and Tattersall, M.H.N. Cancer Res. 40, 1718-1721 (1980).

(15) Goulian, M., Bleile, B., and Tseng, B.Y. *Proc. Natl. Acad. Sci. U.S.A. 77,* 1956-1960 (1980).
(16) Hall, Z.W., and Lehman, I.R. *J. Molec. Biol. 36,* 321-333 (1968).
(17) Hanawalt, P.C., Cooper, P.K., Ganesan, A.K., and Smith, C.A. *Ann. Rev. Biochem. 48,* 783-836 (1979).
(18) Haynes, R.H., and Kunz, B.A. In: *The Molecular Biology of the Yeast Saccharomyces,* (J.N. Strathern, E.W. Jones & J.R. Broach, eds.), chapter 28, Cold Spring Harbor Laboratory Monograph, Cold Spring Harbor, New York (1981).
(19) Hibner, U., and Alberts, B. *Nature 285,* 300-305 (1980).
(20) Hopkins, R.L., and Goodman, M.F. *Proc. Natl. Acad. Sci. U.S.A. 77,* 1801-1805 (1980).
(21) Kaufman, E.R., and Davidson, R.L. *Proc. Natl. Acad. Sci. U.S.A. 75,* 4982-4986 (1978).
(22) Kunkel, T.A., and Loeb, L.A. *J. Biol. Chem. 254,* 5718-5725 (1979).
(23) Kunz, B.A., Barclay, B.J., Game, J.C., Little, J.G., and Haynes, R.H. *Proc. Natl. Acad. Sci. U.S.A. 77,* 6057-6061 (1980).
(24) Kunz, B.A., Barclay, B.J., and Haynes, R.H. *Mutation Res. 73,* 215-220 (1980).
(25) Kunz, B.A., and Haynes, R.H. *Ann. Rev. Genetics 15,* 57-89 (1981).
(26) Levan, A., Levan, G., and Mitelman, F. *Hereditas 86,* 15-30 (1977).
(27) Little, J.G., and Haynes, R.H. *Molec. Gen. Genetics 168,* 141-151 (1979).
(27a) Milunsky, A., Graef, J.W., and Gaynor, M.F. *J. Pediatric 72,* 790-795 (1968).
(28) Pall, M.L. *Proc. Natl. Acad. Sci. U.S.A. 78,* 2465-2468 (1981).
(29) Reynolds, E.D., Harris, A.W., and Finch, L.R. *Biochim. Biophys. Acta 561,* 110-123 (1979).
(30) Sinha, N.K., and Haimes, M.D. *J. Biol. Chem. 256,* 10671-10683 (1981).
(31) Toper, R., Fäth, W.W., and Brendel, M. *Molec. Gen. Genetics 182,* 60-64 (1981).
(32) Wickner, R.B. *J. Bacteriol 117,* 252-260 (1974).
(33) Witkin, E.M. *Bacteriol. Rev. 40,* 869-907 (1976).
(34) Zimmermann, F.K., Kern, R., and Rasenberger, H. *Mutation Res. 28,* 381-388 (1975).

Mechanisms of Induced Mutagenesis in Yeast

Christopher Lawrence

Department of Radiation Biology and Biophysics, University of Rochester Medical Center, Rochester, New York, U.S.A.

ABSTRACT Agents that induce mutations in Saccharomyces cerevisiae are of three kinds: specific mutagens that are RAD6 - repair independent (misreplication type), less specific mutagens that are RAD6 - repair dependent (misrepair type), and specific mutagens that are RAD6 - repair dependent (a new category). Most powerful mutagens, therefore, act via mechanisms related to repair or recovery. RAD6 repair does not depend on either excision or recombination, and most is non-mutagenic. Fifteen or more genes are concerned with induced mutagenesis, and these are often involved in the production of only some kinds of mutations by some mutagens; yeast, therefore, possesses several RAD6-dependent mutagenic pathways. In aggregate, these act with very different efficiencies at different genetic sites, even when of identical nucleotide sequence. A high proportion of the mutations induced by misrepair mutagens, such as UV, are untargeted, that is occur in lesion free regions of DNA, presumably indicating a general reduction in replicational fidelity.

INTRODUCTION

Mechanisms of induced mutagenesis in bakers' yeast, Saccharomyces cerevisiae, have been investigated chiefly because they may differ from those in the more intensively studied prokaryotes. Greater opportunities for avoiding or correcting errors are likely to exist in eukaryotes both because DNA is synthesized during only part of the cell cycle and because replication is much slower than in bacteria and their viruses (5). Moreover, as suggested by Drake (3), greater fidelity may be required to replicate and maintain the large genomes characterstic of eukaryotes, a view for which there is some supporting evidence (3,10). The molecular mechanisms responsible for such fidelity are not well understood, however, and thus the ways in which mutagens circumvent, modify or saturate them are also largely unknown. Current knowledge of mutagenesis in yeast comes chiefly from analysis of the types of changes induced by various mutagens and of the genetic control of these processes.

TYPES OF MUTAGENS

Agents that induce mutations in yeast appear to be of three general kinds, perhaps indicating three basic modes of action: those which are RAD6-repair independent and produce fairly specific types of mutations (misreplication type), those which are RAD6-repair dependent and produce a variety of DNA alterations (misrepair type), and those which are also RAD6-repair dependent but produce fairly specific types of mutations (a new category).

Positive identification of misreplication mutagens in yeast has been limited by the scarcity of studies with rad6 mutants, but the acridine half mustard ICR-170, which produces GC addition frameshifts in GC runs with high specificity (2), appears to be of this type (24). The base analogue 6-N-hydroxyaminopurine may also be of this type (1), though its specificity is not yet known. UV and ionizing radiation appear to be misrepair mutagens: they are ineffective in rad6 mutants (15) and induce a variety of different mutational alterations, though they are far from random mutagens. UV, for example, produces six times more transitions than transversions (13), and can show marked specificity at some genetic sites (23). The third group, first discovered by Prakash (20), includes alkylating agents which are specific mutagens (22), though ineffective in stationary phase rad6 mutants. It is possible that such mutagens might also act by misreplication if administered during S-phase, when possible error-free repair mechanisms may have no time to act (10).

RAD6 DEPENDENT ACTIVITIES

Despite their diversity therefore, many -- perhaps most -- powerful mutagens owe their effectiveness in yeast to RAD6 gene function. This locus appears to play a central, possibly regulatory, role in a group of repair or recovery activities that are of great importance to the survival of wild-type cells exposed to any one of a wide range of DNA-damaging agents (12). Such activities do not appear to depend either on excision or recombination (see 10), and for the most part are error-free: mutagenic activities account at most for only a very small part of wild-type resistance (12,10). The molecular nature of RAD6 dependent repair or recovery is not yet known, however.

GENETIC CONTROL OF MUTAGENESIS

The minor component of RAD6 repair or recovery associated with mutagenesis includes several at least partly independent branches. Among the fifteen or more genes (10) specifically concerned with mutagenesis, many have characteristic mutational phenotypes, that is are involved in the production of only certain types of mutations induced by only some mutagens (17,21). The rev mutations isolated by Lemontt (16) for example, block mutagenesis induced by UV, gamma rays and 4-nitroquinoline -1-oxide (16,28,21), but not by ethyl methane sulfonate or nitrous acid (17,21). Moreover, rev1 mutations reduce UV-induced mutation frequencies by 20 to 100 fold at some genetic sites, but not at all at others (11). Similar results were found using gamma rays (18). This discrimination between genetic sites does not appear to depend on any simple feature of these sites, such as local nucleotide sequence or the type of base-pair alteration detected (11): presumably some other feature of chromatin structure is responsible.

In wild-type strains, simple features of chromatin structure, such as local nucleotide sequence, are also incapable of explaining the existence of hotspots or coldspots for UV mutagenesis (13). Hotspots do not contain a higher density than normal of adjacent-pyrimidines; indeed, the frequency with which a given base pair substitution is induced at different sites can vary by more than fifty fold, even when flanking base pairs are identical and excision repair is absent. These data also emphasize the important part features of chromatin structure other than primary sequence play in determining the activity of mutagenic enzymes, though they do not reveal what these features are.

TIME OF MUTATION INDUCTION

Using mitotic pedigree analysis, Kilbey and James (6) have shown that the majority of mutations induced in wild-type cells exposed to moderate fluences of UV occur during the G1 phase, presumably during the course of excision. In contrast, similar experiments with an excision deficient (rad1-1) strain show that mutations occur no earlier than G2 (7,9). A similar conclusion was reached by J. Henriques and E. Moustacchi (personal communication), using a different method. In the excision deficient-strain, a significant proportion of the mutations occurred in subsequent cell generations, indicating that a single inherited lesion can be mutagenesis in yeast (7).

N-methyl-N'-nitro-N-nitrosoguanidine (NG) treatments induce mutations when administered at all stages of the cell cycle, but are particularly effective during S-phase (8). Although a RAD6-dependent mutagen in stationary cells, its status during S-phase is unknown. Possibly NG produces two types of adduct; those effective only during RAD6-dependent replication, and misreplication lesions. In wild type cells, the latter might be removed so rapidly by error-free repair that only S-phase treatments would be effective (10).

In addition to establishing the time of UV mutagenesis, the experiments of Kilbey and James also confirm that heteroduplex, or mismatch, repair frequently occurs during the induction of mutations (6,7). The existence of this type of repair had previously been inferred from the occurrence of pure mutant clones (see 10). Since mismatch correction appears also to take place in excision-deficient cells (4,7), the two processes are presumably at least partly different.

UNTARGETED MUTAGENESIS

Up to 40 percent or more of mutations induced by UV in excision deficient cells of yeast are untargeted, occurring in lesion free regions of DNA (14). They are presumably caused by DNA lesions, but only indirectly. Their high frequency suggest that cellular capacity for accurate replication is much reduced by UV irradiation, even on undamaged templates. It is not clear whether this infidelity is required to permit replication past template lesions, or whether its occurrence is coincidental. Untargeted mutations, like the targeted variety, are not induced in rad6 or rev3 mutants, though both occur normally in rad52 strains, showing that their induction does not require recombination. Although it would be appealing to ascribe their production to the synthesis of a factor that inhibits one of the mechanisms that normally maintains fidelity, no evidence to support this view has yet been found (14).

MULTIPLE BASE-PAIR ALTERATIONS

Most mutations induced by most mutagens probably involve substitutions, additions, or deletions of single bases, but a low frequency of multiple alterations also occurs. Further, this category of mutation is usually detected either poorly or not at all by most test systems, so its frequency is likely to be underestimated.

The data of Sherman, Stewart and colleagues (see 10), concerning mutations in the CYC1 locus, indicate that about 3 percent of the base pair substitutions they analyzed are multiple, and that such alterations can be induced by many mutagens. This is probably an underestimate: over 20 percent of frameshift revertants represent multiple changes, perhaps because functional requirements are less severe in these cases. As pointed out by Sherman and Stewart, multiple alterations are usually localized with a region of about ten base-pairs, suggesting error-prone events are strongly clustered. Tandem double base-pair substitutions are rarely detected, however, even after UV-irradiation (15), so that it is unlikely that UV selectively produces a high frequency of such events.

CONCLUSIONS

Yeast cells possess a multiplicity of mutagenic processes that require the functions of many genes. Among these, the RAD6 gene is particularly important: most of the commonly used powerful mutagens, even though they produce widely differing types of premutational lesions, owe their effectiveness to RAD6 activity. The relatively small number of mutagens that act by simple misreplication in yeast may imply that the greater accuracy with which it replicates the DNA (10) is achieved, at least in part, by a greater capacity to recognize and discriminate against subtle template defects.

A multiplicity of mutagenic processes may be required for two general reasons: to cope with a diversity of premutational lesions, and to cope with chromatin structure. The distinction between RAD6-dependent specific and less specific mutagens may reflect the existence of two basically different strategies for repair or recovery in these cases, though exceptions occur and the situation is not clear. In particular the usual distinction made between mispairing lesions and non-pairing ones seems overstated. A variety of evidence suggests that a pyrimidine dimer, often taken as the paradigm of non-pairing lesions, can be replicated with moderate, even if reduced, fidelity (10). More probably, different lesions present a graded challenge to mechanisms that maintain fidelity, and entail similarly graded risks of error that nevertheless fall short of unity. The existence of features of chromatin structure that are difficult to replicate accurately seems to be implied by the inability to explain hotspots and characteristic mutational phenotypes in terms of primary sequence, though their nature is unknown.

Since all mutations that reduce the frequency of induced mutagenesis are recessive, the normal products of these genes presumably promote this process, either actively or passively. Perhaps these gene products are non-essential components of the replication complex that facilitate the replication of difficult structures, though at the cost of a modest level of error. This penalty may be of relatively small consequence if episodes of error-prone replication are rare. The existence of high levels of untargeted mutagenesis after treatment with at least some mutagens suggests that the selective pressure to maintain high fidelity in these circumstances cannot be very great.

Acknowledgements: This work was supported by US Public Health Service research grant GM21858 from the National Institutes of Health and by the US Department of Energy at the University of Rochester Department of Radiation Biology and Biophysics and has been assigned report No. UR-3490-2069.

(1) Cassier, C., Chanet, R., Henriques, J.A.P., and Moustacchi, E. Genetics 96, 841-857 (1980)
(2) Culbertson, M.R., Charnas, L., Johnson, M.T., and Fink, G.R. Genetics 86, 745-764 (1977)
(3) Drake, J.W. Symp. Soc. Gen. Microbiol. 24, 41-58 (1974)
(4) Eckardt, F., Teh, S.-J., and Haynes, R.H. Genetics 95, 63-80 (1980)
(5) Huberman, J.A., and Riggs, A.D. J. Mol. Biol. 32, 327-341 (1968)
(6) James, A.P., and Kilbey, B.J. Genetics 87, 237-248 (1977)
(7) James, A.P., Kilbey, B.J. and Prefontaine, G.J. Molec. Gen. Genet. 165, 207-212 (1978)
(8) Kee, S.G., and Haber, J.E. Proc. Nat. Acad. Sci. USA 72, 1179-1183 (1975)
(9) Kilbey, B.J., and James, A.P. Mutat. Res. 60, 163-171 (1979)
(10) Lawrence, C.W. Advances in Genetics 21, (in press).
(11) Lawrence, C.W. and Christensen, R.B. J. Med. Biol. 122, 1-21 (1978)
(12) Lawrence, C.W., and Christensen, R.B. J. Bact. 139, 866-876 (1979)
(13) Lawrence, C.W. and Christensen, R.B. Molec. Gen. Genet. 177, 31-38 (1979)
(14) Lawrence, C.W. and Christensen, and Schwartz, A. Oak Ridge Natl. Lab. Symp., "Molec. and Cellular Mechanisms of Mutagenesis" (in press)
(15) Lawrence, C.W., Stewart, J.W., Sherman, F., and Christensen, R.B. J. Mol. Biol. 85, 137-162 (1974)
(16) Lemontt, J.F. Genetics 68, 21-33 (1971)
(17) Lemontt, J.F. Molec. Gen. Genet. 119, 27-42 (1972)
(18) McKee, R.H. and Lawrence, C.W. Genetics 93, 375-381 (1979)
(19) Petes, T.D., and Williamson, D.H. Exptl. Cell Res. 95, 103-110 (1975)

(20) Prakash, L. Genetics 78, 1101-1118 (1974)
(21) Prakash, L. Genetics 83, 285-301 (1976)
(22) Prakash, L., and Sherman, F. J. Mol. Biol. 79, 65-82 (1973)
(23) Sherman, F., and Stewart, J.W. Genetics 78, 97-113 (1974)
(24) Walsh, J., and Fink, G.R., personal communication.

The Importance of Mutant Yield Data for the Quantification of Induced Mutagenesis

R. H. Haynes,[1] F. Eckardt,[2] B. A. Kunz,[3] and W. Göggelmann[2]

[1]York University, Department of Biology, Toronto, Canada; [2]Gesellschaft für Strahlen- und Umweltforschung, Neuherberg, Germany, and [3]National Institute for Environmental Health Sciences, Research Triangle Park, North Carolina, U.S.A.

ABSTRACT Analysis of mutant yield data (*mutants induced per cell treated*) is the most informative way to compare the mutagenicities of different agents in a given cellular system, or to assess the mutabilities of different cells with respect to any given mutagen. Mutant yield is a measure of the net genetic result of the toxic and mutagenic actions of the agent being studied. Here we use yield data to compare both the mutagenic and recombinagenic efficiencies of certain physical and chemical mutagens, and also to contrast the mutagenic responses of different strains of yeast (*Saccharomyces cerevisiae*) and *Salmonella typhimurium* to 254 nm ultraviolet light.

INTRODUCTION

In this paper we describe a practical approach to the quantitative analysis and comparison of mutagenic responses in yeast and *Salmonella* treated with physical or chemical mutagens. Those interested in the mathematical basis of these methods should consult our previous papers where we first showed how mutant yield, the quantity actually measured in most *in vitro* assays for mutagenesis, can be used for such purposes (6, 7, 10, 12). The definitions of the three basic quantities which concern us here, mutant yield (Y), mutation frequency (M) and surviving fraction (S), are most straightforward in in experiments which involve acute mutagen exposures to homogeneous suspensions of single, non-replicating, auxotrophic cells which do not require a period of growth on supplemented minimal medium in order to obtain maximum expression of prototrophic revertants. Standard yeast protocols generally fulfill these requirements (4, 5). Data derived from mammalian cell culture experiments may require a more complex mathematical treatment, especially if mutant and non-mutant cells should multiply at different rates during subculturing, if lethal sectoring occurs during culture growth, or if the number of generations required to express the mutant phenotype should be dose-dependent (17).

BASIC DEFINITIONS

Consider that we begin with a homogeneous suspension of N_O viable cells per unit volume. First, the number of pre-existing mutants in this population must be determined in order to correct for spontaneous background in the experimental data (4, 5). After any mutagen dose x, let the measured number of induced mutants be $N_m(x)$ and the number of surviving cells be $N_S(x)$. Thus, three interrelated quantities can be calculated: (1) surviving fraction of cells, $[S(x) = N_S(x)/N_O]$; (2) induced mutant yield *per cell initially treated*, $[Y(x) = N_m(x)/N_O]$; and (3) mutation frequency [mutants per survivor, $M(x) = N_m(x)/N_S(x)$]. It is mutant yield and surviving fraction which actually are measured, whereas mutation frequency is a derived quantity, given by the ratio of yield to survival [i.e., $M(x) = Y(x)/S(x)$]. The number of "lethal hits" (z) produced by dose x is given by the negative logarithm of the survivng fraction, i.e., $z = -\ln S(x)$.

The meaning of these three quantities is best expressed in terms of probabilities. Thus, surviving fraction is the probability of clone formation for any typical "non-mutant" cell in the population after exposure to the mutagen. However, this may or may not be equal to the probability of clone formation by mutant cells in the particular assay system used. If these probabilities are *not* equal we say there is a "δ-effect" in the system and such effects can have a profound influence on the shapes of dose-frequency curves for induced mutagenesis (5). Mutation frequency is the probability that a particular mutational change occurs in any given cell, irrespective of whether that cell survives to form a clone. Mutant yield is the joint probability for the occurrence of a mutational hit but no lethal hit in any given cell. Thus, mutant yield is a measure of the number of *viable* mutant clones which develop after exposure to the mutagen. At very low mutagen doses, where there is little or no cell killing, mutant yield and mutation frequency are approximately equal; at higher doses, yield is *less* than the frequency because of the toxic effect of the mutagen on mutant as well as the non-mutant cells (10, 12).

SHAPES OF FREQUENCY AND YIELD CURVES

The simplest case to consider is that in which the number of mutational events, or mutation frequency, increases linearly with dose $[M(x) = mx]$ and the survival curve for the population is exponential $[S(x) = e^{-kx}]$. More complex kinetic response patterns are treated in our previous papers (10, 12). The constants of proportionality m and k can be regarded, respectively, as the intrinsic mutagenicity and toxicity of the agent, in the system in question. In Fig. 1 we have plotted yield and frequency (on linear scales) and surviving fraction (semi-logarithmically) for two imaginary mutagens, A and B, whose doses we assume can be measured in comparable units. In this calculation the mutagenicity of the two agents (m) is the same, but their toxicity (k) is different, B being more toxic than A. At very low doses, where killing is negligible, viable mutants accumulate in direct proportion to dose and the initial linear slope of the frequency curve is equal to the constant m. However, as the dose increases, the yield of the mutants reaches a maximum and then begins to decline toward zero as killing dominates over mutation.

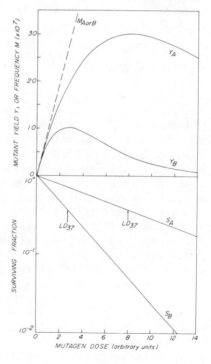

Figure 1

The experimental demonstration of such bell-shaped curves is proof that mutations actually are induced in the experiment, as opposed to the mere selection of pre-existing mutants by exposure to a cytotoxic agent.

The maximum mutant yield for any agent increases with its mutagenicity and decreases with its toxicity. For purely linear kinetics, $Y_{max} = m/ke$, and the dose which gives the maximum yield ($\hat{x}$) corresponds to 37% survival (LD_{37} or one lethal hit). For more complex dose-response kinetics, $\hat{x}$ shifts away from the LD_{37} in generally predictable ways (10, 12). If there should be δ-effects in the system, then the mutation frequency curve becomes non-linear at high doses and the rate of decline of the yield curve (when plotted semi-logarithmically) is not parallel to the final slope of the survival curve. We have found that the non-linear frequency responses observed at high doses for UV-mutagenesis in excision deficient yeast are caused primarily by such δ-effects, whereas similar non-linearities for reverse mutations in repair-proficient strains may reflect the existence of inducible error-prone repair processes in these systems (8, 11).

RELATIVE MUTAGENIC EFFICIENCY (RME) OF DIFFERENT MUTAGENS

Direct comparison of frequency curves for different mutagens (or for different cellular systems treated with the same mutagen) can lead to ambiguities of interpretation because in calculating frequency, the toxicity of the agent (or sensitivity of the cells) is deliberately cancelled out. At seen in Fig. 1, two mutagens could be equally "mutagenic" in terms of induced frequencies, but actually generate different numbers of viable mutants because of differences in cytotoxicity. The importance of taking toxicity into account in environmental mutagenesis has been stressed recently by Margolin et al.(14) and earlier was recognized by Munson and Goodhead (15) who showed how the slope of frequency (M) versus lethal hit (z) curves could be used to measure the relative efficiencies of different mutagens. Unfortunately, this latter method is appropriate only for strictly linear kinetics, for which a plot of $M(z)$ gives a straight line with a slope m/k: if the frequency response contains non-linearities, $M(z)$ is curvilinear and hence no unique slope can be defined.

We argue that mutagenic efficiency should be defined so as to increase with mutagenicity of the agent (m), and decrease both with the toxicity (k) and with the position of the maximum yield along the abscissa ($\hat{z}$). In comparing the efficiencies of different agents, one cannot use mutagen dose as the abscissa, because different chemical exposure doses are incommersurable quantities (a mole of chemical A is as different from a mole of chemical B as is a dozen apples from a dozen oranges). As a further complication, it should be noted that the use of metabolic activation systems could introduce non-linearities into the relation between the genetically effective dose and the measured exposure dose. Thus, we have suggested (6, 7) that relative mutagenic efficiency (RME) be defined in terms of yield versus lethal hit [$Y(z)$] plots, where the use of the lethal hit scale as abscissa, though admittedly not ideal, at least provides a practical and uniform biological measure of mutagen "dose". Such $Y(z)$ plots have the same general shape as the yield curves shown in Fig. 1.

It is useful to select some simple properties of the $Y(z)$ curves upon which to base numerical measures of mutagenic efficiency. Several possibilities exist, all equally arbitrary, though each emphasizes different features of the dose-response relation. Two curve characteristics which satisfy our requirements and are easy to measure, are (1) the slope of the line joining the origin with the maximum yield [$Y_{max}/\hat{z}$], and (2) the area under the $Y(z)$ curve up to some fixed value of z [the "integral yield", I_z] (6, 7). $Y_{max}/\hat{z}$ is a useful quantity because it can be determined uniquely even in non-linear cases where the initial slope of the yield curve is zero; also, it takes account of shifts in the position of $\hat{z}$ associated with different dose-response kinetics. The integral yield is useful because its value will reflect the contribution of any δ-effects in the high dose region.

Figs. 2 and 3 show $Y(z)$ plots for four mutagens [254 nm UV, ^{60}Co γ-rays, 4-nitroquinoline-N-oxide (4NQO) and N-methyl-N'-nitrosoguanidine (MNNG)] for both ILV reversion (Fig. 2) and gene conversion at TRP 5 (Fig. 3) in an excision proficient (RAD wild-type) strain D7 of S. cerevisiae (18). The experimental points, of which there are an average of 11 per curve, were omitted merely to enhance the clarity of the figures. Table I shows the relative mutagenic and recombinagenic efficiencies (against UV as "standard" mutagen) for these agents calculated both in terms of $Y_{max}/\hat{z}$ and integral yield (I_z

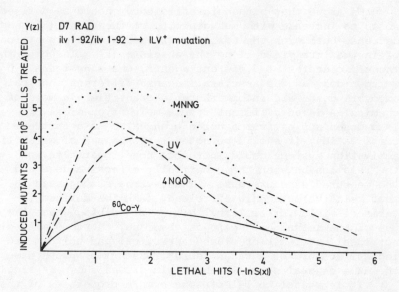

Figure 2: Reversion yields for 2 physical and 2 chemical mutagens in *S. cerevisiae* strain D7 *RAD* wild-type.

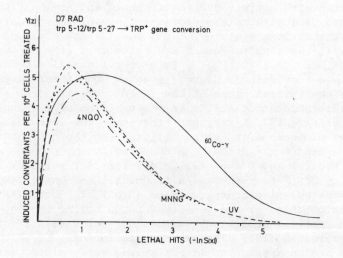

Figure 3: Gene conversion yields for 2 physical and 2 chemical mutagens in *S. cerevisiae* strain D7 *RAD* wild-type.

TABLE 1: Efficiencies for (a) Reversion at *ILV* 1-92 ($\times 10^{-5}$) and (b) Gene conversion at *TRP* 5 ($\times 10^{-4}$) in *S. cerevisiae* strain D7 (*RAD* wild-type).

Mutagen		$Y_{max}/\hat{2}$	RME	I_z	RME
254 nm UV	(a)	2.3	1.0	14.5	1.0
	(b)	7.9	1.0	10.7	1.0
^{60}Co γ-rays	(a)	0.68	0.30	5.2	0.36
	(b)	4.3	0.54	17.2	1.6
MNNG	(a)	4.4	1.9	19.0	1.3
	(b)	6.1	0.78	11.7	1.1
4NQO	(a)	3.7	1.6	11.5	0.79
	(b)	4.9	0.62	10.3	0.96

evaluated over the range 0 to 5 lethal hits). The following points are worthy of note: (1) Recombination yields (measured as gene conversion) are about 10-fold higher than mutation yields; (2) Mutation efficiencies vary over a range two to three times greater than do conversion efficiencies; (3) For these agents there is no indication of any positive correlation between increasing mutagenic and recombinagenic efficiencies; (4) The yield curves for gene conversion are more closely similar than those for mutation, especially at high survival levels; (5) The curves for UV and 4NQO mutagenesis are not identical, as we found previously to be the case for these two agents at a different locus (*LYS* 2) in a haploid strain (6); and (6) Ionizing radiation is a potent recombinagen, but a relatively weak mutagen. This latter observation is of special interest because in other, so far unpublished, experiments we have found that X-rays are about twice as efficient as UV for the production of mitotic crossing-over (in D7 at *ADE* 2); furthermore M. Elkind (personal communication) has shown that ionizing radiation is likewise about twice as efficient as 254 nm UV for the production of neoplastic transformation in C3H/10T½ cells (9). This suggests that there may be a positive correlation between recombinagenicity and carcinogenicity because of a common involvement of DNA strand breaks and genetic exchange events in these two processes (2, 3, 13, 16).

RELATIVE SENSITIVITY AND RESOLUTION OF DIFFERENT CELLS

Plots of yield *versus* lethal hits also can be used to calculate the relative mutabilities of different cells in ways exactly analogous to the calculation of mutagenic efficiencies for different mutagens (6). However, when comparing different cells against the same mutagen, exposure dose can be used as the abscissa and plots of yield *versus* dose $[Y(x)]$ enable one to determine the *mutational sensitivity* of cells in reciprocal units of dose. We use the slope of the line joining the maximum yield to the origin on a $Y(x)$ plot as a measure of mutational sensitivity (6). Another quantity of interest is the *mutational resolution*, which we define to be equal to Y_{max} itself. The maximum yield characterizes the ability of a given cellular system to detect induced mutants over and above the spontaneous background (6). The relative sensitivity and resolution of different cells are important design parameters for short-term test systems. High sensitivity is desirable for the detection of very low mutagen doses, whereas high resolution is necessary for the detection even of rather large doses of weak mutagens. The inclusion of various DNA repair deficiencies in strains used to assay environmental mutagens can be expected to increase mutational sensitivity (1). However, comparison of $Y(x)$ curves for haploid wild-type and an excision-

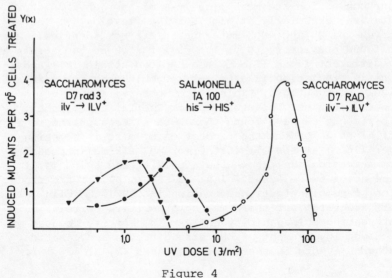

Figure 4

deficient strain of yeast (*rad* 1-1) showed that the sixfold increase in mutational sensitivity to UV of the *rad* 1 strain was accompanied by an eleven-fold *decline* in mutational resolution (6, 7). Figure 4 shows that a similar effects occur as a result of the inclusion of the *rad* 3 excision deficiency in strain D7. Here, the mutational sensitivities ($Y_{max}/\hat{x}$) of the *RAD* wild-type and *rad* 3 strains are 0.076 x 10^{-5} and 1.5 x 10^{-5} $(J/m^2)^{-1}$ respectively, whereas the corresponding mutational resolutions (Y_{max}) are 3.8 x 10^{-5} and 1.9 x 10^{-5}. Thus, a 20-fold increase in mutational sensitivity to UV is achieved by making strain D7 homozygous for *rad* 3, but this is accompanied by a two-fold decline in mutational resolution. Clearly, the inclusion of different repair deficiencies in different strains can have quantitatively different effects on mutational sensitivity and resolution; and, on the basis of available data, it would appear that increases in sensitivity are accompanied by declines in resolution. For purposes of comparison, we have included data for *HIS* reversions in *S. typhimurium* strain TA 100 in Fig. 4. This data was obtained using a liquid assay technique so that both mutation and cell killing by UV could be scored independently as in the yeast protocols. It would appear that the TA 100 system is about 40% less sensitive [0.6 x 10^{-5} $(J/m^2)^{-1}$] than the D7 *rad* 3 system and its resolution is only half that of D7 *RAD* wild-type.

CONCLUDING REMARKS

Analysis of mutant yield data in the yeast *S. cerevisiae* indicates that the increased mutational sensitivity of DNA repair-deficient strains generally is accompanied by a loss of mutational resolution. This loss in resolution is caused by the increased sensitivity to killing conferred by the repair deficiency. Quantitative comparisons, on a yield *versus* lethal hit basis, of the relative efficiencies of UV and ionizing radiation for reverse mutation, mitotic recombination and neoplastic transformation, is consistent with the notion that genetic exchange events, perhaps stimulated by DNA strand breaks, play an important role in carcinogenesis.

REFERENCES

(1) Ames, B.N., Lee, F.D., and Durston, W.E. *Proc. Natl. Acad. Sci. USA 70*, 782-786 (1973).
(2) Barclay, B.J., Kunz, B.A., Little, J.G., and Haynes, R.H. *Can. J. Biochem.* March 1982 (in press).
(3) Cairns, J. *Nature 289*, 353-357 (1981).
(4) Eckardt, F., and Haynes, R.H. *Mutation Res. 43*, 327-338 (1977).
(5) Eckardt, F., and Haynes, R.H. *Genetics 85*, 225-247 (1977).
(6) Eckardt, F., and Haynes, R.H. *Mutation Res. 74*, 439-458 (1980).
(7) Eckardt, F., and Haynes, R.H. In: *Short-Term Tests for Chemical Carcinogens*, (H.F. Stich & R.H.C. San, eds), pp. 457-473, Springer-Verlag, New York (1981).
(8) Eckardt, F., Moustacchi, E., and Haynes, R.H. In: *DNA Repair Mechanisms*, (P.C. Hanawalt, E.C. Friedberg & C.F. Fox,eds), pp. 421-424, Academic Press, New York (1978).
(9) Han, A., and Elkind, M.M. *Cancer Res. 39*, 123-130 (1979).
(10) Haynes, R.H., and Eckardt, F. *Can. J. Genet. Cytol. 21*, 277-302 (1979).
(11) Haynes, R.H., and Eckardt, F. In: *Radiation Research: Proceedings of the Sixth International Congress of Radiation Research*, (S. Okada, M. Imamura, T. Terasima and H. Yamaguchi, eds), pp. 454-461, Japanese Association for Radiation Research, University of Tokyo, Tokyo (1979).
(12) Haynes, R.H., Eckardt, F. In: *Chemical Mutagens*, (F.J. de Serres and A. Hollaender, eds), Vol. 6, pp. 271-307, Plenum Publishing Corp.,New York (1980).
(13) Levan, A., Levan, G., and Mitelman, F. *Hereditas 86*, 15-30 (1977).
(14) Margolin, B.H., Kaplan, N., and Zeiger, E. *Proc. Natl. Acad. Sci. USA 78*, 3779-3783 (1981).
(15) Munson, R.J., and Goodhead, D.T. *Mutation Res. 42*, 145-160 (1977).
(16) Pall, M.L. *Proc. Natl. Acad. Sci. USA 78*, 2465-2468 (1981).
(17) Thompson, L.H., and Baker, R.M. In: *Methods in Cell Biology* (D.M. Prescott, eds) pp. 209-281, Academic Press, New York (1973).
(18) Zimmerman, F.K., Kern, R., and Rasenberger, H. *Mutation Res. 28*, 381-388 (1975).

Examining the Mechanism of Mutagenesis in DNA Repair-Deficient Strains of *Drosophila melanogaster*

P. Dennis Smith, Ruth L. Dusenbery, Sheldon F. Cooper, and Clifford F. Baumen

Department of Biology, Emory University, Atlanta, Georgia, U.S.A.

ABSTRACT Nearly 30 mutagen-sensitive (mus) loci have been identified in the *Drosophila* genome on the basis of hypersensitivity to MMS. Mutants at two of these loci, mei-9 and mus(2)201, have also been shown to be sensitive to ultraviolet light and defective in the removal of UV-induced thymine dimers from DNA. We report that cell cultures prepared from these mutants are also defective in the repair of alkylation damage induced by MMS and MNU. Mutagenesis experiments using either repair-deficient females or males and several mutagenesis assay systems indicate that these mutants are hypermutable by EMS, MMS, and MNU. Future experiments will address whether this hypermutability results from the persistence of particular alkylated bases such as O^6 methyguanine and subsequent miscoding during normal semi-conservative replication or the interruption of replication and the induction of error-prone repair processes.

ALKYLATION-SENSITIVE MUTANTS

Drosophila melanogaster has served for many years as the model system for the study of mutagenesis in complex multicellular organisms. The ability to study mutation induction in both somatic and reproductive cells, to employ in vivo direct acting mutagens as well as those requiring metabolic activation and to detect a wide variety of genetic endpoints represent particular advantages not readily available in other mutagenesis systems.

Because Drosophila is a complex, highly differentiated animal, continued development of this system offers the potential for a well-characterized short-term mutagenicity assay program which can provide substantial information of relevance for genetic risk assessment in human populations.

Recent experimental approaches have focused on the development of this system for the coordinated study of DNA repair and mutagenesis. These approaches have relied on the isolation of mutagen-sensitive mutants to identify loci involved in DNA repair functions (1) and the development of in vitro tissue culture procedures to characterize specific DNA repair defects in these mutant strains (2).

Nearly 30 mutagen-sensitive (mus) loci have been identified on the basis of larval hypersensitivity to methyl methanesulfonate (MMS) (1, 3) and the pattern of mutagen sensitivity conferred by these mutations is summarized in Table 1.

Table 1. Mutagen Sensitivity Phenotypes

Mutant	MMS	X-Ray	UV	HN2
101	+	+	+	+
102	+	+	-	-
103	+	-	-	-
105	+	+	-	-
106	+	+	-	-
108	+	NT	NT	NT
109	+	+	+	+
mei-9	+	+	+	+
mei-41	+	+	+	+
201	+	-	+	+
202	+	+	+	+
203	+	+	+	+
204	+	-	+	-
205	+	-	+	-
206	+	-	NT	+
207	+	-	+	+
301	+	NT	NT	+
302	+	NT	NT	+
304	+	NT	NT	-
305	+	NT	NT	+
306	+	NT	NT	-
307	+	NT	NT	+
303	-	NT	NT	+
309	+	NT	NT	-
310	+	NT	NT	-
311	+	NT	NT	-
312	+	NT	NT	+

A subset of these mutants which displays sensitivity to UV light has been examined for ability to repair UV-induced thymine dimers and strains defective in both excision (4, 5) and postreplication (6, 7) repair have been identified. Among the excision-defective strains, alleles of the mei-9 and mus(2)201 loci have been shown to be extremely sensitive to UV light and totally defective in the repair of UV damage.

REPAIR OF ALKYLATION DAMAGE

Our research is presently focused on the nature of DNA lesions induced by monofunctional alkylating agents, the mechanisms by which these lesions are repaired and the mutational consequences which occur when alkylation repair is impaired.

We have examined the ability of primary cell cultures prepared from homozygous embryos to perform replicative DNA synthesis when treated with various monofunctional alkylating agents. In Figure 1, the ability of the repair-proficient Oregon-R and cn^{35} control strains to synthesize DNA in the presence of MMS is compared with that of alleles of the mei-9 and mus(2)201 loci.

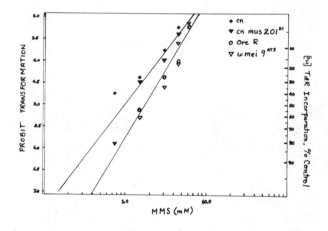

Figure 1. Inhibition of Replicative DNA Synthesis by MMS Treatment

These experiments show that the inhibition of DNA synthesis during the first 2 hours of mutagen treatment is dose-dependent and does not differ between mutant and control cells. We conclude that the introduction of alkyl lesions and their immediate effect on semi-conservative replication is identical in both the proficient and deficient cell cultures. Current studies are in progress to compare the proficient and deficient strains for ability to recover replicative synthesis at varying periods following mutagen treatment.

Cell cultures prepared from mutant and control strains were compared for ability to perform unscheduled DNA synthesis (UDS) following treatment with UV, MMS and MNU (Table 2).

Table 2. Unscheduled DNA Synthesis (UDS) in Primary Cell Cultures

Genotype	Mutagen		
	UV	MMS	MNU
Oregon-R	+	NT	NT
y w	NT	+	NT
cn^{35}	+	+	+
y w mei-9^{AT1}	−	−	NT
cn^{35} mus 201^{D1}	−	−	−
sn^3 mei9^{D2}; cn mus 201^{D1}	−	−	−

+ = REPAIR-PROFICIENT
− = REPAIR-DEFICIENT
NT = not tested

Following UV treatment, the Oregon-R and cn^{35} strains performed dose-dependent synthesis while the y w mei-9^{AT1}, cn^{35} mus 201^{D1} and sn^3 mei-9^{D2}; cn^{35} mus 201^{D1} strains failed to do so. These studies demonstrate that UDS can be employed to monitor excision repair capacity of Drosophila cell cultures and confirm earlier studies (4) that mei-9 mutants are deficient in UV-induced excision repair. Mutant cell cultures treated with MMS and methylnitrosourea (MNU) similarly exhibited an inability to perform unscheduled DNA synthesis. These latter studies provide the first direct evidence that the mei-9 and mus(2)201 loci are defective in the repair of alkylation-induced damage.

ALKYLATION MUTAGENESIS

Repair-deficient mutants have been incorporated into experiments designed to examine the role of these loci in alkylation-induced mutagenesis during spermatogenesis.

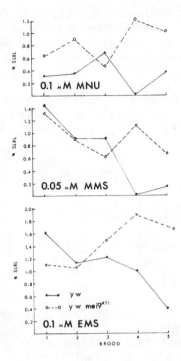

Figure 2. Frequency of Sex-Linked Recessive
Lethals induced by Alkylating Agents
in excision-proficient and excision-deficient males

Figure 2 presents the results of these experiments for mutation induction during spermatogenesis in the repair-proficient y w and repair-deficient y w mei-9^{AT1} strains by MNU, MMS and ethyl methanesulfonate (EMS). Data were pooled from broods 1, 2 and 3 ("postmeiotic") and broods 4 and 5 ("premeiotic") and the strains were compared by statistical analysis. The data demonstrate, for all three agents, that no significant difference exists between the strains for the "postmeiotic" broods but that a highly significant statistical difference exists between the strains for the "premeiotic" broods. This hypermutability observed for the y w mei-9^{AT1} strain represents the first direct demonstration of an altered mutation induction rate during spermatogenesis in a DNA repair-deficient male and supports earlier suggestions (8, 9) that DNA repair functions are not active following meiosis in males.

Alternatively, experiments have been reported which

incorporate repair-deficient mutations in females and assess repair of damaged spermatozoa following fertilization (9, 10). We have employed a similar design to compare spontaneous and alkylation-induced mutation rates in mei-9 and mus(2)201 females

Table 3. Spontaneous Mutation Rates with Excision-Defective Females

FEMALE GENOTYPE	NUMBER OF CHROMOSOMES			% SLRL
	LETHAL	LETHAL+	TOTAL	
y w (control)	15	9799	9814	0.15
y w mei-9^{AT1}	57	6375	6432	0.89
y w mei-9^{AT2}	59	5567	5626	1.05
y w mei-9^{AT3}	31	3149	3180	0.97
y w mei-9^a	54	2788	2842	1.90
cn^{35} (control)	1	2701	2703	0.04
cn^{35} mus(2)201^{D1}	34	4011	4045	0.84

Table 3 summarizes data on the rate of spontaneous sex-linked recessive lethals which occur in Basc chromosomes when Basc males are mated to either excision-proficient (y w, cn^{35}) or excision-deficient (y mei-9, cn^{35} mus(2)201) females. For the mei-9 locus, we confirm earlier studies (9) which have reported the influence of the maternal genome on spontaneous mutagenesis in mature spermatozoa. Interestingly, for the mus(2)201 mutant which, like the mei-9 mutant, appears to be blocked at a pre-incision step of excision repair (11), females show a similar influence on the mutation rate of male X chromosomes.

Experiments which monitor the ability of excision defective females to process alkylation damage introduced into mature spermatozoa are outlined in Table 4.

Table 4. Influence of Excision-Defective Females on Alkylation-induced SLRL Mutation Rates

AGENT	FEMALE GENOTYPE	% INDUCED SLRL	DEFICIENT PROFICIENT RATIO
EMS	y w	6.5	----
	y w mei 9^{AT1}	15.8	2.4
	cn^{35}	2.3	----
	cn^{35} mus 201^{D1}	5.0	2.2
MMS	y w	1.53	----
	y w mei 9^{AT1}	7.80	5.1
	y w mei 9^{AT3}	13.12	8.6
	y w mei 9^a	15.00	9.8
MNU	y w	0.17	----
	y w mei 9^{AT1}	0.94	5.5
	y w mei 9^{AT2}	0.60	3.5
	y w mei 9^{AT3}	2.04	12.0
	y w mei 9^a	2.08	12.2

For 0.5 mM EMS, both types of excision-defective females show a greater than two-fold increase in the SLRL mutation frequency. For the mei-9^{AT1} allele, this compares favorably with the earlier results of Graf, Green and Wurgler (9) who reported a 1.5-fold increase with mei-9^{L1} females mated to Basc males treated at 2.3 mM EMS. For the methylating agents, a more dramatic response is observed. Basc males, treated with either 0.05 mM MMS or 0.1 mM MNU and mated to various mei-9 females, yield a 5-12 fold increase in mutation frequency.

Although the SLRL test has been shown to be an extremely sensitive monitor of alterations in the mutation rate, we have extended our studies on the influence of excision-defective loci on alkylation mutagenesis to include an additional genetic system (Table 5).

Table 5. Influence of Excision-defective Females on Alkylation-induced Mutation Rates at the rosy (ry) Locus

INFLUENCE OF EXCISION-DEFECTIVE FEMALES ON ALKYLATION-INDUCED MUTATION RATES AT SINGLE LOCI

SYSTEM	AGENT	FEMALE GENOTYPE	TOTAL PROGENY	MUTANT FREQUENCY per 10^5 F1	DEFICIENT TO PROFICIENT RATIO
rosy (ry)	EMS	y w	17,611	17.0	----
		y w mei 9^{AT1}	2,730	439.6	25.8
	MMS	y w	32,253	9.3	----
		y w mei 9^{AT1}	1,586	252.2	27.1

The rosy (ry) locus, which is the structural gene for xanthine dehydrogenase, offers the opportunity to monitor for alterations not only in the frequency of mutation production but also in the nature of changes induced in the DNA. Preliminary experiments comparing the induction of ry mutants by EMS and MMS in excision-proficient (y w) and excision-deficient (y w mei-9^{AT1}) females have indicated a dramatic increase in mutant induction at this locus. No data are yet available for possible changes in the molecular nature of the newly-induced mutants.

CONCLUSION

The mei-9 and mus(2)201 mutants are defective in alkylation repair and hypermutable by alkylating agents. Whether this hypermutability results from the persistence of particular alkylated bases such as O^6-methylguanine and miscoding during replication or the interruption of replication and the induction of error-prone repair processes remains to be seen.

ACKNOWLEDGMENTS

Supported by P.H.S. grants ES-01101 and ES-02037 and E.P.A. grant R807538010. Our thanks for technical assistance to Kelly Burnett, Jeffrey Jefferson, Shannon McCormick and Sylvia Smith.

(4) Boyd, J.B., Golino, M.D. and Setlow, R.B. Genetics 84, 527-544 (1976)
(3) Boyd, J.B., Golino, M.D., Shaw, K.E.S, Osgood, C.J. and Green, M.M. Genetics, in press
(5) Boyd, J.B. and Harris, P.V. Chromosoma 82, 249-257 (1981)
(2) Boyd, J.B., Harris, P.V., Osgood, C.J. and Smith, K.E. DNA Repair and Mutagenesis in Eukaryotes, 209-221 (1980)
(6) Boyd, J.B. and Setlow, R.B. Genetics 84, 507-526 (1976)
(7) Boyd, J.B. and Shaw, K.E.S. Molec gen. Genet., in press
(11) Boyd, J.B., Snyder, R.D., Harris, P.V., Presley, J.M., Boyd, S.F. and Smith, P.D., in preparation

(9) Graf, U., Green, M.M. and Wurgler, F.E., Mutat. Res. 63, 101-112 (1979)
(1) Smith, P.D., Snyder, R.D. and Dusenbery, R.L. DNA Repair and Mutagenesis in Eukaryotes, 175-188 (1980)
(8) Sobels, F.H. Rev. Suisse Zool. 79, 143-152 (1972)
(10) Wurgler, F.E. and Graf, U. DNA Repair and Mutagenesis in Eukaryotes, 233-240 (1980)

Mechanisms of Mutagenesis in Cultured Mammalian Cells

Roger Cox

MRC Radiobiology Unit, Harwell, Didcot, Oxon, U.K.

ABSTRACT The mechanisms of mutagenesis in cultured mammalian cells are discussed with reference to:
a) the induction of potentially mutagenic lesions,
b) the role of DNA-repair processes in mutagenesis and c) the biochemical and genetic characteristics of mutants induced by various mutagens.

Potentially mutagenic DNA lesions have been identified after treatment of cells with some mutagens and, within the limits of current methodology, it is concluded that the repair of such lesions is largely error-free. Evidence is presented to support the contention that point mutation is not the dominant mutagen-induced event at some genetic loci.

INTRODUCTION

Experimental studies on the complex process of mutagenesis are conducted in the belief that they will increase our understanding of the effects of environmental agents on human populations. Cultured mammalian cells offer distinct advantages over other biological systems for studying this problem. Mutagenesis in cultured mammalian cells is unquestionably more relevant to man than that in bacterial, fungal or insect species. At the same time, compared to whole animal studies, cultured cells offer the advantages of ease of handling and rapidity of assay. It may also be argued that since mutagenesis is essentially a cellular process, growing cells in culture frees them from normal *in vivo* physiological restraints which, in whole animal studies, may mask some of the more subtle elements of the mutagenic process. In considering the mutation process in cultured cells it is important to be aware of some of the technical and conceptual limitations in our understanding of the problem: our inability to manipulate the most relevant cells; the paucity of suitable mutation assay systems; uncertainties on the genetic nature of some mutagen induced phenotypic changes (3) and our limited knowledge of the identity of potentially mutagenic lesions and their modification by DNA repair systems (5). Despite these limitations, recent advances do allow some general and specific comments on the mechanisms of induced mutation in cultured mammalian somatic cells. For convenience of presentation I shall divide the process of mutagenesis into three parts: 1) the induction of specific DNA lesions; 2) the modification of these lesions by DNA-repair processes and 3) the expression of the mutational change with regard to the structure of the gene and the properties of its product(s).

1) THE INDUCTION OF SPECIFIC DNA LESIONS

Physical and chemical mutagenic agents obviously differ in their mode of action on DNA. Ionising radiations (X and γ-rays) produce relatively non-specific molecular changes in DNA, mediated directly or indirectly through chemical radicals generated by energy deposition in the cell; UV light modifies DNA principally through photon absorption and excitation in particular molecular configurations; the vast majority of chemical mutagens act directly on DNA to produce predictable structural modifications e.g. base analogue substitution, intercalation, cross-linking, alkylation and other various forms of DNA adduct. Since however there is evidence of indirect (radical mediated) attack on DNA after exposure to UV and some chemical agents (4), the distinctions made above are probably over simplistic.

The multiple modes of attack on DNA exhibited by many agents makes it difficult to structurally identify the major potentially mutagenic lesions. For ionizing radiation this situation is at its worst, and whether the relevant lesion is a DNA strand scission, base or pentose phosphate modification is uncertain. For UV, the bulk of

evidence supports the contention that pyrimidine dimers
are a major form of potentially mutagenic lesion, but a
significant contribution from other DNA photoproducts
cannot be excluded. In the case of chemical mutagens I
will cite only two examples of provisionally identified
potentially mutagenic lesions. The relative mutagenicity
of alkylating agents of different activities and
specificities tends to support the view that O^6 alkylation
of guanine is the major potentially mutagenic lesion in
cells (19). However, the enormous excess of induced
alkylations compared to mutational events, generates a high
level of uncertainty in these experiments and it has been
suggested that O^4 alkylation of thymine may be an important
lesion in cell mutagenesis (13). The second example of a
chemical mutagen that produces an identifiable potentially
mutagenic lesion is the diol epoxide of benzo (a) pyrene.
For this mutagen there is evidence (28), summarised in the
next section, which strongly suggests that the major
potentially mutagenic lesion in DNA is an N^2-deoxyguanosine
adduct.

Under normal physiological conditions the majority
of mutagen-induced DNA damage does not appear to be
expressed as mutations in surviving cells. If this is
indeed the case, then even a total knowledge of lesions
induced in DNA by a given agent plus precise data on the
mutagenicity of that agent would be insufficient to allow
us to identify the principal potentially mutagenic lesions.
A second dimension has to be considered, namely, the
efficiency and fidelity of the DNA-repair systems that act
to eliminate or modify mutagen induced molecular lesions
in DNA.

2) THE ROLE OF DNA-REPAIR PROCESSES

DNA-repair processes clearly play a crucial role in
cell mutagenesis. Without entering into detailed
biochemical descriptions of the comparative aspects of
DNA-repair in mutagen treated cells, one relatively
simple criterion may be used to distinguish the way in
which different mutagen-induced lesions are processed in
cells. All observable repair of X or γ-ray induced
cellular damage is completed within post-irradiation
periods of up to 6 hours (8, 27) and all unrepaired damage,
as measured by chromosomal changes, is expressed during
the cell cycle in which the damage is sustained (4).
In contrast, for UV and the vast majority of chemical

mutagens, induced damage may persist in cells for many
hours (e.g. 4) and results in the generation of chromosomal and, perhaps, phenotypic changes over a number of
cell cycles (12). This is strongly suggestive of a basic
difference in lesion type and mode of repair after ionizing
radiation and after UV or chemical exposure. Lesions
induced by UV and chemicals, which are principally of the
base damage type, must therefore be ignored or bypassed
during DNA replication. This is not unexpected for lesions
which do not disturb the gross integrity of the DNA duplex
and will pass through a DNA replication fork, albeit with
a certain probability of replication errors, such as base
mismatch or, 'gapping', in the daughter strand (16).

These differences, between the repair of ionising
radiation induced lesions and those induced by UV and
chemicals, must have implications for the mechanisms of
mutagenesis operating after treatment with these agents
and may influence the phenotype of induced biochemical
mutants (see §3).

The role of DNA repair processes in mutagenesis may
be more critically investigated by comparing mutation
induction in repair-competent and repair-deficient strains
of cells. The best example of this approach may be found
in studies on the sun-sensitive human genetic syndrome,
xeroderma pigmentosum (XP). Cells cultured from classical
XP patients are more sensitive than normal to both the
lethal and mutagenic effects of UV. However, on the basis
of mutations per lethal event, classical XP cells and
normal cells do not differ. This observation implies that
the thymine dimer excision process, that is deficient in
classical XP, operates in a largely error-free manner (16).
However, cells from XP variant patients show a considerable
increase over normal in the number of mutations per lethal
event after UV (15, 18), suggesting that the post-replication
repair system, that is deficient in XP variants, normally
operates to correct potentially mutagenic lesions in DNA
daughter strands resulting from the presence of photoproducts that escape pre-replicative excision. The hypermutability, but near normal sensitivity, of XP variant cells
to the lethal effects of UV would then require these cells
to repair daughter strand damage by a second, error-prone,
route.

The removal, or modification, of altered DNA bases or
chemical adducts from DNA by repair processes has been
demonstrated in cultured cells treated with a variety of
different chemical mutagens (21). However, as implied

earlier (§1), it is necessary to be able to identify those
lesions with a significant mutagenic potential in order
to comment on the role of DNA repair in mutagenesis. Such
comment has been possible in the case of the mutagen,
benzo (a) pyrene-diol epoxide (Bza P-diol epoxide). The
excision repair defect in XP human cells renders them
sensitive to a variety of adduct forming chemical mutagens
including Bza P-diol epoxide. Recent studies (28) show
that the mutagenicity of this agent in XP cells correlates
with the failure to excise an N^2-deoxyguanosine adduct
from DNA. As well as indicating that this adduct is the
principal potentially mutagenic lesion after Bza P-diol
epoxide exposure the data also imply that its excision
in normal cells is largely error-free.

The role of DNA repair in ionising radiation
mutagenesis is unclear. The relatively constant value
for mutations per lethal event after irradiation of normal
cell strains derived from different species (24) and of
V79 Chinese hamster cells under conditions which enhance
cellular repair and recovery (2, J. Thacker, unpublished),
is suggestive of a ubiquitous and possibly error-free
repair function. Since, however, the molecular form of the
relevant DNA lesion after ionising radiation remains
obscure, it is not yet possible to comment on the mode of
repair. Studies on radiosensitive, repair deficient, human
ataxia telangiectasia cells (1) may provide a means of
elucidating radiation mutagenesis but, unfortunately,
limitations in current techniques do not yet permit
definite statements on the mutability of these cells.

Physiological factors other than DNA repair are also
likely to influence the mutability of cells e.g. the
presence of natural clastogenic substances (10), or changes
in the availability of DNA precursors for repair or
replication (17). Consequently, changes in the
mutability of a cell strain should not be interpreted as
a direct DNA-repair effect without sound biochemical
evidence. Nevertheless, quantitative mutagenesis
experiments with somatic cell strains, having characterised
defects in DNA-repair or metabolism, have great potential
for elucidating mutagenic mechanisms. Cell cultures from
mutagen-sensitive human patients with clinically defined,
but generally poorly understood, genetic diseases, are
often difficult to work with and may only provide a
narrow view of the whole problem. The *de novo* isolation
and characterisation of DNA repair mutants from
established cell lines (22) should circumvent these

problems and considerably extend our knowledge of the mutagenic process.

3) THE EXPRESSION OF MUTATIONAL CHANGES

If different mutation pathways operate after treatment with different mutagens (see §2) then it may be anticipated that such differences would be reflected in the spectra of induced mutant phenotypes. Evidence of such mutagen specificity first came from studies of mutations at the ouabain-resistance locus (OUA^R). Whereas UV and the majority of chemical mutagens efficiently induce OUA^R mutations, ionising radiations and the frameshift mutagen, ICR 191, are extremely inefficient mutagens (14, 25). Since the OUA^R phenotype is associated with changes in the essential Na/K ATPase (EC3.6.1.3) function of the cell, it may be, that the poor mutagenicity of ionising radiations and ICR 191 reflects their inability to generate changes at the ouabain binding site of the ATPase without lethally affecting catalytic activity. More direct evidence for such mutagen specificity should be available from studies of mutant phenotypes induced at the locus of a non-essential enzyme, such as that for hypoxanthine-guanine phosphoribosyl transferase (HGPRT, EC2.4.2.8) where, in principle, all mutational events should be tolerated.

In order to compare the mutagenic action of ionising radiation and the chemical mutagen, ethyl methane sulphonate (EMS), we have recently characterised thioguanine-resistant (TG^R), HGPRT-deficient, mutants of human fibroblasts and V79 hamster cells using biochemical, immunological and cytogenetic techniques. The hamster cell studies (R. Brown - unpublished) may be summarised in the following way: TG^R mutants induced by ionising radiation all had zero HGPRT activity, and none showed the presence of any antigenic material (CRM) which cross reacted with an anti-serum raised against purified normal V79 HGPRT. Of the TG^R mutants induced by EMS, ∼80% had zero HGPRT activity and, of these, only a minority were CRM^+. The remaining EMS induced mutants showed residual HGPRT activity of between 2 and 100% of normal and, amongst these, were mutants with electrophoretically altered HGPRT. Amongst spontaneously arising mutants there were equal proportions of mutants with zero and residual HGPRT activities and, again, mutant HGPRT with altered electrophoretic properties was detected in some mutants with residual activity.

Residual mutant enzyme activity, enzyme with altered electrophoretic properties and antigenic cross-reaction are properties expected of the protein product of a gene containing relatively small DNA changes (point mutations or small deletions). These properties were observed amongst spontaneous and EMS induced mutants but, were completely absent in mutants induced by ionising radiation. The zero HGPRT activity and absence of CRM in the radiation induced mutants is suggestive of complete loss of gene product which could most simply result from a large deletion, early frameshift or chain terminating codons, or from a shut down of normal gene expression. However, these extreme phenotypes (null mutations) were not limited to the radiation induced mutants since ~60% of EMS induced mutants were also HGPRT$^-$, CRM$^-$. It appears, therefore, that the major mutagenic mechanisms at the HGPRT locus after both radiation and the alkylating agent, EMS, generate null mutations rather than point mutations.

Karyotypic analysis of radiation-induced HGPRT$^-$ mutants of both human (7), and hamster (23), cells provides further evidence of the gross nature of the final mutational event. HGPRT is specfied by an X-chromosome linked gene, and a significant proportion of radiation-induced mutants show X-chromosome changes (deletions and translocations) consistent with the mapped position of the HGPRT locus. Preliminary evidence for chemically induced HGPRT mutants with X-chromosome changes has also been obtained (6, R. Brown - unpublished). It could be argued, therefore, that the major mutagenic mechanisms operating at the HGPRT locus generate gross genetic changes at the level of the chromosome, resulting in, either deletion of relatively long sequences of DNA or (in the case of X-chromosome translocations), in loss of gene expression though 'position effect'. Whilst there is supporting evidence for gross genetic events, in the form of multi locus mutations or chromosome-associated mutations, from studies with other cellular mutation systems (9, 26), the results obtained at the HGPRT locus should not yet be considered as representative of the whole genome.

Mutant characterisation is clearly a very productive approach to understanding mutagenic mechanisms but, unfortunately, conventional biochemical and chromosomal techniques are limited in their capacity to resolve the nature of mutational events. The new techniques of molecular genetics, however, provide us with

the means of analysing mutations at the level of DNA.
If a cloned complementary DNA sequence is available for
the gene of interest then mutant gene structure may be
analysed by restriction endonuclease digestion of mutant
DNA and visualization of the mutant gene DNA fragments by
hybridization with the cloned complementary sequence (20).
Changes in the DNA restriction patterns of the mutant
gene followed, if necessary, by DNA sequencing techniques
gives total resolution of genetic change.

In principle, application of these methods will
resolve base pair changes, small and large deletions,
transpositions, methylation of DNA sequences and DNA
translocations within or close to the gene in question.
At present, these methods may be applied to only a few
cellular mutation systems, but there is no doubt that their
future impact on mutation research will be considerable.

SUMMARY AND CONCLUSIONS

1) Potentially mutagenic lesions have been identified
for only a few mutagens. A knowledge of initial mutagenic
events in DNA for a range of mutagens is important for
predicting, and ranking, the potency of mutagenic agents.

2) The role of DNA repair in mutagenesis is only
partly understood, but the available evidence supports
the view that the repair of induced damage is largely
error-free. Studies with repair-deficient cell strains
will increase our knowledge of the mutagenic process and
our understanding of the complex clinical manifestation
of DNA repair deficiency in man.

3) Characterisation of somatic cell mutants is
providing evidence of both point mutations and chromosome-associated null mutations after mutagen exposure.
If, as is suggested, chromosome-associated mutations are
a major category of induced mutation at some loci, then,
it may be necessary to re-examine some of the basic
concepts of mutagenesis, which derive from studies of
true gene mutation in micro-organisms, in terms of their
relevance to mammalian cells. Also, on a more practical
point, chromosome-associated null mutations are, on
average, likely to be more deleterious to the offspring
than point mutations. Even though poor 'mutant fitness'
may eliminate many chromosomally associated null

mutations *in vivo*, a knowledge of the relative frequency of these events after exposure to a given mutagen should be a consideration in assessing genetic risk to man.

ACKNOWLEDGEMENTS

I wish to thank R. Brown and J. Thacker for allowing me to cite unpublished work and all my colleagues for their comments on the manuscript.

REFERENCES

(1) Arlett, C.F. in *Proceedings of the 6th International Congress of Radiation Research* (ed. Okada, S. et al.), p. 596-602, JARR, Tokyo (1979).
(2) Asquith, J.C. *Mutation Res. 43*, p. 91-100 (1977).
(3) Bradley, W.E.C. *J. Cell Physiol. 101*, p. 325-340 (1979).
(4) Cerutti, P.A. in *DNA Repair Mechanisms* (ed. Hanawalt, P.C. et al.), p. 1-14, Academic Press, New York (1978).
(5) Cleaver, J.E., Williams, J.I., Kapp, L., and Park, S.D. as ref. 4, p. 85-93 (1978).
(6) Cox, R. in *Progress in Environmental Mutagenesis, Vol. 7* (ed. Alacevic, M.), p. 33-36, Elsevier/North Holland, Amsterdam (1980).
(7) Cox, R., and Masson, W.K. *Nature 276*, p. 629-630 (1978).
(8) Cox, R., Masson, W.K., Weichselbaum, R.R., Nove, J., and Little, J.B. *Int. J. Radiat. Biol. 39*, p. 357-365 (1981).
(9) Edwards, Y.H., and Povey, S. as ref. 6, p. 213-226 (1980).
(10) Emerit, I., and Cerutti, P.A. *Proc. Natl. Acad. Sci. (U.S.A.) 78*, p. 1868-1872 (1981).
(11) Evans, H.J. in *Chromosome and Cancer* (ed. German, J.), p. 191-237, Wiley, New York (1974).
(12) Evans, H.J. in *Progress in Genetic Toxiciology* (ed. Scott, D., Bridges, B.A., and Sobel, F.H.), p. 57-74, Elsevier/North Holland, Amsterdam (1977).
(13) Fox, M., and Brennand, J. *Carcinogenesis 1*, p. 795-799 (1980).

(14) Friedrich, U., and Coffino, P. *Proc. Natl. Acad. Sci. (U.S.A.) 74*, 679-683 (1977).
(15) Maher, V.M., Ouelett, L.M., Curren, R.D., and McCormick, J.J. *Nature 261*, p. 593-595 (1976).
(16) McCormick, J.J., and Maher, V.M. as ref. 4, p. 739-749 (1978).
(17) Meuth, M., L'Heureux-Huard, N., and Trudel, M. *Proc. Natl. Acad. Sci. (U.S.A.) 76*, p. 6505-6509 (1979).
(18) Myhr, B.C., Turnbull, D., and DiPaolo, J.A. *Mutation Res. 62*, 341-353 (1979).
(19) Newbold, R.F., Warren, W., Medcalf, A.S.C., and Amos, J. *Nature 283*, p. 596-599 (1980).
(20) Pellicer, A., Robins, D., Wold, B., Sweet, R., Jackson, J., Lowy, I., Roberts, J.M., Sim, G.K., Silverstein, S., and Axel, R. *Science 209*, p. 1414-1422 (1980).
(21) Roberts, J.J. in *Advances in Radiation Biology* (ed. Lett, J.T. and Adler, H.), 7, p. 211-436 (1978).
(22) Schultz, R.A., Chang, C., and Trosko, J.E. *Environmental Mutagenesis 3*, p. 141-150 (1981).
(23) Thacker, J. *Cytogenet. Cell Genet. 29*, p. 16-25 (1981).
(24) Thacker, J., and Cox, R. *Nature 258*, p. 429-431 (1975).
(25) Thacker, J., and Stephens, M.A., and Stretch, A. *Mutation Res. 51*, p. 225-270 (1978).
(26) Waldren, C., Jones, C., and Puck, T.T. *Proc. Natl. Acad. Sci. (U.S.A.) 76*, p. 1358-1362 (1979).
(27) Weibezahn, K.F., and Coquerelle, T. *Nucleic Acid Res. 9*, p. 3139-3150 (1981).
(28) Yang, L.L., Maher, V.M., and McCormick, J.J. *Proc. Natl. Acad. Sci (U.S.A.) 77*, p. 5933-5937 (1980).

Mutation in Germ Cells

Germ-Cell Sensitivity in the Mouse: A Comparison of Radiation and Chemical Mutagens

Antony G. Searle

Medical Research Council, Radiobiology Unit, Harwell, Didcot, Oxon, United Kingdom

ABSTRACT Sensitivity patterns to mutation induction depend partly on common non-specific factors like germinal selection but are greatly influenced by specific modes of action of different mutagens and by metabolic factors. Radiation induces mutations in all stages (except dictyate oocytes) but the response to chemicals is more restricted. While early spermatids are very sensitive to aberration induction by radiation, only later stages are affected with many monofunctional alkylating agents (AA). With some of these AA, sensitivities to aberration and to mutation induction show inverse correlation, as in Drosophila. So far, chemically induced mutations show less homozygous lethality than with radiation and give more intermediate alleles. Much lower translocation frequencies arise from chemical than from radiation treatment of spermatogonia but very high frequencies may be induced in post-meiotic stages. Expression of aberrations may be delayed with chemicals, which have revealed the importance of maternal repair at the pronuclear stage.

INTRODUCTION

The reasons for which the mouse was chosen for the experimental study of genetic risks from radiation hold also for risks from chemical mutagens. Much information now exists on the latter as well as the former topic, which allows useful comparisons between different mutagens on (i) stages affected for each mutational endpoints, (ii) relative stage sensitivities, (iii) nature of mutational response. Comparisons between the mouse and Drosophila are also valuable in helping to determine whether mammalian sensitivity patterns depend on characteristics of mammalian germ cells and their environment rather than just on the biochemical nature of the mutagen.

The main mutational end-points studied in the mouse after germinal exposure have been (i) dominant lethals, as indications of chromosome breakage, (ii) reciprocal translocations, detected in spermatocytes or in F_1 progeny (thus heritable), (iii) specific locus mutations (56,63) including those at biochemical and histocompatible loci, (iv) various categories of dominant mutation, although almost entirely after irradiation. The somatic equivalent of the specific locus test is the "spot test" in which mutant clones at specific loci are looked for in the mouse coat (55). Comparisons with the germinal test are clearly of interest.

CHROMOSOME ABERRATIONS

The mutagenic behaviour of this category seems distinct from that of gene mutations with respect to monofunctional AA, since, those with high s-values, like methyl methanosulphonate (MMS), break chromosomes readily in Drosophila, while those with low values, like ethyl nitrosourea (ENU) or diethylnitrosamine (DEN), give high frequencies of recessive lethals but only induce translocations at toxic levels of exposure (70). In Tables 1-3 these AA will be given in descending order of s-value, following radiation.

Dominant lethal mutations. The radiation pattern is not followed by any chemical (Table 1). Post-implantation lethality after spermatogonial irradiation is a secondary result of translocation induction (52); the pre-implantation loss with many chemicals comes from failure of fertilization (24), though death of fertilized eggs may be important in postmeiotic stages, e.g. after MMS treatment (24,45). 6-mercaptopurine (6-MP) differs from the others since it induces deletions at premeiotic S-phase. Several monofunctional AA induce lethals in spermatozoa and late spermatids, with little effect thereafter, apart from isopropylmethanesulfonate (IMS) though it differs from propyl(P)MS only in its steric configuration. Dominant lethal induction by ethyl(E)MS shows high correlation with

Table 1. Stage sensitivities for dominant lethals
A,spermatozoa; B,spermatids; C,spermatocytes; D,sp'gonia
CP,cyclophosphamide; EO,ethylene oxide; rest in text

MUTAGEN	A	B	C	D	REFERENCES
Radiation	+	+ → ++ [a]	+	(+) [b]	6,21,34,52
MMS,EMS,PMS	++	++ → - [a]	-	-	21,24,30,53
IMS	++	+ → - [a]	++	-	31
TEM	++	++ → + [a]	-	-	10,67
CP	++	++	-	-	24
EO	++	+	-	-	16,41
Fosfestrol	++	-	-	-	26
Mitomen	++	-	-	-	23
Procarbazine	-	+	+	(+) [c]	23
Mitomycin C	-	+	+	(+) [c]	21
6-MP	-	-	+	+ [d]	38
5-fluouracil	-	-	(+) [c]	(+) [c]	22,24

a Change in sensitivity from late to early spermatids
b Secondary, the result of translocation induction
c Pre-implantation loss through failure of fertilization
d Only in late (premeiotic) gonia, not in stem cells

****

alkylations of sperm protamine (66) which replace histones in mid-to-late spermatids. Possibly these protamines are the target for other monofunctional AA, but early spermatids are more sensitive to radiation than late ones. Oxygen tension differences are thought important here (4). Quite different patterns are found with other chemicals (Table 1).

The response of maturing oocytes to triethylene-melamine (TEM) is very similar to radiation (9) but decidedly lower than that of spermatozoa; this also holds for trenimone (54). Dominant lethal frequencies in female mice are strain-dependent (35). Females of the sensitive T-stock are also better able to repair damage coming in from the male and induced by IMS (40) and EMS (42) than some hybrid females. Thus the egg genotype is important in processing premutational lesions.

Reciprocal translocations. Tests for heritable translocations are responsible for all the data on postmeiotic stages and some from spermatogonia (7,39), the rest is from examination of spermatocytes at metaphase-I (49). Radiation and chemicals again differ with respect to spermatogonia (Table 2). Brewen (8) has explained the very low yield

Table 2. Stage sensitivities for reciprocal translocations

MUTAGEN	GERM CELL STAGE		REFERENCES
	Spermatozoa/'tids	Spermatogonia	
Radiation	++	+	48,65
MMS	+	−	2,50
EMS	++	−	42,50
IMS	+	−	40,50
TEM	++	+	9,14,39
CP	++	−	69
Ethylene oxide	++	?	41
TEPA	++	(+)[a]	39,70
Procarbazine	+	?	2
Nitrogen mustard	+	?	33,50
Mitomycin C	+	−[b]	1,2

[a] conflicting evidence
[b] whole-arm chromatid exchanges found in spermatogonia

****

from chemicals at this stage (as judged by metaphase-I examination) in terms of the induction of chromatid-type rather than chromosome-type interchanges, with lower transmission, and perhaps a sensitivity to a small fraction of the cell cycle. Delayed expression is another possibility (see below). In contrast, postmeiotic stages may give very high frequencies, e.g. a 600-fold difference with TEM (8,39). Thus genetic risks from postmeiotic stages must be considered with chemicals, even though the exposure period is less than 1% that of spermatogonial stem-cells (15).

There are many similarities between findings for translocations and dominant lethals. However, 6-MP induces chromatid deletions rather than interchanges at premeiotic S-phase (36) while IMS induces much higher frequencies of dominant lethals than heritable translocations in spermatozoa. This may be because many IMS-induced lesions reach expression only after pronuclear DNA synthesis, so have less chance of interaction (37). However, some EMS-induced lesions only become fixed at that stage (42) although EMS induces many translocations. Generoso and colleagues have reported that lesions induced by TEM at pachytene may be delayed in their expression as translocations until the F_1 generation (38); such delay in Drosophila is also associated with chromosome-breaking AA (5,70). Heritable translocations are recovered at very

Table 3. Stage sensitivities for specific locus mutations

MUTAGEN	Post-sp'cytes	Spermato-gonia	Maturing oocytes	REFERENCES
Acute X-rays	++	+	++	57,59,62
MMS,EMS	+	−	?	25
PMS	−	+	?	25
IMS	?	−	?	25
TEM	++	+	+	11,12,13
CP	+	?	?	17
ENU	?	++	?	61
DEN	−	−	?	60
MNNG	−	−	?	25
Procarbazine	+	+	−	29
Mitomycin C	−	+	(+)[a]	23,26

[a] not significantly above controls

****

Table 4. Numbers of mutations at specific loci after mutagen treatment of spermatogonia

TREATMENT	LOCI									REF.
	a	b	c	p	d	se	dse	s	Total	
X- or γ-rays	5	45	27	38	42	3	1	108	269	58
Fast neutrons	3	8	9	14	9	6	2	40	91	62
Procarbazine	1	8	4	12	6	3	0	5	39	29
Mitomycin C	1	2	1	6	3	0	0	1	14	28
TEM	2	0	0	1	3	0	0	0	6	11
ENU	>0	>0	>0	>0	5	2	0	2	28	61

****

low frequency by irradiating maturing oocytes (46,64), and by IMS treatment (39).

SPECIFIC LOCUS MUTATIONS (SLM)

Differences between radiation and chemicals are both quantitative and qualitative (Tables 3 and 4); the effects of radiation itself depend greatly on dose-rate and on oocyte stage (62). The dose-response curve for X-irradiation of maturing oocytes is concave (51). Procarbazine, like X-rays, gives a humped curve after gonial treatment (29); the basic responses of both may also have a

dose-squared component. Older spermatogonia may be more
sensitive to mutation induction by TEM (11); older oocytes
may be more sensitive to X-rays (57). There are some signs
of the expected correlation with s-values (70): methane-
sulfonates with high s-values induce few specific locus
mutations, while ENU, with a low s-value, induces more SLM
than any other chemical tested (61). The ENU frequency of
electrophoretic mutations is also high (43) with no sign of
structural aberrations. These results are in line with
Drosophila results (70) and the status of ENU as a
"supermutagen"; however, MNNG is not mutagenic in the mouse
(25), nor is diethylnitrosamine (DEN) in the SL test (60)
though a potent inducer of gene mutations in Drosophila (70).
So far, IMS has given negative results, despite its low
s-value, but Ehling (25) has warned that the mating system
used might have led to false negatives with PMS and IMS.
SLM have been recovered after TEM and mitomycin C exposures
of maturing oocytes (Table 3) but not after a larger test
with procarbazine.

The spectra of SLM found in gonia with the chemicals
tested (Table 4) differ from those with radiation in showing
(i) fewer s locus and more p locus mutations, (ii) no
d + se deficiencies after gonial or post-gonial treatment.
While 66% of SLM recovered from gonial irradiation were
homozygous lethal the figures for mitomycin C and procarbazine were 21% and 24% (28,29). More intermediate alleles
are found with chemicals, suggesting that more of the SLM
may be simple base-pair changes.

Results with the "spot test" for somatic mutations
agree well with those from the specific locus test.
However, both DEN and MNNG seem positive in the spot test
(32, but see ref. 2), although negative in the specific
locus test. There are also some contrasting results for
histocompatibility loci, since these have been negative
for radiation (19) yet positive for diethylsulphate (20)
and TEM (44).

CONCLUSIONS

The most striking difference between germ cell
responses to radiation and chemicals is one of homogeneity
vs. heterogeneity, since radiation can induce genetic damage
at all stages (except the dictyate oocyte) while chemicals
tend to induce particular types of damage at some stages
but not others. Presumably this is related to the more

specific action of particular chemicals which may be less
directly on DNA itself but, for example, on sperm protamines
instead (66). The general greater sensitivity of postmeiotic stages is partly the result of germinal selection,
but the gap seems particularly wide for structural aberrations induced by some monofunctional AA. This may be
connected with the nature of the aberration induced (8) and
possibly with its delayed expression (38). This latter
phenomenon may well turn out to be widespread in the mouse
as in Drosophila. With chemical treatment of post-meiotic
stages, the length of time breaks stay open seems important,
as does the differing capacity of maternal genomes for
repair or restitution of the damage.

The response of Drosophila and the mouse to monofunctional AA have much in common, since in both chromosome
aberration induction tends to be associated with high s-
values and gene mutation induction to low ones, though there
are several discrepancies, e.g. with DEN. This may well be
broken down, before reaching the mouse testis. Direct DEN
injection into the testis might reveal whether the mouse has
a gonadal activation system as found in Drosophila. The idea
that some AA are mainly chromosome breakers and others mainly
gene mutators tends to conflict with Ehling's view (23,28)
that "chemical mutagens are capable of inducing dominant
lethals and specific locus mutations only in the same germ
cell stages". Tables 1 and 3 suggest possible exceptions;
however, it would be very difficult to prove complete
insensitivity of particular stages to mutation induction.
It is clear that there are marked quantitative differences
in this respect. This is just one out of many intriguing
phenomena in this field which deserve further study.

REFERENCES

(1) Adler,I.-D. *Mutat.Res. 23*,369-379(1974).
(2) Adler,I.-D. *Mutat.Res. 74*,77-93(1980).
(3) Adler,I.-D. *Teratogen.Carcinogen.Mutagen. 1*,75-86(1980).
(4) Ashwood-Smith,M.J. et al. *Mutation Res. 2*,544-551(1965).
(5) Auerbach,C. & Kilbey,B.J. *Ann.Rev.Genet. 5,* 163-218
      (1971)
(6) Bateman,A.J. *Heredity 12,* 213-232(1958).
(7) Bishop,J.B. & Kodell,R.L. *Teratogen.Carcinogen.Mutagen.
      1,* 305-332(1980).
(8) Brewen,J.G. *Mutat.Res. 41,*15-24(1976).
(9) Cattanach,B.M. *Z.Vererblehre 90,* (1959)1-6.

(10) Cattanach,B.M. *Int.J.Radiat.Biol. 1*,(1959)288-292.
(11) Cattanach,B.M. *Mutat.Res. 3*,346-353(1966).
(12) Cattanach,B.M. *Mutat.Res. 4*,73-82(1967).
(13) Cattanach,B.M. *Mutat.Res.* in press (1981).
(14) Cattanach,B.M. & Williams,C.E. *Mutat.Res. 13*,371-375 (1971).
(15) Cox,B.D. & Lyon,M.F. *Mutat.Res. 30*,293-298(1975).
(16) Cumming,R.B. & Michaud,T.A. *Envir.Mutagen. 1*,166-167 (1979).
(17) Cumming,R.B. & Walton,M.F. *Genetics 68*, s14(1971).
(18) Dunn,G.R. & Kohn,H.I. *Envir.Mutagen. 1*,158(1979).
(19) Dunn,G.R. & Kohn,H.I. *Mutat.Res. 80*,159-164(1981).
(20) Egorov,I.K. & Blandova,Z.K. *Genet.Res. 19*,133-143 (1972).
(21) Ehling,U.H. *Mutat.Res. 11*,35-44(1971).
(22) Ehling,U.H. *Mutat.Res. 21*,29(1973).
(23) Ehling,U.H. *Mutat.Res. 26*,285-295(1974).
(24) Ehling,U.H. *Arch.Toxicol. 38*,1-11(1977).
(25) Ehling,U.H. In *Chemical Mutagens, vol. 5*,233-256. Plenum, New York (1978).
(26) Ehling,U.H. *Arch.Toxicol. 42*,171-177(1979).
(27) Ehling,U.H. In *Progress in Environmental Mutagenesis*, 47-58,Elsevier/N.H.,Amsterdam (1980).
(28) Ehling,U.H. *Arch.Toxicol. 46*,123-138(1980).
(29) Ehling,U.H. & Neuhäuser,A. *Mutat.Res. 59*,245-256 (1979).
(30) Ehling,U.H. et al. *Mutat.Res. 5*,417-428(1968).
(31) Ehling,U.H. et al. *Mutat.Res. 15*,175-184(1972).
(32) Fahrig,R. In *Chemical Mutagens, vol. 5*,151-176. Plenum, New York (1978).
(33) Falconer,D.S. et al. *J.Genet. 51*,81-88(1952).
(34) Ford,C.E. et al. *Cytogenetics 8*,447-470(1969).
(35) Generoso,W.M. & Russell,W.L. *Mutat.Res. 8*,589-598 (1969).
(36) Generoso,W.M. et al. *Mutat.Res. 28*,437-447(1975).
(37) Generoso,W.M. et al. *Genetics 93*,163-171(1979).
(38) Generoso,W.M. et al. *Genetics 85*,65-72(1977).
(39) Generoso,W.M. et al. In *Advances in Modern Toxicology, vol. 5*,109-129. Hemisphere, Washington (1978).
(40) Generoso,W.M. et al. *Proc.Nat.Acad.Sci., U.S.A. 76*, 435-437(1979).
(41) Generoso,W.M. et al. *Mutat.Res. 73*,133-142(1980).
(42) Generoso,W.M. et al. *Mutat.Res. 91*,137-140(1981).
(43) Johnson,F.M. & Lewis,S.E. *Proc.Nat.Acad.Sci., U.S.A. 78*,3138-3141(1981).

(44) Kohn,H.I. *Mutat.Res.* 20,235-242(1973).
(45) Kratochvil,S. *GSF-Bericht B 565*, 5-17(1975).
(46) Krishna,M. & Generoso,W.M. *Genetics* 86,s36-s37(1977).
(47) Lee,W.R. In *Radiation Research*, 976-983. Academic, New York (1975).
(48) Léonard,A. *Mutat.Res.* 11,71-88(1971).
(49) Léonard,A. In *Chemical Mutagens, vol. 3*,21-56, Plenum, New York (1973).
(50) Léonard,A. *Rad.Envir.Biophys.* 13,1-8(1976).
(51) Lyon,M.F. et al. *Mutat.Res.* 63,161-173(1979).
(52) Lyon,M.F. et al. *Genet.Res.* 5,448-467(1964).
(53) Moutschen,J. *Mutat.Res.* 8,581-588(1969).
(54) Rohrborn,G. *Humangenetik* 2,81-82(1966).
(55) Russell,L.B. & Major,M.H. *Genetics* 42,161-175(1957).
(56) Russell,W.L. *Cold Spring Harb.Symp.Quant.Biol. 16*, 327-336(1951).
(57) Russell,W.L. In *"Repair from Genetic Radiation Damage"*,205-217. Pergamon, Oxford (1963).
(58) Russell,W.L. In *Genetics Today, vol. 3*,257-264. Pergamon, Oxford (1965).
(59) Russell,W.L. *Proc.Nat.Acad.Sci., U.S.A.* 74,3523-3527 (1977).
(60) Russell,W.L. & Kelly,E.M. *Genetics* 91,s109-s110(1979).
(61) Russell,W.L. et al. *Proc.Nat.Acad.Sci., U.S.A. 76*, 5818-5819(1979).
(62) Searle,A.G. *Adv.Radiat.Biol.* 4,131-207(1974).
(63) Searle,A.G. In *Handbook of Mutagenicity Test Procedures*,311-324. Elsevier, Amsterdam (1977).
(64) Searle,A.G. & Beechey,C.V. *Mutat.Res.* 24,171-186 (1974).
(65) Searle,A.G. et al. *Mutat.Res.* 22,157-174(1974).
(66) Sega,G.A. & Owens,J.G. *Mutat.Res.* 52,87-106(1978).
(67) Soares,E.R. & Sheridan,W. *Mutat.Res.* 43,247-254 (1977).
(68) Sotomayor,R.E. & Cumming,R.B. *Mutat.Res.* 27,375-388 (1975)
(69) Sram,R.J. et al. *Folia Biol.* 16,367-368(1970).
(70) Vogel,E. & Natarajan,A.T. *Mutat.Res.* 62,51-100(1979).

ACKNOWLEDGMENTS

I am indebted to Drs. B.M. Cattanach and M.F. Lyon for their critical comments on an earlier draft, also to Drs. F.H. Sobels and E. Vogel for useful discussions.

Dependence of Mutagenesis on Germ Cell Stage and Sex (Introductory Remarks by Symposium Chairman)

F. H. Sobels

Department of Radiation Genetics and Chemical Mutagenesis, University of Leiden, Leiden, The Netherlands

Stage-dependent differences in sensitivity are important for various reasons. First, they present an interesting biological phenomenon, an analysis of which may lead to a better understanding of the process of induced mutation. Second, they are important for an evaluation of test results. Insufficient knowledge of stage-dependent differences in sensitivity may lead to both underestimates or overestimates of the mutagenic potential. Third, they are relevant to risk assessment, because they may point towards special precautionary measures in protecting certain sections of the population, for example pregnant women or children. In individuals of reproductive age they may indicate the special need to avoid procreation at certain stages, for example within short time of exposure or at late parental ages (6).

As an introduction to the topic of stage specificity, I would like to take you back some thirty years. In 1951, Charlotte Auerbach (1) described one of the most striking examples of stage- and sex-dependent sensitivity. Formaldehyde when administered in the food of Drosophila larvae produces mutations only in one part of the cell cycle of one germ cell stage, in one developmental phase and one sex, namely in the early stage of the primary larval spermatocytes. No mutations are produced in adults of either sex, in female larvae and in larval spermatogonia or late spermatocyte stages. The sensitive stage in the larval testis corresponds to the auxocytes, that is the long growth stage that precedes meiosis. It is during this stage that the chromosomes replicate and formaldehyde thus presumably exerts its action during DNA replication (3).

Formaldehyde, when injected into adult flies, however, manifests quite a different sensitivity pattern; mutations are produced in mature sperm and to a lesser extent in spermatogonia (8, 9). In this case the mechanism of action is different. Peroxides are presumably involved, since inhibition of catalase with cyanide enhanced the mutation frequency (10, 13) and dihydroxydimethylperoxide which is formed by the combination of formaldehyde and hydrogen peroxide produced mutations in the same stages (9). Moreover, germ cells in Drosophila females which normally do not respond to formaldehyde could be made to mutate following cyanide pretreatment (10).

Auerbach's findings for formaldehyde and mustard gas led her to a detailed reexamination of the effects of X-rays (2). By mating treated males to a succession of tester females a series of broods is obtained representing germ cells that were successively younger at the time of treatment. In a typical brood pattern early spermatids have the highest sensitivity, yielding 2-3 times as many sex-linked lethals as do mature spermatozoa. As an example of the complications encountered in the analysis of the underlying causes determining these stage-specific differences in response, let me summarize some of our findings.

Part of the difference between sperm and spermatids is due to better oxygenation of the early spermatids. When flies are irradiated in nitrogen followed by posttreatment in air or oxygen, the difference largely disappears and the mutation frequencies observed in early spermatids are roughly similar to those in mature spermatozoa (12).

The degree of oxygenation cannot, however, be the sole cause of the sensitivity difference. If this were so, the difference should also disappear when both stages are fully oxygenated through irradiation in pure oxygen. This, however, was not the case: irradiation in oxygen yielded higher mutation frequencies in early spermatids than in mature sperm. Early spermatids appeared to show a higher oxygen enhancement ratio of 3.3 than sperm, with an oxygen enhancement ratio of about 2. Early spermatids are thus characterized by a greater intrinsic sensitivity to the induction of radiation damage in the presence of oxygen, than mature spermatozoa.

Normally, intrinsic sensitivity and differential repair are often confounded. These two factors could be separated by irradiating the flies in nitrogen or oxygen and then posttreating them with nitrogen or oxygen. Following irradiation in nitrogen contrasting responses were observed for spermatozoa and spermatids, in that recovery from potential damage is favoured by nitrogen in spermatozoa and by oxygen in spermatids. Other experiments with specific inhibitors suggested that repair in sperm is favoured by a glycolytic pathway and in spermatids by oxidative respiration (7, 11).

A puzzling feature remained the observation that post-treatments remained without any effect following irradiation in oxygen. A possible interpretation put forward at the time was that irradiation in oxygen produces more irreparable damage than irradiation in nitrogen. Further confirmation for the notion that irradiation damage in oxygen is qualitatively different from that induced in nitrogen has recently been obtained by Ferro in our department.

During the past ten years evidence has been accumulated that the recovery of genetic damage from irradiated Drosophila spermatozoa is subject to modification by repair processes operating in the female. Thus, depending on the genotype or the physiological environment of the oocytes more or less mutations are obtained from irradiated post-meiotic male germ cells; and there are indications from Generoso's (5) studies that a similar situation is true for the mouse.

Recently Drosophila mutants have become available which are deficient in specific components of the repair process. Thus mei-9 is deficient for excision repair. In Ferro's experiments Muller-5 males were irradiated with 2000 R in N_2 or 1000 R in O_2 and then mated to wild-type females or females homozygous for mei-9^a. Ferro observed that when the spermatozoa had been irradiated in nitrogen, considerably higher yields of sex-linked lethals were recovered from the repair-deficient mei-9^a females than from wild-type females. However, this was not at all the case when irradiation had been administered in oxygen (4). These findings thus give further support to the earlier notion that irradiation in oxygen produces irreparable damage. In view of the fact that oxygen plays such a crucial role in determining stage specific differences in response to the mutagenic action of irradiation, we believe that this is an observation relevant to the subject of this symposium.

References
(1) Auerbach, C.   Hereditas 37, 1-16 (1951)
(2) Auerbach, C.   Z. ind. Abstamm. u. Vererb.lehre 86, 113-125 (1954)
(3) Auerbach, C.   Mutation Research, Chapman Hall, London, 504 pp (1976)
(4) Ferro, W.   Mutation Res. (in press)
(5) Generoso, W.M., Cain, K.T., Krishna, M., Huff, S.W. Proc. Natl. Acad. Sci. (U.S.) 76, 435-437 (1979)
(6) Lyon, M.F.   Mutation Res. (in press)
(7) Mukherjee, R.N. and Sobels, F.H.   Mutation Res. 6, 217-225 (1968)
(8) Sobels, F.H.   Amer. Natural. 88, 109-112 (1954)
(9) Sobels, F.H.   Nature 177, 979-982 (1956)
(10) Sobels, F.H.   Z. ind. Abstamm. u. Vererb.lehre 87, 743-752 (1956)
(11) Sobels, F.H.   Int. J. Radiat. Biol. 2, 68-90 (1960)
(12) Sobels, F.H.   Mutation Res. 2, 168-191 (1965)
(13) Sobels, F.H. and Simons, J.W.I.M.   Z. ind. Abstamm. u. Vererb.lehre 87, 735-742 (1956)

Dependence of Mutagenesis in *Drosophila* Males on Metabolism and Germ Cell Stage

Ekkehart W. Vogel

Department of Radiation Genetics and Chemical Mutagenesis, State University of Leiden, Leiden, The Netherlands

ABSTRACT. An attempt was made to identify some shared general features of the actions of genetically active agents in Drosophila.
(1) From the relationship of cell stage sensitivity to action of promutagens, it is obvious that with this type of mutagens germ cell metabolism is the most significant factor determining their effectiveness in various cell stages.
(2) To obtain a realistic picture of the intrinsic stage mutability throughout the Drosophila germ cell cycle, only autosomal recessive lethal tests can be used to compare stage-sensitivity to a given mutagen.
(3) Response of Drosophila to some, but not all, alkylating mutagens can be modified considerably by introducing into the assay system repair-deficient mutants (e.g. mei-9). There is a tendency to add repair-deficient strains to the battery of tester strains commonly in use and thereby assume that the Drosophila assay has been improved. This can only be determined if each new strain is validated prior to general use with a wide selection of mutagens. From our experience with the mutants mei-9^{L1} and mei-9^a mei-41^{D5}, these strains offer no apparent advantage over repair-proficient tester strains.

1. INTRODUCTION

A central question in chemical mutagenesis is whether general characteristics and relationships can be established for the genetic effects of chemical reagents that are valid beyond the particular species, strain, cell stage, and set of environmental conditions under which the correlations were obtained. It is the purpose of the present paper (1) to seek for such general patterns and common elements of the effects of chemical mutagens in Drosophila and (2) to trace some of the more recent developments that have taken place in this respect during the past years.

A judgement of general characteristics of mutagenic effects in Drosophila germ cells requires special definition of how sensitivity, cell-stage mutability and changes in mutational response are to be measured. Since it is known that the extent of changes in sensitivity depends very much on the kinds of damage under observation, it seems most desirable to look at several kinds (2, 24). But this is not always possible, due to the paucity of the available data. The bulk of information on stage-response in Drosophila has been collected by means of tests for recessive lethal mutations in males, using a wide array of direct-acting agents and promutagens. Thus, in the present case we are going to depend rather heavily on data obtained from experiments using the induction of X-chromosomal and, though to a minor extent, autosomal recessive lethals in male germ cells. Cell stage comparisons for other types of genetic damage will also be included where feasible.

2. RECESSIVE LETHALS

2.1. Direct-acting mutagens

General pattern. Following treatment of the male germ cell cycle with direct-acting reagents, for each mutagen a specific pattern is characteristic in terms of the stage response observed and the type as well as the magnitude of damage produced. This would seem to be a logical consequence of the large number of factors affecting chemical mutagenesis. In spite of this diversity in response, there are some characteristic features direct-acting mutagens have in common. Firstly, as a general rule, highest percentages of X-chromosomal recessive lethals are mostly obtained from cells which have passed the meiotic divisions. This observation has been amply confirmed, e.g. for MMS (13), EMS (10, 13), ethyleneimine (1), epichlorohydrin (reviewed in ref. 37), DEB (27), TEPA (31), TEM and trenimon (reviewed in ref. 37). The differences in mutational response found between postmeiotic and premeiotic stages are generally large for those agents. Browning (10) reported for EMS a 4- to 5-fold higher yield in postmeiotic cells of recessive lethals compared with the yield in meiotic and premeiotic cells, whereas in experiments by Fahmy and Fahmy (13) the difference between postmeiotic and premeiotic cells was even more than one order of magnitude. Analysis of the trifunctional alkylating agent TEM (reviewed in ref. 37) revealed a reaction pattern which is typical of alkylating agents with more than one reactive site. At high concentrations ($> 2.0 \times 10^{-1}$ mM) postmeiotic stages are the only ones which can survive the treatment, showing a high rate of recessive lethals, while meiotic and premeiotic stages are destroyed by TEM as indicated by complete sterility of the later broods. In the concentration range from 1.0×10^{-1} to 3.0×10^{-2} mM TEM, younger cells survive, but recessive lethal yield was 5-10 times below that determined for postmeiotic cells similarly treated.

Mutagens which do not follow the general pattern described here are ICR 170 (11) and Mitomycin C (20). For both reagents, no major differences in response of premeiotic and postmeiotic stages are seen when X-linked recessive lethal induction is determined.

<u>Germinal selection</u>. The reason which has been advanced to mainly account for these large changes in sensitivity within the male germ cell cycle is known under the term "germination selection", which was introduced by Muller (21). It refers to the fact that X-linked deletions that include a gene required for development into a spermatozoon are removed by germinal selection. Thus, to obtain a realistic picture of the intrinsic stage mutability throughout the cell cycle, only autosomal lethals can be used to compare the sensitivity to a given mutagen, and the ratio of autosomal to X-linked lethals in postmeiotic cells as compared with that in spermatogonia, is a measure of germinal selection (27). If this concept is valid, it follows that the relative difference in yield between X-chromosomal and autosomal lethals should be greater in spermatogonia as opposed to postmeiotic cells, if the mutagen in question is capable of inducing multi-locus deletions and gross structural aberrations.

Mutagens for which this approach has been followed, are
DEB, DEN, EMS, and 2-chloroethylmethanesulfonate (25,
27). The work of Shukla and Auerbach (27) on DEB and DEN
has shown that strong germinal selection takes place in
the case of DEB, known to produce chromosome breaks in
Drosophila. With DEN, the ratio of autosomal to X-linked
lethals also tended to be somewhat higher in premeiotic
cells as opposed to that in spermatozoa; but the
differences observed were only very small. DEN is a
potent mutagen in terms of point mutations, but seems
fairly ineffective to break the chromosomes of Drosophila (34, 35).

Experiments in our Institute (Blijleven, unpublished)
indicated that following treatment with EMS and, somewhat
unexpected, also with DEN, a substantial portion of
spermatogonia carrying X-linked lethals did not pass
through meiosis. EMS showed maximum activity in spermatozoa and spermatids, while DEN preferentially acted in
spermatids and spermatogonia stages. With DEB, there
appears no stage-specificity for autosomal recessive
lethals (27); this picture is quite different from that
obtained with other multifunctional agents in the X-
linked recessive lethal test, indicating the significance
of strong germinal selection for this type of mutagen.

2.2. Stage-specificity of promutagens

In the preceding section, recent data on induced
recessive lethals have been summarized and examined from
the point of view of relationship to cell stage of induction. It is now instructive to review the results of
those experiments that have provided evidence for sensitivity differences within the male germ-cell cycle after
treatment with mutagens which require metabolic activation.

A very striking feature of promutagen action in
Drosophila is the significance of intragonadal metabolism
for their genetic effectiveness. It is known that insects
do not possess a specific organ for detoxification, but
several tissues appear to be involved. These are the fat
bodies, various parts of the digestive tract (12) and,
crucial for the explanation of the pattern induced by
promutagens, certain parts of the gonadal tissue. There
is considerable evidence, derived from experiments with
alkaryltriazenes and other promutagens, that release of
mutagenic metabolites from the parent promutagen takes
place directly in gonadal tissue, the cells that are

tested in the recessive lethal assay. In the majority of
cases triazenes show a persistent peak mutagenic activity
in metabolically active stages, i.e., in various spermatid stages, while spermatocytes are highly susceptible to
killing (33, 37, 38). Tates (32) has reported that these
stages have a highly developed endoplasmic reticulum.

A pattern of sensitivity similar to that described
for alkaryltriazenes has also been recorded for other
classes of promutagens: nitrosamines, oxazaphosphorines,
aflatoxins, azoxyalkanes, halo-olefins, haloalkanes,
pyrrolizidine alkaloids (33). The observation that highest
mutation frequencies were observed in spermatids and spermatocytes, also accounts for DEN (Blijleven, unpublished)
and aflatoxin B_1 (17), the only promutagens extensively
studied in the autosomal recessive lethal assay. It
should be realized, however, that the occurrence of peak
mutagenic activity in spermatids or earlier stages alone
is no proof that the chemical under study is a promutagen.

The second piece of evidence supporting the significance of intragonadal activation for most classes of promutagens comes from concentration-effect studies with
procarbazine (5), DMN (14), and DEN (34). With those
mutagens, concentration-effect relations are different
for metabolically inert sperm versus metabolically active
spermatids. In spermatids, a steep increase is found for
recessive lethals between 0.5 mM and 10.0 mM procarbazine
(with a plateau above 20.0 mM) and between 1.0 mM and
10.0 mM DMN. This stands in marked contrast to the flat
curves found following treatment of spermatozoa. In
spermatozoa, there is no single proportionality between
the concentration applied and the response obtained in
the recessive lethal test.

The metabolically inert spermatozoa, then, must have
been attacked by more stable material from the surrounding
tissue, and the degree to which this happens may determine
the extent of sensitivity differences seen between sperm
and spermatids. Whether this interpretation is correct is
not known, but the consequences for mutagenicity testing
of an intragonadal activation are obvious - particular
attention must be paid to metabolically active germ cell
stages. A further implication from these results is that
there is a strongly marked concentration dependence of
stage-sensitivity and of stage-sensitivity differences,
which are presumably linked to differences in metabolic
activation. This is a good example of the care that has
to be taken when interpreting on the basis of only one

dose differences between the responses of germ cells to mutagens.

With direct-acting alkylating agents, however, this pattern in mutagenic response is no longer present (36). This type of mutagens over a wide range of doses generally shows the expected linear concentration-effect relationship.

Dose-effect relationships of the type reported for postmeiotic cells have yet to be established for spermatogonia stages.

2.3. Role of repair

In recent years, a number of recombinant-defective and repair-defective mutants have been isolated in Drosophila to study the genetic control of mutagenesis in this eukaryot (3). One large class of mutagen-sensitive mutants has been isolated on the basis of hypersensitivity to killing by MMS (6, 28, 29). Complementation tests suggest that there are approximately 60 loci in the Drosophila genome which control sensitivity to MMS (19, 30). Some of these mutants have been reported to be defective in meiotic recombination and female fertility (4, 6, 7, 29), excision repair (7), and postreplication repair (8).

In attempts to explore the in vivo function of repair mutants on chemically induced mutagenesis, and the possibility that these mutants could enhance the sensitivity of Drosophila for detecting environmental mutagens, several of these mutants were analysed for their sensitivity to AAF, BP, aflatoxin B_1, nitrogen mustard and to a series of monofunctional alkylating agents. Most work has been carried out with the single-mutants mei-9^a, mei-9^{L1}, mus(1)101, mei-4^{D5} and the double mutagen-sensitive mutant mei-9^a mei-45^{D5} (15, 16, 22, 23, 33). The picture obtained is far from being clear, some observations are even controversial. One experimental approach has been to treat either wild-type or mutagen-sensitive males and compare mutation induction in postmeiotic and meiotic cells. Since mature sperm do not possess the biochemical machinery to repair premutational lesions or to convert these into mutations (18, 26, 39), differences between repair-proficient and repair-defective strains should appear only in broods corresponding to the treatment of spermatids or spermatocyte stages. Nix et al. (23) observed a significant increase in the frequency of X-linked recessive lethals, when mei-9^a males were treated

with several alkylating agents. Treatment of males from
strains Oregon R (wild-type), mei-9^a, mei-9^{L1}, mei-41^a
and mei-41^{D5} with EMS and DEN gave no significant differences between the five genotypes. DMN and MMS treatments
produced even fewer lethals in mei-9^a and mei-9^{L1} than
in mei-41^a, mei-41^{D5}, or wild-type males.

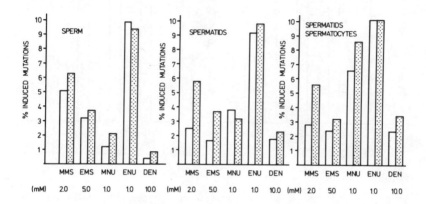

Fig. 1. Influence of the genotype on mutation induction
(rec. leth.) by alkylating agents. Treatment of
either Berlin K males (white columns) or mei-9^{L1}
males (dotted columns).

The results with mei-9^{L1} and Oregon R for EMS and
MMS by Nix et al. (23) differ from the picture obtained
in similar experiments in our laboratory with mei-9^{L1} and
Berlin K, the strain from which mei-9^{L1} was isolated. In
Fig. 1 summarized data are presented from results to be
published elsewhere. Treatment of mei-9^{L1} males with MMS
or EMS, at varying concentrations, leads to a consistent
and significant enhancement in the lethal frequency, but
only in progeny corresponding to treated spermatid or
spermatocyte stages. By contrast, the effects of ENU and
DEN treatments are virtually the same in the two types
of males.
    We also performed the reciprocal cross, i.e., repair-
proficient males were treated and mated to either mutagen-

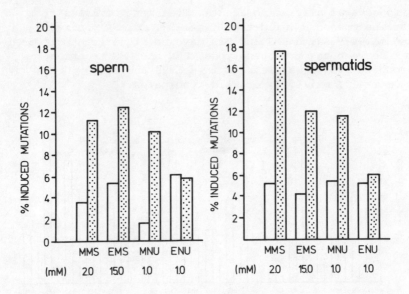

Fig. 2. The role of maternal repair on the induction of recessive lethals after mutagen treatment of Basc males for 2 days. Crosses performed with either Berlin K females (white columns) or mei-9^{L1} females (dotted columns).

sensitive or wild-type females, to explore the effects on mutation induction of repair-deficient females. Graf and Würgler (16) and Graf et al. (15) reported that DNA alkylation by MMS, EMS, MNU or ENU in spermatozoa always led to increased percentages of X-linked recessive lethals in oocytes with an excision repair deficiency (mei-9) compared to wild-type oocytes. This pattern is consistent with the picture observed in similar experiments in our laboratory with MMS, EMS and MNU (Fig. 2). There is, however, no agreement concerning the effects produced by ENU. Although a whole series of ENU experiments was performed, utilization of mei-9^{L1} females did not alter the yield of lethals determined at several ENU concentrations which ranged from 5×10^{-3} mM to 2.5×10^{-1} mM. At high concentrations of 0.5 mM and 1.0 mM ENU, a marked decline in male and female fertility and mutation induction was noted for the cross mei-9^{L1} ♀♀ X Basc ♂♂, as compared to the combination Berlin K ♀♀ X Basc ♂♂. Therefore, one of the questions that can be asked is whether selective killing of certain cell stages can explain part of the rather puzzling interlaboratory discrepancies. To clarify the whole issue, we feel that there is a stringent need for extensive concentration effect studies.

3. Other types of genetic damage

In the preceding sections, data on induced recessive lethal mutations were presented and considered from the point of view of relationship to cell stage of induction, metabolic activation and repair. In the last section, an attempt will be made to summarize information on chromosome aberrations (chromosome loss, translocations, deficiencies) for an over-all view of relative sensitivities of the various germ-cell stages of Drosophila to the induction of chromosome aberrations. However, as already pointed out, we shall be concerned again entirely with male germ cell stages.

Comparisons of spermatozoa, spermatids, spermatocytes and spermatogonia give generally consistent results in various studies. Thus, on the basis of induced sex-chromosome aberrations and translocations, post-meiotic cells yield higher frequencies of aberrations than do spermatogonia (1, 9, 31). Spermatocytes are readily eliminated by cell killing at moderate and higher concentrations. To illustrate this aspect with an example, in a detailed brood fractionation analysis with ten 3-day broods following treatment with MNNG, 4.3% translocations (14/412 gametes) were detected in brood 1, 2.5% (5 in 201) in brood 2, 0.5% (1 in 211) in brood 3, but none thereafter in about 1400 tests (9).

There are, however, basic difficulties in comparing spermatocytes and spermatogonia with postmeiotic stages. The possibility of strong segregational elimination during meiosis makes it impossible to draw any conclusion concerning initially induced breakage frequencies from the final yield observed.

This leaves one with the question whether there are differences in relative sensitivity between the post-meiotic spermatozoa and spermatid stages. In an analysis with a series of monofunctional alkylating agents (AA), two parameters were applied for comparing in these two stages the chromosome-breaking efficiency of some AA: the lowest exposure concentration producing both chromosome aberrations (translocations and ring-X loss) and recessive lethal mutations, and the proportion of translocations to mutations (T : M) at equal or at least similar mutation frequencies (35). In some experiments with MMS, the T : M ratios tended to be somewhat higher in spermatids, but the differences observed were only very small. Experiments with ring-X chromosomes (36) seemed to provide more evidence of an essential difference in sensitivity to chromosome aberration induction between spermatozoa and spermatids. Again, the comparisons were made on the basis of similar percentages of recessive mutations. These comparisons revealed a somewhat lower response of spermatozoa relative to spermatids for DMN, MNU, EMS, and MMS.

ACKNOWLEDGEMENT

This work was supported by the National Institute of Environmental Health Sciences (U.S.A.), Contract ESO 1027-04/06, and the "Stichting Koningin Wilhelmina Fonds" (The Netherlands), Contract 81.90. Part of this investigation also received support from the Association Contract 139-77-1 ENV N between the European Communities (Environmental Research Programme) and the State University of Leiden.

REFERENCES

(1) Alexander, Mary L., and Evalea Glanges. Proc.Natl. Acad.Sci. 53, 282-288 (1965)
(2) Auerbach, C. Mutation Research: Problems, Results and Perspectives. Chapman and Hall, London, 504 pp (1976)
(3) Baker, B.S., Boyd, J.B., Carpenter, A.T.C., Green, M.M., Nguyen, T.D., Ripoll, P., and P.D. Smith. Proc.Nat.Acad.Sci. (USA) 73, 4140-4144 (1976)
(4) Baker, B.S., and A.T.C. Carpenter. Genetics 71, 255-286 (1972)
(5) Blijleven, W.G.H., and E. Vogel. Mutation Research 45, 47-59 (1977)
(6) Boyd, J.B., Golino, M.D., Nguyen, T.D., and M.M. Green. Genetics 84, 485-506 (1976a)
(7) Boyd, J.B., Golino, M.D., and R.B. Setlow. Genetics 84, 527-544 (1976b)
(8) Boyd, J.B., and R.B. Setlow. Genetics 84, 507-526 (1976)
(9) Browning, Luolin S. Mutation Res. 8, 157-164 (1969)

(10) Browning, L.S. Drosophila Information Service 45, 75-76 (1970)
(11) Carlson, E.A., and J.L. Southin. Genetics 48, 663-675 (1963)
(12) Casida, J.E. In: Microsomes and Drug Oxidations, Academic Press, New York, p. 517 (1969)
(13) Fahmy, O.G., and M.J. Fahmy. Genetics 46, 361 (1961)
(14) Fahmy, O.G., and M.J. Fahmy. Chemical.-Biol.Interact. 11, 395-412 (1975)
(15) Graf, U., Green, M.M., and F.E. Würgler. Mutation Res. 63, 101-112 (1979)
(16) Graf, U., and F.E. Würgler. Abst. 10th Ann.Meet. of the EEMS, Athens, Greece, Mutation Res., in press
(17) Lamb, M.J., and L.J. Lilly. Mutation Res. 11, 430-433 (1971)
(18) Maddern, R.H., and B. Leigh. Mutation Res. 41, 255-268 (1976)
(19) Mason, J. Mutation Res. 72, 323-326 (1980)
(20) Mukherjee, R. Genetics 51, 947-951 (1955)
(21) Muller, H.J. In: A. Hollaender (ed.), Radiation Biology, Vol. 1, McGraw Hill, New York, pp. 475-626 (1954)
(22) Nguyen, T.D., Boyd, J.B., and M.M. Green. Mutation Res. 63, 67-77 (1979)
(23) Nix, C.E., McKinley, C., and J.L. Epler. Genetics 94, Supplement, s. 77 (1980)
(24) Parker, D.R. In: Repair from Genetic Radiation Damage and Differential Radiosensitivity in Germ Cells, F.H. Sobels (ed.), Pergamon Press, Oxford-London, 11-19 (1963)
(25) Purdom, C.E. Nature (London) 180, 81-85 (1957)
(26) Sankaranarayanan, K., and F.H. Sobels. In: M. Ashburner and E. Novitski (eds.), The Genetics and Biology of Drosophila, Academic Press, London-New York-San Francisco, Vol. 1c, 1083-1250 (1976)
(27) Shukla, P.T., and C. Auerbach. Mutation Res. 72, 231-243 (1980)
(28) Smith, P.D. Mutation Res. 20, 215-220 (1973)
(29) Smith, P.D. Molec.Gen.Genetics 149, 73-85 (1976)
(30) Smith, P.D., Snyder, R.D., and R.L. Dusenberg. In: F.J. de Serres, W.M. Generoso and M.D. Shelby (eds.), DNA Repair and Mutagenesis in Eukaryotes, Plenum Press, New York, in press
(31) Sram, R.J. Folio Biol. (Prague) 18, 139-148 (1972)
(32) Tates, A.D. Thesis, University of Leiden (1971)
(33) Vogel, E. In: Short-Term Tests for Chemical Carcino-

gens, H.F. Stich and R.H.C. San (eds.), Springer, New York, 379-398 (1981)

(34) Vogel, E., and B. Leigh. Mutation Res. 29, 383-396 (1975)

(35) Vogel, E., and A.T. Natarajan. Mutation Res. 62, 51-100 (1979a)

(36) Vogel, E., and A.T. Natarajan. Mutation Res. 62, 101-123 (1979b)

(37) Vogel, E., Schalet, A., Lee, R.L., and F. Würgler. Chapter 9 in: Mutagenicity of selected chemicals (F.J. de Serres, ed.), in press

(38) Vogel, E., and F.H. Sobels. In: Chemical Mutagens, Principles and Methods for their Detection (ed. A. Hollaender), Plenum Press, New York-London, Vol. 4, 93-142 (1976)

(39) Würgler, F.E., and H. Ulrich. In: M. Ashburner and E. Novitski (eds.), The Genetics and Biology of Drosophila, Vol. 1c, Academic Press, New York, pp 1269-1298 (1976)

Mutagenesis and Carcinogenesis

Comparison of Mutagenesis and *in vitro* Transformation Models in Detecting Potential Carcinogens

Verne A. Ray

Medical Research Laboratories, Pfizer, Inc., Groton, Connecticut, U.S.A.

ABSTRACT The correlation between mutagenicity of chemicals and carcinogenicity in rodents is high when genotoxic carcinogens are examined in multiple gene or point mutation assays. However, most non-genotoxic or epigenetic carcinogens would not be detected by these procedures which prescribes a role for in vitro transformation tests that are a more direct measure of carcinogenic potential. A group of 61 chemicals were selected based on data available in Ames Salmonella, mouse lymphoma (L5178Y) or Chinese hamster ovary gene mutation models, the Syrian hamster embryo clonal in vitro transformation assay and compared to oncogenicity results in rodents. This group of chemicals contained 42 carcinogens, 14 non-carcinogens and 5 indeterminate compounds. The battery of 3 assays identified correctly 41 out of 42 carcinogens and 10 out of 14 non-carcinogens. The high sensitivity of this small battery of assays in detecting compounds with carcinogenic potential recommends its consideration in the safety evaluation of chemicals.

INTRODUCTION

The evaluation of chemicals for irreversible types of toxicity at an early stage of their development into important items of commerce has been advanced markedly by the introduction of short-term tests for mutagenicity and carcinogenic potential. In fact, there is presently no other satisfactory way to acquire perspective on carcinogenic potential than by performing long-term oncogenicity studies in rodents. Analysis of structure and physical properties of chemicals are helpful, especially in prioritizing lists of chemicals for testing, but such efforts cannot substitute for information gained from complex biological systems. It is also quite clear that a number of carcinogens such as 3-aminotriazole, thioacetamide, thiourea and ethionine which are either non-genotoxic or extremely weak in existing assay systems, would not be detected as having carcinogenic potential by tests which have a purely genetic endpoint. This defines a role for in vitro transformation models that do detect such substances in batteries of assays commonly employed for screening purposes.

Among the short-term assays that are used to detect carcinogenic potential of chemicals, gene or point mutations in microbial and mammalian cells plus in vitro transformation studies have provided correlations with oncogenicity determinations which are as good or better than a number of other test batteries identified to date (51), (7). It should be mentioned in this context that substances which are positive in chromosomal-level assays such as urethane and diethylstilbestrol but negative or extremely weak in gene or point mutation models, can also be detected in in vitro transformation tests.

COMPARISON OF IN VITRO TRANSFORMATION MODELS

The relationship between mutagenesis and in vitro transformation has been discussed in several recent reviews (34), (5). Those studies which measure both mutation and in vitro transformation frequency in the same cell populations have provided the most perspective on this relationship. Barrett and Ts'o (6) compared mutation and transformation in diploid Syrian hamster embryo cells induced by benzo(a)pyrene and MNNG. Spontaneous transformation in these cells was determined to be approximately 10^{-4} (6 colonies in approximately 62,000 control colonies) whereas the spontaneous mutation frequency at the HGPRT (8-azaguanine,6-thioquanine resistance) and the Na^+/K^+ ATPase loci (quabain resistance) was determined to be in the vicinity of 10^{-6}. Induced frequencies of transformation with 1-5 μM MNNG or 1-10 μg/ml of benzo-(a)pyrene was in the range of 10^{-3} to 10^{-2} whereas induced mutation frequency at these concentrations was in the range of 10^{-5} to 10^{-4}.

Although variation was encountered in the transformation frequency of different pools of SHE cells, transformation frequency always exceeded mutation frequency (25-450X) and an average ratio was 100-1. Huberman, et. al., (20) had previously reported a ratio of 20-1 with benzo(a)pyrene and its 7,8 dihydrodiol in primary cultures of golden hamster embryo cells. It is also interesting to note that the expression time for both mutation and morphological transformation was 6-8 population doublings. However, anchorage independendent cellular growth or growth in soft-agar required a period of 32-75 population doublings. This characteristic, closely associated with the ability to establish tumors in animals, represents a period of phenotypic expression that is quite different from single-gene mutational processes. Further, initial experiments indicate that less than 10 percent of morphologically transformed colonies can progress to the anchorage-independent state. From this evidence, the authors have concluded that neoplastic transformation is more complex than a single gene mutational event and may well be a combination of genetic and epigenetic factors.

In contrast to these studies, those of Bouck and di Mayorca (8) concluded that chemical induction of malignant transformation in the quasidiploid hamster line BHK21/C113 seemed to result from a somatic mutation. Such cells have a low spontaneous frequency of transformation as measured by the ability to grow in soft agar (Selected a supernormal line 10, BHKSNC110, for these determinations), are induced to high frequencies of transformation by mutagens, transformants are stable and have a low frequency of reversion, but can be induced to revert to normal phenotype by a mutagen and often display temperature-restricted phenotypes. These characteristics are similar to those achieved by somatic cell mutants.

The frequency of spontaneously arising transformants of the quasi-diploid BHR21SNC110 cells was determined at 1.3×10^{-7} and corrected to be 8.8×10^{-7} per viable cell. This differs markedly from the spontaneous frequency of transformants in the diploid SHE cells at 10^{-4}. Furthermore, the BHK21 derived cells can grow in soft agar whereas the SHE cells require many doublings after morphological transformation to achieve this state. The BHK spontaneous transformation frequency is similar to a spontaneous mutation frequency which distinguishes this line from a diploid line such as SHE. Such differences may be a reflection of an alteration from the true diploid state. The ability to grow as a continuous cell culture and the ability of induced transformants to grow directly in soft-agar indicates a number of changes have occurred and the line could be regarded as partially transformed or preneoplastic. Barrett and Elmore (5) have pointed out that DES which does not induce mutation at the HGPRT or Na^+/K^+ ATPase loci in SHE cells, does not induce BHK21 cells to grow in soft agar. This can be a reflection of several processes being involved in achieving the neoplastic condition, one can be mutation, others can be epigenetic modification or the induction of aneuploidy. Cairns, (10) has suggested recently that transpositions in DNA may be more important in cancer induction than localized changes in sequence.

The influence of ploidy state on the <u>in vitro</u> transformation process is worthy of additional comment. Transformation of diploid SHE cells to the neoplastic condition results in an alteration in chromosome number to that of an aneuploid line. Currently, there are several mouse derived cell-lines in use for transformation work which are not diploid. The BALB/C3T3 and $C_3H/10T\frac{1}{2}$ lines are near tetraploid. These lines are already in a ploidy state which can be described as aneuploidy and therefore resemble many tumor cell types in this regard.

Barrett and Elmore (5) have concluded that aneuploid lines, which are not tumorigenic, have acquired some properties of neoplastic cells and progress to neoplastic cells more readily than normal diploid cells and should be considered as preneoplastic cells. Further, they propose that transformation of normal, diploid cells to aneuploid, preneoplastic cells may occur by a mechanism different from the transformation of preneoplastic cells to neoplastic cells.

A comparison of the SHE, BALB/C-3T3 and RLV-Infected Fischer 344 rat embryo transformation models using 49 chemicals has been reported recently, Dunkel et. al. (15). Most of the chemicals, which included polycyclic aromatic hydrocarbons, aromatic amines, alkylating agents, nitrosamines, heterocyclics, inorganics and hormones, were tested as coded samples. Thirty-seven of the chemicals were tested in all three assays. The most uniform test responses were obtained with PAH and inorganic compounds. Data from each assay were analyzed using defined criteria which resulted, with certain chemicals, in a conclusion different from previously published results. Responses of the SHE clonal assay in which one transformed colony was recorded or transformed colonies occurred in non-consecutive doses were regarded as inconclusive.

Examples of chemicals giving one transformed colony included uracil mustard and cyclophosphamide, whereas diethylstilbestrol, 1,2 epoxybutane and N-nitrosodimethylamine had transformed colonies in non-consecutive doses. Questionable responses were also seen in the BALB/C-3T3 assay with benz(a)anthracene, DENA and progesterone. Both SHE and RLV-RE models were found to be more responsive to heterocyclics than BALB/C-3T3.

Differences in intrinsic metabolism of these cellular models could provide a reason for the variation observed. All tests in this study were conducted without exogenous activation and the authors properly identify the need to supply such activation systems before concluding a chemical is indeed incapable of producing transformation. However, in spite of the absence of exogenous metabolic activation, many carcinogens of diverse structure, including non-genotoxic chemicals, are detected by these assays which indicates a considerable metabolic capability.

The utility of specific criteria in analyzing data from short-term tests is sorely needed and this paper makes a definite contribution toward standardizing the process of declaring a result positive, negative or inconclusive.

Pienta, (46) has reported a 92% correlation of cell transformation with carcinogenicity in the SHE clonal assay using cryopreserved pooled cells. Over 100 chemicals have been examined in this assay without added metabolic activation. Although Dr. Pienta has reported no false-positives were observed in this assay, several compounds reported negative in rodent carcinogenicity studies have produced positive results. These same compounds are also positive in Salmonella and mammalian cell gene mutation assays. It was also reported that false negatives could be eliminated by providing metabolic activation from liver homogenates or cultured hepatocytes. Considering the difficulties which can be encountered in using primary cell cultures for routine in vitro transformation screening of chemicals, the accomplishments of Dr. Pienta's group are especially noteworthy and have provided a stimulus to the utility of primary diploid cultures in this field.

IN VITRO TRANSFORMATION, AMES SALMONELLA AND MAMMALIAN CELL GENE MUTATION ASSAYS AS A BATTERY TO INDICATE CARCINOGENIC POTENTIAL

The primary utility of short-term tests for many pharmaceutical companies is the detection of carcinogenic potential of chemicals. A minimal or core group of assays for the detection of mutagenic and carcinogenic potential of chemicals has been under investigation at Pfizer for the past 4 years. Components of this battery are: (1) Bacterial point mutation assay with four strains of Salmonella typhimurium, (1a) Assay of urine from mice treated with test chemical on Salmonella strains, (2) Mammalian cell point mutation assay with mouse lymphoma (L5178Y), (3) Cytogenetic analysis in vivo with rat or mouse bone marrow cells, (4) A cytogenetic assay in human lymphocytes in vitro with mitotic indices as an indicator for dose selection, and (5) Transformation in hamster embryo cells in vitro, Ray, (52). Wherein comparative data exist, a subgroup of this battery composed of the Syrian hamster embryo clonal assay for in vitro transformation, the Ames Salmonella assay and the mouse lymphoma (L5178Y) or Chinese hamster ovary cell gene mutation assays has shown considerable promise as a valuable group of tests for assessing the carcinogenic potential of chemicals. It should be stated that although the Ames Salmonella and mouse lymphoma or Chinese hamster ovary cell assays may be performed routinely, the Syrian hamster embryo assay is difficult to develop to a point where routine application is feasible. A computerized selection of compounds within a data base of 517 substances on which comparative results were available, is shown in Table 1. Of the 61 compounds listed, 51 showed agreement between rodent carcinogenicity determinations and the battery of 3 assays. A positive agreement here means a positive result in one or more of the short-term tests and the rodent bioassay. Agreement on negative compounds means negative in all 3 short-term tests and the rodent bioassays. This group of 61 compounds contained 42 carcinogens (68.8%), 14 non-carcinogens (23.3%) and 5 indeterminate substances (8.2%).

Of the remaining 10 compounds, one substance, phenobarbital, was negative in the short-term tests but had a declaration of positive in rodent bioassays. Another, methylcarbamate, was negative in short-term tests but indeterminate in rodent bioassays. The remaining 8 compounds that are positive in short-term tests can be divided on the basis of carcinogenicity results into two groups; positive and indeterminate. The four compounds whose carcinogenicity is indeterminate are as follows:

	COMPOUND	STT RESULT
1.	Acridine orange	SHE(+),Sal(+),ML(?)
2.	Hycanthone	SHE(+),Sal(+),ML(+),CHO(+)
3.	Phenylenediamine, M	SHE(+),Sal(+),ML(+)
4.	Succinic anhydride	SHE(+),Sal(?),ML(N)

The remaining four compounds that have negative carcinogenicity determinations but are positive in the short-term tests are:

	COMPOUND	STT RESULT
1.	Benzo(e)pyrene	SHE(+),Sal(+),ML(+),CHO(+)
2.	Epoxybutane, 1,2	SHE(+),Sal(+),ML(+)
3.	Methotrexate	SHE(-),Sal(-),ML(+)W
4.	Phenylenediamine, 4 Nitro O	SHE(+),Sal(+),ML(+)

It is interesting to note that of the four compounds declared positive in short-term tests and indeterminate in carcinogenicity assays, 2 out of the four were positive in all 3 short-term tests. Three of the four compounds which had declarations of negative in rodent bioassays also were also positive in all 3 short-term tests.

The overall performance of the 3 test battery on compounds that were designated as carcinogenic or non-carcinogenic was 51 detected correctly out of 56 compounds or 91 percent. The detection rate of carcinogens was 41 out of 42 compounds or 97.6 percent. Ten out of 14 non-carcinogens were labelled correctly by STT for a 71.4% rate. Because of the high sensitivity of these 3 assays for detecting carcinogenic potential, a negative result appears to be highly significant and of considerable value in the safety evaluation process.

Although the number of compounds is relatively small on which comparative data in all 3 short-term tests and carcinogenicity are available, it does indicate that emphasis should be given to this small battery in screening and collaborative studies. Further, when a battery of tests declares a substance as mutagenic and capable of producing _in vitro_ transformation, its toxic potential has to be regarded with more than a modicum of concern. At the very least, additional documentations of this

type are needed to establish confidence in the utility
of short-term tests as indicators of carcinogenic potential and to serve as a focus for reexamination of rodent
bioassays as the ultimate indicator of carcinogenicity
declarations in toxicology.

REFERENCES

(1) Amacher, D. E., Paillet, S. C., Turner, G., Ray, V. A. and Salsburg, D. S. Mutation Res. 72, 447-474 (1980)
(2) Amacher, D. E. and Paillet, S. C. Mutation Res. 78, 279-288 (1980).
(3) Amacher, D. E. and Turner, G. N. Mutation Res. (in press) (1981).
(4) Ames, B. N., Kammen, H. O. and Yamasaki, E. Proc. Nat. Acad. Sci. USA 72, 2423-2427 (1975).
(5) Barrett, J. C. and Elmore, E. In "Mutagenesis and Carcinogenesis" edited by L. S. Andrews, R. J. Lorentzen and W. G. Flamm. Handbook of Experimental Pharmacology, Springer-Verlag, Berlin, in press. (1981)
(6) Barrett, J. C. and Ts'o. P.O.P. Proc. Natl. Acad. Sci. USA, 75, 3297-3301 (1978).
(7) Bartsch, H., Malaveille, C., Camus, A.-M., Martel-Planche, G., Breen G., Hautefeuille, A., Sabadie, N., Barbin, A., Kuroki, T., Drevon, C., Piscoli, C., and Montesano, R. IARC Scientific Publications No. 27, 179-241 (1980).
(8) Bouck, N. and di Mayorca, G. Nature, 264, 722-727 (1976).
(9) Burki, H. J. Rad. Res. 70, 650 (1977)
(10) Cairns, J. Nature 289, 353-357 (1981)
(11) Casto, B. C., Pieczynski, W. J., Nelson, R. L. and Dipaolo, J. A. Proc. Amer. Ass. Can. Res. 17, 12 (1976)
(12) Casto, B. C., Janosko, N. and Dipaolo, J. A. Can. Res. 37, 3508-3515 (1977)
(13) Clive, D., Johnson, K. O., Spector, J. F. S., Batson, A. G. and Brown, M. M. M. Mutation Res. 59, 61-108
(14) Dunkel, V. C. In Molecular and Cellular Aspects of Carcinogen Screening Tests. (R. Montesano, H. Bartsch, L. Tomatis, W. Davis EDS.) IARC Scientific Publications #27, International Agency for

Research on Cancer, Lyon, 1980
(15) Dunkel, V. C., Pienta, R. S., Sivak, A. and Traul, K. JNCI, in press (1981)
(16) Heddle, J. A. and Bruce, W. R. In Origins of Human Cancer. Book C. Human Risk. Assessment (H. H. Hiatt, J. D. Watson, J. A., WI Eds.) Cold Spring Harbor Laboratory, Cold Spring Harbor, New York (1977).
(17) Hetrick, F. H. and Kos, W. L. J. Toxicol. Environ., Health 1, 323-327 (1975)
(18) Hollstein, M and McCann, J. Mutation Res. 65, 133-226 (1979)
(19) Hsie, A. W., O'Neill, J., San-Sebastian, Jr., Couch, D. B., Brimer, P. A., Sun, W. N. C., Fuscoe, J. C., Forbes, N. L., Machanoff, R., Riddle, J. C. and Hsie, M. H. Report: EPA-600/9-78-027 PP. 293-315 (1978)
(20) Huberman, E., Mager, R. and Sachs, L. Nature 264, 360-361 (1976)
(21) IARC Monographs Volume 1, 1972
(22) IARC Monographs Volume 3, 1973
(23) IARC Monographs Volume 4, June, 1973
(24) IARC Monographs Volume 6, February, 1974
(25) IARC Monographs Volume 7, 1974
(26) IARC Monographs Volume 8, 1975
(27) IARC Monographs Volume 9, April, 1975
(28) IARC Monographs Volume 10, 1976
(29) IARC Monographs Volume 11, February, 1976.
(30) IARC Monographs Volume 12, 1976
(31) IARC Monographs Volume 13, 1977
(32) IARC Monographs Volume 15, February, 1977
(33) IARC Monographs Volume 16, 1978
(34) IARC Monograph, Supplement 2, 295-308 (1980)
(35) Kalina, L. M., Polukhina, G. N. and Lukasheva, L. I. Genetika 13, 1089-1092 (1977)
(36) McCann, J., Choi, E., Yamasaki, E and Ames, B. N. Proc. Nat. Acad. Sci. USA 72, 5135-5139 (1975)
(37) NCI Technical Report Series NCI-CG-TR-84
(38) NCI Technical Report Series NCI-CG-TR-130
(39) NCI DHEW Publication No. (NIH) 78-1329
(40) NCI Technical Report Series NCI-CG-TR-19
(41) Technical Report Series NCI-CG-TR-164
(42) NCI Technical Report Series NCI-CG-TR-180
(43) OTC Office of Technology Assessment (OTA) Report 1977 #052-003-00471-2, U.S. Gov't Printing Office Washington, D.C. 20402

(44) Palmer, K. A., Denunzio, A and Green, S. J. Environ, Pathol. Toxicol. 1, 87-91
(45) Pienta, R. J. Personal communication
(46) Pienta, R. J. In Chemical Mutagens, Vol 6, edited by F. J. de Serres and A. Hollender, Plenum Press, 175-202 (1980)
(47) Pienta, R. J., Poiley, J. A. and Lebherz III, W. B. Int. J. Can. 19, 642-655 (1977)
(48) Pienta, R. J., Lebherz, III, WB. and Takayama, S. Can. Lett. 5, 245-251 (1978)
(49) Pienta, R. J. Presentation at Symposium No. 12 Bioassay Systems for Carcinogens at the XIITH International Cancer Congress, Buenos Aires, Argentina, October 6, 1978
(50) Purchase, I. F. H., Longstaff, E., Ashby, J., Styles, J. A., Anderson, D., Lefevre, P. A. and Westwood, F. R. Br. J. Cancer 37, 873-903 (1978)
(51) Purchase, I. F. H., Longstaff, E., Ashbey, J., Styles, J. A., Anderson, D., Lefevre, P. A. and Westwood, F. R. Nature 264, 624-627 (1976)
(52) Ray, V. A. Pharmacological Reviews 30, 537-546 (1979)
(53) Simmon, V. F., Kauhanen, K. and Tardiff, R. G. Dev. Toxicol. Environ. Sci. 2, 249-258 (1977)
(54) Simmon, V. F. J. Natl. Can. Inst. 63, 893-899 (1979)

TABLE I

COMPOUND	ANIM	IVT	AMES	MC	REFERENCES			
ACETONE	−	N	−	N	36	47	36	1
ACETOXY-2-AAF,N	+	P	P	P	36	47	36	1
ACETYLAMINOFLUORENE,2	+	P	+	+	36	47	36	13
ACETYLAMINOFLUORENE,4	−	N	?	−	36	50	36	13
ACRIDINE ORANGE	I	P	+	?	36	45	36	1
AF-2	+	P	P	P	36	48	36	13
AFLATOXIN B1	R	P	+	P	28	47	36	18
AMINOAZOBENZENE,4	R	P	+	+	26	47	36	3
AMINOFLUORENE,2	+	P	+	+	36	47	36	3
ANILINE	R	N	−	+	38	50	36	1
ANTHRACENE	−	N	−	−	36	47	36	1
AURAMINE	B	+	+	−	21	47	36	1
BENZ(A)ANTHRACENE	M	P	+	+	22	47	36	1
BENZO(A)PYRENE	B	P	+	+	22	47	36	13
BENZO(E)PYRENE	−	+	+	+	22	47	36	13
BERYLLIUM SULFATE	R	P	+	P	21	47	53	19
BROMODEOXYURIDINE,5	−	N	−	?	47	47	16	9
CADMIUM CHLORIDE	B	P	?	P	29	11	35	19
CAFFEINE	−	N	−	N	36	50	36	1
CAPROLACTONE,E	−	N	−	N	36	47	36	13
CYCLOPHOSPHAMIDE	B	P	+	+	27	50	36	13
DIAMINOANISOLE,2,4	B	N	+	P	37	45	4	44
DIETHYL NITROSAMINE	B	+	+	+	21	47	36	13
DIETHYLSTILBESTROL	B	P	?	P	24	50	36	13
DIMETHYL CARBAMYLCHLORIDE	M	P	P	+	30	47	36	3
DIMETHYL NITROSAMINE	B	P	+	+	21	47	36	13
DIMETHYL SULFOXIDE	−	N	−	N	36	47	36	1
DIMETHYL-B(A)A,7,12	+	P	+	+	36	47	36	19
DIPHENYL NITROSAMINE	R	P	−	−	41	47	36	13
EPOXYBUTANE,1,2-	−	P	P	P	36	49	36	1
ETHANOL	−	N	−	N	36	47	36	1
ETHYL METHANESULFONATE	B	P	P	P	25	50	36	13
GLYCIDALDEHYDE	B	P	P	+	29	47	36	3
HYCANTHONE METHANESULFONATE	I	P	P	P	31	17	36	13
HYDRAZINE	B	P	+	N	23	47	36	1
HYDROXY-2-AAF,N	+	P	+	P	36	47	36	19
LEAD ACETATE	B	P	−	N	21	47	53	2
METHANOL	−	N	−	N	47	47	52	1
METHOTREXATE	−	N	−	P	39	47	53	13

TABLE I (CONTINUED)

COMPOUND	ANIMT	IVT	AMES	MC	REFERENCES			
METHYL CARBAMATE	I	N	–	–	30	49	36	3
METHYL IODIDE	R	P	P	P	32	47	36	13
METHYL METHANESULFONATE	B	P	P	P	25	12	36	13
METHYLCHOLANTHRENE, 3	+	P	+	+	36	47	36	1
MNNG	B	P	P	P	23	47	36	13
NATULAN	+	P	–	P	40	47	36	13
NICKEL CHLORIDE/SULFATE	+	+	+	P	47	47	47	2
NITRITE, SODIUM	+	N	+	–	36	45	36	19
NITRO-O-PHENYLENEDIAMINE, 4	–	P	+	P	42	49	14	44
NITRO-P-PHENYLENEDIAMINE, 2	M	P	+	P	33	49	14	45
NITROBIPHENYL, 4	+	P	P	+	23	47	36	3
NITROQUINOLINE-1-OXIDE, 4	+	P	P	P	36	47	36	1
NITROSOETHYLUREA, N	B	P	P	P	21	47	36	19
PHENOBARBITAL, SODIUM	B	N	–	N	31	50	36	1
PHENYLENEDIAMINE, M-	I	P	+	P	33	49	4	44
PROPIOLACTONE, BETA	B	+	P	P	23	50	36	13
PYRENE	–	N	–	–	36	47	36	1
SACCHARIN	R	N	–	P	43	43	43	13
SUCCINIC ANHYDRIDE	I	P	?	N	32	47	36	13
THIOACETAMIDE	B	P	–	–	25	47	36	3
URACIL MUSTARD	B	P	P	P	27	49	36	13
URETHANE	B	+	–	–	25	47	36	3

Animal Carcinogenicity
  M_1R and B designate positive in mouse, rat, or both mouse and rat studies. I designates indeterminate study. M_1R and B also indicate study was evaluated by IAEC or was an NCI(U. S.) study.

Mutagenicity and in vitro transformation
  P and N designate positive and negative results without exogenous metabolic activation, + and – with exogenous metabolic activation.

Correlations between Mutagenesis and Neoplastic Transformation in *in vitro* Systems

Leonard M. Schechtman and Richard E. Kouri

Microbiological Associates, Bethesda, Maryland, U.S.A.

ABSTRACT Criteria for selection of mammalian target cell systems for use in mutation assays include: (1) high plating efficiency, (2) short cell-generation time, (3) expression of mutations at ≥ 1 locus, and (4) near diploid chromosome complement. Criteria for selection of a cell system useful in transformation assays include: (1) endogenous mixed-function oxidase activity, (2) a stable source of target cells, (3) low cell saturation density, and (4) a readily discernible transformed phenotype. Mutation and transformation assays are not usually comparable because of differences in: (1) the assays themselves, (2) target cells used, (3) the type selective pressures applied, and (4) the definitions of such parameters as expression period, cells at risk, mutations and transformation frequency (MF and TF). Generally one can perform mutation assays in systems designed for transformation but not vice versa. To date, only a limited number of mammalian cell systems have been found useful for both mutagenesis and transformation, e.g. (1) BALB 3T3 CL.A31-1, (2) C3H 10T 1/2 CL.8, (3) Syrian hamster embryo, (4) Fischer rat embryo cells infected with Rauscher leukemia virus, and (5) certain human diploid fibroblast lines. For quantitive correlations between mutation and transformation, monitoring the simultaneous induction of both endpoints in a single cell system is preferable. However, limitations associated with both types of assays make interpretations of TF, MF and TF/MF ratios difficult. Nevertheless, such systems provide suitable models for ascertaining the molecular relationship between mutagenesis and carcinogenesis and for screening potential genotoxins.

I. COMPARISON OF MUTATION AND TRANSFORMATION SYSTEMS

Among the available in vitro mammalian cell bioassay systems, there exist but a select few which are suitable for monitoring multiple genetic (and epigenetic) endpoints associated with cytotoxicity, mutagenicity and/or carcinogenicity. Moreover, those systems which are suitable for monitoring such endpoints simultaneously in the same target cell system are even more limited. Generally mammalian cell mutation assays have been performed using established quasidiploid, functionally hemizygous target cell lines, many of which are already transformed and yield tumors in vivo. Cell systems employed for transformation assays include both diploid and aneuploid cells and, by definition, are non-tumorigenic in vivo. The properties of these various systems which make them desirable for either mutation or transformation assays are summarized in Table 1 and 2, respectively.

TABLE 1

DESIRED PROPERTIES FOR CELLS USED IN MUTATION ASSAYS

1. NEAR DIPLOID CHROMOSOME COMPLEMENT
2. HIGH CLONING EFFICIENCY
3. SHORT GENERATION TIME
4. EXPRESSION OF MUTATIONS AT $\geq$ 1 LOCUS
5. CAPACITY TO ACCEPT EXOGENOUS METABOLIC ACTIVATION SYSTEM

TABLE 2

DESIRED PROPERTIES FOR CELLS USED IN TRANSFORMATION ASSAYS

1. LOW CELL SATURATION DENSITY
2. READILY DISCERNIBLE DIFFERENCES BETWEEN NORMAL AND TRANSFORMED PHENOTYPES
3. STABLE SOURCE OF TARGET CELLS
4. LOW AND REPRODUCIBLE SPONTANEOUS TRANSFORMATION FREQUENCY
5. TRANSFORMED PHENOTYPE RELATES TO ABILITY TO FORM TUMORS IN VIVO
6. ENDOGENOUS MICROSOMAL MONOOXYGENASE ACTIVITY
7. CAPACITY TO ACCEPT EXOGENOUS METABOLIC ACTIVATION SYSTEMS

Obviously, those properties which make systems desirable as mutation assays vs. transformation assays are quite disparate. Other dissimilarities reside in the manner in which such assays are performed, the respective definitions of mutation frequency (MF) and transformation frequency (TF), and the differences in expression time and selective pressures applied. It is therefore apparent that most assay systems suitable for mutagenesis are unsuitable for transformation; however, transformation systems are available which are amenable to mutation. With the availability of such systems, a variety of basic biological questions relating to mutagenesis and carcinogenesis can be addressed, such as (a) Is transformation a mutagenic event?, (b) What is the gene target size responsible for neoplastic transformation?, (c) Are all mutagens carcinogens and/or vice versa? However, due to inherent limitations associated with both mutation and transformation assays (not the least of which is that for the most part, mutagenesis studies and transformation studies have been performed in different target cell systems), comparisons and correlations between MF and TF are difficult.

II. SIMULTANEOUS MUTAGENESIS AND TRANSFORMATION

In order to accurately correlate mutagenicity and neoplastic transformation it is desirable to investigate both endpoints in the same target cell system simultaneously. Table 3 gives examples of assay systems which have been used for such endpoints. Model systems have employed both rodent and human cells. Typical genetic loci monitored include resistance to ouabain, 6-thioguanine, and 8-azaguanine. The endpoints observed for transformation include altered growth characteristics, morphological aberrations of constituent cells, etc. For studies in which both assays were performed in the same system, TF/MF ratios (Table 3) generally ranged from 10-32 for mouse cells, ~17-100 for hamster embryo cells and are probably in the order of ~20 for human cells. Variations have been observed in mouse (15) and human (11) systems; but these discrepancies are not understood.

It is apparent that the weaknesses related to mutation and transformation assays necessitate caution when considering TF/MF ratios. In this respect, there are some very specific assumptions which must be satisfied in order to permit legitimate comparison between TF and MF. These assumptions are as follows: (a) each transformation results from a single mutation; (b) there are no artifacts or inaccuracies involved in measuring either transformation or

Table 3

SIMULTANEOUS MUTAGENESIS AND TRANSFORMATION
IN VARIOUS CELL TYPES

CELLS	LOCUS[a]	TREATMENT[b]	"TF/MF"[c]	REFERENCE
MOUSE				
C3H 10T 1/2	OUA^R	UV	~10	3
	OUA^R	BaP, N-AC-AAF	10-20	12
		AA	NO TF	15
BALB/3T3	OUA^R	BaP, MNNG	10-15	16,18
	OUA^R	UV	~10	11
M2	OUA^R	MNNG	~30	13
		PAHD	100-1000	
HAMSTER	OUA^R	BaP, BaP-7,8-DIOL	~17	10
	OUA^R, $6\text{-}TG^R$	MNNG, BaP	~100	1
RAT	OUA^R, $8\text{-}AG^R$	4-NQO	NA[d]	14
HUMAN	OUA^R	UV	<<1	11
	6-TG	PS	~20	20

[a] OUA^R = ouabain resistance; $6\text{-}TG^R$ = 6-thioguanine resistance; $8\text{-}AG^R$ = 8-azaguanine resistance.

[b] UV = ultraviolet; BaP = benzo(a)pyrene; N-AC-AAF = N-acetoxy-acetylaminofluorene; AA = alkylating agents; MNNG = N-methyl-N'-nitro-N-nitrosoguanidine; PAHD = polycyclic aromatic hydrocarbon derivatives; 4-NQO = 4-nitroquinoline-1-oxide; PS = propane sultone.

[c] TF/MF = Transformation to mutation frequency ratio.

[d] NA = not applicable.

mutation frequencies; (c) neither the transformation nor mutation genes are separately located at "hot spots;" and (d) the size of targets on the genes involved in transformations and mutations are equal.

III. PARAMETERS INFLUENCING MF AND TF

To date, among the available mammalian cell systems, in virtually no case have these assumptions been adequately evaluated. Moreover, there are situations in which these conditions have not been met, thereby tending to limit the feasibility of comparing TF to MF. Thus, there are instances in which genome target sizes may differ from that of the gene itself. For example, mutations to ouabain resistance have been shown to involve a specific subunit (97,000 daltons) of the Na^+/K^+ ATPase (3). Hence, the target size for mutation to ouabain resistance is smaller than that of the whole gene (molecular weight of 250,000 daltons) for the Na^+/K^+ ATPase. It follows, then, that any extrapolations as to the size of the target responsible for neoplastic transformation (based upon TF/MF ratios) will reflect such "misleading" factors, and will result in inaccurate conclusions.

Cross-feeding via cell-cell gap junctions (intercellular communication) has been shown to occur in certain types of cells, e.g. primary human skin fibroblasts and certain clones of BALB/3T3 mouse cells (5). Such factors can likewise influence the survival of non-mutant cells, thereby altering the expected mutation frequencies. As a result, calculations of MF, and as a consequence TF/MF likewise become inaccurate.

MF and TF can also be influenced by the chemical class employed. For example, N-methyl-N'-nitro-N-nitrosoguanidine (MNNG) and ethyl methanesulfonate more often induce ouabain resistance mutations in S49 mouse lymphoma cells, whereas ICR-191 and X-ray induce 6-thioguanine resistance mutations for the most part (4). In addition MNNG is a potent transforming agent of BALB/3T3 cells (6) but induces little or no transformation of C3H 10T 1/2 cells without proper assay modification (2,17). Thus, different chemicals induce different mutational responses (and at different rates), and may or may not transform cells, depending upon the methodologies employed. Accordingly, TF's are subject to wide variations, and the different sensitivities for different chemical mutagens exhibited by different genetic loci can yield marked variations in MF. Consequently, TF/MF ratios may reflect abnormally high or low induction of either endpoint.

Transformation frequencies can vary up to 200-fold depending upon the number of cells available to chemical treatment (7-9; Tu, A.S. and Sivak, A., personal communication; Schechtman, L.M. et al., unpublished results), and subculture regimen post-treatment (17). The resulting increase or decrease in calculated transformation frequencies will alter the TF/MF ratio accodringly.

Further influences on TF/MF ratios may reside in the differences in expression time and in the selective pressures applied toward identifying mutants and transformants. For example, expression time in a mutation assay (i.e. that time required for fixation of the genotypic lesion which will become manifest as the mutant phenotype) can be terminated by the addition of the selective agent or drug employed to select for a specific mutational marker. On the other hand, the expression of the transformed phenotype may be a continuous event until the termination of the assay. Should restrictive conditions inherent in a particular assay limit development of the transformed phenotype, TF/MF can be inaccurately low.

IV. CONCLUSIONS

From the foregoing discussion, it appears that it may be premature to attempt to relate neoplastic transformation and gene mutation. Due to inherent limitations associated with each type of assay and the failure of both mutation and transformation assays to meet certain specific requirements necessary for such correlations to be meaningful, it is even less likely that the gene target size for transformation can be defined based upon TF/MF ratios derived from currently available test systems.

Nevertheless, for practical purposes, the monitoring of multiple genotoxic endpoints concurrently in the same target cell system can be of significant value with respect to the screening and identification of potential genotoxins (6,16). Such an approach offers certain advantages, including (a) enhancing the credibility of results obtained through supportive data, (b) allowing accurate evaluation of chemical potency (e.g. discriminating potent and weak chemical mutagens/carcinogens), and (c) limiting the possibility of false negative results.

Thus, although much more work is necessary in both mutation and transformation assay systems (e.g. in terms of parallel methodologies, data handling and interpretation) their value is well recognized. Such systems are an absolute requirement for establishing and understanding the molecular mechanisms of chemically induced carcinogenesis. They also provide an invaluable tool for the identification of potentially biohazardous compounds.

REFERENCES

1. Barrett, J.C. and Ts'o, P.O.P. *Proc. Natl. Acad. Sci. USA 75*, 3297-3301 (1978)
2. Bertram, J.S. and Heidelberger, C. *Cancer Res. 34*, 526-531 (1974)
3. Chan, G. and Little, J.B. *Proc. Natl. Acad. Sci. USA 75*, 3363-3366 (1978)
4. Coffino, P. and Friedrich, U. *Proc. Natl. Acad. Sci. USA 74*, 679-683 (1977)
5. Corsaro, C.M. and Migeon, B.R. *Nature 268*, 737-739 (1977)
6. Curren, R.D., Kouri, R.E., Kim, C.M. and Schechtman, L.M. *Environment Int.* (in press) (1982)
7. Fernandez, A., Mondal, S. and Heidelberger, C. *Proc. Natl. Acad. Sci. USA 77*, 7272-7276 (1980)
8. Haber, D.A., Fox, D.A., Dynan, W.S. and Thilly, W.G. *Cancer Res. 37*, 1644-1648 (1977)
9. Haber, D.A. and Thilly, W.G. *Life Sciences 2*, 1663-1674 (1978)
10. Huberman, E., Mager, R. and Sachs, L. *Nature 264*, 360-361 (1976)
11. Kakunaga, T., Crow, J.D. and Augl, C. In: Radiation Research. S. Okada, M. Imamura, Y. Terashima and H. Yamaguchi, eds. Jap. Assoc. Rad. Res. p.589-595 (1979)
12. Landolph, J.R. and Heidelberger, C. *Proc. Natl. Acad. Sci. USA 76*, 930-934 (1979)
13. Marquardt, H. *Carcinogenesis 1*, 215-218 (1980)
14. Mishra, N.K., Pant, K.J., Wilson, C.M. and Thomas, F.O. *Nature 266*, 548-550 (1977)
15. Peterson, A.R., Landolph, J.R., Peterson, H., Spears, C.P., and Heidelberger, C. *Cancer Res. 41*, 3095-3099 (1981)
16. Schechtman, L.M., Beard, S.F., Curren, R.D. and Kouri, R.E. In: Proc. 3rd Int. Conf. Envir. Mutagens, Tokyo p. 54 (1981)
17. Schechtman, L.M., Kiss, E., McCarvill, J., Gallagher, M. Kouri, R.E. and Lubet, R.A. In: Proc. Am. Assoc. Cancer Res. (submitted) (1982)
18. Schechtman, L.M. and Kouri, R.E. In: Progress in Genetic Toxicology D. Scott, B.A. Bridges and F.H. Sobels, eds. Elsevier/North-Holland Biomedical Press, N.Y. p. 307-316 (1977)
19. Schechtman, L.M., Lubet, R.A., Kouri, R.E. Sivak, A. and Tu, A.S. In: Proc. 2nd NCI/EPA/NIOSH Collaborative Workshop: Progress on Joint Environmental and Occupa-

tional Cancer Studies (in press) (1982)
20. Silinskas, K.C., Kateley, S.A., Tower, J.E., Maher, V.M., and McCormick, J.J. *Cancer Res. 41*, 1620-1627 (1981)

Free Oxygen Radicals — Modulator of Carcinogens

Walter Troll

Institute of Environmental Medicine, New York University Medical Center, New York, New York, U.S.A.

ABSTRACT Agents inactive as mutagens in microbial systems modulate the response of active mutagenic carcinogens. These promoting agents may exert their action by the formation of free oxygen radicals. The evidence for this has come from the observation that promoting agents (e.g., phorbol-myristate-acetate and teleocidin) cause formation of superoxide anions (O_2^-). Inactive analogues (e.g., phorbol diacetate) fail to produce O_2^-. Slaga has demonstrated that organic peroxides are specific tumor promoters. This mechanism provides a possible tool for identifying promoting agents and enabling one to devise chemopreventive agents.

INTRODUCTION

The identification of carcinogens by microbial mutagenesis has become more difficult because of the discovery of nonmutagenic agents which modify the cancer response. The mechanism of action of these substances is of particular interest for measuring carcinogenic agents in our environment and devising preventive measures against their activity. The following types of compounds will be elucidated:

Tumor Promoters: These materials produce tumors by repeated application after a single subcarcinogenic exposure to a carcinogen (initiation). The tumor promoters act any time during the lifetime of the animal even if a period of time elapses between exposure of the animal to the initiator and to the promoter.

Cocarcinogens: These materials produce increased tumors when applied simultaneously with the initiating carcinogens. These include all promoters plus multihydroxylated compounds (e.g., catechol and pyrogallol).

Chemopreventive Agents: These materials prevent the action of promoters and cocarcinogens. They include protease inhibitors, anti-inflammatory cortical steroids, retinoids and antioxidants.

We are proposing that the action of tumor promoters depends on the formation of free oxygen radicals by the cell membrane. The chemopreventive agents counteract this action.

FREE RADICALS AS TUMOR PROMOTERS

Carcinogenesis is a multistage process causing biologically irreversible toxication and biologically reversible, preventable toxicity. This has been most clearly shown in two-stage skin carcinogenesis in the mouse. Berenblum and Shubik (3) clearly identified two stages: initiation caused by a variety of primary carcinogens; and promotion caused by the tissue irritant croton oil. The active principle of croton oil has been identified by Hecker and Van Duuren to be phorbol-myristate-acetate (PMA) (8, 18). The action of the promoter was blocked by synthetic protease inhibitors, leupeptin, cortical steroids and retinoids (2, 9, 14, 16). Tumor promotion may be a consequence of formation of superoxide anions caused by cell membrane perturbation of certain cells, PMNs and macrophages. This effect is counteracted by all promotion inhibitors, such as protease inhibitors, retinoids and steroids (19). Experiments supporting the role of oxygen intermediates have come from the demonstration by Slaga that benzoyl peroxide is a tumor promoter and that antioxidants block promotion (15). Ito *et al.* (10) showed that hydrogen peroxide produces duodenal tumors in mice by oral administration.

PROTEASE INHIBITORS AND OXYGEN RADICALS

The α_1-protease inhibitor (α_1-PI) may play a role in preventing cancer caused by tumor promotion, in addition to its established effect in blocking emphysema in man.

The destruction of α_1-PI by particles in cigarette smoke occurs by activating polymorphonuclear leukocytes and macrophages. The activated leukocytes and macrophages form hydrogen peroxide and together with myeloperoxidase oxidize the methionine of α_1-PI (11). The methionine in the inhibitor appears to be an essential element for inhibiting proteases (5). Inactivation of the trypsin inhibitor, observed in cigarette smokers with emphysema, may also be a necessary condition of lung cancer. The recent finding that the skin tumor promoters, PMA and teleocidin, form superoxide and hydrogen peroxide with PMNs and macrophages, which may destroy α_1-PI, has given additional support to the pivotal role of this endogenous inhibitor.

The following steps may be involved in tumor promotion:

1. Formation of O_2^- by perturbation of the cell membrane of PMNs or macrophages.

2. Formation of organic and inorganic peroxides. Synthetic benzoyl peroxide and lauryl peroxide presumably enter at this stage.

3. Oxidation of the methionine in α_1-PI by hydrogen peroxide or organic peroxide together with myeloperoxidase (5). This action will inactivate the protease inhibitor.

4. Increased protease levels due to the destruction of a major protease inhibitor causes a number of changes advantageous to the cancer cell such as increased mitosis and destruction of connective tissue.

Thus, protease inhibitors applied exogenously interact in two areas in this scheme of promotion: 1) in the formation of O_2^- which starts the process by inactivating the endogenous α_1-protease inhibitor; and 2) in replacing the endogenous inhibitor that has been destroyed by an oxidation-stable protease inhibitor. The formation of superoxide can be blocked by a variety of substances, in addition to protease inhibitors, which have been shown to inhibit tumor promotion, for example dexamethasone, retinoids and antioxidants. The protease inhibitors most effective in the first step are different from those in the second step. Thus, α_1-PI appears to be ineffective in preventing superoxide formation as is the competitive protease substrate tosylarginine methylester (TAME) (7); yet TAME has been shown to inhibit tumor promotion in mouse skin (16).

Protease inhibitors block tumorigenesis when applied

via a variety of routes. Applied topically, tumor promotion in mouse skin is inhibited by TPCK or TLCK (16). Aprotinin, a broad-spectrum protease inhibitor, administered i.p. blocks methylcholanthrene-induced squamous cell carcinomas in mice (12). Troll et $al.$ (17) demonstrated that rats fed a soybean diet were protected against x-ray induced breast tumors. Studies using diets supplemented with the Bowman-Birk or the Kunitz inhibitor demonstrated that protease activity in the small intestine was decreased only in diets supplemented with the Bowman-Birk inhibitor (6).

The occurrence of breast and colon cancers throughout the world is directly proportional to meat and fat consumption (4). Seventh-Day Adventists, a vegetarian population, have a lower incidence of breast, colon and other cancers than the general population (13). Epidemiological studies also demonstrate that high legume and cereal consumption decreases breast cancer incidence (1). Thus, it appears that not only may cancer incidence decrease if less fat (meat) is consumed but that diets rich in legumes contain protective factors, among them protease inhibitors, which may protect against some cancers.

CONCLUSION

The tentative identification of free radicals as the proximal agent of tumor promotion offers new opportunities for identifying tumor promoters and chemopreventive agents. All substances causing formation of superoxide anions by PMNs may be considered potential tumor promoters. These are the known tumor promoters, PMA and teleocidin. Other substances (e.g., Concanavalin A and calcium ionophore) also causing O_2^- generation are under investigation as tumor promoters. Agents which prevent this reaction or destroy O_2^- are potential chemopreventive agents. Clearly, much work is required to confirm these generalizations.

ACKNOWLEDGMENTS

This research was supported by U.S. Public Health Service Grant CA 16060 and is part of a New York University Medical Center Program supported by the

National Institute of Environmental Health Sciences, National Institutes of Health, Grant ES 00260.

REFERENCES

(1) Armstrong, B., and Doll, R. *Int'l. J. Cancer 15*, 617-631, 1975.
(2) Belman, S., and Troll, W. *Cancer Res. 32*, 450-454, 1972.
(3) Berenblum, I., and Shubik, P. *Toxicology and Occupational Medicine*. W.B. Deichmann (Ed.), Elsevier North Holland Press, New York (1979) pp. 243-252.
(4) Carroll, K.K. *Cancer Res. 35*, 3374-3383, 1975.
(5) Clarke, R.A., Stone, P.J., Hag, A.E., Calore, J.D., and Franzblau, C. *J. Biol. Chem. 256*, 3348-3353, 1981.
(6) Gertler, A., Birk, Y., and Bondi, A. *J. Nutrition 91*, 358-370, 1967.
(7) Goldstein, B.D., Witz, G., Amoruso, M., and Troll, W. *Biochem. Biophys. Res. Comm. 88*, 854-870, 1979.
(8) Hecker, E. *Methods of Cancer Research, Vol. 6.* H. Busch (Ed.) Academic Press, New York-London (1971) pp. 439-484.
(9) Hozumi, M., Ogawa, M., Sugimura, T., Takeuchi, T., and Umezawa, H. *Cancer Res. 32*, 1725-1729, 1972.
(10) Ito, A., Watanabe, H., Naito, M., and Naito, Y. *Gann 72*, 174-175, 1981.
(11) Janoff, A., Carp, H., Lee, D.K., and Drew, R.T. *Science 206*, 1313-1314, 1979.
(12) Ohkoshi, M. *Gann 71*, 246-250, 1980.
(13) Phillips, R.L. *Cancer Res. 35*, 3515-3522, 1975.
(14) Slaga, T.J., Fischer, S.M., Nelson, K., and Gleason, G.L. *Proc. Natl. Acad. Sci. U.S.A. 77*, 3659-3663, 1980.
(15) Slaga, T.J., Klein-Szanto, A.J.P., Triplett, L.L., Yotti, L.P., and Trosko, J.E. *Science 213*, 1023-1025, 1981.
(16) Troll, W., Klassen, A., and Janoff, A. *Science 169*, 1211-1213, 1970.
(17) Troll, W., Wiesner, R., Shellabarger, C.J., Holtzman, S., and Stone, J.P. *Carcinogenesis 1*, 469-472, 1980.
(18) Van Duuren, B.L. *Prog. Exp. Tumor Res. 11*, 31-68, 1969.

(19) Witz, G., Goldstein, B.D., Amoruso, M., Stone, D.S., and Troll, W. *Biochem. Biophys. Comm. 97*, 883-888, 1980.

Role of DNA Damage and Repair in Carcinogenesis

Taisei Nomura

Faculty of Medicine, Osaka University, Osaka, Japan

ABSTRACT Two experimental approaches were used for assessing a role of DNA damage and repair in *in vivo* carcinogenesis.
(1) Treatment of parental germ cells of mice with X-rays, 4-nitroquinoline-1-oxide (4NQO), or urethan increased the incidence of tumors in the offspring significantly higher (up to 6 times) than controls, clearly indicating that heritable tumorigenic changes are induced in the genetic materials of germ cells. The tumor incidence dramatically decreased after fractionated X-irradiation of spermatogonia and mature oocytes, as an indication of DNA repair, whereas it did not after that of the postmeiotic sperm.
(2) The incidence of lung tumors was greatly diminished with increasing doses of caffeine which was given immediately after 4NQO treatment to either Day 15 fetuses or young adults. The lung of young adult mice remained sensitive to the antineoplastic action of caffeine for more than 20 days after 4NQO treatment, suggesting that DNA damage by 4NQO remains unexcised in cancer progenitor cells (stem cells) for a long time. Stem cells may be repair-deficient, while excision repair is known to be active in cultured cells.

INTRODUCTION

A role of DNA damage and repair in carcinogenesis

has been studied mainly with microorganisms and cultured mammalian cells, but very rarely with whole animal systems. This is due to the complexity of whole body animals, which often disturbs basic study. However, there are some advantages in the whole body experiment. I designed to test the role of DNA damage and repair in *in vivo* carcinogenesis by the two independent methods. The first method is to use germ cells. If treatment of parental germ cells with known mutagenic agents can increase the incidence of cancer in the offspring produced from these cells, this clearly indicates that damage induced in DNA results in the heritable tumorigenic DNA alterations. The second is to use stem cells, which are uniquely available only in whole body animals.

DNA DAMAGE AND REPAIR IN PARENTAL GERM CELLS

To test the idea that DNA damage induced in parental germ cells by mutagenic agents can increase cancer incidence in the offspring, measurements have been made of translocations, dominant lethals, and anomalies as well as tumors in the offspring after exposure of parental ICR mice to X-rays, urethan, and 4NQO. The data were obtained from 23086 mice and fetuses (9).

Induction of Heritable Tumors by Mutagens.

The incidence of tumors in the F_1 offspring for

Table 1. Incidences of tumors in the F_1 offspring after parental exposure to X-rays and urethan.

Treatment to parent			Tumors in the offspring examined 8 months after birth		
Agents	Sex	Dose (rad or mg/g)	Incidence (%)	P	Details
X-rays	Male	36 - 504	153/1529 (10.0)	< 0.002	138 LT, 19 OC, 11 L, 1 ST.
X-rays	Female	36 - 504	101/1155 (8.7)	< 0.02	91 LT, 7 OC, 4 L.
Urethan	Male	1.5	136/1254 (10.9)	< 0.001	118 LT, 7 OC, 12 L, 1 Lip, 1 G, 1 Th.
Urethan	Female	1.5	139/963 (14.4)	<< 0.001	117 LT, 17 OC, 7 L, 1 Th, 2 LH, 1 He.
Urethan	Female	1.0	115/772 (14.9)	<< 0.001	94 LT, 13 OC, 9 L, 1 G.
Untreated		0.0	29/548 (5.3)		26 LT, 2 L, 1 OC.

* LT, lung tumor; OC, ovarian cystadenoma; L, lymphocytic leukemia; ST, stomach tumor; Lip, lipoma; G, granulosa cell tumor; Th, thyroid gland tumor; LH, liver hemangioma; He, hepatoma.

all stages induced by the exposure of male and female germ cells to X-rays and urethan are summarized in Table 1. The incidences were significantly higher than the controls. Most tumors were found in the lung (papillary adenoma).

When X-rays were given in a single dose, tumor incidence increased with increasing doses of X-rays up to 6 times higher than controls (Fig. 1). Spermatids were about 2 times more sensitive than spermatogonia. Oocytes at late follicular stages (1 to 7 days before ovulation) were resistant to 36 to 216 rad of X-rays, but very sensitive to higher doses. Maturated eggs may have high repair ability in an intermediate dose range.

The pattern of stage dependent sensitivity of germ cells for the induction of "tumor mutations" are similar to those for dominant skeletal mutations (2) and 7 specific locus mutations (11). This suggests that "heritable tumorigenic factors" induced by X-rays in germ cells are not distinguishable from accepted mutations in germ cells (2, 11).

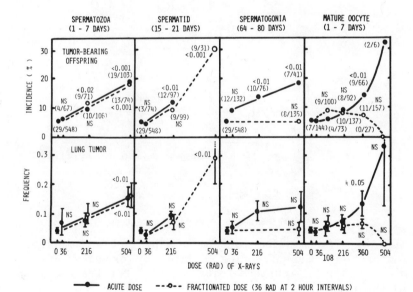

Fig. 1. Stage- and dose-related induction of tumors in the F_1 offspring after parental exposure to acute and fractionated doses of X-rays. X-rays were given at a dose rate of 72 rad/min.

To confirm that "heritable tumorigenic factors" induced by X-rays are damage induced in the genetic materials of germ cells, inheritance of X-ray induced "heritable tumorigenic factors" was tested following the scheme outlined in Fig. 2. Male mice were exposed to X-rays and then mated with untreated females. From the F_1 offspring of X-ray-treated male parents, some males and females were mated randomly, and F_2 offspring were examined for tumors. Tumors in the parents of F_2 offspring were retrospectively surveyed. The incidence of lung tumors in F_2 animals were significantly higher than controls, when either of the parents F_1 had tumors. The similar results were obtained after paternal exposure to urethan and 4NQO. Although, in the case of chemicals, we can not exclude extra-chromosomal factors which may transmit to the next generation, similar results obtained in the X-ray study strongly suggest that "heritable tumorigenic factors" are induced by mutagens in the genetic macromolecules of germ cells.

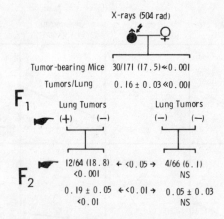

Fig. 2. Transmission pattern of "heritable tumorigenic factors" in the offspring induced by exposure of male mice to X-rays at spermatozoa stage. Details are given in the text.

Repair Capacity of Germ Cells.

For assessing a role of DNA repair in "tumor mutations", various stages of germ cells were treated with fractionated doses of X-rays. For fractionated doses, 36 rad of X-rays were given at 2 hour intervals. Results are summarized in Fig. 1. Fractionation effects of X-rays on "tumor mutations" seem to exist in spermatogonia and mature oocytes ($p<0.05$ between acute and fractionated doses at 504 rad), whereas there was no difference in the tumor incidence between acute and fractionated doses when postmeiotic stages were treated. X-ray-induced DNA damage responsible for "tumor mutations" seems to be repairable in spermatogonia and in oocytes at the late follicular stage, but not in the postmeiotic sperm. These results agree with the Russell's finding (11) that the sperm lose their repair capacity after the meiotic division.

Present observations clearly indicate a significant role of DNA damage and repair in carcinogenesis. However, we need an enormous number of animals for further analytical studies.

DNA DAMAGE AND REPAIR IN SOMATIC STEM CELLS.

The mouse lung is one of the simplest model-organ to study *in vivo* carcinogenesis (6). A single alveolus of the mouse lung is composed of only two cells. One is type 2 cell (stem cell) and the other is type 1 cell (mortal cell). Type 2 cell divides asymmetrically to produce daughter type 2 cell and mortal type 1 cell in the mature lung. Type 2 cells, cancer progenitor stem cells, are considered to appear or to be differentiated in the mouse embryo at late organogenesis stage of the lung (6), and at the rapidly growing stage, stem cells divide symmetrically so that they can increase their numbers enough to form adult organ, while at the adult age (stationary stage), stem cells divide asymmetrically only to maintain the organ in the normal condition. Using the chracteristics of the fetal and adult lung of the mouse, a role of DNA damage and repair in somatic stem cells is tested (7, 8).

Timing of Chemically Induced Neoplasia in Stem Cells.

When and how DNA damage induced by carcinogens are fixed in stem cells? Is it the same as mutations in *E. coli*? Such study may be possible by using the antineoplastic action of caffeine. As a carcinogen, 4NQO is most suitable, because the nature of 4NQO-induced DNA lesions and their relationship to mutation

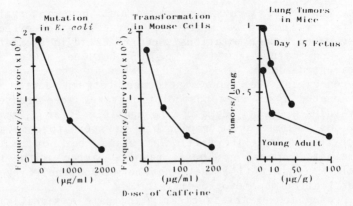

Fig. 3. Inhibiting effects of caffeine on 4NQO-induced mutagenesis, transformation, and lung tumorigenesis. Excisionless strain of *E. coli* and cultured mouse cells were treated with caffeine immediately after 4NQO treatment (3, 4). Young adult mice and mouse fetuses were treated with caffeine at 6 hr intervals for 36 hr immediately after 4NQO treatment (7, 8).

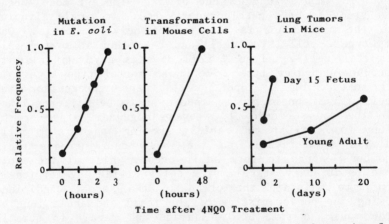

Fig. 4. Changes in antimutagenic, antitransformational, and antitumorigenic effects of caffeine with time after 4NQO treatment. The ratio of mutants, transformants, and lung-tumor nodules in the caffeine treated groups to those in caffeine-less groups was plotted against the time starting caffeine treatment. Caffeine was given at 6 hr intervals for 36 hr to mouse fetuses and for 120 hr to young adult mice. Data are from Kondo (4), Kakunaga (3), and Nomura (7, 8).

and transformation are well studied at the molecular level (5). 4NQO-purine adducts seem to be a major cause of mutation in *E. coli* (4), and transformation in cultured mouse cells (3), both of which are inhibited by caffeine (Fig. 3), seemingly due to the suppression of inducible error-prone repair of 4NQO-induced DNA lesions.

When caffeine was given to ICR mice after 4NQO treatment, yields of tumor nodules in the lung were greatly reduced with increasing doses of caffeine, to a maximum of about 20% of the 4NQO alone controls (Figs. 3). The lung of Day 15 mouse fetuses treated with 4NQO was sensitive to the antineoplastic action of caffeine for about 1.5 days, while that of young adult females (21 days old) was still sensitive to the antineoplastic action of caffeine even 21 days after 4NQO treatment (Fig. 4). The difference in the caffeine sensitive period parallels the difference in the generation time of stem cells in the lung between Day 15 fetuses (1.4 days) and young adults (20 to 25 days). Antineoplastic action of caffeine on 4NQO-initiated transformation in cultured mouse cells (3), as well as antimutagenic action on 4NQO-initiated mutagenesis in *E. coli* (5), is also limited mostly to the first post-4NQO cell division, as in the case of 4NQO-initiated lung tumorigenesis in mouse fetuses (Fig. 4). Thus, it seems to be an important conclusion that 4NQO-induced pretumorigenic damage in the lung of mouse fetuses, pretransformational damage in cultured mouse cells, and premutational damage in *E. coli* are similarly fixed during the first post-4NQO DNA replication period.

Not only in the mouse lung, caffeine post-treatment also suppressed tumors in the rat pancreas induced by 4-hydroxy-aminoquinoline-1-oxide (4HAQO), an active form of 4NQO, when 4HAQO was given after partial pancreaticotomy (1).

Contrary to the lung tumors induced by 4NQO, those by alkylating agents (ethylnitrosourea, methylnitrosourea, and N-hydroxyurethan) were not reduced by caffeine post-treatments. However, this was the case in the mutagenesis in *E. coli* (Kondo, S., Ryo, H., and Ishidate, K. personal communication), because caffeine does not affect mispairing activity. Consequently, it seems to be another important conclusion that caffeine effects on neoplasia depend on the kinds of treated carcinogens, probably kinds of DNA damage and repair.

Are Cancer-prone Stem Cells Repairless?

One problem remains unsolved. The lung of young adult mice was still sensitive to caffeine even 21 days after 4NQO treatment (Fig. 4). If antineoplastic action of caffeine is caused by the suppression of error-prone post-replication repair of 4NQO-induced DNA damage, as reported for 4NQO-induced mutations in *E. coli* (5), 4NQO-induced pretumorigenic damage in DNA of the stem cells in the lung must remain unexcised for more than 20 days. However, cultured mouse cells possess excision repair capacity for 4NQO-induced DNA damage (5). This contradiction is explained by supposing that cancer-prone stem cells in the lung is repairless, while cells in other components of the lung show efficient repair ability. This proposal lacks direct evidence, but recently, Potten (10) has reported that the stem cells in the crypts of small intestine are hypersensitive to radiation killing, suggesting that they may lack some kinds of repair capacities.

An alternative explanation is that antineoplastic action of caffeine is due to the selective killing of 4NQO-damaged cancer-susceptible stem cells by caffeine. This mysterious finding is open for future research.
(supported by the Ministry of Education, Science, and Culture, The Asahi Science and Art Promotion Foundation, Princes Takamatsu Cancer Research Foundation, Nissan Scientific Foundation, and Prix Isabella Decazes de Noüe)

References
(1) Denda, A., and Konishi, Y. submitted to *Carcinogenesis*.
(2) Ehling, U. H. *Genetics 63*, 1381-1389 (1966).
(3) Kakunaga, T. *Nature 258*, 248-250 (1975).
(4) Kondo, S. *Br. J. Cancer 35*, 595-601 (1977).
(5) Kondo, S. in *Carcinogenesis Vol. 6* (ed., T. Sugimura), pp. 47-64, Raven Press, New York (1981).
(6) Nomura, T. *Cancer Res. 34*, 3363-3372 (1974).
(7) Nomura, T. *Cancer Res. 40*, 1332-1340 (1980).
(8) Nomura, T. *Nature 260*, 547-549 (1976).
(9) Nomura, T. in *Tumors of Early Life in Man and Animals* (ed. L. Severi) pp. 873-891, Perugia Univ. Press, Perugia (1978).
(10) Potten, C. S. *Nature 269*, 518-521 (1977).
(11) Russell, W. L. in *Repair from Genetic Radiation* (ed. F. H. Sobels) pp. 205-217, Pergamon Press, Oxford (1963).

Increased Cancer Incidence in the Progeny of Male Rats Exposed to Ethylnitrosourea before Mating[1]

L. Tomatis, J. R. P. Cabral, A. J. Likhachev, and V. Ponomarkov*

*International Agency for Research on Cancer, Lyon, France, and *Cancer Research Centre, Academy of Medical Sciences of the USSR, Moscow, U.S.S.R.*

ABSTRACT The possible role of prezygotic events in determining cancer risk was investigated in the progeny of male rats treated with ethylnitrosourea (ENU) before mating with untreated females. Eight BDVI male rats were given a single i.p. dose of 80 mg/kg bw ENU and each rat was then caged at weeks 1, 2, 3 and 4 after treatment with three untreated females. Fertility was lower and preweaning mortality higher in the experimental group, as compared to controls, particularly at the 4th-week mating. Survival rates after weaning were similar in the progeny of treated males and controls, as was the total incidence of tumours. However, analysis of tumour incidence at the various organ sites showed an increased incidence of neurogenic tumours in the progeny of ENU-treated males, as compared to that of controls.

INTRODUCTION

Results obtained in previous experiments indicate that the exposure of rodents to chemical carcinogens during pregnancy may result not only in a high incidence of tumours in the progeny of the first generation, but also in an increased tumour incidence in those of subsequent untreated generations (21-23). In the preliminary experiment reported here, we investigated the possible

[1] This article, reproduced here with minor modifications, has already appeared in Int. J. Cancer, 28, 475-478, 1981

role of prezygotic events in determining cancer risks, using a different experimental model, namely, treatment of male rats with ethylnitrosourea (ENU) before mating with untreated females.

MATERIALS AND METHODS

An inbred BDVI rat strain, obtained in 1970 from Professor Druckrey and maintained by strict brother-to-sister mating in the IARC animal house, was used. The animals were kept in Makrolon no. III cages and fed a pelleted diet (Aliment Extralabo Biscuits, Pietrement, France) and water *ad libitum*. ENU (Schuchardt, Munich, Federal Republic of Germany) was prepared as a 0.8% solution in buffered saline (pH 6) within half an hour of its use, and a single i.p. dose of 80 mg/kg bw was given to eight male rats aged 9 weeks.

At weeks 1, 2, 3 and 4 after treatment, each of the treated males was caged with three untreated females for three consecutive nights. The females were controlled for conception every morning, and those with positive vaginal smears were caged separately; only those with positive vaginal smears were kept under observation for a long period.

In order to compare breeding records and preweaning mortality, eight males treated with buffered saline were each mated for three consecutive nights with two untreated females at one-week intervals. By error, at the 4th-week mating, the males were caged with the females for one night only; this resulted in a lower rate of pregnancies in that week.

Experimental and control rats were killed only when moribund; survivors were sacrificed at 120 weeks of age. All animals were necropsied, and the brain, hypophysis, spinal cord, lung, thymus, liver, kidney and adrenal glands, spleen and all other organs showing macroscopic lesions were fixed in 10% buffered formalin and examined histologically.

RESULTS

Breeding records, fertility and fecundity records, as well as data on preweaning mortality are summarized in Table 1. The fertility of females mated with treated males was slightly lower than that of females mated with control males for all mating intervals, except at the 1st-week mating. The proportion of females with a positive vaginal smear that actually delivered was lower than that of controls for females fertilized at the 1st and 4th weeks; the average litter size was the same as in controls for females mated at the 1st and 2nd weeks, but decreased considerably for the 3rd- and 4th-week matings. Preweaning mortality followed the same pattern, being similar to controls for offspring of 1st- and 2nd-week matings and progressively higher for those of the 3rd and 4th weeks.

Table 1. Breeding Records of ENU-Treated and Control BDVI Male Rats Mated With Untreated Females, And Preweaning Mortality Among Their Progeny

NO. OF MALES	MATING INTERVAL (WEEKS)*	NO. OF FEMALES MATED	NO. OF FEMALES WITH A POSITIVE VAGINAL SMEAR	NO. OF FEMALES THAT DELIVERED	NO. OF OFFSPRING AT BIRTH	AVERAGE NO. OF OFFSPRING IN LITTER	NO. (%) OF OFFSPRING DEAD BEFORE WEANING
EXPERIMENTAL GROUP (80 MG/KG BW ENU I.P.)							
	1	24	12	9	62	6.9	1 (1.6)
	2	24	8	8	55	6.9	0 -
	3	24	10	9	52	5.8	6 (11.5)
	4	24	8	5	22	4.4	5 (22.7)
CONTROL GROUP							
	1	16	7	6	44	7.3	1 (2.2)
	2	16	8	7	52	7.4	2 (3.8)
	3	16	8	8	63	7.8	1 (1.05)
	4	16	3**	3	26	8.6	1 (3.4)

* FOR TREATED MALES, WEEK(S) AFTER TREATMENT. ** ONLY ONE NIGHT'S MATING, INSTEAD OF THREE

Survival rates after weaning were similar in the progeny of treated males in all groups and in that of controls: over 50% of the animals were still alive at 100 weeks of age (Table 2). The lifespan of the eight treated male rats was considerably shorter, and none of these animals survived beyond 98 weeks.

Table 2. Survival Rates of the Progeny of ENU-Treated BDVI Male Rats Mated 1-4 Weeks After Treatment With Untreated Females, And of Controls

GROUP	SEX	TIME IN WEEKS													
		0	5	10	20	30	40	50	60	70	80	90	100	110	115
PROGENY OF 1ST-WEEK MATING	F	32	31	31	31	31	31	31	31	30	28	24	22	14	13
	M	29	29	29	29	29	29	29	29	26	26	26	24	22	21
PROGENY OF 2ND-WEEK MATING	F	30	30	30	30	30	30	30	30	29	26	26	22	18	17
	M	25	25	25	25	25	25	25	24	23	21	20	18	13	12
PROGENY OF 3RD-WEEK MATING	F	25	25	25	25	25	24	24	24	22	19	18	17	14	12
	M	21	21	21	21	21	21	21	21	20	20	20	17	15	15
PROGENY OF 4TH-WEEK MATING	F	6	6	6	6	6	6	6	6	6	6	6	6	3	2
	M	11	11	11	11	11	11	11	11	10	10	9	7	7	7
CONTROLS	F	71	71	71	71	71	71	71	70	68	65	60	49	37	33
	M	81	81	81	81	78	75	73	73	73	71	65	59	51	43

Of the eight treated males, five had one or multiple nervous-tissue tumours, one had a haemangioma of the spleen and two died with no tumours. Data on tumour incidence in their progeny and in controls are summarized in Table 3.

The total incidence of tumour-bearing animals was similar in the progeny of ENU-treated males and in that of their controls: 42.2% of treated females and 28% of treated males and 42% and 32% of controls, respectively. However, an analysis of tumour incidences at various organ sites showed an increased incidence of nervous-tissue tumours in the progeny of ENU-treated males as compared with that in controls: 10 nervous-tissue tumours occurred in 179 progeny (5.5%) of ENU-treated male rats and 3 (1.9%) in the 152 controls. This difference was of borderline significance ($p = 0.08$). The incidence of nervous-tissue tumours varied among the groups of progeny from different weekly mating intervals and was highests in the group obtained at the 2nd-week mating. The incidence in this group - 5 nervous-tissue tumours in a total of 55 rats (9.9% in males and females combined) was significantly

Table 3. Tumour Incidence in Progeny of BDVI Male Rats Mated At 1-4-Week Intervals With Untreated Females, And in Controls

GROUP	SEX	NO. AT START	NO. OF TBA*	NO. OF TBA>1**	NO. OF ANIMALS WITH TUMOURS				
					CENTRAL AND PERIPHERAL NERVOUS TISSUE	LYMPHO-RETICULAR	MAMMARY	PITUITARY	OTHER
PROGENY OF 1ST-WEEK MATING	F	32	12	2	1 (BRAIN GLIOMA)	1	7	3	1
	M	29	5	1	1 (SPINAL CORD NEURINOMA)	2	0	0	3
PROGENY OF 2ND-WEEK MATING	F	30	16	5	2 (PLEXUS BRACHIALIS NEURINOMA; HEART NEURINOMA)	2	7	5	5
	M	25	6	0	3 (BRAIN EPENDYMOMA; SCIATIC NERVE NEURINOMA; VAGUS NERVE NEURINOMA)	1	0	0	2
PROGENY OF 3RD-WEEK MATING	F	25	7	0	1 (SCIATIC NERVE NEURINOMA)	0	5	0	1
	M	21	8	0	1 (SCIATIC NERVE NEURINOMA)	2	0	0	5
PROGENY OF 4TH-WEEK MATING	F	6	3	1	0	0	2	0	2
	M	11	2	0	1 (BRAIN GLIOMA)	0	0	0	1
CONTROLS	F	71	30	2	0	0	16	4	12
	M	81	27	2	3 (2 HEART NEURINOMAS; 1 TAIL NEURINOMA)	8	0	3	14

* TUMOUR-BEARING ANIMALS; ** ANIMALS WITH MORE THAN ONE TUMOUR

increased compared with that observed in control rats (1.9% of males) (p = 0.018). (In males only, the incidence was 12% in the treated group and 3% in the controls).

The nervous-tissue tumours in the progeny of ENU-treated males occurred in the brain (2 gliomas and 1 ependymoma), in the spinal cord (1), in the sciatic nerve (3), in the plexus brachialis (1), in the vagus nerve (1) and in the heart (1). Of the 3 nervous-tissue tumours observed in the controls, one was a neurinoma of the tail and 2 were neurinomas of the heart. All neurogenic tumours were malignant, according to histological criteria; they were seen in animals that died or were killed at or after 100 weeks of age, except for a heart neurinoma that was found in a female in the experimental group which died at 71 weeks of age. No significant differences were observed in tumour incidences at other organ sites.

DISCUSSION

After the first report by Druckrey et al. (3) of the teratogenicity and carcinogenicity of ENU in rats following prenatal exposure, this compound has become one of the most widely used chemicals in experimental carcinogenesis. ENU is a direct alkylating agent with a pronounced organ specificity for the nervous tissue of rats. It has been found to be mutagenic in a variety of tests, including those with *Salmonella typhimurium*, *Escherichia coli* and *Drosophila melanogaster*, and produces chromatid and chromosomal aberrations in cultured human cells (8).

ENU has been reported to be the most effective known chemical in producing locus mutations in mice (19), and the hypothesis has been advanced that the high mutagenic activity of ENU is related to the high rate of alkylation at the O^6, as compared with the N7, position of guanine (12,19). It is not yet known whether the rates of alkylation at the various positions of other DNA bases, as well as the rates of loss of the different DNA adducts, is comparable in spermatogenic cells to those which have been reported for somatic cells (5,14,17).

The results of the experiment reported here indicate that the mating of male rats with untreated females at various intervals after i.p. administration of 80 mg/kg bw ENU resulted in an increase in nervous-tissue tumours in the progeny.

Central and peripheral nervous-tissue tumours, although rare, do occur in untreated animals: in an earlier experiment, the incidence of such tumours was 1.1% in rats of the same strain as that used in the present experiment; and the incidence recorded in various other laboratories varies from 0.8% to 1.2% (24).

The present findings point to a possible role of prezygotic events in determining an increased cancer risk in successive generations. They also appear to be in keeping with those obtained previously in this laboratory (22-24) and in others (15,16; for a review see also 21), indicating the persistence of an increased cancer risk in subsequent, otherwise untreated generations following prenatal exposure to a chemical carcinogen.

Additional evidence that toxic effects of mutagens can be transmitted through the male germ line is provided by recent results that indicate that the untreated progeny of cyclophosphamide-treated male rats mated with untreated females exhibit behavioural deficits (1).

Only a limited, and insufficient number of studies are available at present to assess the significance of these findings for the human situation. A possible association between adverse effects observed in progeny and exposure of the father to a carcinogen and/or mutagen prior to conception has been documented in some reports (2,4,7,9-11,18), but not confirmed by others (6,13,20,25).

ACKNOWLEDGEMENTS

The authors wish to thank Drs R. Montesano and H. Bartsch for their comments, Mrs B. Euzeby and Miss M. Laval for their technical help, Mrs E. Heseltine for editing and Miss J. Mitchell for typing the manuscript.

REFERENCES

(1) Adams, P.M., Fabricant, J.A. and Legator, M.S. *Science 211*, 80-82 (1981)
(2) Corbett, T.H., Cornell, R.G., Endres, J.L. and Leiding, K. *Anesthesiology 41*, 341-344 (1974)
(3) Druckrey, H., Ivankovic, S. and Preussmann, R. *Nature (Lond.) 210*, 1378-1379 (1966)
(4) Fabia, J. and Thuy, T.D. *Br. J. Prev. Soc. Med. 28*, 98-100 (1974)
(5) Goth, R. and Rajewsky, M.F. *Proc. Natl. Acad. Sci. USA 71*, 639-643 (1974)
(6) Hakulinen, T., Salonen, T. and Teppo, L. *Br. J. Prev. Soc. Med. 30*, 138-140 (1976)
(7) Hemminki, K., Saloniemi, I., Salonen, T., Partanen, T. and Vainio, H. *J. Epidemiol. Community Health 35*, 11-15 (1981)
(8) IARC. *IARC Monographs on the Evaluation of the Carcinogenic Risk of Chemicals to Humans 17*, International Agency for Research on Cancer, Lyon (1978)

(9) Infante, P.F., Wagoner, J.K., McMichael, A.J., Waxweiler, R.J. and Falk, H. *Lancet i*, 734-735 (1976)
(10) Infante, P.F., McMichael, A.J., Wagoner, J.K., Waxweiler, R.J. and Falk, H. *Lancet i*, 1289-1290 (1976)
(11) Kantor, A.F., McCrea Gurnen, M.G., Meigs, J.S. and Flannery, J.T. *J. Epidemiol. Community Health 33*, 253-256 (1979)
(12) Lawley, P.D. *Mutat. Res. 23*, 283-285 (1974)
(13) Li, F.P., Fine, W., Jaffe, N., Holmes, G.E. and Holmes, F.F. *J. Natl. Cancer Inst. 62*, 1193-1197 (1979)
(14) Montesano, R., Pegg, A.E. and Margison, G.P. *J. Toxicol. Environ. Health 6*, 1001-1008 (1980)
(15) Nomura, T. In: *Tumours of Early Life in Man and Animals*, ed. L. Severi, Perugia Quadrennial International Conferences in Cancer, Perugia, pp. 873-891 (1977)
(16) Nomura, T. *Cancer Res. 35*, 264-266 (1975)
(17) Pegg, A.E. *Adv. Cancer Res. 25*, 195-269 (1977)
(18) Peters, J.M., and Preston-Martin, S.Yu.M.C. *Science 213*, 236-237 (1981)
(19) Russel, W.L., Kelly, E.M., Hunsicher, P.R., Banham, J.W., Maddur, S.C. and Phipps, E.C. *Proc. Natl. Acad. Sci. USA 76*, 5818-5819 (1979)
(20) Sew-Leong, K.W.A. and Fine, J. *J. Occup. Med. 22*, 792-794 (1980)
(21) Tomatis, L. *Natl. Cancer Inst. Monogr. 51*, 159-184 (1979)
(22) Tomatis, L. and Goodall, C.M. *Int. J. Cancer 4*, 219-225 (1969)
(23) Tomatis, L., Hilfrich, J. and Turusov, V. *Int. J. Cancer 15*, 385-390 (1975)
(24) Tomatis, L., Ponomarkov, V. and Turusov, V. *Int. J. Cancer 19*, 240-248 (1977)
(25) Zack, M., Cannon, S., Loyb, D., Heath, C.W.Jr., Falletta, J.M., Jones, B., Housworth, J. and Crowley, S. *Amer. J. Epidemiol. 111*, 329-336 (1980)

Toxicology of Environmental Mutagens

Introduction to the Symposium on Toxicology of Environmental Mutagens

René Truhaut

Member of French Academy of Sciences and French Academy of Medicine, Professor Emeritus of Toxicology at University René Descartes, Paris, France

First of all, I want to express my deep gratitude to the Japanese Organizing Committee of this third international Conference on environmental mutagens and especially to the President, my eminent colleague and friend Professor Sugimura, for inviting me to chair this Symposium in cooperation with my good friend Professor Kawashi. It is for me, at the same time, a great honour and a deep pleasure, because it is a nice opportunity to meet again colleagues of whom I appreciate the contribution to the advancement of our scientific knowledge and also the friendship and courtesy.

I. GENERAL REMARKS ON THE SUBJECT OF THE SYMPOSIUM "TOXICOLOGY OF ENVIRONMENTAL MUTAGENS"

I will deal successively with two points:
A. General remarks on toxicological evaluation of environmental chemicals.
B. Specific considerations on mutagenic effect.

A. <u>General remarks on toxicological evaluation of environmental chemicals.</u> One of the main characterial of our time is the great development of chemical industry and the use of an increasing number of <u>chemicals</u> in many fields of human activity:
1. drugs in therapeutics
2. pesticides in agriculture and in programs of

public health in the fight against vectors of infectious diseases
3. additives to human food as well as to animal food
4. cosmetics
5. industrial chemicals for many purposes, which, in this way, are added to natural chemicals sometimes, very toxic.

As a consequence, there is an increasing chemical pollution of our environment in the wide sense (soil, waters, food, air, biota ...) and, as a result, exposure of man and other living organisms being in dynamic equilibrium with their milieu and between themselves. In this regard, one must not forget that every chemical, natural or man made, has potential toxicity pending on the exposure dose. For these reasons, the chemical age poses problems of harmful, acute, short-term and overall, insidious long-term effects on health of man and of noxious effects on other living organismes (plants, bacteria and animals, with, very often, impact on man.

The task of the toxicologist is to study these effects with the ultimate objective of preventing them, obviously in ensuring safety in use. This is what I called "Salvi-toxicology". One can not prevent dangers one doesn't know and, as a consequence, the best way of prevention is to disclose risks by adequate methods relying on many scientific disciplines: biological sciences of course (Pathology, Histology, Biochemistry, Genetic, and Immunotoxicology ...) and even non biological sciences: (Chemistry, Physic, Mathematics applied to statistical evaluation and to risk assessment).

Toxicology is typically a multidisciplinary science. There are 3 stages in toxicological evaluation:
1. Collection of facts
a) from animal experimentation and diverse in vitro systems. The main limiting factor is uncertainty of extrapolation to man, because of big differences in sensitivity to the toxic actions of chemicals.
b) Observations in exposed human subjects by epidemiological studies. Because of the importance of long-term effects, the main limiting factor is the length of life in man compared with experimental animals.

2. **Interpretation of data.** Paradoxically, with the improvement of methodological approaches in toxicological evaluation, it becomes more and more difficult to interpret the data, because their toxicological significance is not easy to establish. The interpretation of the data should be done in an <u>integrated</u> context, that is in conjunction with each other. For this reason, I don't like to apply rigid schemes of routine testing procedures and prefer to take into account:
- <u>physico chemical panorama</u> of the substance under test
- <u>conditions of use</u> and, as a consequence, <u>of exposure</u>
- <u>first results of preliminary toxicity studies</u>
- <u>biochemical behaviour</u> in the organism of man, as well as fate in the physical (abiotic) and biological environment,

and to orientate intelligently exploration in depth of effects on certain organs or tissues, on certain sub-cellular targets such as endoplasmic reticulums, metochondria lysosomes, membranes and on <u>biochemical targets</u>, mainly enzymatic systems.

One should not forget the fundamental importance of the dose-effect relationship. This is a golden rule in pharmacoloy and toxicology. Unfortunately, at present, in our Society, when something is found to be toxic at very high doses, there is a tendancy to think that it is also toxic even at very low doses.

In these conditions, it is not <u>the dose which makes the poison</u> ("Dosis facit venenum" said Paracelsus about five centuries ago), but, because of the increasing sensitivity of analytical methods, it is <u>the analyst who makes the poison</u>.

This is completely wrong, in my opinion, and, in interpreting the toxicological findings, one should exert a <u>value judgment</u>. Each substance should be evaluated on its own merit.

3. <u>Action to be taken</u>, with the aim of achieving the protection of public health and welfare.

The toxicologist must be present, but should not be alone. Specialists of the interested parties, including consumers, should be present to indicate, in applying the concept of <u>benefits versus risks</u>, the benefits for the collectivity as a whole. I will not elaborate on that which was stressed and examined critically by many national and international Expert Committees.

B. **Specific considerations on mutagenic effect**. My last general remark will apply to this specific toxic effect, that is production of mutation. Mutation refers to those changes in the genetic material of somatic or germ cells brought about spontaneously or by chemicals or radiations whereby their successors differ in a permanent and heritable way from their predecessors.

The detection of chemicals that may be potential human mutagens has recently been a rapidly expanding field. This is because, apart from the eventually serious implications for future generations if supplementary genetic diseases are added to the far from negligable current burden, there is evidence that somatic cell mutation, as opposed to gene cell mutation, may be associated with the development of certain types of cancer and, according to many authors, to embryotoxic effects including some teratogenic effects and transplacental carcinogenesis.

Damage of the genetic apparatus may be at the level of individual genes (gene mutation) or the interference may be a grosser type, in which the structure of the chromosome (structural chromosomal aberrations) or their number (numerical chromosomal aberrations) is altered.

I don't want to overlapp with what would be said by my colleagues who will present their views after my introduction, but I hope they will forgive me, if I stress some points relating to the study of mutagenic potential of environmental chemicals.

1. Mutagenic effects are to be classified among the most severe toxic effects, because they are generally irreversible. Disclosing of mutagenic effects in the broad sense is therefore of a central importance in toxicological evaluation of chemicals. In the two or three recent decades, the knowledge of the ability of certain chemicals to alter certain genetic materials (genotoxicity) has constituted a corner stone for the experimental development of modern genetics.

2. It is clearly <u>impractical</u> for any <u>routine testing</u> to cover the <u>entire spectrum of genetic toxicity tests</u> currently available. To have significant results, it is necessary to apply a <u>battery of tests</u>, in view to study both gene and chromosome damage and to complete in a sequential approach testing in <u>microorganisms</u>, mainly bacteria, with and without mammalian metabolism activation, by testing on systems which utilizes higher eucaryotes, preferably mammals, both <u>in vitro</u> and <u>in vivo</u>.

3. Results of mutagenicity testing may differ, <u>qualitatively and quantitatively</u>, widely, pending of the type of tests applied and, in the case of <u>in vivo</u> testing, of the kind of species or strains used. This constitutes a very important limiting factor for extrapolation to man and confers a special value to studies made on human cells, such as studies of chromosomal aberrations in circulating human blood lymphocytes.

4. Interpretation of the results should, <u>as a consequence</u>, be made very carefully, in conjonction with other parameters, such as the structure of the compound under test, its metabolic behaviour and its eventual association with other factors, such as endogenous chemicals, as well as man made or natural substances which could lead to antagonistic, additive or potentiating effects.

5. In regard of the predictive value of positive results of mutagenicity testing for carcinogenic potential, I leave the presentation of the problems to my eminent colleague and friend, Professor Schmähl. I want only to say that, in my opinion, positive results constitute <u>warning signs</u> in a preliminary screening. They should serve as criteria for establishing priorities for exploration in depth of the mutagenic potential and for a classical longterm testing on carcinogenicity.

One should not forget, in this regard, that there are several categories of carcinogens:

a) those which act, directly or after metabolic activation, on certain sites of macromolecules, mainly DNA, and are, consequently, genotoxic, and

b) those which are active by other mechanisms, generally by a two stage process and which are not genotoxic. I called them "secondary carcinogens".

They are generally called "epigenetic carcinogens". John Weisburger will elaborate on this important issue.

6. In a review written in 1975 in Mutation research, Mrs. Auerbach wrote: "The estimation of the risk of genetic hazards is a dangerous procedure, because it will create the impression that our conclusions are meaningfull, whereas, in reality, they are full of uncertainties".
I hope not to be provocative in sharing this opinion and emphasizing the still tenuous nature of current procedures for the precise estimation of hazards from genotoxic substances to man.
This means that, in spite of the usefulness, as research tools, of mutagenicity tests, there is a need for further research in this very important field of toxicology, especially on the mechanisms of action and dose-effect relationships. In this field, as in many others, Scientists are speciacialists who have some gaps in their ignorance ("Un savant est un specialiste qui a quelques lacunes dans son ignorance") and they must make every effort to increase their number.

II. ORGANIZATION OF THE SYMPOSIUM

In full agreement with the co-chairman of this Symposium, my good friend Professor Kawashi, we selected as speakers for this Symposium four famous scientists well known in the field over the world.

They will speak in the following order:

1. My countryman <u>Professor Chouroulinkov</u>, Director of Research Department on Chemical Carcinogenesis in Institut de recherches scientifiques sur le cancer CNRS, Villejuif, France, will present his views on

    "Place of mutagenicity tests in toxicological evaluation of chemicals likely to be present in the environment".

2. **Professor Loprieno**, the famous italien geneti-
cian from Istituto di Biochimica, Biofisica e
Genetica, University of Pisa, will present a
review on

"Mutagenic hazard and genetic risk evaluation
of environmental chemical mutagenic substances".

3. **Dr. John Weisburger**, famous specialist in chemical carcinogenesis, who had a fruitful carrier
in U.S. National Cancer Institute and who is,
at present, conducting an efficiant research
team in Institute for disease prevention of
American Health Foundation, Valhalla, New York
State, will present, in cooperation with a very
well known scientist at international scale,
Dr. Gary M. Williams, a review on

"Classification of carcinogens as genotoxic
and epigenetic as basis for improved toxicologic bioassay methods".

4. An other internationally famous specialist in
chemical carcinogenesis, my eminent friend,
Dietrich Schmähl, Director of the Institute of
Toxicology and Chemotherapy of the German Cancer
Research Center, will present, in cooperation
with his collaborator Louise Beatrice Pool,
views on

"The predictive value of short term mutagenicity
testing to determine the carcinogenic potential
of compounds".

Time allocated for presentation of each
speaker is __20 minutes__ followed by __5 minutes__ of
discussion.

At the end of their presentations, I thought
and Dr. Kawashi agreed, that it would be desirable
to offer to the audience a general panel discussion
on Toxicology of environmental mutagens and avenues
of research to be explored, __if there is enough time
available before 11 H 30__.

Finally, Professor Kawashi will conclude.

Thank you for your attention.

Place of Mutagenicity Tests in Toxicological Evaluation of Chemicals Likely to Be Present in the Environment

Ivan Chouroulinkov

Institut de Recherches Scientifiques sur le Cancer, Villejuif, France

INTRODUCTION

The topic on which I have been asked to speak seems to have been solved already and this presentation, if I may say so, unnecessary. I had myself this impression till the moment when I was confronted by the real problems concerning chemical mutagenesis and carcinogenesis : significance and interpretation of the results, extrapolation and decision. The subject was inviting but fantastic and hazardous. Fantastic because it includes all life science knowledge, and it is hazardous to treat such a subject in limited time and limited space. A brief recall of Toxicology and mutagenesis definitions can give an idea of the complexity of the problem. *Toxicology* is the science dealing with poisons or toxic substances, namely with their origin (industrial or natural), their physico-chemical and biological properties, their modes and mechanisms of action, their detection and dosage, and, with the elaboration of prophylactic and preventive measures. In complement, a chemical is called toxic when after its penetration into the organism, by any route, whether in one high dose (or closely repeated doses), or in several low doses, it induces immediate or delayed, transitory or lasting alterations of one or several functions which may lead to death. So, the toxicology is an experimental multi-disciplinary science at the cross roads between many fundamental disciplines such as physico-chemistry and biology. Therefore, it should borrow their methods, it should take into acount their progress, and apply their findings to the study of its own problems. Finally the Toxicology is also a science of action with ecological and social characters, one of the essential objectives being health protection (1).

Mutagenesis is a biological phenomenon during which chemicals interact with the genetic material of the cells (DNA, chromosomes) inducing qualitative and quantitative changes which can be followed by immediate and heritable modifications (mutations) of the genotype and the phenotype. Consequently a chemical is mutagenic or mutagen when, directly or indirectly (after metabolic activation), it induces mutations or increases significantly the mutation rates above the spontaneous background level.

Obviously, mutagenesis as a discipline studying mutations in relation with the character transmission, is an integral part of genetics. On the other hand, when mutagens and health risks are concerned, it belongs to toxicology. Moreover the mutagens being poisons, toxic chemicals with genetic effect, their study is therefore assumed by general toxicology as was indicated in the definition.

TOXICOLOGICAL ASPECT OF MUTAGENESIS

The approaches and the methods for mutagenic potential of chemicals are the same applied in toxicology for acute and chronic toxicity evaluations. The only difference is that in mutagenesis the target molecule is known and some mechanisms are elucidated. DNA being the target molecule, all organisms having DNA are susceptible to be used for mutagenicity studies. A large perspective for research was opened. Dealing with toxicology the presentation is in concern with mammals like mice, rats, hamsters, species accepted and used in toxicological evaluation. Even though different, these species are more relevant for extrapolation to man.

Table 1 shows schematically the ways through which the mutagens can arrive in contact with the target cells and the factors which can play a role on the final response. From this scheme we can see that contrary to the somatic (epidermal, lung, etc ...) cells, the germinal cells can not be exposed directly. The chemicals should borrow one of the three mentioned routes, pass through different organs and/or cells before arrival in the germinal cell. On the way they can be submitted to metabolic detoxification or metabolic activation.

TABLE 1. Toxicological aspect of mutagenesis (Scheme)

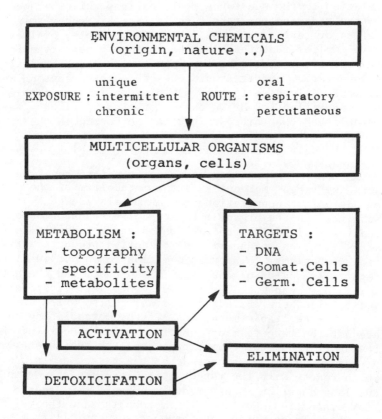

The exposure, closely related to the dose problem can be unique, intermittent or continuous as, respectively, in acute, subacute and chronic toxicity exposures. The unique exposure is the most frequently used in mutagenicity assays. The nature, the relative rapidity of the effect and the dynamic character of the targets (cells and molecule) justify the unique exposure. The unique or two closely applied exposures permit the dose-effect evaluation on cell levels (number of chromosomal aberrations) or on cell population level (incidence of gene mutation). The intermittent or the chronic exposures eems not to be adequate for this purposes the cells being in permanent renewal which however does not exclude the accumulation of mutations in germ cells after repeated dose.

Route exposure : The route of exposure is important in relation to the substance modifications which can occur before reaching the target cells (somatic and germinal cells) or target molecule (DNA). For example, oral exposure implies that the substance before penetrating into the organism, should be in contact with many chemicals and bacteria, and it should pass through various pH conditions. All these factors may play an important role in the detoxification or in the activation. It is now well established that nitrosamines can be formed in the stomach in given conditions which may occur in nature.

The same comments can be made for respiratory and percutaneous routes.

Metabolism. The most important point before reaction with the target molecules is the metabolism of the compounds. After penetration, the chemicals are generally submitted in the organism to metabolic transformation leading to inactive products which can be identified as detoxification mechanisms, or to active products-activation mechanisms. This transformation can occur in different cells and/or organs, specific or not, and the active products can exercice the mutagenic activity locally and/or elsewhere including germinal cells, which are involved in heritable mutations. If the metabolising cells are not the target, many questions arise concerning the transportation, the elimination, and the real quantity of active compound which reaches the target cells and which reacts with the target molecules. The answer to the last question is of great importance for genetic risk evaluation. Moreover the metabolism and the response are strongly dependent on the host susceptibility (species, strains, organs etc ...), on the dietary habits (alcool, tobacco, drugs) which can modify the enzymatic capacity (detoxifying or activating), and on the factors susceptible to modify the DNA-repair mechanism.

Normally I would have illustrated my presentation. I did not try. Many points will be discussed by our collegues during this meeting. Moreover most of the references concern carcinogenesis. I would like to emphasise that, with regard to health problems the hereditable mutations are as important as the cancer, even more, may be more important, but for the future generations.

Dealing with toxicology and genetic effect I have in mind the problems with germinal mutations and with the germinal cells as target in mammals. This approach can give a better idea of the real problems of genetic risk evaluation and raise questions on the relevance of different short term mutagenicity test for prediction of risks for genetic heritable mutations. I am convinced that we are able to find a model sensitive for each substances. But for what ? I separate clearly scientific interest which increases knowledge and the toxicological interest which, based on the scientific data, should indicate preventive or prophylactic measures. Evaluation of genetic risk to man from short term test only,

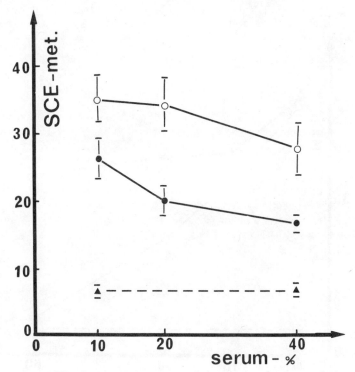

FIGURE 1. SCEs induced by Benzo(a)pyrene (o—o - 0.1, •—• - 0.01µg/ml) in Syrian hamster embryo cells as function of serum concentration (2).

present problems. We may over estimate or underestimate
the mutagenic effects. For example, increased serum concentrations decrease significantly the SCE induced with
benzo(a)pyrene (BaP) in Syrian Hamster embryo cells
(Figure 1). These results indicate that serum prevent
BaP penetration into the cells, modify its metabolism
and decrease its biological effects (2). In this case
the use of organic solvent and low level of serum in
cell culture, which is not representative for the whole
organism, overestimate the biological effect of BaP. With
methyl nitrosourea (MNU) the results are different. The
MNU activity decreases very rapidly when is in medium
with low serum concentration (Figure 2). Using serum as
vehicle their activity remains for many hours (3). Consequently the MNU activity may be understimated in certain
in vitro testing conditions.

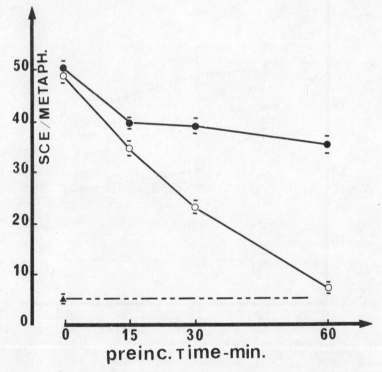

FIGURE 2. SCEs induced by (MNU) (3) preincubated at
37°C in MEM (o—o) (3) or in serum (●—●) (unpublished
data). Ordinate : mean value of SCEs per metaphase (n =
50). Dashed line : control. (V-79 Cell line)

PLACE OF MUTAGENICITY TESTS IN TOXICOLOGICAL EVALUATION

From the point of view of applied toxicology, the place of mutagenicity tests is in acute and sub-acute toxicity evaluation as *tests* for genetic effects. The nature of the reaction (dynamic) between the chemicals and the genetic material (DNA), the nature of the effect at molecular and chromosomal level (definite structural and quantitative modifications), the relative rapidity of the events between the compounds penetration and the genotype change, and the use of only one exposure like in acute toxicity studies indicates that mutagenic evaluation can be assimilated to the acute or sub-acute toxicity tests even if in some cases the observation of the new phenotype is delayed.

This proposal presents two positive aspects, preventive and psychological and raises problems of applications.

The preventive aspect is concerned with the premarked dossier for evaluation of mutagenic potential of chemicals and of the environmental pollutants. In respect to the premarked dossier we can impose *routine assays for toxicity with genetic effect* in the same manner as for acute toxicity. The information furnished by such assays will be beneficial for health protection by *primary preventive* measures. In the case of environmental pollution (air, water, foods) the same tests can be used but with specific approach : collection, extraction, mutagenic effect evaluation, chemical analysis for identification and risk evaluation.

The psychological aspect is that the toxicological evaluation is a part of the habitual activity. Including mutagenic effect evaluation in this activity will contribute to dedramatise the mutagens problems.

It was written some years ago, I quote "mutagenic effects deserve attention not only for reasons of protecting future generations, but also for the need to avoid deleterious effects in the present one in view of the striking correlation between the carcinogenic and mutagenic potential of most chemicals. A task of immediate concern thus becomes one of how such genetic and carcinogenic hazard can be avoided or reduced" (4). It is a generous proposal. The actual conference on Environmental mutagens confirms the importance of the concern. Nevertheless, I should note, that in different papers, in different conferences, scientific or not, the terms mutagens and carcinogens were and are used without any discernment, in such a way that uninformed man can not find his bearings. The word mutagen identified as carcinogen, feels dangerous and insidious. A simple anonimous paper can provoke a real public psychosis as happened in France two years ago, and discredit the research on the subject. This situation was illustrated well by a journal picture presenting a man in a bar who, after reading on the wall : *"alcohol is noxious, tobacco is dangerous, coffee contains mutagens, sausages are with conservatives"* said to the waiter : *"a glass of cyanide please !"*.

Consequences : Obviously, to impose for routine evaluation tests for *toxicity with genetic effects* for all substances likely to be in the environment or likely to be in contact with human beings raises problems of application, economics and toxicological.

The economic aspect I keep out of the matter. It should not play an important role when it is a question of health. In the other hand we must pay now for the benefit of the future generations.

The *toxicological problems* have already been mentioned. Nevertheless, from the pratical point of view I like to repeat one proposed procedure (5) : 1) primary identification or detection of mutagenic activity ; 2) verification ; 3) quantification ; 4) extrapolation. Some regulatory guidelines are elaborated for the primary identification and for the confirmation.

This procedure seems to resolve the problems of mutagenic effect evaluation. However, in concern to the decisions it gives no satisfaction even using numerous short term tests. In most of the cases the use of multiple systems complicate the situation, contradictory results, marginally active compounds, confirmation in higher organisms not easy etc... In these conditions it is less easy to evaluate the risks for humans.

CONCLUSION

What then must we do ? In terms of toxicology I feel that we should come back to the animal, the "black box" of the toxicologists, for routine evaluation of *toxicity with genetic effects*, like in classical toxicity evaluation.

For this proposal some of Mailing's recommandations
(5) may be used as working bases such as :
(1) - Study the mutation induction after sub-acute and
chronic exposure to chemical mutagens.
(3) - Make the present test systems easier and more economical.
(6) - Develop new point mutation systems for whole animals, in which we are sure to measure a broad spectrum
of genetic events.

The following scheme should be as objective for
mutagenic effect evaluation.
- gene mutation *in vivo*
- cytogenetic in bone marrow cells
- cytogenetic in male germ cells and/or
- germ cell mutation in mice.

This scheme may seem unrealistic and inacceptable.
I do not ask agreement. I known that toxicologists need
systems more close to the human and with perceptible
significance. I know that legislations need informations
not only for carcinogenic risk, but also for genetic
(heritable) risk. The last one seems not easy to evaluate. May be. But we should keep in mind that the health
of future generations is in question and think how to do
our best for environmental chemicals mutagenic potential.

REFERENCES

1. Truhaut, R., 1979, La Toxicologie : Aperçus sur les
   buts, ses problèmes et ses méthodes. Sciences, Tome
   $\underline{V}$, N°2, 35-49.
2. Coulomb, H., Gu, Zuwei, Audu, S., Chouroulinkov, I.,
   1981, The Uptake and release of BaP and BeP *in vitro*
   by Syrian Hamster embryo cells as a function of serum
   concentration. Carcinogenesis, $\underline{2}$, 523-527.
3. Stahl, K.W., Cheng, S.J., Bayer, U., and Chouroulinkov, I., 1981, Genetoxicity of N-methyl-N-nitrosourea
   in the Presence of Amphiphilic Membrane-Active Compounds, Toxicology and Applied Pharmacology, $\underline{60}$, 16-25.
4. Sobels, F.H., 1977, Some problems Associated with the
   Testing for Environmental Mutagens and a Perspective
   for Studies in "Comparative Mutagenesis". Mutation
   Research, $\underline{46}$, 245-260.
5. Mailling, H.V., 1976, Mutagenesis Testing-Mammalian
   Systems, Mutation Research, $\underline{41}$, 171-172.

Mutagenic Hazard and Genetic Risk Evaluation on Environmental Chemical Substances

Nicola Loprieno

Istituto Biochimica Biofisica e Genetica, University of Pisa, Pisa, Italy

ABSTRACT The presence among environmental anthropogenic chemicals of substances which have resulted to produce in different experimental organisms genetic damages, and the development which has occurred during the last decade in the knowledge of chemical mutagenesis and human genetics have supported the concept of a high mutational load in the human population as a consequence of the environmental contamination.

The suggestion that a relationship between environmental mutagenic contamination and the frequency of human diseases does exist, depends at the present on the assumption that mutagenic hazards and genetic risks are common to all so far tested chemicals, i.e. on the predictive value of short-term mutagenicity tests employed singularly in regard to the ability to produce heritable damages in mammals.

This latter relationship has been discussed, by examining some environmental chemicals so far investigated such as ethylene oxide and formaldehyde.

INTRODUCTION

The predominant problem of most of the geneticists is, at present, the need to reduce the mutation load of human population, that is the result of the genetic variability. The mutation load contributes, in a way not yet clearly established, to the number of genetic diseases, (17) although up to now its degree has not been defined, because of the lack of a definitive mutation rate's value of many recessive and dominant genes in man. We are not able, today, to evaluate the variation in mutation rate among different human populations, as national and international programs devoted to this biological problem are completely lacking in it.

We can only depend on a few direct or indirect values on the present load of congenital defects, most of which are supposed to have a genetical basis. As a matter of fact, according to A.G. KNUDSON, J.R., most of prenatal deaths in man (spontaneous abortions) represent 20% of all pregnancies and have a genetic basis both chromosomal and genic (11). It is documented that in 50% of spontaneous abortions, the fetal tissues show chromosomal anomalies (Tab. 1); for the remaining 50%, it is possible to

TABLE 1
TYPES AND RELATIVE FREQUENCIES OF CHROMOSOMAL ANOMALIES OBSERVED IN FETAL TISSUES OF SPONTANEOUS ABORTIONS.

CHROMOSOMAL ANOMALY	N° OBSERVED AND PERCENTAGE	
ANEUPLOIDY	1.400	32.7
POLIPLOIDY	456	10.7
STRUCTURAL CHROM. RIARRANGEMENTS	62	1.4
MOSAICISM	41	0.95
TOTAL	1.959	45.8
TOTAL N° ANALYZED	4.274	100.0

hypothesize dominant mutations due to small deletions (11).

On these premises the prenatal genetic load to be considered amounts approximately to 0.2, half of which chromosomal and half genic. A large part of it, consists of fresh mutations, since only a small fraction of all chromosomal anomalies is represented at birth; in fact 0.6% of live born or 1:146 present congenital anomalies due to chromosomal disorders (Tab. 2). It is even more difficult to evaluate the postnatal genetic load due to gene mutations. The present estimates made by international agencies (f. i. UNSCEAR) according to NEEL (17) are an underestimates of the total number of recessive mutations, whose consequencies on human health in the heterozygotic condition are still unknown. Moreover a present estimate suggests

TABLE 2
CHROMOSOME ANOMALIES IN LIVE BIRTHS, PER 1,000

NUMERICAL	SEX CHROM. ANEUPLOIDS OR MOSAICS	2.69
	AUTOSOMAL	1.45
	TRIPLOIDY	0.02
STRUCTURAL	BALANCED	1.93
	UNBALANCED	0.28
UNCLASSIFIED		0.51
TOTAL		6.88

that almost 10% of all live born will present at the birth or during the life severe genetic diseases (17): the values indicated by UNSCEAR are reported in Table 3 (21).

TABLE 3
ESTIMATES OF FREQUENCY OF GENETIC DISORDERS PER 100 LIVEBORN (UNSCEAR, 1977).

DISEASE - CATEGORY	PERCENTAGE
1. DOMINANT	0.95
2. X-LINKED	0.05
3. RECESSIVE	0.10
4. CHROMOSOMAL	0.4 (*)
5. CONGENITAL MALF. & MULTIFACT.	9.0
TOTAL	10.5

(*) This value has been underestimated by UNSCEAR (see Table 2).

The social dramatical aspects of these estimates are well represented in their own gravity by the recent data collected by TRIMBLE and SMITH: they have indicated that children affected by genetic diseases are hospitalized 5-10 more times than average children (20); DAY and HOLMES have moreover shown that 17% of children in hospitals (1:5) are affected by genetic diseases (4).

It seems therefore possible to consider the estimate of 0.2 of postnatal genetic load, suggested by KNUDSON (Tab. 4) as a valid one.

TABLE 4
EXPRESSION OF THE GENETIC LOAD

	CHROMOS.	GENIC	TOTAL
PRENATAL	.1	.1	.2
POSTNATAL	.005	.2	.2
TOTAL	.1	.3	.4

After A.G. KNUDSON, Jr., 1979 (11).

It is a general conviction in most geneticists that chemical mutagenic substances so widely spread on human environment (food contaminant and additive, cosmetics, agricultural and industrial contaminants, etc.) do contribute to a considerable extent to mutations which are directly responsible of part of the genetic load.

This opinion is not supported by any epidemiological human evidence; in fact we do not have data on the change of the mutation rate in two different groups of persons one of which could have been exposed to some environmental mutagenic contaminant; moreover the present data are only related to small groups such as the one made up of the children whose parents were exposed to γ and neutron irradiation during the atomic explosions in Hiroshima and Nagasaki (18).

It is not possible, either as it has been demonstrated for human carcinogens, to collect evidences on the presence of human mutagens due to epidemiological difficulties, in spite of some suggestions which have been reported in the literature (23, 2, 10).

The only evidence which is possible at present consists of an extensive scientific literature on the mutagenic properties of thousands of chemical substances, such as food additives, cosmetics, drugs, pesticides, industrial compounds which represent a total of 70,000 molecules (Table 5): these substances have been demonstrated to

TABLE 5
CLASSIFICATION OF CHEMICALS IN COMMON USE IN THE POPULATION

CLASS OF COMPOUNDS	N° ESTIMATE
CHEMICALS IN EVER DAY USE	50,000
ACTIVE INGREDIENTS IN PESTICIDES	1,500
ACTIVE INGREDIENTS IN DRUGS	4,000
EXCIPIENTS	2,000
FOOD ADDITIVES	2,500
CHEMICALS TO PROMOTE LIFE	3,000
TOTAL	63,000

interact with the DNA of several organisms and to produce a series of biological events related to the mutational process of genes and chromosomes (Tab. 6).

TABLE 6
ACTION OF CHEMICAL MUTAGENS

DIRECT	
ON RESTING DNA	ON REPLICATING DNA
1. DEAMINATION OF BASES	1. INCORPORATION OF BASE ANALOGUES
2. ALKYLATION OF BASES	2. INTERCALATION BETWEEN BASES
3. CROSSLINKS OF DNA CHAINS	3. INTERACTION WITH MEMBRANES
4. DELETION AND TRANSLOCATION OF DNA MOLECULES	4. INTERACTION WITH ENZYMES

INDIRECT
1. MISTAKES OF DNA REPLICATING ENZYMES
2. MISTAKES OF DNA REPAIR ENZYMES
3. DIFFERENT EVENTS DURING RECOMBINATION
4. STIMULATION OF DNA-REPAIR PROCESSES
5. ANOMALOUS SEGREGATION OF CHROMOSOMES

From the knowledge of these facts it derives for the geneticist the need to evaluate the present data on environmental mutagens in terms of genetic toxicity for man, in order to define the hazard for the environment and for human beings due to the presence of mutagenic substances and for some of them to assess, their genetic risk, or their ability to increase the present mutation load of human population.

MUTAGENICITY TESTS AND THEIR VALUE IN THE ASSESSMENT OF GENETIC RISK

The short-term mutagenicity tests already validated to a different extent on several classes of chemicals are intended to provide the demonstration that chemical substances produce irreversible changes in the genetic structure of experimental organisms, i.e. mutations; or else, some of them can be used to demonstrate, as well, that chemical substances stimulate biological reactions which are related to genotoxic processes (binding with the DNA, thus producing lethal effects, chain breaks or induction of repair synthesis in the DNA molecule). All these help to demonstrate the above mentioned effects which are reported in Table 6.

The presently available short-term tests for the evaluation of genotoxic effects of chemical substances are reported schematically in Table 7 (13, 14, 16), where all tests have been classified according to the genetic end point (gene, or chromosomal mutation; indicator biological effect), or the cellular type (somatic or germinal cell), or the methodologies and organisms employed (animal organisms, or in vivo; unicellular organisms, or in vitro).

Such a classification also contributes to understand the validity of the data so far obtained and the way to interpret them in term of hazard or genetic risk, which represents the final task of mutation researches.

A positive result in whichever test in vitro or in vivo (somatic cell) allows us to identify it as a substance with a potential for mutagenicity.

This type of result represents only a qualitative information to be used to predict a potential hazard for man; it cannot be used to define the genetic risk for man, even in the case of a well documented human exposure. The identification of substance having a genetic risk is based only on those data provided by methodologies which employ germ cells as a target, and which are represented by some of the in vivo assays (small animals or insects).

Therefore we can make use, at present, of just two different types of methodology: those consisting of somatic

TABLE 7
DIFFERENT MUTAGENICITY TEST SYSTEMS (IN VITRO AND IN VIVO)

SOMATIC CELL		GERMINAL CELL	
IN VITRO	IN VIVO	IN VITRO	IN VIVO

A. GENE MUTATION

SOMATIC CELL IN VITRO	SOMATIC CELL IN VIVO	GERMINAL CELL IN VITRO	GERMINAL CELL IN VIVO
1. Bacterial test	1. Mouse spot test		1. Drosophila reces. lethal test
2. Yeast test			2. Mouse specific locus test (morph., skelet., domin., biochemical)
3. Fungal test			3. Sperm abnormality in F 1
4. Mammalian test			

B. CHROMOS. MUTATION

SOMATIC CELL IN VITRO	SOMATIC CELL IN VIVO	GERMINAL CELL IN VITRO	GERMINAL CELL IN VIVO
1. Mammalian cells	1. Mouse & rat micro nucleus test		1. Drosophila dominant lethal and chromosome loss tests
2. Peripheral blood lymphocytes	2. Mouse & rat bone marrow cytogenetic test		2. Mouse heritable translocation test
			3. Mouse & rat dominant lethal test
			4. Mouse & rat germ cell cytogenetic test
			5. Chromosome non—disjunction test

C. INDICATOR BIOLOGICAL EFFECTS

SOMATIC CELL IN VITRO	SOMATIC CELL IN VIVO	GERMINAL CELL IN VITRO	GERMINAL CELL IN VIVO
1. Mammalian cells (UDS)	1. Hepatocytes & other tissues (UDS)		1. Mouse germ cell (UDS; DNA Adducts & Breaks Sce)
2. Mammalian cells (DNA Adducts & Breaks)	2. Different tissues (DNA Adducts & Breaks)		
3. Mammalian cells (Sce)	3. Hemoglobin alkylation		
4. Gene conversion & recombination in yeast	4. Different tissues (Sce)		

effects, and able to evidenziate the potential genotoxicity of chemical substances, and those consisting of germinal effects, and able to predict the existence of a human hazard. In the latter case we have not yet the demonstration of the presence of a genetic risk, for these are also other relevant factors to be considered, such as the extent and the duration of the human exposure. This is proved for gene mutations and for chromosomal mutations as well.

On Table 7, the tests included in the last class, allow us the collection of results indicating complementary informations useful to correctly analyze the data derived from the previous assay systems.

Recently a group of OECD*experts have classified the

---

* Organization for Economic Cooperation and Development.

present available mutagenicity tests to be used for national and international regulations, as the following (14):
(a) tests to be used for the identification of potential mutagenic chemical substances: they include both preliminary and verification tests as well (in vitro and in vivo);
(b) tests to be used to elucidate the genetic hazard;
(c) indicator tests of genetic toxicity (Tab. 8).

TABLE 8
DIFFERENT CLASSES OF SHORT-TERM MUTAGENICITY TESTS (OECD'S WORKING GROUP)

1. MUTAGENICITY TEST SYSTEMS TO PROVIDE INFORMATION ON THE POTENTIAL OF CHEMICALS TO CAUSE GENETIC DAMAGE
   1.A. PRELIMINARY TESTS
   1.A.1. GENE MUTATION IN PROKARYOTIC CELLS (SALMONELLA; E.COLI)
   1.A.2. CHROMOSOMAL ABERRATIONS IN MAMMALIAN CELLS GROWN IN VITRO; MICRONUCLEUS TEST AND METAPHASE ANALYSIS OF BONE MARROW CELLS.

   1.B. VERIFICATION TESTS
   1.B.1. GENE MUTATION TESTS IN (i) EUKARYOTIC MICROORGANISMS; (ii) MAMMALIAN CELL GROWN IN VITRO; (iii) DROSOPHILA (S.L.R.L.); (iv) THE MOUSES's EMBRYO.
   1.B.2. CHROMOSOME ABERRATION TESTS IN (i) IN VIVO CYTOGENETIC ON SOMATIC CELLS, OR ON GERMINAL CELLS; (ii) DOMINANT LETHALS IN RODENTS.

2. MUTAGENICITY TESTS SYSTEMS TO ELUCIDATE THE POTENTIAL GENETIC HAZARD OF CHEMICALS
   2.A. MOUSE SPECIFIC LOCUS MUTATION TEST
   2.B. MOUSE HERITABLE TRANSLOCATION TEST

3. MUTAGENICITY TEST SYSTEMS WHICH ARE INDICATOR OF THE POTENTIAL GENOTOXIC ACTION OF CHEMICALS
   3.A. (i) DNA DAMAGE IN BACTERIA; (ii) DNA DAMAGE AND REPAIR IN MAMMALIAN CELLS; (iii) SISTER CHROMATID EXCHANGE; (iv) SPERM ABNORMALITY TEST; (v) MITOTIC RECOMBINATION AND GENE CONVERSION IN YEAST.

Tables 7 and 8 classify the mutagenicity test systems on the basis of their use and on the value of the results provided.

The present discussion refers only to the genetic effects analyzed by the mutagenicity test methodologies and to their significance in the definition of genetic risk and does not take into consideration the predictive value of the mutagenicity tests for the cancerogenic potential of chemical substances: a predictive value for cancerogenicity of the short-term mutagenicity tests has been universally recognized (13), but it is not of interest in the present discussion.

The importance of the genetic risk as a result of human exposure to mutagenic chemical susbtances, i.e. the possible increase of mutation load of the human population, is completely different from that related to the carcinogenic risk, which indeed represents one of the most frequent cause of human death, because of the distinct significance of the two diseases. The cancer affects only the exposed individual, whereas the genetic disease affects the exposed individuals, and produces almost in all cases a severe alteration in the individual affected, alteration which is seldom a fatal one.

We believe that the research for the estimation of the contribution of contaminated environment to the development of genetic disease is to be indipendently evaluated from those on cancer causes. According to a statement made recently by S. ABRAHAMSON (1), a study commissioned by EPA* has shown that the cost paid by the society for each genetic diseases is five times more (in dollars) than the one for a cancer disease.

RELEVANT FACTORS FOR THE DETERMINATION OF GENETIC RISK

In the determination of the genetic risk of a mutagenic compound, it is highly relevant the estimation of the dose of the substance which reaches the gonads of experimental animals, depending the amount of DNA damages in the germ cells and of the genetic transmissible damages on the concentration in the gonadal tissues.

According to LEE and DIXON (12) it is generally assumed that the test chemical under study for a possible

---

* Environmental Protection Agency.

definition of its potential genetic risk readily reaches all types of gonadal tissues including the germ cells, because it spreads in these tissues with the same rate as it does in the somatic tissues. Such an assumption, however, does not take into account all kind of pharmacokinetic concepts regarding the distribution of chemicals in the testis and male accessory glands.

Fig. 1 illustrates schematically the present problems

FIG. 1

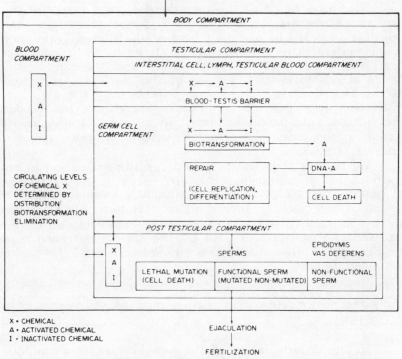

After Lee & Dixon (12).

concerning the comprehension of the parmacokinetic of toxic chemical mutagens in the male gonad system: a chemical substance has to penetrate the blood-testis barrier (BTB), then it might be subjected to the enzymatic activation and/or detoxification system present in different compartments of gonads; thereafter the chemical should interact with the DNA and the resulting damage may be involved in different DNA repair processes, from which the final production of a stable genetic alteration depends.

At present only by indirect evidences we can prove that the distribution of the mutagenic chemical in the gonadal tissues is different from the distribution observed in the other tissues and that the enzymatic systems present in the gonadal tissues are quantitatively different from what is known to occur in the f.i. liver tissues.

According to EHRENBERG (5) it is proved that alkylating agents, such as Methylmethanesulfonate (MMS), which remain for some minutes in the body of a mouse, show the same dose in every organs including the gonads; on the contrary, those mutagenic compounds which undergo metabolic conversion, spread in the gonads at a different level. The dose of vinyl chloride (VCM) found in the gonads is 10 times lower than in the liver and N-Nitrosodimethylamine (NDMA) in the gonads is about 100 times lower than in the liver and in the blood (5). Such a difference in the concentration of these types of chemicals in the liver and in the gonads might be due to several factors: (i) the distribution in different tissues; (ii) the elimination rate in different tissues; (iii) the number of possible active metabolites and their pharmacokinetics; (iv) the gonadal dose of the compound itself and of each possible mutagenic species the compound gives rise to by metabolic reactions (in the case of ethilene oxide (EO) f.i., the compound gives rise to ethylene chlorohydrin, chloroacetaldehyde, and ethylene glycol and we know that they possess a different mutagenic specific activity).

These information, however, are far from being exaustive: we have some other types of indirect evidences from the results obtained by R. FAHRIG with the host mediated assay (8): they show a quantitative mutagenic difference in some chemicals when tested in testes, liver, or lung of rats.

Table 9 presents mutagenicity data obtained with 4 different chemical mutagens both direct and indirect administered to animals by different ways and tested on yeast cells which have been inoculated into the liver and lung tissues by an intravenous injection, or injected directly into the testis tissues.

In all cases the mutagenic activity found has been lower or negative, in the testes, than the one in the liver or in the lungs, and this is resulted true also for the direct alkylating mutagen, MMS.

TABLE 9
COMPARATIVE MUTAGENIC ACTIVITY OF DIFFERENT SUBSTANCES IN THE LIVER, LUNG, AND TESTES OF RAT (GENETIC END POINT: GENE-CONVERSION IN YEAST; METHODOLOGY: HOST MEDIATED ASSAY).

SUBSTANCE	CONCENT. (MM)	ADE CONVERT/5×10^6 CELLS IN(1)		
		TESTES	LIVER	LUNG
4.NQO (I)	0.25	0	11.5	107.6
	(44)	(0)	(10)	(100)
ENDOXAN (I)	2	3.7	13.6	10.2
	(44)	(27)	(100)	(75)
NDMA (I)	10	2.9	66.0	32.0
	(44)	(4)	(100)	(48)
MMS (D)	4	5.9	21.8	17.9
	(14)	(27)	(100)	(82)

(1) THE CONTROL VALUE HAS BEEN SUBSTRACTED
I: INDIRECT MUTAGEN; D: DIRECT MUTAGEN
IN BRACKETS THE PERCENT VALUES
AFTER R. FAHRIG, MODIF. (8)

Moreover, the different data could be the result of different mechanisms as well, such as those already mentioned.

Information on the efficiency of metabolism for converting promutagenic chemicals into ultimate active metabolites present in testes or spermatogenic cells of rats in comparison with that of liver tissue of the same animals are provided by the conclusions made by LEE and DIXON (12) reported in tables 10 and 11.

The present data show that in general different types of activation (c, d) detoxification (a, b) are present in testes as in the liver, but their amounts in testicular tissues are low about 10% of those present in the liver. The differences are still more definable when we compare amount of enzymes in two different compartments of the testis. The distribution of the enzymes and the cytochrome P-450 in the interstitial and germ cell compartment indicates that AHH activity and cytochrome P-450 content of the microsomes from the interstitial cells were nearly twice bigger than that in the tubule; in contrast with it, the

TABLE 10
ENZYME'S ACTIVITY IN TESTIS AND LIVER OF ADULT RATS

ENZYME	SPECIFIC ACTIVITY	
	TESTIS	LIVER
(a) Glutathione S-transferase n mole product/min-mg protein with B(a) P 4,5 oxide substrate (176,000 x g supernatant)	19.99 (48)	41.29 (100)
(b) Epoxide hydrase n mole product/min-mg protein with B(a) P 4,5 oxide substrate (microsomes)	0.77 (7)	10.85 (100)
(c) Aryl hydrocarbon hydroxylase p mole 3-HO-B(a) P formed/min -mg protein (microsomes)	5.17 (5)	106 (100)
(d) Cytochrome P-450 n mole/mg protein (microsomes)	0.125 (15)	0.85 (100)

IN BRACKETS PERCENT VALUES
AFTER LEE & DIXON MOD. (12)

specific activities of the detoxification enzymes, EH and GSH.T. in tubules were as much as the one found in the interstitial cells.

The reliability of extrapolation in data from laboratory animals to man and the resulting estimates of genetic risk depends also on the possibility to define, on a quantitative base the different sensitivity of germ cells during the spermatogenic process: contrary to X-irradiation which is able to induce mutations in all stages of germ cells, many chemical mutagens show a differential response to the induction of mutations during the spermatogenic process. A well documented set of data is represented by the induction of dominant lethal mutations in germ cells of mice: Table 12 reports some of these data collected by V. EHLING (7).

It has been suggested that these differential results could be due to the existence of different repair systems

TABLE 11
ENZYME'S ACTIVITY IN DIFFERENT COMPARTMENT TISSUES OF RAT'S TESTES

ENZYME	SPECIFIC ACTIVITY	
	INTERSTITIAL CELLS	SPERMATOGENIC CELLS
(a) Glutathione S-transferase n mole/min-mg protein	65.3 (55)	119 (100)
(b) Epoxide hydrase n mole/min-mg protein	1.09 (46)	2.36 (100)
(c) Aryl hydrocarbon hydroxylase p mole/min-mg protein	5.98 (100)	3.18 (53)
(d) Cytochrome P-450 n mole/mg protein	0.196 (100)	0.084 (43)

a: in 176,000 x g supernatant; b, c, d: in microsomes
IN BRACKETS PERCENT VALUES
AFTER LEE & DIXON, MOD. (12)

It has been suggested that these differential results could be due to the existence of different repair systems during spermatogenic development: as to MMS, f.i., it has been shown that the lack of dominant lethal effects of spermatogonia and spermatocytes (premeiotic cells) is due to their ability to repair DNA, while spermatids and spermatozoa are unable to repair DNA, therefore lethal mutations can be expressed (12).

The knowledge of the dose of the chemical mutagen in the gonads, the rate of metabolic reactions to which the substance is subjected into the gonadal tissues, the sensitivity of different germ cell stages to the mutagenic activity of the substance, the role of DNA repair systems operating in the germ cells may all of them contribute to a better estimation of the potential genotoxic effect of a chemical mutagen to be used in the definition of genetic risk existing for the human population exposed to that substance.

TABLE 12

SPERMATOGENIC RESPONSE OF MICE TO THE INDUCTION OF DOMINANT LETHALS BY CHEMICAL MUTAGENS

COMPOUND	SPERMATOGENIC CELL STAGE			
	S.zoa	S.tids	S.ytes	S.onia
FOSFETROL	●	○	○	○
MITOMEN	●	○	○	○
MMS	●	●	○	○
CYCLOPHOSPHAMIDE	●	●	○	○
5-FLUROURACIL	○	○	●	○
NATULAN	○	●	●	●
MITOMYCIN C	○	●	●	●
ISOPROPYL-METHANESULFONATE	●	●	●	●
BUSULFAN	●	●	●	●
X-RAYS	●	●	●	●

● : POSITIVE RESULTS
○ : NEGATIVE RESULTS

AFTER U. EHLING, MODIF. (7)

CRITERIA FOR THE EVALUATION OF THE GENETIC RISK OF ENVIRONMENTAL CONTAMINANTS

The evaluation of the genetic risk as a consequence of human exposure to a mutagenic substance comprises a series of tests able to accumulate relevant information.

The right approach to the risk assessment for environmental contaminants should include at least 5 phases (fig. 2): 1. the production of data showing the ability of

FIG. 2

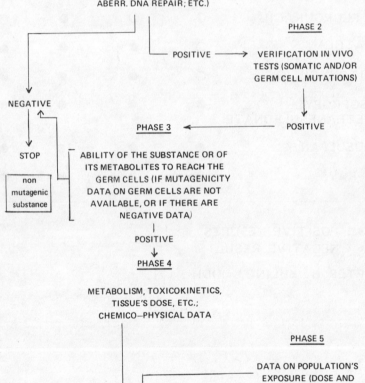

the substance in question to produce all kinds of genotoxic effects, including point mutation, chromosome aberrations, DNA repair stimulation; 2. in the affermative, that is in case of positive results in phase 1, it is necessary to verify the mutagenic activity of the substance in an in vivo condition, in somatic and or germ cells tissues; 3. when the latter information cannot be available, or in case of negative results on germ cells, data should be provided on the ability of the substance or of some of its metabolites to reach the germ cells; 4. production of data on the metabolic fate of the substance, its toxicokinetic and tissues's distribution and dose, the stability of the substance in the environment should also be assessed; 5. the knowledge of the human population's exposure (dose and duration) is also relevant.

On that basis it is possible to evaluate the mutagenic hazard of the substance and to assess its genetic risk.

Phase 1, 2, and 3 are clearly represented in the schema of figs. 3 and 4 (or 5 and 6), relating to the sequential tests to be put in operation for assessing the mutagenic hazard: they have been developed on the basis of the Guidelines for the Mutagenic Risk Assessment proposed by US-EPA on Federal Register (22).

An attempt has been made to evaluate if the proposed approach for the definition of phases 1-4 may be satisfied by the data so far available for two environmental contaminants, namely Ethylene oxide (EO) and Formaldehyde (FA).

EO has been proved to produce gene mutation on all types of microorganisms tested in vitro (viruses and phages, S.typhimurium, E.coli, N.crassa, A.nidulans, S.pombe); these in vitro data have been corroborated by the results already obtained from analyses on the induction of recessive lethal mutations in Drosophila melanogaster (6), by treating flies in different ways. CUMMING et al. (3) have moreover established that EO administered to mice by inhalation for 4th at 600 and 800 ppm stimulates the DNA repair synthesis in the spermatids 10-20 folds the untreated animals (fig. 3). In spite of the lack of information on the mutagenic activity of EO in vivo mammalian tissues, we way conclude that EO presents a potential for the production of gene mutations. The data on the clastogenic activity of EO are more conclusive than the previous ones (fig. 4).

EO has been shown to produce chromosome aberrations in vitro and in vivo; moreover the induction of dominant lethals (9), and heritable translocations in mice

FIG. 3
ETHYLENE OXIDE

GENE MUTATIONS

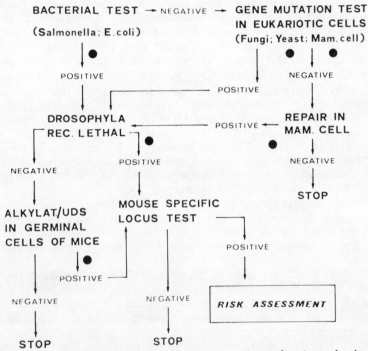

(9) has been demonstrated; data on tissue's dose in human beings and animals are also available. According to some of these data, EHRENBERG and HUSSAIN have recently made a calculation of the genetic risk for human occupational exposure to EO, expressed in rad-equivalent (6); we think that it should be possible at present to express the genetic risk in terms of increasing induction of translocations and dominant lethals.

The data published by GENEROSO et al. (9) point out that an exposure of mice to a dose of EO of 800 ppm. h produces dominant lethal effects: this dose corresponds to an occupational exposure of two days (50 ppm×8h×2=

FIG. 4
ETHYLENE OXIDE

CHROMOSOMAL MUTATIONS

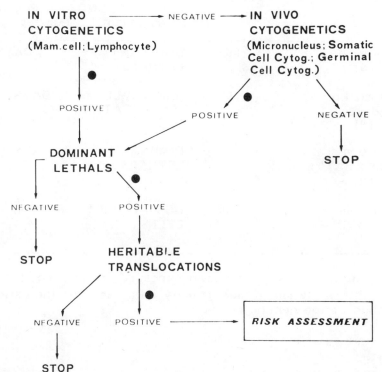

= 800 ppm.h.) in the working place; the same AA have demonstrated that a dose of 4,000 ppm.h. in mice, corresponding to an occupational exposure of 10 days, produces heritable translocations (Table 13).

A similar example may be provided by another environmental contaminants: i.e. Formaldehyde (FA) (figs. 5 and 6), by which several genetic end points have been produced in in vitro and in vivo mutagenicity tests and which is proved to reach the gonads (15, 19).

In conclusion, the evaluation of the genetic risk for human groups of environmental substances should be based on an extended analysis of its mutagenicity potential both in vitro and in vivo; additional data on its metabo-

TABLE 13
GENETIC EFFECTS PRODUCED BY ETHYLENE OXIDE IN MICE AND CORRESPONDING HUMAN OCCUPATIONAL EXPOSURE

MICE'S EXPOSURE	ppm.x.h	GENETIC END POINTS FOUND POSITIVE	OCCUPATIONAL EXPOSURE (MAC=50 ppm)
150 mg/kg (9)	800	DOMINANT LETHALS	2 DAYS
30 mg/kg x5d.x5w. (9)	4,000	HERITABLE TRANSLOCATION	10 DAYS
300 ppm x8h.x5d. (9)	12,000	DOMINANT LETHALS	30 DAYS
600 ppm x 4 h (3)	2,400	UDS IN GERM CELLS	6 DAYS

lism and dose levels of human exposure are relevant for the interpretation of mutagenicity data, and for the estimation of the genetic risk.

FIG. 5

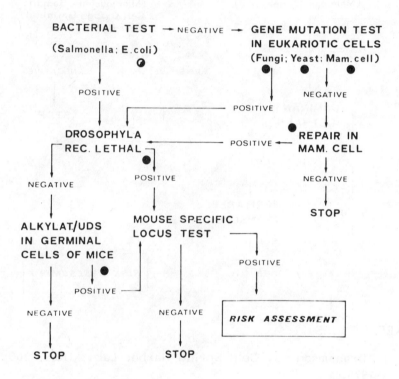

FIG. 6
FORMALDEHYDE

CHROMOSOMAL MUTATIONS

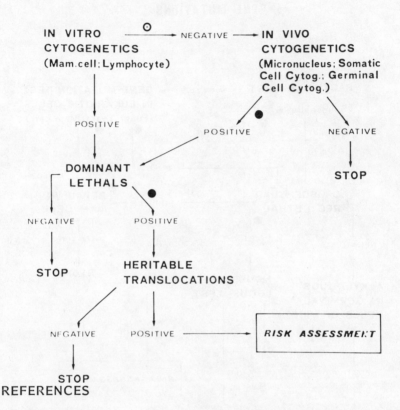

REFERENCES

(1) Abrahamson, S. Cold Spring Harbor Lab., USA, 305, 1979.
(2) Bridges, B.A., J. Clemmesen and T. Sugimura. Mutation Res. 65, 71-81, 1979.
(3) Cumming, R.B., R.A. Michaud, L.R. Lewis, and W. H. Olson. Mutation Res. (in press).
(4) Day, N. and L.D. Holmes. Am. J. Hum. Genetic. 25, 237, 1973.
(5) Ehrenberg, L. Cold Spring Harbor Lab. USA, 243-245, 1979.
(6) Ehrenberg, L. and S. Hussain. Mutation Res. 86, 1-

113, 1981.
(7) Ehling, U.H. Elsevier North Holland Biomedical Press, Amsterdam, 47-56, 1980.
(8) Fahrig, G. Elsevier Scient. Publ. Co., Amsterdam, 135-147, 1977.
(9) Generoso, W.M., K.T. Cain, M. Krishna, C.W. Sheu, and R.M. Gryder. Mutation Res., 73, 133-142, 1980.
(10) Kapp, R.W.Jr., D.J. Picciano, C.B. Jacobson. Mutation Res., 64, 47-51, 1979.
(11) Knudson, A.G. Jr. Am. J. Hum. Genet. 31, 401-413, 1979.
(12) Lee, I.P. and R.L. Dixon. Env. Health Perspect. 24, 117-127, 1978.
(13) Loprieno N. Biomedical Press, Amsterdam, 107-124, 1980.
(14) Loprieno N. European Environmental Mutagen Society, Budapest, Hungary, July, 6-11th, 1981.
(15) Loprieno N. (in press).
(16) Loprieno N. Biology Internat. 3, 2-14, 1981.
(17) Neel, J.V. Cold Spring Harbor Lab., USA, 7-26, 299, 1979.
(18) Neel, J.V. Proc. Nat. Acad. Sci. USA, 77, 4221-4225, 1980.
(19) Stott, W.T. and P.G. Watanabe. Toxicol. Appl. Pharmac. 55, 411-416, 1980.
(20) Trimble, B.K. and M.E. Smith. Can J. Genet. Cytol. 19- 375, 1973.
(21) Unscear. UN 1977.
(22) U.S. Federal Register. 45, 74984-74988, 1980.
(23) Wagoner, J.K. et al. Ed. E. Bingham, 100-113, 1976.

Classification of Carcinogens as Genotoxic and Epigenetic as Basis for Improved Toxicologic Bioassay Methods

John H. Weisburger and Gary M. Williams

American Health Foundation, Naylor Dana Institute for Disease Prevention, Valhalla, N.Y., U.S.A.

ABSTRACT Neoplastic transformation may involve a somatic mutation. Thus, environmental chemicals which can change the normal genetic apparatus of a cell, and which have been found genotoxic, can convert a normal cell to a tumor cell. Certain alkylating agents can interact with DNA directly, but procarcinogens or promutagens require metabolic activation to reactive electrophiles interacting with nucleophilic DNA. Such reactive electrophiles are also mutagenic in prokaryotic and eukaryotic systems and can thereby be detected. Some carcinogenic metal ions, such as cobalt or nickel, alter DNA, probably not by direct reaction, but through modification of the fidelity of the DNA polymerases.

In addition, there are chemicals which participate in the carcinogenic process by controlling the growth and development of transformed cells. Such chemicals do not interact with DNA, but exert their effects through epigenetic mechanisms. This kind of chemical may be important in carcinogenesis and may exert a decisive, crucial role as to whether or not neoplastic cells develop into overt tumors.

The risk assessment and, thus, regulatory controls applicable should be distinct for genotoxic carcinogens and mutagens versus epigenetic agents as regards dose-response, the ease of establishing a low-risk threshold, and importantly, the reversibility of their action. Based on these concepts, improved, more economical testing approaches for the detection and estimation of chemical carcinogens have been developed.

CANCER--A SET OF DISEASES CAUSED BY ENVIRONMENTAL FACTORS

A large fraction, 70-90%, of human cancers, have been attributed to environmental causes (9,14,15,45). However, the environment is complex. The causative elements are often misunderstood and related mainly to ubiquitous chemicals, and, more specifically, to modern technology and industrial development. Certainly, a number of food additives, pesticides, insecticides, and industrial chemicals introduced commercially in the last 40 years have exhibited carcinogenic properties in animal models (14,18,27), and extensive exposure to chemicals at the workplace or as drugs has led to human cancers (14, 17). However, most of the main human cancers in the Western world do not stem from intentional or even inadvertent chemical contaminants in the environment. Cancer denotes many different diseases, and an understanding of the complex causes of cancer requires a detailed study of the factors associated with each specific kind of cancer. The major elements can be traced to lifestyle as practiced in many parts of the world (14,23,38). It is important to identify the real, actual causes of cancer, as a rational, effective basis for prevention.

MECHANISMS OF CANCER

Cancer Causation - Historic Definitions. Traditionally, chemical carcinogens were usually characterized by their ability to induce tumors. The classic operational definition of carcinogen is that of any chemical compound or mixture that yields an increased incidence of cancer over that seen in control groups (1,39,41,44). Regulatory agencies in many countries determine whether a product is safe or unsafe by this definition. This definition requires refinement in the light of contemporary knowledge as to the mechanisms of action of carcinogens.

Cancer Causation - Modern Definition. The neoplastic state is heritable at the cellular level, meaning that the progeny of the division of a neoplastic cell inherit the neoplastic potential. The mechanisms whereby chemicals convert normal cells to malignant ones must eventually account for the permanence of the transformation.
The important discovery that many types of carcinogens interact with DNA resulting in a direct alteration

of the genotype has been the key to explain the permanence of the neoplastic state (5,17,20). At present, the mechanisms by which direct intervention of a carcinogen with DNA could lead to malignant conversion are much better understood than those that postulate a role for an indirect protein or RNA interaction.

Many types of carcinogens can be metabolized to electrophilic moieties, the ultimate carcinogenic forms of these carcinogens (24) which can react with cellular macromolecules, and especially with DNA. Those who favor nongenetic mechanisms of carcinogenesis visualize that such electrophilic reactants could react with protein or RNA to produce the changes involved by quite indirect mechanisms, such as faulty differentiation. Yet, no ultimate electrophilic carcinogen is known that fails to react with DNA but attacks only RNA and protein. Those who subscribe to genetic mechanisms of carcinogenesis regard the ability of carcinogens to interact with DNA as critical and have developed a further generalization that such carcinogens are mutagens (2).

These generalizations on the action of carcinogens, however, have not been demonstrated to apply to all chemical carcinogens as defined by the classic operational definition. Chemical carcinogens comprise a highly diverse group of agents, and thus it would be truly remarkable if they all acted on cells in the same manner. In fact, there are carcinogens such as plastics, asbestos, hormones or immunosuppresant drugs that have not been demonstrated to alter DNA, are not mutagenic, and whose structure does not suggest an obvious reactive electrophilic form (14). Thus, it would seem that several mechanisms can be involved in the overall carcinogenic effects of different chemicals. This is the reason we have proposed a mechanistic classification of carcinogens into two major categories, genotoxic and epigenetic (39,41,44)

<u>Genotoxic Processes</u>. New knowledge on the mechanisms of carcinogenesis has been developed in recent decades. Neoplasia can be considered to result from a somatic mutation, namely, a change in the fundamental structure at the level of the genetic material in normal cells (10,24,32). A change in the genetic material can arise through several mechanisms. The genome can be altered by a direct attack with radiation, chemicals, or viruses. Radiation or chemicals can damage the genetic material at a number of points along the DNA chain lead-

ing to permanent alterations by mispairing of bases during replication of the damaged regions. With viruses, a more specific insertion of DNA segments or, through reverse transcriptase, of RNA sequences takes place through the operation of specific polymerases and hence yields an abnormal DNA containing new information (3).

Other mechanisms include faulty operation of DNA polymerase during DNA synthesis, resulting in inaccurate transcription of the parent DNA segment and yielding abnormal DNA. Certain carcinogenic metal ions may have this effect (22).ABnormal DNA can also ensue, especially during postreplicative DNA synthesis, through errors introduced by specific DNA polymerases concerned with DNA repair (25,43).

The infidelity of DNA polymerases may lead to further abnormalities in DNA, produced during replication of early tumor cells, and thus may represent the means whereby tumor cells progress to less differentiated, more malignant cancer types, during their growth and development.

Epigenetic Phenomena. The production of abnormal DNA by any of the above mechanisms is only the first step in a long sequence of events terminating in a malignant invasive neoplasm. An important element is the ability of an abnormal cell population to achieve a selective growth advantage in the presence of surrounding normal cells. The cell duplication process is highly dependent on a number of endogenous and exogenous controlling elements. Two of the key controlling elements are promoters or inhibitors of growth which either enhance or retard the process.

As numerous experiments documenting this phenomenon indicate, promoters do not lead to the production of an invasive cancer in the absence of an antecedent cell change (15,19,28,40). Thus, in exploring the causes of any specific human cancer, consideration must be given both to the agents leading to an abnormal genome and to any other agents possibly involved in the growth and development of the resulting abnormal neoplastic cells, and their further progression to malignancy.

Chemical carcinogens, accounting for the majority of human cancers, have been classified by Weisburger and Williams (39,41,44) into eight classes that, in turn, belong to two main groups: genotoxic carcinogens and agents, including promoters, operating by epigenetic path-

Table 1. Classes of Chemical Carcinogens

Type	Mode of Action	Example
GENOTOXIC		
1) Direct-acting or primary carcinogen	Electrophile, organic compound, genotoxic, interacts with DNA.	Ethylene imine bis(chloromethyl)ether
2) Procarcinogen or secondary carcinogen	Requires conversion through metabolic activation by host or in vitro to type 1.	Vinyl chloride, benzo-(a)pyrene, 2-naphthyl-amine, dimethylnitros-amine
3) Inorganic carcinogen	Not directly genotoxic, leads to changes in DNA by selective alteration in fidelity of DNA replication.	Nickel, chromium
EPIGENETIC		
4) Solid-state carcinogen	Exact mechanism unknown; usually affects only mesenchymal cells and tissues; physical form vital.	Polymer or metal foils, asbestos.
5) Hormone	Usually not genotoxic; mainly alters endocrine system balance and differentiation; often acts as promoter	Estradiol, diethyl-stilbestrol
6) Immuno-suppressor	Usually not genotoxic; mainly stimulates "virally induced", transplanted, or metastatic neoplasms.	Azathioprine, anti-lymophocytic serum
7) Cocarcinogen	Not genotoxic or carcinogenic, but enhances effect of type 1 or type 2 agent when given at the same time. May modify conversion of type 2 to type 1	Phorbol esters, pyrene, catechol, ethanol, n-dodecane, SO_2
8) Promoter	Not genotoxic or carcinogenic, but enhances effect of type 1 or type 2 agent when given subsequently.	Phorbol esters, phenol, anthralin, bile acids, tryptophan metabolites, saccharin.

ways (table 1). This classification, in relation to an understanding of the relevant mechanisms, is important in dissecting the complex causes of diverse kinds of cancer and arriving at a specification of the role of each agent - genotoxic carcinogen, cocarcinogen, or promoter - in the overall carcinogenic process for each kind of cancer.

This classification is important not only for theoretic, fundamental reasons, but also because different types and classes of carcinogens require distinct regulatory actions in order to ensure safety and public protection (39).

CLASSES OF CHEMICAL CARCINOGENS

Carcinogens that interact with and alter DNA would be classified as genotoxic. This category contains the "classic" organic carcinogens that are electrophilic reactants either in their parent form or after metabolism (24). Such carcinogens can be identified by their genotoxic effects through in vitro short-term tests that reliably measure DNA damage. It seems highly likely that DNA alteration is the basis for the carcinogenicity as well as the mutagenicity of these compounds. Furthermore, the observations of Loeb and colleagues (22) that inorganic chemicals affect the fidelity of DNA polymerases suggest that these might yield abnormal DNA by a mechanism distinct from the electrophilic genotoxic compounds.

The category of epigenetic carcinogens contains all types of carcinogens for which no evidence of genotoxicity has been found through appropriate tests. The lack of an interaction with DNA for these chemicals would obviously exclude this as a mechanism of action. On the other hand, possible mechanisms may involve chronic tissue injury, hormonal imbalance, immunologic effects, or promotional activity on cells that are either genetically altered or have been independently altered by genotoxic carcinogens. Less likely, but also deserving consideration, are epigenetic effects mediated through covalent interactions with protein or RNA, yielding to faulty differentiation (10). This category contains solid-state carcinogens, hormones, immunosuppressants, cocarcinogens, and promoters (39).

It is important to note that this concept does not preclude that genotoxic carcinogens or more likely certain of their metabolites could also have epigenetic effects. Indeed, the potency of some carcinogens may reside in their combined promoting actions together with their genotoxicity.

This classification, which has gained support since its introduction, has major implications for risk extrapolation to humans from data on experimental carcinogenesis. Genotoxic carcinogens, because of their effects on genetic material, may be a distinct qualitative hazard to humans, albeit they do demonstrate dose-response features in their action (44). Nonetheless, these carcinogens are occasionally effective after a single exposure, act in a cumulative manner, and operate together with other genotoxic carcinogens having the same organotropism (26,30). Hence, it follows that the level of human exposure acceptable as a "no risk" environment needs to be evaluated most stringently in the light of existing data and relevant mechanisms.

In contrast, with various classes of epigenetic agents, the expression of carcinogenic effects requires high and sustained levels of exposure leading to prolonged physiologic abnormalities, hormonal imbalances, or tissue injury. Consequently, the risk due to exposure may be of a strictly quantitative nature, and in addition, involves a time element. This is almost certainly the case with estrogens which are carcinogenic in animal studies at high chronic exposure levels; otherwise every individual would develop cancer. Thus, with epigenetic carcinogens, it may be possible to establish a "safe" threshold of exposure, once their mechanism of action is elucidated.

TESTS FOR GENOTOXIC CARCINOGENS

The relevant tests for the detection of chemical carcinogens need to consider our classification. Tests suitable to reveal agents affecting DNA, or genotoxic carcinogens, are based on the fact that such agents, or other active metabolites, do react with DNA. In the last ten years, a number of such assays have been developed in prokaryotic and eukaryotic systems where the endpoint is a change in DNA. Whether or not a given test is reliable often depends less on the capability of detecting the alteration in DNA and more on the ability of the overall system to metabolize a given chemical to the reactive electrophilic species. This area has been reviewed in some detail (2,7,19,20,25,28,40,43,44).

Effects of Epigenetic Agents. Chemicals operating on cell systems as epigenetic agents act by diverse mechanisms that are definitely different from those involving genotoxic pathways. The immediate corollary of this concept is that test systems for genotoxic agents are not suitable, in general, for the detection of agents with epigenetic modes of action. Except for the well studied example (4,8,12,13,31,36) of the phorbol esters, the active ingredients in croton oil, there are few data on mechanisms and systems to detect epigenetic agents. In view of the potential importance of such agents in human carcinogenesis, this is a gap that requires urgent attention (6,23,29). Considering the specificity of such agents for select tissues, and sometimes even for certain species, it may be difficult to develop a generally valid and applicable system that can detect the spectrum of epigenetic carcinogens. In any case, the quantitative aspects of their action are different from those applying to genotoxic carcinogens since their mechanisms are so distinct. Thus far, the known agents have been tested mainly at high dose levels and seem to exhibit a no-effect threshold. Also, their action is reversible. For chemicals acting by such mechanisms that have a demonstrated role of cancer causation in man, an effective control mechanism to reduce cancer risk would be to adjust the levels of such agents in the human environment to lower levels.

Two contrasting examples will be noted. Tobacco smoke is a most complex material including genotoxic carcinogens such as polycyclic aromatic hydrocarbons, certain nitrosamines, and perhaps heterocyclic amines (14, 16,21,35,46). The major effect due to tobacco smoke, however, is due to the presence of co-carcinogens, and especially of promoting agents, acting by epigenetic mechanisms. Since the action of these agents is reversible, this accounts for the fact that in individuals who have not smoked for too long (typically less than 15-20 years) and who stop smoking, the postitive pressure of promotion is removed. Hence, such former smokers eventually resume the low risk of lung cancer of a nonsmoker (14,16,35). On the other hand, except for a minor promoting effect of salt (34), the agents leading to gastric cancer appear to be powerful genotoxic carcinogens, the action of which is basically irreversible (33,37). Migrants from a high-risk region to a low-risk region do not lose their propensity for gastric cancer (11,38).

In conclusion, the classification of chemical carcinogens not only yields an understanding of the precise mode of action of specific agents but also suggests the rational development of appropriate public health actions to reduce disease risk. This is especially necessary in the context of the complex human environment where exposure is rarely to a single agent but more likely involves complex mixtures of genotoxic carcinogens, epigenetic agents, and accelerators or inhibitors for each kind of agent. It is important to sort through these complex situations in order to have an understanding of the relative importance of specific elements and thus be able to take the appropriate steps, which are not only scientifically sound but are realistically feasible, to lower disease risk.

EXTENSION TO THE DEVELOPMENT OF SYSTEMATIC BIOASSAYS

Our classification of carcinogens has been used for their systematic detection through short-term in vitro and in vivo bioassays [see 39,44). In addition, the recognition of diverse classifications permits a critical evaluation of the effect of chemicals yielding ambiguous data in already completed bioassays to detect carcinogenicity. Such bioassays can be interpreted intelligently and logically if corollary data on the possible presence or absence of genotoxic properties are obtained through appropriate test systems.

Carcinogenicity of a chemical in an in vivo animal bioassay without in vitro evidence of genotoxicity, may mean that it acts as an epigenetic agent. The reliability of this conclusion depends upon the relevance of the in vitro bioassays. For example, certain organochlorine pesticides do not exhibit genotoxic effects in liver cell systems identical to the in vitro target cell for these agents (42). This finding constitutes a sound argument for the concept that these chemicals may act by epigenetic mechanisms. These agents, therefore, may represent only quantitative hazards to humans. The appropriate toxicologic dose-response studies would need to be performed to acquire data as to safe levels.

Acknowledgements: Research described in this paper was supported by USPHS grants and contracts, and contracts from the U.S. Environmental Protection Agency.
This publication is dedicated to the founder of the American Health Foundation, Dr. Ernst L. Wynder, on the occasion of the 10th anniversary of the Naylor Dana In-

stitute for Disease Prevention.
We thank Mrs. C. Horn for excellent editorial assistance.
(1) Althouse, R., Tomatis, L., Huff, J., Wilbourn, J., eds. IARC Monogr., Vol. 1-20, Suppl. 1, 1-69, 1979
(2) Ames, B.N. Science 204, 587-593, 1980
(3) Berg, P. Science 209, 296-303, 1981
(4) Boutwell, R.K., in, Carcinogenesis, vol. 2, Mechanisms of Tumor Promotion and Cocarcinogenesis, T.J. Slaga, A. Sivak, and R.K. Boutwell, Raven Press, New York, 1978
(5) Brookes, P. British Med. Bullet. 36, 1-104, 1981
(6) Burchenal, J.H. and Oettgen, H.F., eds. Cancer: Achievements, Challenges, and Prospects for the 1980's, vol. 1, Grune & Stratton, New York, pp 1-685, 1981
(7) Coon, M.J., Conney, A.H., Estabrook, R.W., Gelboin, H.V., Gillette, J.R. and O'Brien, P.J., eds. Microsomes, Drug Oxidations, and Chemical carcinogenesis, vol. 1, 2, Academic, New York, 1980
(8) Diamond, L., O'Brien, T.G., and Baird, W.M. Adv. in Cancer Res. 32, 1-75, 1980
(9) Doll, R. and Peto, R. J. Nat. Cancer Inst. 66, 1191, 1981
(10) Emmelot, P. and Kriek, E., eds. Environmental Carcinogenesis: Occurrence, Risk Evaluation and Mechanisms, Amsterdam, Elsevier/No. Holland, 1979
(11) Haenszel, W., in Persons at High Risk of Cancer, J. Fraumeni, ed., Academic, New York, pp 361-372, 1975
(12) Hecker, E., in, Naturally Occurring Carcinogens-Mutagens and Modulators of Carcinogenesis, E.C. Miller, J.A. Miller, I. Hirono, T. Sugimura, and S. Takayama, eds., Japan Scientific Societies Press, Tokyo, pp. 263-286, 1979
(13) Hecker, E., in, Cocarcinogenesis and the Biological Effects of Tumor Promoters, E. Hecker, N. Fusenig, F. Marks and W. Kunz, eds., Raven, New York, 1981
(14) Hiatt, H.H., Watson, J.D., and Winsten, J.A., eds. Origins of Human Cancer, Cold Spring Harbor Lab., New York, 1977
(15) Higginson, J. In, Environmental Carcinogenesis Occurrence, Risk Evaluation and Mechanisms, P. Emmelot and E. Kriek, eds. Elsevier/North Holland Biomed. Press., Amsterdam, 1979
(16) Hoffmann, D., Tso, T.C., and Gori, G.B. Prev. Med. 9, 287-296, 1980

(17) Hollaender, A. and DeSerres, F.J., eds. Chemical Mutagens, Principles and Methods for Their Detection, vol. 1-6, Plenum, New York, 1971-1980
(18) IARC Monographs (series), On the evaluation of the carcinogenic risk of chemicals to humans, International Agency for Research on Cancer, Lyon, France
(19) Jollow, D.J., Kocsis, J.J., Snyder, R., Vaino, H., eds. Biological Reactive Intermediates. Formation, Toxicity, and Inactivation, Plenum, New York, 1977
(20) Kawachi, T., Nagao, M., Yahagi, T., Takahashi, T., Sugimura, T., Matsushima, T., Kawakami, T. and Ishidate, M. in, Naturally Occurring Carcinogens-Mutagens and Modulators of Carcinogenesis, E.C. Miller, J.A. Miller, I. Hirono, T. Sugimura, S. Takayama, eds., Japan Scientific Societies Press, Tokyo, 1979
(21) LaVoie, E., Hecht, S.S., Hoffmann, D. and Wynder, E.L., in, Banbury Rept. No. 3, A Safe Cigarette?, Cold Spring Harbor Pub., Cold Spring Harbor, N.Y., pp 251-160, 1980
(22) Loeb, L.A. and Zakour, R.A., in, Nucleic Acid-metal Ion Interactions, G.T. Spiro, ed., John Wiley & Sons, New York, pp. 115-144
(23) Miller, A.B., Gori, G.B., Kunze, M., Graham, S., Reddy, B.S., Hirayama, T. and Weisburger, J.H. Prev. Med. $\underline{9}$, 189-196, 1980
(24) Miller, E.C. and Miller, J.A. Cancer $\underline{47}$, 2327-2345, 1981
(25) Montesano, R., Bartsch, H., Tomatis, L. and Davis, W., eds. IARC Sci. Pub. $\underline{27}$, IARC, Lyon, France, 1980
(26) Nakahara, W., in, Chemical Tumor Problems, W. Nakahara, Ed., Japanese Society for the Promotion Science, Tokyo, pp. 287-330, 1970
(27) Nelson, N. J. Nat. Cancer Inst. $\underline{67}$, 227-231, 1981
(28) Norpoth, K.H. and Garner, R.C., eds. Short-Term Test Systems for Detecting Carcinogens, Springer-Verlag, Berlin, 1980
(29) Reddy, B.S., Cohen, L.A., McCoy, G.D., Hill, P. and Weisburger, J.H. Adv. in Cancer Res. $\underline{32}$, 237-345, 1980
(30) Schmähl, D., in, Critical Reviews in Toxicology, vol. $\underline{6}$, CRC Press, Boca Raton, Fla., pp. 257-281, 1979
(31) Slaga, T.J., Sivak, A., Boutwell, R.K., eds. Mechanisms of Tumor Promotion and Cocarcinogenesis, vol.

II, Raven Press, New York, 1977
(32) Straus, D.S. J. Nat. Cancer Inst. 67, 233-241, 1981
(33) Sugimura, T. and Kawachi, T., in Gastro-intestinal Tract Cancer, M. Lipkin and R. Good, eds., Plenum, New York, pp 327-342, 1978
(34) Tatematsu, M., Takahashi, M., Fukushima, S., Hananouchi, M. and Shirai, T. J. Nat. Cancer Inst. 55, 101-104, 1975
(35) Van Duuren, B.L., in, Chemical Carcinogens, ACS Monogr. 173, C.E. Searle, ed., American Chemical Society, Wash. D.C., pp 24-51, 1976
(36) Weinstein, I.B., Lee, L.S., Fisher, P.B., Mufson, A. and Yamasaki, H., in, Environmental Carcinogenesis, P. Emmelot and E. Kriek, eds., Elsevier, Amsterdam, pp 265-286, 1979
(37) Weisburger, J.H., Marquardt, H., Mower, H., Hirota, N., Mori, H. and Williams, G.M. Prev. Med. 9, 297-304, 1980
(38) Weisburger, J.H., Reddy, B.S., Hill, P., Cohen, L.A., Wynder, E.L. and Spingarn, N.E. Bullet. N.Y. Acad. Med. 56, 673-696, 1980
(39) Weisburger, J.H. and Williams, G.M., in, Casarett and Doull's Toxicology, 2nd ed., J. Doull, C. Klaassen, M. Amdur, eds., Macmillan, New York, 1981
(40) Weisburger, J.H. and Williams, G.M., in, Cancer: A Comprehensive Treatise, vol. 1, 2nd ed., F.F. Becker, ed., pp. 241-333, Plenum, New York, 1981
(41) Weisburger, J.H. and Williams, G.M. Science, in press, 1981
(42) Williams, G.M. Ann. N.Y. Acad. Sci. 349, 273-282, 1980
(43) Williams, G.M., Kroes, R., Waaijers, H.W., van de Poll, K.W., eds., The Predictive Value of Short-Term Screening Tests in Carcinogenicity Evaluation, Elsevier, Amsterdam, New York, 1980
(44) Williams, G.M. and Weisburger, J.H. Ann. Rev. Pharmacol. Toxicol. 21, 292-416, 1981
(45) Wynder, E.L. and Gori, G.B. J. Nat. Cancer Inst. 58, 825-832, 1977
(46) Wynder, E.L. and Hoffmann, D. New Eng. J. Med. 300, 894-903, 1979

What Is the Predictive Value of Short-Term Mutagenicity Testing to Determine Carcinogenic Potential of Compounds?

Dietrich Schmähl and Beatrice Luise Pool

Institute of Toxicology and Chemotherapy, German Cancer Research Center, Heidelberg, F.R.G.

ABSTRACT The carcinogenic potential of a substance can only be assessed when dose response relations have been studied in as many animal species as possible. Only then is it possible to predict the carcinogenic risk of the substance for man, all the more when regarding the organotropy which is essential for the evaluation of human carcinogens. Mutagenicity tests, however, do not reveal organotropic effects.
It is well established that these and other short-term tests measuring diverse biological activities of compounds are of major importance for qualitative evaluation on the carcinogenic potential of many substances. In our opinion, however, they are mainly applicable for predicting activities of genotoxic substances. Their predictive value for epigenetic agents remains to be investigated.

When regarding the activity of chemical carcinogens, two different classes of compounds may be distinguished from a toxicologist's point of view: Compounds which are local carcinogens (e.g. inhalation of tobacco smoke producing bronchus cancer in man) and compounds which act systemically (e.g. aromatic amines, producing bladder cancer in man).

Furthermore a distinction must be made for genotoxic substances and other compounds with an "epigenetic" mode of action.

It is the aim of the following essay to discuss the possibility of predicting the carcinogenic potential of chemicals with the aid of mutagenicity tests. However, the prediction of the carcinogenic potential of a given compound is already very difficult to achieve using data from animal experiments, or even from human pathology (16,23,26). The carcinogenic potential of a compound is its "power" of inducing tumors in man. This ability is determined with its potency of effectiveness in animal experiments, unless adequate human epidemiological data is available. "Potency" of a carcinogen is experimentally determined on the basis of dose needed for tumor induction, incidence and type of tumors, their biological and histological behaviour and the latent periods of tumors induced. So-called potent carcinogens induce cancer in various animal species and their organotropy is dependant on species and dose (24). Such "multipotent" carcinogens are strongly suspected to be also carcinogenic in man. "Weak" carcinogens induce tumors only in one or two animal species, in low incidence after applying high doses. The distinction of potent (high risk for man) and weak (low risk for man) carcinogens is generally only possible when exact dose effect relationships of a given compound have been established in several animal species. Of course, an enormous number of experiments must be performed to produce the necessary data, and therefore only very few compounds have been studied so extensively (22,24).

For evaluating a chemical carcinogen it is very important to regard its organotropy of action. The carcinogenic effectiveness of vinylchloride, for instance, might not have been recognized had it not induced a very rare type of tumor, the angiosarcoma mainly of the liver. This is similarly true for ß-naphtylamine, which induced urinary bladder tumors in man. Had the major activity of these compounds been limited to inducing stomach or intestinal cancer, their carcinogenic effectiveness would not have been detected, since the high background of the so-called spontaneous tumors would not allow for recognition of the few additional compound-induced tumors. These two examples imply how essential the organotropic effectiveness of compounds is, when evaluating chemical carcinogens for their potential in inducing tumors in humans. At present, however, the prediction of carcinogenic organotropy of chemicals can not be resolved with mutagenicity assays.

This presentation will briefly deal with the evaluation of soluble epigenetically active substances. Recently we were able to induce skin carcinoma after topically administering hydrochloric acid and potassium hydroxide solution to the mouse skin. We expect an epigenetic and not a genotoxic effect to be responsible for the induction of these tumors. A similar example is the Barret syndrome in man, where there is a constant reflux of hydrochloric acid and chyme in the lower esophageal region. The observed hydrochloric acid induced precancerous stages are probably due to an epigenetic mode of action.

In this context the problematics of testing insoluble materials such as asbestos, glass fibers or metals with submammalian mutagenicity assays should also be covered. To quote a basic principle of pharmacology and toxicology: "Corpora non agunt nisi soluta". However, as we know, especially from data on human epidemiology, these compounds may induce tumors at high potency even though they are not dissolved in body fluids. A carcinogenic potential of these compounds is to date evidently not predictable with assays for mutagenicity.

There are numerous other examples of compounds which do not act via electrophiles. Therefore, it is important to remember that, besides the genotoxic acting compounds which will further be discussed, epigenetic mechanisms of action are not uncommon.

For genotoxic compounds, it is now well accepted that batteries of "short term tests" generate valuable information on a possible qualitative carcinogenic potential (2,5,9,17, 21,25). Correlation figures of 90% have been published for individual tests and combinations of tests to be able to predict potential carcinogens (14,20). Nevertheless, even for a 90% correlation a 10% non agreement must still be accounted for. It is these 10% false positives and false negatives which are the object of concern. Two general examples obtained in our laboratory were chosen to discuss possible reasons for the non-correlation: N-nitroso(2,2,2-trifluorethyl-ethyl)amine (F_3NDEA) is a structural analogue of N-nitrosodiethylamine, originally synthesized to study the metabolism of N-nitrosodialkylamines by blocking metabolism at specific molecular sites (19). With S-9 mix it was predominantly dealkylated at the unfluorinated alkyl group, suggesting the generation of a high proportion of a trifluoroethylating and a small proportion of an ethylating species (11). It was at most only marginally active in Salmonella typhimurium TA 1535 and even negative in Escherichia coli, measuring lethal DNA damage in repair deficient strains, when compared to N-nitrosodiethylamine. In in vivo experiments, it was found to be a carcinogen, inducing tumors in the nasal cavity and oesophagus after oral application to rats (19a), while NDEA acts as a liver and oesophagus carcinogen. The weak activity of F_3NDEA in bacterial systems therefore does not correlate with its carcinogenic activity. One possible explanation for the low activity of F_3NDEA in the bacterial systems could be that trifluoroethylation of the bacterial DNA did not occur in sufficient amount. This could be due to the high reactivity of the respective trifluoroethylating species prior to reaching the target DNA (11). Also a different metabolic capacity of intact cells versus S-9 mix (6, 8) is possible. From these results it is evident that mutagenicity and carcinogenicity of substances are not always comparable, even if the compounds are structurally closely related and well metabolized in vitro.

An opposite example is obtained from results of N-nitrosocimetidine, which is formed after N-nitrosation of cimetidine in vivo. N-nitrosocimetidine is a potent mutagen in Salmonella typhimurium and causes lethal DNA damage in repair deficient strains of E.coli WP2 without metabolic activation (18). Ongoing carcinogenicity experiments (16 months) show no carcinogenic activity of N-nitrosocimetidine.
This, however, was unexpected, since comparable doses of the closely related potent mutagen N-methyl-N'-nitro-N-nitrosoguanidine (MNNG) (7,12) already yielded 100% tumors. Why MNNG and N-nitrosocimeditine, which both generate the same methylating species upon decomposition in vitro, should vary so much in their ability to induce tumors in vivo is, as yet, unresolved. One postulated reason is that rate and pathway of decomposition of both compounds may be intrinsically different from each other. Thus for N-nitrosocimetidine, a denitrosation not yielding a methylating species may be favored (12). Numerous in vivo pharmacological and toxicological barriers including inactivation mechanisms, resorption, distribution, excretion, as well as protective processes such as repair of DNA lesions may additionally be responsible for this lack of tumor induction by an otherwise biologically active compound.

These two examples were chosen to demonstrate possible reasons for the false negative and false positives. Even when all test systems (including the carcinogenicity assays upon which the evaluations are based) are performed in the most sensitive manner, these discrepancies will probably always occur, since "not all mutagens are carcinogens" and vice versa. The scheme outlines some complicating factors involved during the genesis of a tumor, even after introducing potent genotoxic and epigenetic agents in vivo.

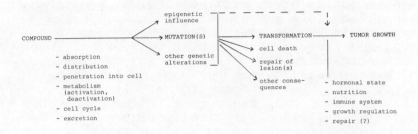

Determinations of potency of chemical carcinogens are very much to be desired in view of risk benefit assessment of human exposure. Numerous contradictory reports concerning the predictability of carcinogenic potency of genotoxic agents with short term tests have been published (1,3,4,5,10,13,15). These authors have closely described limitations of short term tests as well as of the carcinogenesis experiments which are taken as basis to evaluate the results of the short term assay data. A major limitation which these mutagenicity assays have in predicting potency of chemicals is their great variation in response when alterations of experimental parameters are introduced. Furthermore chemicals will respond in different orders of magnitude in different test systems. When sufficient data on carcinogenesis is not available, the choice of the response in a given test system to establish potency is quite arbitrary. In general we support the reservations voiced, and agree that mutagenicity tests are not able to detect actual potency of chemicals. The above detailed examples of our results were also described to substantiate our opinion.

There is no doubt that the introduction of short term tests for predicting biological activities of chemicals has been very useful in the recognition of potential carcinogens. However, the results generated in any of the available tests must be critically evaluated. A summary of considerations to be included in any evaluation to predict carcinogenic potential of unknown compounds with one special category of short term tests, the mutagenicity tests, is as follows:

1. Sufficient evidence is available that numerous carcinogenic compounds are not mutagenic.

2. Test compounds must be soluble to induce genotoxic effects, but not to induce tumors.

3. Commonly employed mutagenicity tests utilizing bacteria, other submammalian organisms or cell cultures, can not produce information on the possible organotropy of carcinogens.

4. The estimation of carcinogenic potency is very difficult to achieve with mutagenicity tests. This is even more apparent when considering limitations encountered in extrapolating data obtained in animals to the human situation.

5. High inter- and intralaboratory variability may be encountered during performance of these tests. Therefore only highly experienced investigators should participate in the critical evaluation of the results generated for determination of potential carcinogenicity.

6. To date it does not seem possible to evaluate the absolute potential of carcinogens solely with short term mutagenicity tests.

We do appreciate the necessity of future research programs in elucidating these aspects. Further, we recognize the invaluable information on the activities of compounds generated in short term tests and their importance for the general field of chemical carcinogenesis.

REFERENCES

1. Ames, B.N., and Hooper, K., Nature 274, 19-20 (1978)
2. Ames, B.N., Durston, W.E., Yamasaki, E. and Lee, F.D. Proc. Nat. Acad. Sci. 70, 2281-2285 (1973)
3. Andrews, A.W. Mutation Res. 62, 396-398 (1979)
4. Ashby, J. and Styles, J.A., Nature 271, 452-455 (1978)
5. Bartsch, H., Malaveille, C., Camus, A.-M., Martel-Planche, G., Brun, G., Hautefeuille, A., Sabadie, N., Barbin, A., Kuroki, T., Drevon, C., Piccoli, C. and Montesano, R. Mutation Res. 76, 1-50 (1980)
6. Bigger, C.A.H., Tomaszewski, J.E., Dipple, A., and Lake, R.S., Science 209, 503-505 (1980)
7. Forster, A.B., Jarman, M. and Manson, D. Cancer Letters 9, 47-52 (1980)
8. Glatt, H.R., Billings, R., Platt, K.L. and Oesch, F. Cancer Res. 41, 270-277 (1981)
9. Hollstein, M., McCann, J., Angelosanto, F.A., and Nichols, W.W. Mutation Res. 65, 133-226 (1979)
10. Hooper, N.K., Harris, R.H. and Ames, B.N. Science 203. 602-603 (1979)
11. Janzowski, C., Pool, B.L., Preussmann, R. and Eisenbrand, G., in preparation
12. Jensen, D.E. and Magee, N. Cancer Res. 41, 230-236 (1981)
13. Maugh, T.H. Science 202, 27-41 (1978)
14. McCann, J. and Ames, B.N. Proc. Nat. Acad. Sci. 73, 950 (1976)
15. Meselson, M. and Russel, K. In: Origins of human cancer - Book C. (Hyatt, Watson and Winston, Eds.) Cold Spring Harbor Laboratory pp. 1473-1481, 1977
16. Pitot, H.C. J. Environm. Pathol. Toxicol. 1 3, 431-450 (1980)
17. Poirier, L.A. and de Serres, F.J. J. Natl. Cancer Inst. 62, 919-926 (1979)
18. Pool, B.L. ,Eisenbrand, G. and Schmähl, D. Toxicology 15, 69-72 (1979)
19. Preussmann, R., Habs, M., Pool, B.L., Stummeyer, D., Lijinsky, W. and Reuber, M.D. Carcinogenesis, in press

19a. Preussmann, R. et al., in preparation
20. Purchase, I.F.H., Longstaff, E., Ashby, L., Styles, J.A., Anderson, D., Lefevre, P.S. and Westwood, F.R. Br. J. Cancer 37, 873-959 (1978)
21. Sankaranarayanan, K. J. Cancer Res. Clin. Oncol. 99, 87-102 (1981)
22. Schmähl, D., Crit. Rev. Toxicol. 6, 257-281 (1979)
23. Schmähl, D. and Habs, M. In: The Evaluation of Toxicological Data for the Protection of Public Health (W.J.Hunter and J.G.P.M.Smeets, Eds.) Proc of the Int. Colloq. Luxembourg Pergamon Press pp. 59-63, (1976)
24. Schmähl, D. and Habs, M. Oncology 37, 237-242 (1980)
25. Sugimura, T., Sato, S., Nagao, M., Yahagi, T., Matsushima, T., Seine, Y., Takeuchi, M. and Kawachi, T. In: Fundamental in Cancer Prevention (P.N. Magee, Ed.) University of Tokyo Press, Tokyo/Univ. Park Press, Baltimore, pp. 191-215 (1976)
26. Tomatis, L., Agthe, C., Bartsch, H., Huff, J., Montesano, R., Saracci, R., Walker, E. and Wilbourn, J. Cancer Res. 38, 877-885 (1978)

Mutagens and Carcinogens in the Environment

Overview of Mutagens and Carcinogens in the Environment

Lawrence Fishbein

National Center for Toxicological Research, Jefferson, Arkansas, U.S.A.

ABSTRACT This overview focuses on aspects of the burgeoning chemical industry, primarily over the last several decades and its sequalae, development of many new organic chemical products and increased product applications. Although tens of thousands of chemicals are in common use, those for which there is substantial evidence concerning human carcinogenicity number about two dozen and those for which we have substantial evidence concerning experimental carcinogenicity number in the hundreds. Emphasis is placed on occupational exposure and two major classes of environmental agents, e.g., chlorinated hydrocarbons and alkylating agents. For the vast majority of environmental agents (including perhaps a large number of potential carcinogens and/or mutagens) to which humans are exposed to, it is not now possible to determine the extent to which chemical exposure will influence future cancer rates. We can help prevent human cancers by preventing or minimizing exposures to these agents.

1. <u>Introduction</u>: The chemical industry has grown exponentially since the late 1930's and early 1940's as exemplified by the doubling every 7 to 8 years of production levels of synthetic organic chemicals with the total production accounting for over 175 billion pounds per year (1) and the introduction of many new organic chemical products and increased product applications (2-5). The rate of growth in the production of several classes of bioavailable substances (e.g., plastics and resin

materials, plasticizers, flavors and perfumes (benzenoid and naphthalenoid) and food, drug and cosmetic dyes has been particularly explosive. All have the common characteristic of being present in products designed for uses in, on, or very near to humans, thus it can be assumed that any toxic or potentially toxic chemical substance contained therein are likely consumed by or absorbed by the product user to some degree (1).

Estimates vary concerning the number of chemicals in existence, the number in significant production in the United States, and the number introduced yearly in the United States and worldwide. Approximately 3.5 million to 4.3 million chemicals are known to be in existence (1,4-6), and the number is believed to be growing by about 10% per year (7). Approximately 25,000 to 50,000 chemicals are in significant production in the United States alone (1,3,4,8), and about 50,000 to 65,000 chemicals (not including pesticides, pharmaceuticals and food additives) are believed to be in everyday use (6). About 44,000 chemical substances are listed in the EPA inventory of substances in commercial use in the United States (9). The European chemical industry puts in excess of 30,000 compounds on the market today, retailed in mixtures and preparations exceeding several hundreds of thousands. On a world-wide basis, more than 100 million tons of organic compounds are sold annually and in the Federal Republic of Germany more than 10,000 cleaning preparations are currently retailed (10).

In the United States alone, 300 - 700 to as many as 3000 new industrial chemicals are introduced annually into economic use (4,8). It is a reasonable assumption that more than 1000 new chemicals are introduced each year worldwide. In the past decade, production of synthetic organic chemicals in the United States has expanded 255%.

Some 2500 chemicals or mixtures are reported to be in current use by the plastics industry alone as antioxidants, antistatics, blowing agents, catalysts, colorants, flame retardants, lubricants, plasticizers, stabilizers, and ultraviolet (u.v.) absorbers (11). The EPA estimates that there may be as many as 1500 active ingredients in pesticides, while the Food and Drug Administration (FDA) estimates that there about 4000 active ingredients in drugs, 2500 additives for nutritional value and flavoring and 3000 chemicals used to promote product life (12). The National Academy of Science stated that 1 billion pounds of toxic matter was being introduced yearly for pest control and that the knowledge of the potential harm was superficial (13).

While the number of individuals directly involved in the preparation of these chemicals and their by-products (e.g., plastics and polymers) is relatively small compared to the many industrial segments and processes, their exposure to potentially hazardous (carcinogenic and mutagenic) substances can be very substantial. Many more individuals may be indirectly exposed to these potential carcinogens and mutagens through (1) use applications that contain entrained materials and (2) inhalation, ingestion, or absorption of these agents from air, water, and food, in which they are present because of escape into the atmosphere, leaching, and so on.

A NIOSH survey conducted in 1972 to 1974 showed that more than 7 million workers in the United States were exposed to 20,000 trace name products containing toxic substances regulated by OSHA. Another 300,000 workers may be exposed to similar products containing one of the 16 cancer-causing agents regulated by OSHA. We do not know precisely what percentage of chemicals may be hazardous. Only a very small fraction of the approximately 50,000 synthetic chemicals which are produced and used in significant quantities have been tested for carcinogenicity or mutagenicity before their use and even very high-production chemicals with extensive human exposure were produced for decades before adequate carcinogenicity or mutagenicity tests were performed (3). For example, before ethylene dichloride was found to be a carcinogen, 100 billion pounds of it was produced while vinyl chloride reached a production rate of 5 billion pounds a year before it was recognized as a carcinogen (13).

Of all the chemicals on the market, a relatively small proportion, approximately 7000, have been tested to determine their cancer-causing potential; approximately 1500 have been found to be tumorigenic in test animals. It has been suggested that at least half of these studies were completely inadequate for their purpose, so that the actual number of compounds tested for carcinogenicity may be closer to 3500 and the respective number found to be carcinogenic closer to 750 (2,12). The number of "known carcinogens" also varies considerably depending on the degree of stringency adopted for accepting evidence of carcinogenicity (2). To date, 26 substances or workplace exposures have been assessed by IARC to cause cancer in humans and through laboratory studies, 221 substances are considered as carcinogenic (14).

2. __Epidemiological Considerations__: Although the etiology of human neoplasia is with rare exceptions unknown, there have been various estimations of the percentage of cancers which could be directly or indirectly attributable to environmental factors (using the term in its widest sense and including life style) (15-17). These have ranged from 75% (15) to 80% (16) to 90% (17). [This has been translated to mean the occupational environment (18)]. One can dispute these percentages and reduce them to much lower and more acceptable proportions, but the fact remains that a certain proportion of human cancers are certainly due to environmental factors (19).

Although much emphasis in the past decade has been on occupationally induced cancers and their etiology, there is a growing recognition that these cancers, while largely preventable through proper engineering and safety controls and education of personnel and management account for only a small fraction of cancers, by most recent estimates to range between 1-5% (2,20-23) (Table 1). The proportion of current U.S. cancer deaths attributed to occupational factors was provisionally recently estimated by Doll and Peto (20) to range between 2 and 8% (best estimate 4%) with lung cancer being the major contributor. This is far smaller than has recently been suggested by various U.S. government agencies. Exposure to some agents has been only on a small scale as in the case of a drug introduced briefly for the treatment of a rare disease where exposure to others has been intensive and widespread, as in the case of tobacco smoking and hundreds of thousands of cancers have been caused each year (20).

The hazards that have been recognized thus far tend to be those which increase the relative risk of some particular type(s) of cancer very substantially (e.g., liver angiosarcoma among PVC manufacturers). However, important occupational hazards may quite possibly exist that have not yet been detected because the added risk is small compared with that due to other causes, or because only relatively few people have been exposed, or simply because a hazard has not been suspected and hence not looked for.

Doll and Peto (20) suggested that although different figures may apply to other countries, the __minimum__ proportion of all current U.S. cancer deaths attributable to occupation can hardly be less than 2 or 3%. Even this relatively small percentage provides little solace. Thus,

Table 1. Causes of Human Cancer in the United States (20)

	% of all cancer deaths	
	Best estimate	Range of acceptable estimates
Tobacco	30	25 – 40
Alcohol	3	2 – 4
Diet	35	10 – 70
Food additives	<1	–5 – 2
Reproductive/sexual behavior	7	1 – 13
Occupation	4	2 – 8
Pollution	2	<1 – 5
Industrial products	<1	<1 – 2
Medicines/medical procedures	1	0.5 – 3
Geophysical factors	3	2 – 4
Infection	10 ?	1 – ?
Unknown	?	?

although the proportion is small, cancer is so common and the U.S. population so large that even 2% of cancer deaths amounts to more than 8,000 deaths in the United States every year. Occupational cancer tends to be concentrated among relatively small groups of people among whom the risk of developing the disease may be quite large and such risks can usually be reduced, or even eliminated once they have been identified.

It should also be noted that cancer in humans seldom develops until one or more decades after beginning exposure to a carcinogen and it may well be too soon to be certain whether agents are human carcinogens if they were introduced into industry only during the last 20 years. Ames (3) suggested that the flowering of the chemical age may be followed by genetic birth defects and a significant increase in human cancer during the 1980 decade because of the 20-30 year lag if many of these chemicals with widespread human exposure are indeed powerful mutagens and carcinogens.

3. **Major Class of Carcinogens and Mutagens**: Chemical carcinogens and mutagens represent a spectrum of agents that vary in activity by a factor of at least 10^7 and have strikingly different biological activities, ranging from highly reactive molecules that can alkylate macromolecules and cause mutations in many organisms, to compounds that are hormonally active and have neither of these actions.

While a large number of chemicals are suspected to be carcinogenic on the basis of animal assays, their role in human cancer remains undetermined and possibly undeterminable. To obtain statistically significant data with a reasonable number of mice or rats, cancer incidences of 5 to 10% must be obtained necessitating the use of large doses. Unless there are very high occupational exposures of limited numbers of men, there are few instances of carcinogen exposure which affect such a high proportion of the human population. Thus, assessment of risk in man requires extrapolation of these data to a human population approximately 10^6 times larger, with exposure levels likely to be approximately 10^4 times lower, and with a lifespan 30 times longer (24). It would appear to be extremely difficult, considering both the range of potency and variability of exposure, to predict the proportions of a society that would be at potential risk to a known carcinogen over extended time periods (18).

Two major categories of carcinogenic and mutagenic agents that humans can be substantially exposed to are the halogenated hydrocarbons and alkylating agents since numerous members of these classes are produced and used in substantial volumes in a broad spectrum of applications.

Toxicologial concern is now manifest to practically all of the major commercial halogenated hydrocarbons particulary the lower molecular weight C_1-C_4 chlorinated alkanes and alkenes, numerous members of which have extensive utility as solvents (including extraction of components from foods), fumigants, aerosol propellants, degreasing agents, dry-cleaning fluids, refrigerants, flame-retardants, synthetic feedstuffs, cutting fluids and as intermediates in the production of textiles and plastics and other chemicals, etc. Production figures indicate a growth rate of the halocarbons industry in excess of 6% per annum. The chlorinated aliphatic saturated and unsaturated hydrocarbons (methyl chloride, chloroform, carbon tetrachloride, methylene chloride, ethylene dichloride; 1,1,2-trichloroethane; methyl chloroform; trichloroethylene; perchloroethylene; 1,1,2,2-tetrachloroethane; epichlorohydrin; allylchloride; BCME; Freons F-11 and F-12 and halothane) represent one of the most important categories of industrial chemicals from a consideration of volume, use categories, environmental and toxicological consideration and potentially populations at risk which can include significant segments of occupational, consumer and general public. Available data also suggests that a number of chlorinated hydrocarbons, particularly solvents and active ingredients contained in aerosol products are an important class of air pollutants in the indoor environment (25).

A number of chlorinated hydrocarbons may also occur in drinking water via contamination of water supplies by effluents from chemical industry, or in part, as in the case of chloroform or other trihalomethanes, via disinfection of water supplies by chlorination thus representing another area of large potential human exposure.

There are 15 major structural categories of direct-acting alkylating agents that have been reported to be carcinogenic and mutagenic. These include: epoxides, lactones, aziridines, aliphatic sulfuric acid esters, cyclic aliphatic sulfuric acid esters, alkyl fluorosulfuric acid esters, sultones, diazoalkanes, aryl dialkyltriazenes, phosphoric acid esters, aldehydes, alkane halides, alkyl halides, haloethers and haloalkanols.

The most important alkylating agents, based on numbers of potential carcinogens and/or mutagens, production volume, use applications and potential populations at risk are the epoxides which include: ethylene oxide, propylene oxide, butylene oxide, epichlorohydrin, glycidol, glycidaldehyde, and diglycidyl resorcinol ether and glycidyl ethers.

References:

1. Davis, D. L., and Magee, B. H., Science, 206 (1979) 1956-1957
2. Maugh, T. H., II, Science, 201 (1978) 1200-1205
3. Ames, B. N., Science, 204 (1979) 587-593
4. Fishbein, L., Potential Industrial Carcinogens and Mutagens, Elsevier, Amsterdam (1979) p. 534
5. Fishbein, L., J. Toxicol. Environ. Hlth., 6 (1980) 1133-1137
6. Carter, L. J., Science 201 (1979) 1198-1199
7. CEQ, Toxic Substances, Council on Environmental Quality, Washington, DC (1971)
8. Ames, B. N., Am. Chem. Soc. Symp., 94 (1979) 1-11
9. Roderick, H., Ann. NY Acad. Sci., 363 (1981) 233-244
10. Schmidt-Bleek, F., and Muhs, P., In: The Use of Biochemical Specimens for the Assessment of Human Exposure to Environmental Pollutants, (eds) A. Berlin, A.H. Wolff and Y. Hasegawa, Martinus Nijhoff Publ., The Hague, Boston, and London (1979) 313-319

11. Withey, J. R., In: Origins of Human Cancer (eds) H. H. Hiatt, J. D. Watson, and J. A. Winsten, Book A, Cold Spring Harbor Lab., Cold Spring Harbor, New York (1977) 219-242
12. Maugh, T. H., II, Science, 199 (1978) 162
13. Ramo, S., Science, 213 (1981) 837-842
14. Tomatis, L., Agthe, G., Bartsch, H., et al., Cancer Res., 38 (1979) 877-885
15. WHO Tech. Rep. Ser. 276, Geneva (1964)
16. Boyland, E., Prog. Exp. Tumor Res., 11 (1969) 222-234
17. Higginson, J., Proc. Can. Cancer Res., Conf., 8 (1968) 40-75
18. Council of Scientific Affairs of American Medical Assoc. Carcinogen Regulation, J. Am. Med. Assoc., 246 (1981) 253-256
19. Tomatis, L., Ann. NY Acad. Sci., 271 (1976) 396-409
20. Doll, R., and Peto, R., J. Natl. Cancer Inst., 66 (1981) 1193-1308
21. Weisburger, J. H., Reddy, B. S., et al., Bull. NY Acad. Med., 56 (1980) 674-690
22. Higginson, J., and Muir, C. S., In: "Cancer Medicine", Holland, J. F., and Frei, III (eds) Lea & Febiger, Philadelphia (1981)
23. Higginson, J., J. Environ. Pathol. Toxicol., 3 (1980) 113-125
24. Weinhouse, S., In: "Origins of Human Cancer" Book C, "Human Risks Assessment" (ed) Hiatt, H. W., Watson, J. D. and Winsten, J. A., Cold Spring Harbor Laboratory, Cold Spring Harbor, New York (1977) pp. 1307-1310
25. Bridbord, K., Brubaker, P. E., et al., Env. Hlth. Persp., 11 (1975) 215-220

Styrene—A Widespread Mutagen: Conclusions from the Results of Testing

Gösta Zetterberg

Department of General Genetics, University of Uppsala, Uppsala, Sweden

ABSTRACT Styrene is mutagenic in a variety of tests that include metabolic activation. The active metabolite has been identified as styrene-7,8-oxide, an olefinic epoxide, shown to react with nucleic acid bases and to be mutagenic in prokaryotic and eukaryotic systems. It also induces chromosome aberrations and sister chromatid exchanges (SCE). Styrene oxide has been found to be carcinogenic in rats. Persons occupationally exposed to styrene have an increased number of chromosome aberrations in their lymphocytes. A significant increase in the incidence of cancer in exposed persons has not been reported. Styrene shares properties with vinyl chloride in use, metabolism, and positive outcome in genotoxic tests. The procedures of risk assessment of these two plastic monomers are compared.

PRODUCTION AND COMMERCIAL USE OF STYRENE

Styrene has been produced commercially for more than 40 years. In 1977 the worldwide production is estimated to have been 7 000 million kg. U.S.A.(40 %), West-european countries (33 %), and Japan (16 %) produced the main part. The predominant use of styrene is in the production of plastics and polyester resins. About 10 % is used for making styrene-butadiene rubber. Plastics containing styrene are now used in the manufacture of a large variety of articles, ranging from bulky constructions like boats and big containers to small items like toys and thin tissues for food packaging.

EXPOSURE TO STYRENE

Humans are exposed to styrene monomer mainly in factories for the production of reinforced plastics. The highest concentrations of styrene have been detected in the air of factories making plastic boats and in the curing of tyres. Low concentrations of styrene monomer have been registered in the ambient air, in rivers and in drinking water in industrial regions where styrene is produced or processed. Small amounts of styrene have also been found in food kept in polystyrene containers. Thus a large population is exposed to styrene but the level of exposure is very low for most people. It is believed that it is only the occupational exposure to styrene that can cause health problems. The number of workers exposed is unknown but for small countries like Finland and Sweden it has been estimated to be more than 2 000 in each.

Standards of hygiene in the work environment have been determined in countries that manufacture styrene. The figures range from 1.2 ppm which is the maximum allowable concentration in the USSR to 100 ppm as an eight--hour time-weighted average in FRG and U.S.A. factories. Neurological and psychological disturbances are toxic effects often encountered among persons exposed to styrene. It is likely that the TLVs have been determined according to estimates of the risk for such toxic effects rather than the risk for long term effects such as cancer and heritable defects, since the information from epidemiological investigations of the genotoxic effects of styrene is still very limited.

METABOLISM OF STYRENE

People are exposed to styrene mainly by breathing contaminated air. Styrene is easily absorbed through the lungs (60-90 % of the inhaled amount) and also through the skin and intestines. It is soluble in blood and accumulates in human adipose tissue where the half-life is 2-5 days. It is likely that styrene exerts its toxic effects through its metabolites. At high doses the metabolic clearance rate of styrene is nonlinear and the effects from lower concentrations may be underestimated if extrapolated from observations of the effects at saturation levels.

In the first step styrene is metabolized to styrene--7,8-oxide by mixed-function oxidases and the metabolism proceeds from styrene oxide to styrene glycol, catalyzed by epoxide hydrolase. Styrene oxide can also be conjugated with glutathione. Both styrene and styrene oxide are metabolized in the liver, kidney, intestines, lung, skin, and blood. The major urinary metabolites of styrene are mandelic acid and phenylglycoxid acid, the amounts of which have been used in measurements of recent exposure to styrene. It is likely that the active metabolite causing genotoxic effects in different test systems is styrene-7,8-oxide. Another substance, styrene-3,4-oxide, has also been suggested as the active metabolite but quantitatively the metabolic route via this compound is considered less important.

TESTING RESULTS OF STYRENE AND STYRENE-7,8-OXIDE

Reactions with nucleic acid components Styrene-7,8--oxide has a reactive oxirane ring and can bind covalently to macromolecules such as proteins and nucleic acids. Its alkylating activity has been determined to be a third of that of epichlorohydrin, a carcinogenic and mutagenic epoxide. The nucleic acid adducts formed by styrene oxide were identified as N-3 alkyl cytosine and N-7 alkyl guanine.

Data from mutagenicity tests (Summarized in Table 1) Styrene-7,8-oxide is mutagenic to base-pair substitution strains of Salmonella and E. coli. In the presence of mammalian liver microsomes, the mutagenic effect of styrene oxide is usually reduced because epoxide hydrolase and glutathione enhance the breakdown of the epoxide. When styrene has been tested with bacteria, both in the presence and in the absence of a liver microsome preparation, the results are contradictory. Differences in the activating systems have been accused for this discrepance. If it is assumed that styrene oxide is the active metabolite, its formation from styrene by mixed-function oxidases in the S9-mix must be greater than its breakdown by epoxide hydrolase. Differences in the relative activities of these enzymes are likely to occur in different preparations.

Table 1 Mutagenicity of styrene and styrene-7,8-oxide

Test system	Styrene		Styrene-7,8-oxide	
	Metabolic activation			
	+	−	+	−
Bacteria	±	−	+	+
Yeast	−	−	+	+
Drosophila	+		+	
V79 Hamster cells	−	−	+	

Results from tests using yeast cells and V79 Chinese hamster cells agree with those form bacteria: styrene oxide is mutagenic while styrene is usually nonmutagenic even in the presence of liver microsomes. In host-mediated assays styrene was mutagenic to yeast and E. coli. In Drosophila both styrene and styrene-7,8-oxide increased the frequency of X-linked recessive lethals in the offspring of males fed with the compounds. Pretreatment of the males with phenobarbitone doubled the effect.

Data from tests on chromosome damage (Summarized in Table 2) In root tip cells of Allium, styrene caused chromosome breaks, as did styrene oxide in Vicia. In a Chinese hamster cell line (CHL), styrene induced chromosome aberrations in presence of an activating system from methylcholanthrene-induced rats (5, 8). In a different cell line (CHO) styrene induced SCE only when an epoxide hydrolase inhibitor was added to the S9-mix (16).

Norppa et al. (12) have shown that human lymphocytes cultured and treated *in vitro* can metabolize styrene to styrene-7,8-oxide in amounts that can be detected analytically. A dose-dependent increase of SCE was obtained at concentrations of styrene that could be expected from occupational exposure. Styrene oxide was as effective as methylmethanesulphonate in the induction of SCE. Among four epoxides known to be mutagens acting directly in microbioal tests, styrene oxide was the most effective in inducing chromosome aberrations and SCE in cultured human lymphocytes (13). Styrene and vinyltoluene were about equally effective (10).

Table 2 Chromosome damage by styrene and styrene-7,8-oxide

Test system	Type of damage	Styrene		Styrene-7,8-oxide	
		Metabolic activation			
		+	-	+	-
Allium	Aberr.	+			-
Vicia	"				+
Hamster cells					
V79	Aberr.				+
CHL	"	+	-		
CHO	SCE	+	-	+	+
Human lympho-	Aberr.	+		+	
cytes, *in vitro*	SCE	+		+	
	Micronuclei	+		+	

The results from experimental *in vivo* studies of styrene and styrene-7,8-oxide are conflicting but tests with both Chinese hamsters and mice are essentially negative, even when styrene oxide has been administered. Norppa (11) has shown that styrene in high doses can induce micronuclei in bone marrow cells of mice. It is assumed that mice are more sensitive than Chinese hamsters to styrene exposure because they inactivate styrene oxide at a slower rate.

EPIDEMIOLOGICAL INVESTIGATIONS OF PERSONS EXPOSED TO STYRENE AT WORK

Chromosome aberrations and SCE Several cohorts of persons exposed to styrene have been investigated and found to have an increased frequency of chromosomal aberrations. Usually no dose response has been observed. However, when the exposure to styrene was expressed as the accumulated dose over a long time, several years, Andersson et al. (2) observed a positive dose response for the less exposed personel in a plastic-boat factory but not for the more exposed. An interpretation of this result is that the lymphocytes can accumulate chromosomal damage over long time and that for the high dose level saturation has been reached for the metabolic processes required to render styrene clastogenic.

Little or no increase in SCE has been observed in the epidemiological investigations. This is not to be expected from the results of *in vitro* exposure of lymphocytes, where SCE was a more sensitive measure of the effects of styrene oxide than chromosome aberrations (13). It may be that the damage leading to SCE can be repaired by the nondividing peripheral lymphocytes while the damage later observed as chromosome aberrations is less repairable.

Carcinogenicity of styrene exposure Only a few studies have been published on the possible carcinogenicity of styrene in man. Ott (14) observed a significant increase in the number of leukemias among employees exposed to styrene compared to nonexposed in the same company. The increase was not statistically significant when calculated on the basis of national figures for the incidence of leukemias in the U.S.A. An investigation of Swedish workers exposed to styrene revealed no increase in the incidence of cancer (1). Styrene-butadiene rubber workers had a higher risk for lymphatic and hematopoietic cancers but the exposure pattern was too complicated to allow any conclusions about the carcinogenic effect of styrene alone (9).

Carcinogenicity tests of styrene in experimental mammals showed that high doses administered orally to mice increased the incidence of lung tumours in the offspring (15). The incidence of leukemia and lymphosarcoma types of tumour was increased in rats (6). Styrene-7,8-oxide has been reported to be strongly carcinogenic in rats (7).

EMBRYOTOXICITY AND TERATOGENICITY

Styrene is embryotoxic in rats, hamster and chicken, but its teratogenic effect is low, except in chicken. In Finland an increased frequency of spontaneous abortions was found among women in the styrene industry. Sorsa et al. (17) found indications of an increase in the frequency of chromosomal aberrations in the lymphocytes of children exposed *in utero*.

COMPARISON BETWEEN STYRENE AND VINYL CHLORIDE

Vinyl chloride is mutagenic, teratogenic and carcinogenic in man. It is likely that the active metabolite is

its peroxide, chloroethylene oxide, generated by mixed-
-function oxidases. As shown in Table 3, vinyl chloride
has been found positive in all tests on genotoxicity.

Table 3 Summary of data from tests and epidemiological
studies of the genotoxicity of vinyl chloride and styrene.

Chemical	Experimental data			Exposed persons			
	Muta-genic	Car-cino	Tera-togen	Chrom. aberr. or SCE	Cancer	Abor-tions	Malfor-mations
Vinyl chloride	+	+	+	+	+	+	+
Styrene	+	±	+	+	±	±	±

All these results were obtained in tests started after 1974 when it was reported that vinyl chloride induced angiosarcoma in man. Technical improvements in the factories for vinyl chloride production have made it possible to reduce the air contamination of vinyl chloride below detection level and the TLV for vinyl chloride has been drastically reduced in all countries.

Much valuable information on the mode of action has been gained through the testing of vinyl chloride. However, we must admit that the testing results came too late. It was already an established fact that vinyl chloride had genotoxic effects in humans when the testing results were produced. So the actions were taken in the wrong order.

With the vinyl chloride story as background to the present information on styrene, our test results as well as those from epidemiological studies challenge us to take preventive actions earlier this time. If we consider these results valid and if we believe in the value of testing for predicting a human risk we must be prepared to draw the conclusions. It is my opinion that the present levels of exposure in the manufacture of styren should be considerably reduced in order to avoid a repeat of the vinyl chloride story.

REFERENCES

Reviews on styrene (3, 4, 18) include references to

data published before 1979.
(1) Ahlmark, A. Styrene investigation. Epidemiological report. (Swedish.) Sveriges Plastförbund (1978)
(2) Andersson, H.C., Tranberg, E.Å., Uggla, A.H., and Zetterberg, G. *Mutat. Res. 73*, 387-401 (1980)
(3) Fishbein, L. Potential Industrial Carcinogens and Mutagens, Elsevier, Amsterdam (1979)
(4) IARC Monographs on the Evaluation of the Carcinogenic Risk of Chemicals to Humans, Vol. 19, Lyon (1979)
(5) Ishidate, M.Jr., and Yoshikawa, K. *Arch. Toxicol. 4*, 41-44 (1980)
(6) Jersey, G.C., Balmer, M.F., Quast, J.F., Park, C.N., Scheutz, D.J., Beyer, J.E., Olson, K.J., McCollister, S.B., and Rampy, L.W. Dow Chemical U.S.A. MCA NO: 1.1-Tox-INH (2 yr), Dec 6, (1978)
(7) Maltoni, C., Failla, G., and Kassapidis, G. *Med. Lavoro, 5*, 358-362 (1979)
(8) Matsuoka, A., Hayashi, M., and Ishidate, M.Jr. *Mutat. Res. 66*, 277-290 (1979)
(9) McMichael, A.J., Spirtas, R., Gamble, J.F., and Tousev, P.M. *J. Occup. Med. 18*, 178 (1976)
(10) Norppa, H. *Carciongenesis 2*, 237-242 (1981)
(11) Norppa, H. *Toxicol. Lett. 8*, 247-251 (1981)
(12) Norppa, H., Sorsa, M., Pfäffli, P., and Vainio, H. *Carcinogenesis 1*, 357-361 (1980)
(13) Norppa, H., Hemminki, K., Sorsa, M., and Vainio, H. *Mutat. Res. 91*, 243-250 (1981)
(14) Ott, M.G. Detailed summary of a mortality survey of employees engaged in the development or manufacture of styrene-based products, Dow Chemical, U.S.A. July 31 (1979)
(15) Ponomarkov, V., and Tomatis, L. *Scand. J. Work. Environ. Health 4*, Suppl. 2, 127-135 (1979)
(16) de Raat, W.K. *Chem.-Biol. Interact. 20*, 163 (1978)
(17) Sorsa, M., Hyvönen, M., Järventaus, H., and Vainio, H. Symposium on Toxicology, University of Turku, May 29-30, Abstracts, p. 16 (1979)
(18) Vainio, H., Norppa, H., Hemminki, K., and Sorsa, M. Metabolism and Genotoxicity of Styrene, *In* Snyder, R., Parke, D.V., Kocsis, J., Jollow, D.J., and Gibson, G.G. (Eds.) Biological Reactive Intermediates 2: Chemical Mechanisms and Biological Effects. Plenum Press, New York. *In press*.

Detection of Worker Exposure to Mutagens in the Rubber Industry by Use of the Urinary Mutagenicity Assay

Marja Sorsa, Kai Falck, and Harri Vainio

Institute of Occupational Health, Helsinki, Finland

ABSTRACT The mutagenic activity was measured by the bacterial fluctuation assay from concentrated urine samples of workers representing different job categories in a rubber plant. In spite of a few deviant individual values, the urinary mutagenic activity of the workers was higher than among the control subjects, also employed by the same company. The highest values, detected by E.coli WP2 uvrA, the base-pair substitution strain, were observed among workers employed as weighers and mixers of rubber chemicals.

INTRODUCTION

Elevated cancer risks have been reported in a number of epidemiological studies performed among workers in rubber industry (2,13,14,15,17,18,19). After the detection of some aromatic amines as causative agents especially for excess bladder cancer cases in the rubber industry (4), and the necessary withdrawal of them, no real progress has been made to identify any single chemicals as etiological factors for cancer risk in this field of industry (10).

The reason for the lack of rapid success is obviously in the extremely complex exposure situation varying in different job categories and depending on the type of rubber products manufactured. In rubber operations, several hundred different raw materials and chemicals are handled (9), and their further processing produces an array of toxicologically unknown mixtures, vulcanization fumes and exhausts.

The aim of the present research efforts has been to identify the possibly hazardous chemical exposures occurring in various operations and job categories, by measuring the urinary mutagenicity as estimate for exposure. The analysis of somatic chromosome damage in lymphocytes has been used to estimate the biological effect of such exposure. The preliminary results of the urinary mutagenicity assays are reported here; the findings of sister chromatid exchange (SCE) frequencies among rubber workers are presented elsewhere (20).

MATERIALS AND METHODS

The subjects of this study were workers in a rubber factory, manufacturing mainly technical rubber products. The urine samples (minimum 100 ml) were collected at the end of the work shift, deepfrozen at $-70^{\circ}C$ and stored until analyzed. The control persons, male employees from other than manufacturing duties in the same rubber plant, were matched for age and smoking habits prior to sampling. All subjects were interviewed for their drinking and smoking habits, possible leasure time exposures, drug intake, special diet, and health condition. Also the medical records of the plant were available for check up. Nothing special was observed and all subjects were considered healthy.

The concentration of urine was performed with the XAD-2 resin method described by Yamasaki and Ames (23). The mutagenicity assay was carried out with the bacterial fluctuation test, originally developed by Green et al. (7), using Escherichia coli WP2 uvrA and Salmonella typhimurium TA 98 as detector organisms for base pair and frame shift type mutagens, respectively (see 6 for details of the method).

RESULTS

We have shown earlier, from another rubber plant, that the mutagenicity in the urine samples of workers in some job categories can be detected by both the base pair strain E.coli WP2 uvrA and the frame shift tester strain S.typhimurium TA 98. The effect of smoking on the urinary mutagenicity is only revealed by the frame shift tester strain (6). Thus the occupational effect on the urinary mutagenicity can be clearly seen with the E. coli WP2 uvrA strain, also on individual level, when subsequent urine samples are taken from the same workers after vacation and after work shift (22).

The present results confirm the earlier findings. The mutagenic activity[1] of the urine samples of the group of 26 rubber workers was 509.5 $\pm$ 562.1 (mean $\pm$ S.D.) while it was 115.7 $\pm$ 97.2 among the 15 control subjects. When the individual mutagenic activities among workers and controls are compared (Fig. 1), it is clearly seen that the highest mutagenic activities were among the workers. None of the control persons showed mutagenic activity over 400.

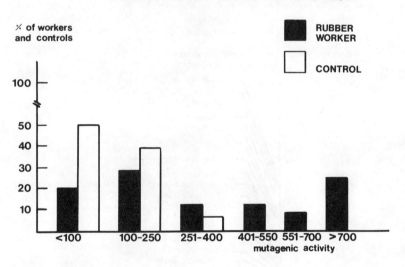

Fig. 1. Distribution of individual mutagenic activity values among rubber workers and controls.

---

[1]mutagenic activity = $\dfrac{\text{revertants in sample} - \text{revertants in solvent control}}{\text{amount of creatinine (mmol) in urine sample}}$

Comparing the different job categories among the rubber workers, the weighers and mixers of chemicals appear to show especially high mutagenic activities. When the highest urinary mutagenic activities are listed, among the 20 top values there are 11 men working in the chemicals' weighing and mixing operations. However, among the 20 highest values obtained, also three control persons appear (Fig. 2). Such individual high values do occur occasionally among persons with no obvious mutagen exposures.

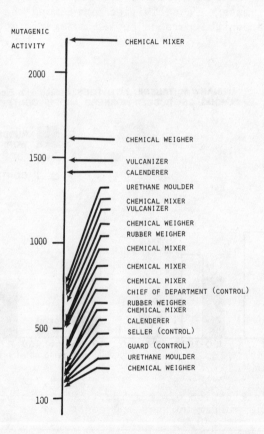

Fig. 2. Job categories of 20 persons showing the highest urinary mutagenic activities among 26 rubber workers and 15 control subjects tested.

Even though the interindividual variation within the job categories, also among mixers and weighers, is quite high, obviously depending on personal working habits, extent of exposure and individual metabolic capacities, the urinary mutagenicity method still seems applicable in estimating exposures to mutagenic chemicals in occupational settings.

DISCUSSION

The analysis of mutagenic constituents in urine as a measure of ambient exposure to mutagens has recently gained increasing interest (1,5,11,12,16,21,23). The method is unspecific in regard to the chemical structure of the exposing agent, but specific concerning its mutagenic character. Consequently, it may be especially useful in monitoring exposure to complex chemical mixtures, such as occurs in many occupational environments.

The analysis is based on detection of the activity due to excreted mutagens, mutagenic metabolites or liberated conjugates. Other short-term test systems, e.g. mammalian cells have also been used (3). In most cases the metabolic activation system seems to be needed for splitting of the conjugates or activation of the possible premutagens present in urine. Obviously much work still needs to be done to define the mutagenic components responsible for the urinary mutagenicity in specific exposures, both lifestyle and occupational.

The present results point to a potential exposure to mutagenic compounds in rubber industry, especially during weighing and mixing operations. This exposure probably occurs through particulate dust of various chemicals easily inhaled during these operations. Some of the single chemical entities handled have been shown to be mutagenic in bacterial mutagenicity assays (8). A careful exposure analysis during the workday needs to be combined to the biological exposure tests on urinary mutagenicity before there is hope to pinpoint individual causative agents.

ACKNOWLEDGEMENTS

This work was made possible by the good collaboration of the personnel in the Nokia Corp., Savio Rubber Plant.

The excellent preparatory assistance of Ms. Ritva Toivanen, Health Center of the Savio factory, and Ms. Vuokko Kytöniemi and Ms. Hilkka Järventaus from our own laboratory, is greatfully acknowledged. Financial support for the project was obtained from the Swedish Work Environment Fund (contract no. 79/189:2).

(1) Aeschbacher, H.V. and Chappuis, C. *Mutation Res.* **89,** 161-177 (1981)
(2) Andjelkovich, D., Taulbee, J. and Blum, S. *J. Occup. Med.* **20,** 409-413 (1978)
(3) Beek, B., Aranda, I. and Thomson, E. *Mutation Res.* in press (1981)
(4) Case, R.A.M., Hosker, M.E., McDonald, D.B. and Pearson, J.T. *Brit. J. Indust. Med.* **11,** 75-104 (1954)
(5) Falck, K., Gröhn, P., Sorsa, M., Vainio, H., Heinonen, E. and Holsti, L.R. *Lancet, i,* 1250-1251 (1979)
(6) Falck, K., Sorsa, M., Vainio, H. and Kilpikari, I. *Mutation Res.* **79,** 45-52 (1980)
(7) Green, M.H.L., Bridges, B.A., Rogers, A.M., Horspool, G., Muriel, W.J., Bridges, J.W. and Fry, J.R. *Mutation Res.* **48,** 287-294 (1977)
(8) Hedenstedt, A., Rannug, U., Ramel, C. and Wachtmeister, C.A. *Mutation Res.* **69,** 313-325 (1979)
(9) Holmberg, B. and Sjöström, B. Investigation Report 1977: 19. National Board of Occupational Safety and Health, Stockholm
(10) IARC Monographs on the evaluation of the carcinogenic risk of exposures in the rubber industry. IARC, Lyon, France, 1982, in press.
(11) Kilian, D.J., Pullin, T.G., Conner, T.H., Legator, M.S. and Edwards, H.N. *Mutation Res.* **53,** 72 (1978)
(12) Legator, M.S., Truong, L. and Connor, T.H. In Hollaender, A., de Serres, F.J. (eds): "Chemical Mutagens: Principles and Methods for Their Detection." Vol. 5, New York, Plenum Press, 1978, pp. 1-23
(13) Mancuso, T.F. *Environm. Health Persp.* **17,** 21-30 (1976)
(14) Mancuso, T.F. and Brennan, M.J. *J. Occup. Med.* **12,** 333-341 (1970)
(15) Mancuso, T.F., Ciocco, A. and El-Attar, A.A. *J. Occup. Med.* **10,** 213-233 (1968)
(16) McCoy, E.C., Hankel, R., Robins, K., Rosenkrantz, H.S., Ginfrida, J.G. and Bizzari, D.V. *Mutation Res.* **53,** 71 (1978)

(17) McMichael, A.J., Andjelkovic, D.A. and Tyroler, H.A. *Ann. New York Acad. Sci.* **271**, 125-137 (1976)
(18) Monson, R.R. and Nakano, K.K. *Am. J. Epid.* **103**, 284-296 (1976a)
(19) Monson, R.R. and Nakano, K.K. *Am. J. Epid.* **103**, 297-303 (1976b)
(20) Sorsa, M., Mäki-Paakkanen, J. and Vainio, H. *Cytogenet. Cell Genet.*, in press (1982)
(21) Sorsa, M., Falck, K., Norppa, H. and Vainio, H. *Scand. J. Work Environ. Health* **7**, in press (1981)
(22) Vainio, H., Falck, K., Mäki-Paakkanen, J. and Sorsa, M. In Aitio, A. et al. (eds) "Host factors in carcinogenesis". IARC, Lyon, France, in press (1982)
(23) Yamasaki, E. and Ames, B. *Proc. Natl. Acad. Sci. (U.S.A.)* **74**, 3555-3559 (1977)

Mutagenicity Screening Studies on Pesticides

Yasuhiko Shirasu, Masaaki Moriya, Hideo Tezuka, Shoji Teramoto, Toshihiro Ohta, and Tatsuo Inoue

Institute of Environmental Toxicology, Tokyo, Japan

Microbial mutagenicity screening studies were done on 228 pesticides and related chemicals including 88 insecticides, 60 fungicides, 62 herbicides, 12 plant-growth regulators, 3 metabolites and 3 other related compounds. The results of our earlier screening studies were previously published (3, 4). The reverse mutation tests using *S. typhimurium* and *E. coli* WP2 hcr with and without metabolic activation were employed.

Out of 228 compounds (Table 1) 50 chemicals (22%) showed mutagenicity. They were 25 insecticides (28%), 20 fungicides (33%), 3 herbicides (5%), 1 plant-growth regulator and 1 other compound. Organic phosphates, halogenated alkanes and dithiocarbamates contained mutagens at a higher ratio. Forty-five of the mutagenic pesticides were direct-acting mutagens and five were indirect ones, which require S-9 mix for their activity (Table 2).

The fate of mutagenicity of some positive pesticides was investigated by *in vitro* metabolic activation system employing S-9 mix (1). In this study mutagenicity of captan, captafol, folpet and NBT disappeared, whereas mutagenic activity of NNN increased. S-9 fraction, cysteine or rat blood was also effective for disappearance of their mutagenicities, indicating the effect of SH groups.

Cytogenetic and dominant lethal studies were conducted on captan which showed the most potent activity in the microbial assays (9). The cytogenetic analysis of human diploid fibroblast cells treated with captan revealed no chromosomal aberrations. Captan also could not induce chromosomal aberrations in the bone marrow cells of male

Table 1 Pesticides tested for microbial mutagenicity

INSECTICIDE (88)

Acephate, Chlorfenvinphos, Chlorpyrifos, Chlorpyrifosmethyl, Cyanofenphos, Diazinon, Dichlofenthion, Dichlorvos, Dimethoate, Dimethylvinphos, Disulfoton, EPBP, EPN, ESP, Ethion, Etrimfos, Fenitrothion, Formothion, Isofenphos, Isoxathion, Malathion, Mecarbam, Monocrotophos, Naled, Phenthoate, Phosmet, Phoxim, Pirimiphosmethyl, Propaphos, Prothiophos, Pyridaphenthion, Salithion, Tetrachlorvinphos, Thiometon, Trichlorfon, Vamidothion, Chlorfenethol, Chlorobenzilate, Chloropropylate, p,p'-DDT, Dicofol, Phenisobromolate, Tetradifon, Chloropicrin, DBCP, D-D, EDB, EDC, Methyl bromide, Aldrin, α-BHC, Dieldrin, Dienochlor, Endosulfan, Endrin, Heptachlor, BPMC, Butoxycarboxim, Carbaryl, Carbofuran, Isoprocarb, Methomyl, MPMC, MTMC, Oxamyl, Pirimicarb, Propoxur, XMC, Metamammonium, Allethrin, Nicotine sulfate*, Petroleum oil*, Piperonyl butoxide*, Permethrin, Polynactins, Pyrethrins, Rotenone*, Diflubenzuron, Amitraz, Benzomate, Binapacryl, Chlorfenson, Cyhexatin, DCIP, Methylisothiocyanate, Propargite, Quinomethionate, Vendex

FUNGICIDE (60)

Edifenphos, IBP, Phosdifen, Pyrazophos, Amobam, DDC, EMSC, Ferbam, Mancozeb, Maneb, Milneb, Polycarbamate, Thiram, TTCA, Zineb, Ziram, Benomyl, Blasticidin S, Griseofulvin, Kasugamycin, Oxytetracycline, Polyoxin B, Polyoxin D·Zn, Validamycin A, Anilazine, Calcium polysulfide*, Captafol, Captan, Carboxin, Chloroneb*, Chlorothalonil, Copper sulfate*, DAPA, Dazomet, Dichlofluanid, Dimethirimol, Dinocap, Dithianon, Echlomezol, Fentin hydroxide, Folpet, Fthalide, Guazatine, Hymexazol, Isoprothiolane, MAF, NBT, NNN, Oxine-copper, Oxycarboxin*, Phenazine oxide, o-Phenylphenol, Probenazole, Quintozene, Sulfur*, Thiabendazole, Thiophanatemethyl, Tricyclazole, Triforine, Zinc sulfate*

HERBICIDE (62)

Amiprofos-Methyl*, Bensulide, Glyphosate, Krenite*, Piperophos, Asulam, Chloropropham, Cycloate, EPTC, Orthobencarb, Phenmedipham, Swep, Bifenox, Chlomethoxynil, Chlorthal-dimethyl, CNP, 2,4-D, Dicamba*, Dinoseb acetate, MCPA, MCPB, Mecoprop, Nitrofen, PCP, Phenothiol, Bromacil, Dimuron, Diuron, Lenacil, Linuron, Methabenzthiazuron, Methyldimuron, Siduron, Terbacil, Thiochlormethyl, Cyanazine, Dimethametryn, Metribuzin, Alachlor, Benefin, Butachlor, Chlorthiamid*, Diphenamid, Napropamide, Trifluralin, Propanil, Propyzamide, ACN, Amitrole, AMS*, Bentazon, Dalapon*, Diquat dibromide, Ioxynil octanoate, Methoxyphenone, Molinate, Oxadiazon, Paraquat dichloride, Sodium chlorate, Sodium fluoride*, TCA*, Triclopyr*

GROWTH REGULATOR (12)

Ethephon, Gibberellin A_3, Gibberellin A_4, 4-CPA, 4-CHMPA, Aminozide, HEH, Maleic hydrazide, Benzyladenine, Indolebutyric acid, NAA^3, OED,

OTHER (3)

Ferric oxide*, Sodium polyacrylate, ETU

METABOLITE (3)

p,p'-DDD, p,p'-DDE, Heptachlor epoxide

*: Pesticide was tested with only TA98 and TA100.

Table 2 Mutagenic potency of pesticides

No.	Pesticide	Strain	Revertants/nmole
Direct Mutagen (− S-9 mix)			
1	Captan	TA100	93.67
2	NBT	TA100	58.28
3	Folpet	TA100	15.00
4	Dienochlor	TA100	12.04
5	NNN	TA100	9.39
6	DAPA	TA98	5.22
7	Captafol	WP2 hcr	3.11
8	Nitrofen	TA100	2.63
9	EMSC	TA100	2.00
10	Ferbam	TA100	1.33
11	Polycarbamate	TA100	1.33
12	Ziram	TA100	1.24
13	Dichlofluanid	WP2 hcr	1.10
14	Thiram	TA100	0.84
15	Chlomethoxynil	TA100	0.79
16	DDC	TA100	0.79
17	TTCA	TA100	0.58
18	Echlomezol	TA100	0.35
19	Phosmet	TA100	0.34
20	Salithion	TA100	0.14
21	EDB	TA100	0.092
22	DBCP	TA100	0.085
23	Fenitrothion	TA100	0.074
24	HEH	WP2 hcr	0.044
25	Chlorfenvinphos	TA100	0.038
26	Dichlorvos	TA100	0.027
27	MAF	TA98	0.020
28	Polyoxin D·Zn	TA100	0.018
29	Thiometon	TA1535	0.013
30	D-D	TA100	0.0087
31	Dinocap	TA98	0.0073
32	Trichlorfon	TA100	0.0067
33	Monocrotophos	TA100	0.0064
34	ESP	TA100	0.0060
35	Vamidothion	TA100	0.0046
36	Dimethoate	TA100	0.0045
37	Pirimiphosmethyl	TA100	0.0040
38	Binapacryl	TA100	0.0036
39	Allethrin	TA100	0.0031
40	Formothion	TA100	0.0029
41	Carbofuran	TA98	0.0023
42	Disulfoton	TA1535	0.0014
43	Acephate	TA100	0.0013
44	ETU	TA1535	0.00065
Indirect Mutagen (+ S-9 mix)			
1	Oxine-copper	TA100	1.30
2	Chloropicrin	TA100	0.37
3	Phenazine oxide	TA1537	0.13
4	Butachlor	TA100	0.046
5	EDC	TA1535	0.0023

rats. Dominant lethal studies on captan failed to produce a positive response.

We conducted a series of mutagenicity testing on ethylenethiourea (ETU), a decomposition product of fungicidal ethylene-bis-dithiocarbamates (5). ETU was weakly mutagenic for TA1535 without metabolic activation, inducing about 4-fold increase in the number of revertants with the higher doses. Neither the in vitro cytogenetic studies using a Chinese hamster cell line (Don) nor the in vivo chromosomal analysis of the rat bone marrow cells could reveal any abnormalities attributable to ETU treatment. Dominant lethal studies in mice also gave negative results.

Interactive mutagenicities of ETU plus sodium nitrite were investigated (4). The reaction mixture of ETU and sodium nitrite at acidic conditions was mutagenic toward TA1535, TA100 and WP2 hcr. The host-mediated assay with S. typhimurium G46 also revealed interactive mutagenicity of ETU and sodium nitrite simultaneously administered into the mouse stomach. Chromosomal aberrations were induced in the bone marrow cells of mice given simultaneous treatment with ETU and sodium nitrite. A single or 5-day oral administration of these two chemicals induced dominant lethality, increasing pre-implantation losses in weeks 5 and 6 (7). N-nitroso-ETU, a possible derivative from ETU and sodium nitrite, also caused similar effects. Furthermore, N-nitroso-ETU was revealed to be a carcinogen (2).

Cytogenetic studies were performed on 10 pesticides and N-nitroso-ETU which were positive in the microbial reversion assays (8). A good correlation was observed between the ability to induce sister chromatid exchanges or chromosomal aberrations in the Chinese hamster V79 cells and the mutagenic potency in bacteria.

Dominant lethal studies on DBCP and EDB revealed that DBCP gave positive results in rats but not in mice, although EDB was negative in both species (6). DBCP induced post-implantation losses at week 4 to 5, exerting its adverse effect on early spermatids.

Our latest studies revealed that DBCP induced sex-linked recessive lethal mutations in Drosophila melanogaster (Fig. 1).

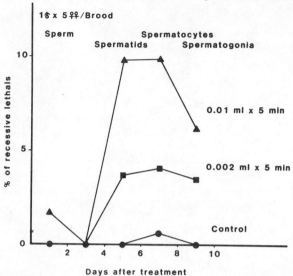

Fig. 1 Induction of sex-linked recessive lethals in Drosophila males by 5-min treatment with gaseous DBCP

REFERENCES
(1) Moriya, M., Kato, K., and Shirasu, Y. Mutation Research 57, 259-263 (1978)
(2) Moriya, M., Mitsumori, K., Kato, K., Miyazawa, T., and Shirasu, Y. Cancer Letters 7, 339-342 (1979)
(3) Shirasu, Y., Moriya, M., Kato, K., Furuhashi, A., and Kada, T. Mutation Research 40, 19-30 (1976)
(4) Shirasu, Y., Moriya, M., Kato, K., Lienard, F., Tezuka, H., Teramoto, S., and Kada, T. "Origins of Human Cancer", Cold Spring Harbor Conferences on Cell Proliferation, Vol. 4 pp. 267-285, Cold Spring Harbor Laboratory (1977)
(5) Teramoto, S., Moriya, M., Kato, K., Tezuka, H., Nakamura, S., Shingu, A., and Shirasu, Y. Mutation Research 56, 121-129 (1977)
(6) Teramoto, S., Saito, R., Aoyama, H., and Shirasu, Y. Mutation Research 77, 71-78 (1980)
(7) Teramoto, S., Shingu, A., and Shirasu, Y. Mutation Research 56, 335-340 (1978)
(8) Tezuka, H., Ando, N., Suzuki, R., Terahata, M., Moriya, M., and Shirasu, Y. Mutation Research 78, 177-191 (1980)
(9) Tezuka, H., Teramoto, S., Kaneda, M., Henmi, R., Murakami, N., and Shirasu, Y. Mutation Research 57, 201-207 (1978)

Knowledge Gained from the Testing of Large Numbers of Chemicals in a Multi-Laboratory, Multi-System Mutagenicity Testing Program

Errol Zeiger

Toxicology Research and Testing Program, National Institute of Environmental Health Sciences, Research Triangle Park, North Carolina, U.S.A.

ABSTRACT A mutagenicity data base comprised of Salmonella, Drosophila and in vitro cytogenetics (CHO cells) test results has been developed and is being expanded. This data base has been used to compare results between laboratories using the same test system and between mutagenicity and carcinogenicity results. The chemicals in the data base can be ordered by chemical structure so that chemical structure can be correlated with biological activity.

The total number of carcinogens that were also mutagenic in Salmonella was low, probably reflecting the distribution of chemicals tested. The number of mutagens that were carcinogenic was relatively high, demonstrating that a positive result in Salmonella is indicative of carcinogenicity; however, 52% of the nonmutagens were carcinogens. This implies that nonmutagenicity in Salmonella is not an adequate criterion for considering a substance to be non-carcinogenic.

INTRODUCTION

The Environmental Mutagenesis Test Development Program began as an attempt within the NIEHS to identify, develop and validate short-term tests for genetic toxins and to develop a large-scale chemical testing program using short-term tests (1). The majority of test development and validation and all testing is performed under contract in non-government laboratories, and the program is directed by scientific personnel in the Toxicology Research and Testing Program in NIEHS.

The descriptor "short-term test" as it is used in

the field of genetic toxicology is a generic one covering a wide range of test systems, both in vitro and in vivo, having a variety of genetic endpoints. The phrase was originally used to imply that the test could be run in less time than a carcinogenicity bioassay (two or more years). As such, the term applies equally well to tests that take 1-2 days as to tests that take 1-2 months. In the same manner, the term is used in apposition to multigenerational tests for heritability of genetic effects in rodents. The main feature of short-term tests with respect to carcinogenicity is that they measure endpoints that have been shown in some, but not all, cases to correlate with carcinogenicity.

PROGRAM DESCRIPTION

In this program large numbers of chemicals are being tested for mutagenicity in three tests:
(1) A preincubation Salmonella mutagenicity test. Strains TA98, TA100, TA1535 and TA1537 are used with both Aroclor 1254-induced rat and hamster liver S9 and without exogenous metabolic activation.
(2) The induction of sex-linked recessive lethal (SLRL) mutations and reciprocal translocations in Drosophila. All chemicals are tested for induction of SLRL mutations by feeding; if negative results are obtained they are retested by injection. Only those chemicals that are positive in the SLRL test are tested for their ability to induce translocations.
(3) CHO cells in culture for the induction of chromosome aberrations and sister chromatid exchanges. All chemicals are tested with Aroclor 1254-induced rat liver S9 and in the absence of exogenous metabolic activation.
All chemicals are tested initially in Salmonella. Those that give a positive or equivocal response are considered for testing in Drosophila. In addition, selected Salmonella-negative chemicals and chemicals for which additional toxicology testing is planned are also tested in Drosophila. Because of the limited CHO cytogenetics capacity not all chemicals can be tested in this system. The chemicals tested in this system are those which give weakly positive or equivocal results in Salmonella, chemicals negative in Salmonella which are known or suspect carcinogens, selected

Salmonella negatives, and chemicals chosen for additional (nongenetic) toxicology testing.

All chemicals are tested and the data evaluated under code. This precludes bias by the testing laboratory or NIEHS personnel during testing or evaluation of the data.

Because all chemicals are coded prior to their shipment to the test laboratories an empirical quality control procedure can be used. That is, 10-15% of all samples sent to _Salmonella_ laboratories are selected from a list of positive and negative controls. Also, another 10-15% of the test chemicals are sent to more than one laboratory or to the same laboratory under a different code at a later date. This allows monitoring of the laboratory's effectiveness and the degree of inter- and intra-laboratory variability. _Drosophila_ and cytogenetics testing laboratories are also under the same scrutiny but with fewer control chemicals.

All test data and data summaries are, or will be, stored in a computer data base readily accessible by the NIEHS project officers. The summary data base contains such information as chemical name and CAS #, WLN (Wiswesser Line Notation: a system for describing a molecule in a linear, alpha-numeric formula), chemical code, testing laboratory, test type, results, highest dose tested, carcinogenicity (where known), chemical purity, date the test was completed, and status of the chemical in the program.

Such information enables searches of the data base to be made on any of the factors mentioned, ordering of the data to allow detection of unique or interesting patterns of results, and comparisons between chemicals or laboratories.

The _Salmonella_ data base is expanding by over 200 chemicals per year, and many of the chemicals under test in _Salmonella_ are those which have been tested in the NCI Carcinogenesis Bioassay Program. Increasing numbers of chemicals are being tested in _Drosophila_ and the CHO system so that in the future these tests will also be used for comparative mutagenicity, inter-laboratory and carcinogen correlations.

The remainder of this article will center on the _Salmonella_ test and, without discussing specific chemicals, will describe the results of testing to date and some relationships and correlations observed.

INTERLABORATORY AGREEMENT

The testing progress is given in Table 1. Of the 373 chemicals tested in Salmonella, 22 were also tested in Drosophila and 21 in CHO cells. A total of 72 chemicals were tested in Salmonella in two different laboratories and 18 were tested in all three.

TABLE 1: TESTING PROGRESS (As of Aug.10, 1981)

	Sal.[1]	Dros.[1]	Cyto.[1]	Total
No. of Tests	496	24	33	553
No. of Chemicals	373	22	21	373

[1] Sal: Salmonella pre-incubation test; Dros: Drosophila SLRL; Cyto: aberrations & SCE's in cultured CHO cells.

The agreement between laboratories is presented in Table 2. There was complete agreement on 83.3% of the chemicals tested in more than one laboratory, 7.8% showed disagreement between laboratories and 8.9% were indeterminate; that is, one laboratory reported an equivocal result while the other laboratory reported an unequivocal (+) or (-).

TABLE 2: INTERLABORATORY AGREEMENT (Salmonella)

	2 LABS	3 LABS	Total
Agree	59(81.9%)[1]	16(88.9%)	75(83.3%)
Indeterminate[2]	7 (9.7%)	1 (5.5%)	8 (8.9%)
Disagree	6 (8.3%)	1 (5.5%)	7 (7.8%)
Total	72	18	90

[1] Number of chemicals (% of total).
[2] Mutagenic (+) or non-mutagenic (-) response in one laboratory vs. an equivocal (?) response in another.

Since there is relatively little Drosophila and cytogenetic data at this time, results of these tests will not be summarized in this report.

CORRELATIONS

A number of studies have addressed the question of what proportion of carcinogens are mutagens as determined by the Salmonella test, and have arrived at correlations ranging from 68% to 92% (2-6). The number of noncarcinogens that were not mutagenic was also high. These exercises have served to justify the use of the Salmonella test as a preliminary screen for identification of carcinogens.

The application of the Salmonella test to screen chemicals of unknown carcinogenicity so that a decision can be made for further testing or for regulatory purposes presents a different problem. The knowledge that the majority of carcinogens are mutagens and the majority of noncarcinogens are not mutagenic does not help to interpret the results from testing a series of chemicals for mutagenicity. The important questions to ask in this situation are not "what proportion of (non)carcinogens are (non)mutagens?", but "if a substance is a mutagen, what is the liklihood of its being carcinogenic?" and, conversely, "if a substance is not a mutagen should we be concerned about its potential carcinogenicity?"

To examine these questions, chemicals tested for mutagenicity in Salmonella and for which a carcinogenicity determination has been made by the NCI (7) or IARC (8) were selected. The 99 chemicals for which these data were available included 12 positive control mutagens. Because inclusion of these controls would bias the correlations, all calculations were performed both with and without them.

The resulting list of 87 chemicals (minus controls) is unlike those previously used for mutagenesis-carcinogenesis correlation determinations in that it contains few, if any, direct alkylating agents, polycyclic aromatic hydrocarbons, or nitrosoamines and amides. Table 3A presents the correlations between carcinogenicity and mutagenicity - the proportion of carcinogens that are mutagens and nonmutagens. When the control chemicals were eliminated from the calculations only 44% of the carcinogens were mutagens, whereas 52% of the carcinogens were judged nonmutagenic. (Inclusion of the positive controls increased the correlation to 55%). When the suspect carcinogens were combined with the carcinogens, the correlation dropped to 42%.

TABLE 3: PERCENT AGREEMENT BETWEEN RESULTS OF SALMONELLA MUTAGENICITY (M) AND RODENT CARCINOGENICITY (C) TESTS

M \ C	A. Mut./Carc.			B. Carc./Mut.		
	+[1] (33)	− (50)	? (4)	+ (33)	− (50)	? (4)
+(50)[2]	44	52	4	67	52	50
−(23)	26	70	4	18	32	25
?(14)	36	57	7	15	16	25

[1] M. +: mutagenic; −: non-mutagenic; ?: equivocal response; C. +: carcinogen; −: non-carcinogen; ?: suspect carcinogen.
[2] Number of chemicals

When the question of what proportion of mutagens are carcinogens was asked, it can be seen that 67% of the mutagens were carcinogenic (Table 3B). Combining the carcinogens with suspect carcinogens increased the correlation to 82%. (With inclusion of the 12 positive controls, the correlations increased to 76% for the carcinogens only and to 87% for the carcinogens plus suspect carcinogens.) Most interestingly, however, 52% of the substances that were not mutagenic were carcinogenic (this percentage increases to 68% if suspect carcinogens are included). These correlations imply that while a Salmonella mutagen is probably a carcinogen, the absence of mutagenicity cannot be interpreted as a probable absence of carcinogenicity.

In an attempt to further resolve the carcinogenicity-mutagenicity correlations, chemicals were sorted, using WLN, into a few broad classes: chlorinated hydrocarbons, nitro-containing chemicals, chemicals containing a free amine and azo-group-containing chemicals. The chlorine-nitro- and amine-containing chemicals were further subdivided into those where the functional group was attached to a benzene or conjugated benzene ring, an aliphatic moiety or some other group. The results of this exercise are presented in Table 4.

TABLE 4: STRUCTURAL CORRELATIONS BETWEEN SALMONELLA MUTAGENICITY AND CARCINOGENICITY

Chemical Class	Carc.[1]	Mutagenicity			Mut/Carc[3]
		+[2]	-	?	
$-\emptyset-Cl$[4]	10	2	8	0	5/16 (31%)
$-C-Cl$	6	3	3	0	
$-\emptyset-NO_2$	6	6	0	0	8/8 (100%)
$-X-NO_2$	2	2	0	0	
$-\emptyset-NH_2$	18	12	6	0	
$-C-NH_2$	3	2	1	0	14/22 (64%)
$-X-NH_2$	1	0	1	0	
$-N=N-$	5	2	2	1	2/5 (40%)

[1] Number of carcinogens in chemical class.
[2] +: mutagenic; -: non-mutagenic; ?: equivocal mutagenic response.
[3] Mutagens/Carcinogens.
[4] $\emptyset-$: phenyl; $-C-$: aliphatic; $X-$: other moiety.

It can be seen that 2/10 chlorinated phenyl carcinogens were also mutagens, 6/6 nitro phenyl carcinogens were mutagens and 12/18 carcinogenic phenyl amines were mutagens, as were 2/5 azo chemicals. While these numbers suggest specific structural correlations, there are too few chemicals in any of the classes examined to support a definitive statement. Based on the calculations in Tables 3 and 4 it can be seen that correlations of carcinogenicity and mutagenicity are not sufficient by themselves for assessing the potential carcinogenicity of a chemical. The chemical structure must also be considered, since there appears to be a wide disparity in the correlations within different chemical classes. At the present time, the data base for the individual chemical subclasses is not large enough for making definitive statements about correlations from class to class.

REFERENCES

1. Zeiger, E. and Drake, J. in *Molecular & Cellular Aspects of Carcinogen Screening Tests. IARC Publication No. 27*, Montesano, R., Bartsch, H., and Tomatis, L., eds., Lyon, p. 303-313 (1980).
2. Bartsch, H., Malaveille, C., Camus, A.-M., Martel-Planche, G., Brun, G., Hautefeuille, A., Sabadie, N., Barbin, A., Kuroki, T., Drevon, C., Piccoli, C., and Montesano, R. in *Molecular and Cellular Aspects of Carcinogen Screening Tests. IARC Publication No. 27*, Montesano, R., Bartsch, H., and Tomatis, L.; eds., Lyon, 179-241 (1980).
3. Heddle, J.A. and Bruce, W.R. in *Origins of Human Cancer, Book C*, Hiatt, H.H., Watson, J.D., and Winsten, J.A., eds., Cold Spring Harbor Laboratory, 1549-1557 (1977).
4. McCann, J., Choi, E., Yamasaki, E., and Ames, B.N. *Proc. Nat. Acad. Sci., USA 72*, 5135-5139 (1975).
5. Purchase, I.F.H., Longstaff, E., Ashby, J., Styles, J.A., Anderson, D., Lefevre, P.A., and Westwood, F.R. *Brit. J. Cancer 37*, 924-930 (1978).
6. Rinkus, S.J. and Legator, M.S. *Cancer Res. 39*, 3289-3318 (1979).
7. Griesemer, R.A. and Cueto, C., Jr. in *Molecular & Cellular Aspects of Carcinogen Screening Tests. IARC Publication No. 27*, Montesano, R., Bartsch, H., and Tomatis, L., eds., Lyon, 259-281 (1980).
8. Althouse, R., Huff, J.E., Tomatis, L., and Wilbourn, J. *Cancer Res. 40*, 1-2 (1980).

Modification of Mutagenesis

Enhancement and Suppression of Genotoxicity of Food by Naturally Occurring Components in These Products

Hans F. Stich, Chiu Wu, and William Powrie

Environmental Carcinogenesis Unit, British Columbia Cancer Research Centre, and Department of Food Science, The University of British Columbia, Vancouver, British Columbia, Canada

ABSTRACT *In vivo* target cells for carcinogens and mutagens are exposed to complex mixtures rather than to single compounds. Currently it is unknown whether genotoxicity tests on isolated purified chemicals can really reveal the actual hazard of these compounds when present in a mixture. The mutagenicity (*S. typhimurium* assay), clastogenicity (chromatid exchanges in CHO cells) and recombinogenicity (yeast D7 strain) of caramelized sugars, Maillard reaction products and of naturally occurring phenols were examined in the presence of polymer carmalens and melanoidins which are also formed by browning reactions. Moreover, the modulating actions of chlorophyll, naturally occurring antioxidants and transition metals at concentrations found in food products were tested on the genotoxicity of cooked food products and isolated food products, including pyrazines, furans and flavonoids. It is of particular interest to note that several common food products can show genotoxic activities as well as antimutagenic properties. The results clearly indicate that genotoxicity of single chemicals can be virtually abolished or greatly enhanced when present in a complex mixture and that tests on single compounds can lead to gross over-estimations or underestimations of the actual genotoxicity of a mixture when the results of single tests are simply summed.

We seem to have arrived at a crossroads in the testing programmes for environmental mutagens. First, the idea of identifying genotoxic agents and removing them from man's environment has proven to be an unrealistic one. Thousands of naturally occurring and man-made genotoxic agents cannot just simply be replaced. Second, the amount and number of genotoxic chemicals ingested through food and beverages seems to greatly exceed industrially- and agriculturally-used mutagens. It seems that new priorities for testing programmes must be set, giving food products a high level of concern. Third, estimation of the mutagenicity, clastogenicity and recombinogenicity of isolated, purified single compounds may not suffice to assess the genotoxic hazard of complex mixtures. However, human populations are exposed to complex mixtures rather than to single compounds. Fourth, the importance of "modulating" factors seems to have been underestimated in the past. Their presence may potentiate the genotoxic activity but may even change so-called "non-mutagens" into active "mutagens". Thus the mutagenicity of a system rather than that of a single compound should be considered when genotoxic hazards are evaluated. Finally, the definition of mutagen and non-mutagen does not seem to be as clear and precise as is presented in textbooks.

In spite of apparently insurmountable difficulties, attempts must be made to design methods to assess the genotoxic hazard posed by complex mixtures. Food products or beverages can be submitted to intensive gas chromatographic and high pressure liquid chromatographic separation combined with mass spectrometry. These procedures have resulted in the discovery of thousands of chemicals in virtually all food items. Representatives of different molecular groups can then be selected and tested for genotoxicity using a battery of different bioassays. Finally, a stepwise reconstruction of the food item can be tried. This approach is feasible. Its main disadvantage is that the number of combinations and permutations of mutagens, comutagens and antimutagens becomes extremely large and unmanageable when the number of individual chemicals in the mixture exceeds four, a number which is very small compared to the thousands of compounds found in food products and beverages.

In this paper we would like to show the difficulties encountered when as little as two or three chemicals start to interact. An example of an enhancing (potentiating) effect is shown in Table 1. The combined clastogenic effect of arecoline, a carcinogenic betel nut alkaloid and chlorogenic acid, which is found in relatively large quantities in coffee, exceeds the sum of the activities of individually applied compounds (14). Moreover, the addition of the non-genotoxic transition metal Mn^{++} further enhances the clastogenic activity of the mixture. Finally, the addition of an S9 preparation from the liver of arochlor 1254-pretreated rats greatly reduces the clastogenic activity of the mixture of clastogens. Maybe we should start thinking in terms of mutagenic or clastogenic systems rather than defining mutagenic and clastogenic capacity of individual compounds.

Table 1. Percentage metaphase plates with chromatid breaks and exchanges in CHO cells

Arecoline (50 µg/ml) + chlorogenic acid (25 µg/ml) + Mn^{++} ($10^{-4}M$)	42.9%
Arecoline (50 µg/ml) + chlorogenic acid (25 µg/ml)	13.1%
Arecoline (50 µg/ml)	0.8%
Chlorogenic acid (25 µg/ml)	1.0%
Mn^{++} ($10^{-4}M$)	0.8%
Arecoline + chlorogenic acid + Mn^{++} + S9 rat microsomal mixture	3.5%

The enhancing effect resulting from the joint application of two or more genotoxic agents has great implications for the setting of so-called "no-effect" or "acceptable" levels, concepts so greatly cherished by regulatory agencies. Based on our experience with phenolics and alkaloids, the concentration which gives a certain frequency of chromosome aberrations with two genotoxic agents is approximately one-quarter of the concentrations required when each compound was applied individually. If this type of potentiation occurs among the hundreds of genotoxic agents which occur in a food product, the minimum dose of a compound which contributes to the overall genotoxicity of a mixture will be considerably smaller than the minimum dose which induces a detectable genotoxic effect when tested individually. In such a situation, the soundness of setting "no-effect" doses on the basis of results obtained on individual compounds should be seriously questioned. To avoid confusion, it may even be necessary to distinguish between "genotoxic range" (mutagenic range or clastogenic range), which would be applicable to individually assayed chemicals, and "cogenotoxic range" (comutagenic range or coclastogenic range), which may include concentrations at which a chemical can exert a potentiating effect when combined with other genotoxic compounds or even non-genotoxic but enhancing agents, such as, for example, transition metals (Fig. 1).

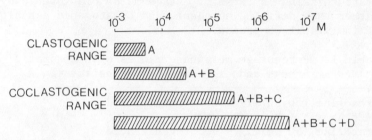

Fig. 1. Genotoxic range versus cogenotoxic range. The range of concentrations which produce a genotoxic (mutagenic, clastogenic, etc.) effect when chemical A is applied is smaller (genotoxic range) as compared to the joint application of A with other agents (B, C and D) (cogenotoxic range).

Considering the relatively large amount and number of genotoxically active compounds in food, including chlorogenic acid, simple phenols, phenolic acids, cinnamic acids and derivatives (11), flavonoids (2), pyrazines (10), furans (12), polycyclic aromatic hydrocarbons (5), protein pyrolysates (15), caramelized sugars (13), alkenylbenzenes (16), 1,2-dicarbonyl compounds (1), hydroxamic acids (3), thiazolidines (6), alkaloids (17), hexachloroacetone (18), thiol derivatives (Stich, unpublished data), ascorbate (8,9), cysteamine and glutathione (7), one is surprised to find a lack of association between food items and human cancer. If one adds up all the mutagens consumed by an individual through food and beverages during one day, one arrives at the almost unbelievable figure of 1-2 g. Obviously a series of efficient protection mechanisms must exist. Since it is beyond the scope of this short paper to survey all these factors, we would like to dwell only on those which already operate within a complex mixture prior to their entry into an organism.

A search for antigenotoxic (antimutagenic, anticlastogenic, etc.) agents can only succeed when simple, reliable detection procedures are available. We have selected two test systems: (i) Nitrosation of methylurea leads to directly-acting mutagens which are readily detectable with the *S. typhimurium* mutagenesis assay. Compounds to be tested can be added during the nitrosation reaction or after its completion; (ii) Aflatoxin B_1 can be activated by S9 mixtures into strong mutagenic metabolites for *S. typhimurium*. Again, the test substance can be added during or after metabolic activation. The results in Figs 2 and 3 exemplify this approach. These two test systems can also be applied to complex mixtures. For example, coffee and tea at concentrations found in a cup of coffee or one cup of tea can virtually block the formation of mutagenic nitroso compounds in the model reaction (Fig. 4).

Once the antimutagenic activities of food chemicals are understood in model systems, naturally occurring conditions can be closely simulated. For example, aqueous extracts of salted Cantonese fish, which seem to be linked to nasopharyngeal carcinomas in south China (4), produce, after nitrosation, potent mutagenic agents. The addition of tannic acid, gallic acid, catechol, etc. suppresses the appearance of mutagenic nitrosation products (Fig. 5). This pattern may actually occur when fish and vegetables are ingested, as is the case with many Chinese meals.

Human population groups continuously ingest with their food thousands of chemicals, many of which are potent mutagens, clastogens and recombinogens. However, they also consume hundreds of agents which show antimutagenic, anticlastogenic and antirecombinogenic activities. Thus the total genotoxicity of a diet is not a simple sum of the genotoxic capacities of its components. A health hazard can result from an increase in genotoxic agents (e.g., smoking and lung cancer; betel nut chewing and oral carcinomas). It could also conceivably be due to a lack of antimutagenic and anticlastogenic agents in deficient diets. The role of modulating factors in carcinogenesis and mutagenesis seems to merit greater attention than it has received in the past.

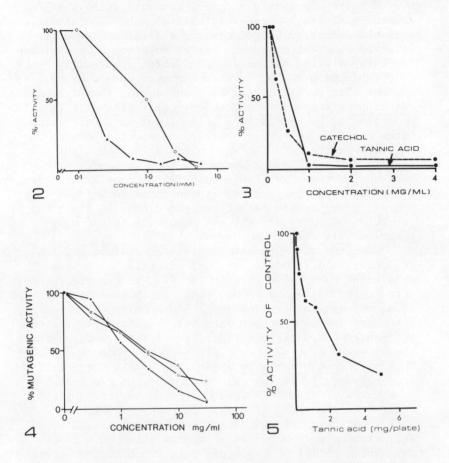

Fig. 2. Inhibitory effect of tannic acid (▲) and chlorogenic acid (O) on the mutagenicity of *S. typhimurium* TA1535 exposed to a nitrosation mixture of nitrite and methylurea at pH 3.6.
Fig. 3. Inhibitory effect of tannic acid and catechol on the mutagenicity (*S. typhimurium* TA98) of aflatoxin B1 (5×10^{-5} M) in the presence of an S9 activation mix.
Fig. 4. Inhibitory effect of three different North American coffees on the mutagenicity (*S. typhimurium* TA1535) produced by the nitrosation of methylurea at pH 3.6.
Fig. 5. Inhibitory effect of tannic acid on the mutagenicity resulting from the nitrosation of an aqueous extract of Cantonese fish.

ACKNOWLEDGMENTS

This study was supported by the Natural Sciences and Engineering Research Council Canada and by the National Cancer Institute of Canada. Dr H. Stich is a Research Associate of the National Cancer Institute of Canada.

REFERENCES

(1) Bjeldanes, L.F., and Chew, H. *Mutat. Res. 67*, 367-371 (1979)
(2) Brown, J.P. *Mutat. Res. 75*, 243-277 (1980)
(3) Hashimoto, Y., Shudo, K., Okamoto, T., Nagao, M., Takahashi, Y., and Sugimura, T. *Mutat. Res. 66*, 191-194 (1979)
(4) Ho, J.H.C., Huang, D.P., and Fong, Y.Y. *Lancet 2*, 626 (1978)
(5) Lo, M-T., and Sandi, E. *Residue Rev. 69*, 35-86 (1978)
(6) Mihara, S., and Shibamoto, T. *J. Agric. Food Chem. 28*, 62-66 (1980)
(7) Stich, H.F., Wei, L., and Lam, P.P.S. *Cancer Lett. 5*, 199-204 (1978)
(8) Stich, H.F., Wei, L., and Whiting, R.F. *Cancer Res. 39*, 4145-4151 (1979)
(9) Stich, H.F., Wei, L., and Whiting, R.F. *Food Cosmet. Toxicol. 18*, 497-501 (1980)
(10) Stich, H.F., Stich, W., Rosin, M.P., and Powrie, W.D. *Food Cosmet. Toxicol. 18*, 581-584 (1980)
(11) Stich, H.F., Rosin, M.P., Wu, C.H. and Powrie, W.D. *Cancer Lett.*, in press (1981)
(12) Stich, H.F., Rosin, M.P., Wu, C.H. and Powrie, W.D. *Cancer Lett. 13*, 89-95 (1981)
(13) Stich, H.F., Rosin, M.P., San, R.H.C., Wu, C.H., and Powrie, W.D. *Banbury Report 7*, 247-266 (1981)
(14) Stich, H.F., Stich, W., and Lam, P.P.S. *Mutat. Res.*, in press (1981)
(15) Sugimura, T. In *Naturally Occurring Carcinogens-Mutagens and Modulators of Carcinogenesis*, Univ. Park Press, Baltimore, pp. 241-261 (1979)
(16) Swanson, A.B., Chambliss, D.D., Blomquist, J.C., Miller, E.C., and Miller, J.A. *Mutat. Res. 60*, 143-153 (1979)
(17) Uyeta, M., Taue, S., and Mazaki, M. *Mutat. Res. 88*, 233-240 (1981)
(18) Zochlinski, H., and Mower, H. *Mutat. Res. 89*, 137-144 (1981)

Mechanisms and Genetic Implications of Environmental Antimutagens

Tsuneo Kada

Department of Induced Mutation, National Institute of Genetics, Mishima, Japan

ABSTRACT   "Antimutagen" means, in the original sense, any factor reducing the apparent frequency of spontaneous or induced mutations. In the case of environmental mutagens a number of agents are known to interact directly with mutagens *in vitro*. These factors are defined as "desmutagens". To search for antimutagens, in a narrower sense acting on the cellular level and interfering with cellular mutagenesis, we carried out systematic screening using microbial systems and found many environmental antimutagens, such as a human placental factor and a tea factor (existing in the Camellia plant), which may be proofing errors in SOS repair.

Detection of environmental mutagens by the microbial system   We are currently carrying out screening of environmental mutagens by the following three methods:
(a) *Bacillus subtilis* rec-assay with and without *in vitro* metabolic activation (4, 10). The original streak method without metabolic activation is most simple and still useful for rapid screening of DNA-damaging agents. More recently, metabolic activation and quantitative analysis of a number of mutagens were accomplished using spores and thin agar.
(b) *Salmonella*/microsome assays that are described by Ames *et al.* (1) with small modifications.
(c) *Salmonella* SD assays (streptomycin dependency to independency)(13). The current *Salmonella*/microsome system utilizes histidine markers and their reversions

are studied. Since samples such as food very often
contain histidine this creates obstacles for use of
the Ames assay. Therefore we isolated a number of
streptomycin-dependent derivatives from the strains
TA98 and TA100, and two of them (SD98 and SD100) are
used for mutagenicity assays of natural products or
foodstuffs.

Examples of environmental mutagens detected   Use of
the rec-assay in combination with the *Salmonella*/microsome
system remarkably speeds up mutagenicity screenings. Cost
and labor for the rec-assays are less than 1/10 of those
of *Salmonella*/microsome assays, and the majority of
positive samples that are picked up by the rec-assay are
usually shown later to be mutagens in the *Salmonella*
assays.

Examples of environmental mutagens which were
detected by means of the combined procedures (*B. subtilis
rec* — *Salmonella his*) include food additives, pesticides,
metal compounds and miscellaneous chemicals including
known carcinogens (5, 6, 11, 15, 16, 18, 21, 26).

Results of the SD assay on 173 foodstuffs which were
purchased from the market revealed that 57 (32.9 %) gave
positive results; that is more than a doubling of the
level of SM-independent colonies were produced by them.
These samples included drinks, cereals, spices, vegetables,
fruits, cakes, etc.

This suggests that our daily food contain a number
of mutagenic components and it is absolutely necessary to
understand the quantitative aspects of mutagenicity. We
propose to express the mutagenic capacity of 1 g of food.
Thus mutations/gram or mut-g indicates the number of
mutations (His$^+$ or SM^{ind} mutants) induced by 1 g of food
sample. Certain examples of mut-g are shown in Table 2.

Desmutagen and antimutagen   Research about antimutagens acting on microorganisms started around 1950 in
certain laboratories and aspects of the basic findings
were recently reviewed by Clarke and Shankel (3).
In these studies, antimutagens are considered to be
acting mostly on the level of cellular mutagenesis.
Recently, modification factors affecting the yeild of
mutations induced by many environmental mutagens have
become known. Some of them are acting directly on mutagens and decompose or inactivate them. We propose the
term "Desmutagen" for these factors directly inactivating

mutagens *in vitro*. Term "Desmutagen" was originally proposed by Dr. John Drake. On the other hand, the term "Antimutagen" was originally defined as any agent which reduces apparently the yield of spontaneous or induced mutations regardless of the mechanisms involved. Therefore the original "Antimutagens" should include "Desmutagens". However, we propose now to define "Antimutagens" as factors acting solely on the cellular level for the suppression of mutations.

Studies on desmutagens   Recently much attention has been given to mutagenic products in charred proteinic food (27). We therefore tried to search for desmutagenic agents in food and found that extracts from certain vegetables such as cabbage, egg plant, burdock and broccoli worked as desmutagens for mutagenic pyrolysis products of tryptophan(7,19). Since the factor is sensitive to heating to 100°C as well as pronase, current purification procedures for proteins were employed utilizing the criterion of reducing the mutagenicity of Trp-P-2. A hemo-protein with a molecular weight of 43,000 was isolated (23). This protein possesses a peroxidase activity depending on Mg^{2+} and resorcinol as well as a NADPH oxidase. The "Cabbage factor" suppresses the mutagenicity of Trp-P-2 in two ways, by inactivating directly the molecular mutagen as well as by suppressing the S9 metabolic activation. It has been reported that a peroxidase prepared from human myeloma cells also inactivates a tryptophan pyrolysis product (29).

Studies on antimutagens   Here, "Antimutagens" are defined as agents acting on cells which have already been exposed to mutagenic agents and reducing incidences of mutations. Agents that reduce spontaneous mutabilities in "normal" and mutator strains are also considered to be antimutagens.

We examined different kinds of tissues of mammals and found that the placenta contains a factor antagonizing ultraviolet light-induced reversions of *Escherichia coli* B/r WP2 *trp* (14). Results of typical experiments using human placenta indicated that presence of similar activities was confirmed in placental tissues of mice, dog, rabbit, rat and cow.   However, in the case of rat, no similar activity was found in tissues other than placenta, such as liver, pancreas, stomack, brain, etc.

Recently Komura and his collaborators (17) studied

the chemical nature of the active agent in human placenta and found that cobaltous ions are a major factor. This was a surprising concidence for us since we discovered the antimutagenicity of cobaltous chloride several years ago and have been engaged in studies of the mechanisms involved.

The frequencies of mutations induced by NTG (N-methyl-N'-nitro-N-nitrosoguanidine) or gamma-rays in E. coli B/r trp were remarkably reduced by the presence of cobaltous chloride ($CoCl_2$) in the post-treatment medium of these bacteria (8, 9). It was also shown that the frequency of induced mutations towards 8-azaguanine resistance was significatively reduced by the presence of cobaltous chloride at a concentration of 1 µg/ml during the expression period of in vitro cultured cells of Chinese Hamster (Yokoiyama, in preparation).

In order to analyze the mechanisms involved, action of the metal compound was examined in a mutator strain of Bacillus subtilis NIG1125 possessing an altered DNA-polymerase III, a chromosome replicating enzyme of this bacterium (22). It was found that, at concentrations of cobaltous chloride which scarcely affected cellular growth, cells grown in the presence of the metal compound had definitively lower mutagenicity compared to those grown in its absence, though the numbers of cell-division events are very similar in both cultures. Further studies were carried out using strains of E. coli K12 such as SG902 (dnaE) possessing a genetic alteration in a DNA-replicating enzyme (25) or DM1187 (spr) having constitutive expression of an SOS DNA-repair enzyme (20). Cobaltous chloride reduced remarkably high spontaneous mutation frequencies in these strains (data to be published elsewhere by T. Inoue et al.).

Since there is increasing evidence that an inducible error-prone repair enzyme is involved in mutation induction (24, 28), and it is suggested that DNA polymerase III is one of the factors influencing mutation induction in UV or gamma-irradiated E. coli (2); we therefore conclude that the metal antimutagen may correct the error-proneness of DNA replicating enzyme(s) by improving fidelity during DNA synthesis.

We then carried out systematic screenings for environmental antimutagens using strain NIG1125 (his met mut-1) of B. subtilis in which the antimutagenicity of cobaltous chloride was clearly shown. The bacteria were grown on minimal glucose agar medium supplemented with a limited amount of required amino acids (the conventional semi-enriched medium used by Witkin, 1956 and Kada et al., 1960) containing, different concentrations of a test sample which was being examined for antimutagenic activity. We showed that crude extracts of juices obtained from certain plants such as Camellia (from which green tea and black tea are produced) reduced drastically the frequency of histidine reversions at concentrations where there is no growth inhibition at all. The potency of antimutagenic capacity can be expressed as that which is equivalent to 1 µg/ml cobaltous chloride.

REFERENCES

(1) Ames, B. N., McCann, J. and Yamasaki, E. *Mutation Res.* 31, 347-364 (1975)
(2) Bridges, B. A. and Mottershead, R. P., *Mol. Gen. Genet.* 162, 35-41 (1978)
(3) Clarke, C. H. and Shankel, D. M., *Bacterial Rev.* 39, 33-53 (1953)
(4) Kada, T., Tutikawa, K. and Sadaie, Y., *Mutation Res.* 16, 165-174 (1972)
(5) Kada, T., *Japan. J. Genetics* 48, 301-305 (1973)
(6) Kada, T., Moriya, M. and Shirasu, Y., *Mutation Res.* 26, 243-248 (1974)
(7) Kada, T., Morita, K. and Inoue, T., *Mutation Res.* 53, 351-353 (1978)
(8) Kada, T. and Kanematsu, N., *Proc. Japan Acad.* 54 (Ser. B), 234-237 (1978)
(9) Kada, T., Inoue, T., Yokoiyama, A. and Russell, L. B., *Proc. 6th Inter. Congr. Radiation Res.*, 711-720 (1979)
(10) Kada, T., Hirano, K. and Shirasu, Y., *"Chemical Mutagens" Vol. 6*, 149-173 (1980)
(11) Kada, T., *"Evaluation of Short-Term Tests for Carcinogens" Vol. 1*, 175-182 (1981)
(12) Kada, T., Inoue, T. and Namiki, M., *"Environmental Mutagenesis, Carcinogenesis and Plant Biology"* (in press)
(13) Kada, T., Aoki, K. and Sugimura, T., (submitted)
(14) Kada, T. and Mochizuki, H., *J. Rad. Res.*, (in press)
(15) Kanematsu, N., Hara, M. and Kada, T., *Mutation Res.* 77, 109-116 (1980)
(16) Kawachi, T., Komatsu, T., Kada, T., Ishidate, M., Sasaki, M., Sugiyama, T. and Tazima, Y., *"The Predictive Value of Short-Term Screening Tests in Carcinogenicity Evaluation"*, 253-267 (1980)
(17) Komura, H., Minakata, H., Nakanishi, K., Mochizuki, H. and Kada, T., *3rd Inter. Conf. Env. Mutagens 3B18*, (1981)
(18) Morimoto, I., Watanabe, F., Osawa, T., Okitsu, T. and Kada, T., *Mutation Res.* (in press)
(19) Morita, T., Hara, M. and Kada, T., *Agric. Biol. Chem.* 42, 1235-1238 (1978)
(20) Mount, D. W., *Proc. Nat. Acad. Sci., USA* 74, 300-304 (1977)
(21) Namiki, M., Udaka, S., Osawa, T., Tsuji, K. and Kada, T., *Mutation Res.* 73, 21-28 (1980)
(22) Inoue, T., Ohta, Y., Sadaie, Y. and Kada, T., *Mutation Res.* 91, 41-45 (1981)
(23) Inoue, T., Morita, K. and Kada, T., *Agric. Biol. Chem.* 45, 345-353 (1981)
(24) Radman, M., Villani, G., Boiteux, S., Defais, M., Caillet-Fanquet, P. and Spadari, S., *"Origins of Human Cancer" Vol. 4*, 903-922 (1977)
(25) Sevastopoulos, C. G. and Glaser, D. A., *Proc. Nat. Acad. Sci. USA* 74, 3947-3950 (1977)
(26) Shirasu, Y., Moriya, M., Kato, K., Furuhashi, A. and Kada, T., *Mutation Res.* 40, 19-30 (1976)
(27) Sugimura, T., Kawachi, T., Nagao, M., Yahagi, T., Seino, Y., Okamoto, T., Shudo, K., Kosuge, T., Tsuji, K., Wakabayashi, K., Iitaka, Y. and Itai, A., *Proc. Japan Acad.* 53, 58-61 (1977)
(28) Witkin, E. M., *Bacteriol. Rev.* 40, 869-907 (1976)
(29) Yamada, M., Tsuda, M., Nagao, M., Mori, M. and Sugimura, T., *Biochim. Biophys. Res. Comm.* 90, 769-776 (1979)

Antimutagens and the Problem of Controlling the Action of Environmental Mutagens

Urkhan Alekperov

Institute of Botany of the Academy of Sciencies of Azerbaijan SSR, Baku, U.S.S.R.

ABSTRACT For the estimation of complex characteristics of the antimutagens by their universality, effectiveness and physiologisity a single system was suggested. For the natural antimutagens the index of correlation between their endogenic content and stability of genetic structure to abiotic stresses was used additionally. Problems of antimutagenesis significance in natural population stability and possible application of this approach for neutralization of the environmental mutagen action are discussed.

INTRODUCTION

Prophylaxis of the possible genetic consequencies of the environmental pollution may be conducted differently. Technological, component and compensational approaches are presently put forward. The technological approach presupposes the industry reorganization according to closed-type cycles. In this case the resulting products of the process shouldn't have the mutagenic properties and as for the intermediate products, they can't produce any genotoxic effect because of their insulation. The component approach presupposes the genetic screening of many factors, they are revealing of the mutagenic products, replacing them by

the permissible analogues and others. Both these
approaches are the most optimal ones, however
in practice their realization presents technical
and economical problems. Being a trend of
compensational approach to the problems of
neutralization of the genetic consequencies of
the environment pollution, the antimutagenesis
is easily realized by the prophylactic application of the antimutagens. Antimutagenesis is
also perspective due to the possible application
of these modificators for the solution of the
mutagenesis general problems, that is a new
approach to the investigation of the mutagenesis.

SYSTEM OF CRITERIEN FOR ANTIMUTAGENS ESTIMATION

Currently we possess the information on
the antimutagenic effect of the synthetic and
natural compounds (1-3). The results of investigation show that this effect extends, practically
to all types of organizms. Besides, the comparative estimation of the antimutagens is complicated by the fact, that the action of these
compounds is studied on different objects, under
the conditions of using diffrerent mutation
inductors. That is why the data of experiments
on antimutagenesis are necessary to systematize.
The systematization may be performed on the
basis of the integrated system of criterien,
involving the complex estimation of the antimutagens according to their universality,
effectiveness and physiologisity.

The universality is one of the most important characteristics of antimutagens. One substance property of decreasing the mutability
of different objects by the mutation induction
with physicochemical factors is important for
the estimation of the premutational parameters
and their modifications. On the other side,
different kinds of the environmental mutagens
and the complexity of teir effect predetermine
the antimutagens estimation according to their
universality characteristic. For the estimation
of this parameter the index of the antimutagen
universality is suggested. Every unit of the

index is characterized by the antimutagenic effect according to one of the mutational tests induced by one type of the mutagens at one kind of the objects. The results of calculations show that α-tocopherol, ascorbic acid and sodium selenite have the highest index of universality - 8 (Table).

Table

CHARACTERISTIC OF UNIVERSALITY, EFFECTIVENESS AND PHYSIOLOGISITY FOR SOME ANTIMUTAGENS

Compound	Universality index	Effectiveness factor	Effect of higher concentrations
ionole	5	0,46	block at different stages of mitotic cycle, polyploidity and aneuploidity of cells
β-carotin	5	0,49	antimutagenic, lack of cytotoxic and mutagenic effects
α-tocopherol	8	0,47	-"-
phylloquinone	5	0,40	-"-
ascorbic acid	8	0,32	mutagenic
sodium selenite	8	0,21	mutagenic, cytotoxic

The well-known antimutagens are considered to be the optimizers of the mutational process. They don't decrease the spontaneous mutation to the level below the optimal one. However, different mutagens are characterized by different influence on the rate of the mutational process. For the quantitative estimation of this process it is necessary to use such a criterion as factor of effectiveness which is determined by a value obtained at one object and characterized by relationship between difference of frequency of the spontaneous and modificated mutations and a level of spontaneous mutability. The most effective of the investigated antimutagens are: β-carotin, α-tocopherol, phylloquinone and others (Table).

The problem of physiologisity in antimutagenesis arose because of the fact that in the first studies of antimutagenesis performed on the eukaryots it was determined that having these properties in low concentrations some of the antimutagens in higher concentrations may also expose the mutagenic and cytotoxic effects. Such a change of the effect was observed for some antimutagens and as for the other (practically natural compounds) the increase of concentration within some orders provide no loss nor change of the effect (Table). Consequently, the property of physiologisity can't be a limiting factor for the application of these antimutagens.

The given above methods of quantitative estimation of antimutagens according to universality, effectiveness and physiologisity criterien provide the objective characteristic for determination among them the perspective ones.

ANTIMUTAGENESIS AND REPARATION

Mechanism of the action of the most antimutagens is not conclusively investigated yet. Fact of universality of many antimutagens shows that these substances may influence on the system reparation. The correlation between the antimutagenesis and the reparation was observed in the experiments with ionole. In these experiments the ionole influence on the DNA additional synthesis was observed. Analysis of the results showed that the antimutagen provides the increased H^3-thymidine induction to DNA in cells being in presynthetical phase (Fig.1).

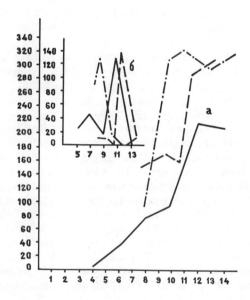

Fig. 1 Ionole action (10 /ml) on H^3-thymidine incorporation (a) and absolute difference of the intensivity of labelling of Crepis capillaris cells being in G_1 phase
Axles: Abscissa - time from beginning of the seeds soaking, hour
Ordinate - number of granes for 1000 cells

As the action of the antimutagen and H^3-thymidine is limited by G_1-phase, the obtained data show the influence of the antimutagen on the additional, probably reparational synthesis of the DNA.

The influence of the antimutagen on the ferment system is observed in the experiments on temperature modification of the induced antimutagens by the recovery process. The obtained data indicate that the antimutagen effect is correlated with the reparation process.

ANTIMUTAGENESIS AND MUTATION OF NATURAL POPULATION

Among the antimutagens the natural compounds, being the products of the plant methabolism take a particular place (1,4,5). It is believed that these compounds associated with the automutagenic group may influence on the rate of spontaneus mutation and determine the species resistance to the extrimal mutagenic effect. We obtained the experimental data confirming this fact in the process of investigation of the spontaneous and induced mutability of Aegilops ovata, growing in nature at different levels of the vertical zonality. In these experiments the investigation of the endogenic content of the tocopherol was conducted. It was detected the correlation between the autoantimutagenic content and general apparatus resistance to the action of the mutagenic factors (Fig. 2).

The autoantimutagenic system may be important for the problem of rare and disappearing species. It is well known that on the territory exposed to an equal technogenic effect some species disappear and the other preserve their population. That is why we studied two species of the oats: Avena eriantha and Avena ventricosa. Both these species are widely spread in one region (Azerbaijan SSR, Apsheron peninsular) and consequently, they are exposed to an equal antropagenic effect. Besides, one of them (A. ventricosa) is the disappearing species and another doesn't reduce the areal. The investigations show that the disappearing species is more sensitive to the effect of the mutagenic factors and content less quantity of the autoantimutagens.

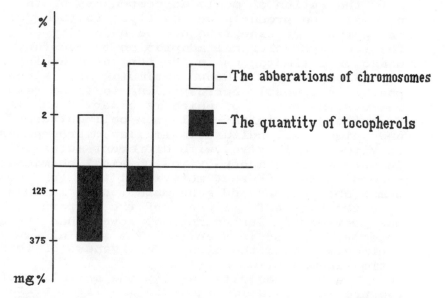

Fig. 2 Correlation of the induced mutability level and content of the endogenic tocopherols at Aegilops ovata

ANTIMUTAGENESIS IS AN ACTIVE PHASE OF GEOFUND PROTECTION

In the problem of geofund protection the antimutagenesis plays an active role and not only in the problem of rare and disappearing species. The antimutagens may be used for many objects which are exposed to the effect of the environmental mutagens. The antimutagens suggested for practical application should correspond to the following requirements:
- high antimutagenic effectiveness with different inductors of mutagenic process and their combinations;
- universality of the antimutagenic action when using the different test systems and mutational tests;
- lack of cumulative properties and consequently, impossible application for a long time;
- possible incorporation of these substances

in the ration of meals as components of the natural food products or additives to them for the purpose of exception of the necessity of the socio-psycological adaptation to constant usage of antimutagens in the form of medicines.

Among the well-known antimutagens, -tocopherol is a mostly corresponding to these requirements, the effect of which is observed on different objects (plant cells, man cell cultures and animal cell cultures, mammalian, microorganisms in vitro and in vivo), with different mutation inductors (spontaneous, virus, physical, chemical) according to different mutational test (chromosomes, aberrations and gene mutations). In this case the high effectiveness of the preparate was provided in concentrations lower than those accepted for use in practice. The special experiments showed the high effectiveness of this antimutagen application (1).

The obtained data testify the actuality of the problem of antimutagenesis and its importance in the course of prevention the remote genetic consequencies of the environmental pollution by mutagens.

REFERENCES:

(1) Alekperov, U. Antimutageny i problema zashchity geneticheskogo apparata (Rus.), Baku, Elm, 133 (1979)
(2) Alekperov, U. Uspekhi sovremennoi genetiki (Rus.), Moscow, Nauka, 8 (1979) 168-181
(3) Clarke, C., Shankel, D. Bacteriol. Rev., 39, 33-53, (1975)
(4) Kada, T., Morita, K., Inone, T. Mat. Res., 53, 351-353, (1978)
(5) Namiki, M., Osawa, T., Tsuji, K., Kada, T. Proc. Jap. Cancer Assoc., 38th Annu. Meet, Tokyo (1979), 67 (Eng.)

Chemical Structure of Mutagens

An Overview of the Structural Features of Mutagenic S-Chlorallylthio and Dithiocarbamate Pesticides and Trichlorfon, Dichlorvos and Their Metabolites

Lawrence Fishbein

National Center for Toxicological Research, Jefferson, Arkansas, U.S.A.

ABSTRACT A small number of pesticides were examined primarily within the classes of herbicidal S-chlorallylthio-and dithiocarbamates, fungicidal dithiocarbamic acid derivatives, and organophosphorus insecticides such as trichlorfon and dichlorvos and their interrelated metabolic, degradation and/or transformation products with a view to relating germane features of their structure with respective mutagenic and/or teratogenic activity.

The principal objective of this overview is to highlight the structural features of several herbicides (e.g., diallate, triallate and sulfallate), ethylene bisdithiocarbamate fungicides and insecticides (e.g., trichlorfon and dichlorvos) in which the compounds themselves, their metabolites and/or degradation and biotransformation products (as well as trace impurities) suggest an <u>a priori</u> mutagenic and/or teratogenic risk.

1. <u>S-Chlorallylthio- and Dithiocarbamates</u>: Herbicides account for an increasing percentage of all pesticides used in the United States. Currently, more than 180 are marketed, in addition to about 6000 formulated products (1). Production in the U.S. of synthetic organic herbicides ranges from 650 - 700 million pounds per year. There has been a rapid growth in usage in preemergence herbicides that are applied to the soil prior to germination of weeds and crops.
    Diallate or Avedex (<u>cis</u>, <u>trans</u> mixture) [S-(2,3-dichloroallyl)diisopropyl-thiocarbamate] (I), triallate or

Avedex BW [S-(2,3,3-trichloroallyl)diisopropylthiocarbamate] (II); and sulfallate, or CDEC or Vegadex (2-chloroallyl-N,N-diethyldithiocarbamate) (III) have been employed as preemergence herbicides for controlling weeds in certain vegetable and field crops for about 15 years (2,3). For example, diallate and triallate are used to control wild oats in wheat and barley crops and vegetable crops and is registered for preemergence use in the U.S. for 11 agricultural crops, largely for age or field crops. Commercial formulations of diallate contain roughly equal amounts of the cis and trans isomers (3). The annual consumption of diallate, triallate, and sulfallate in the U.S. in 1975 was reported to be 0.3, 1.3 and 0.1 million pounds respectively (4).

$$[(CH_3)_2CH]-N-\underset{O}{\overset{\|}{C}}-S-CH_2-\underset{Cl}{\overset{|}{C}}=CHCl \quad [(CH_3)_2CH]-N-\underset{O}{\overset{\|}{C}}-S-CH_2-\underset{Cl}{\overset{|}{C}}=CCl_2$$
$$(I) \qquad\qquad\qquad\qquad (II)$$

$$(CH_3CH_2)_2-N-\underset{S}{\overset{\|}{C}}-S-CH_2-\underset{Cl}{\overset{|}{C}}=CH_2$$
$$(III)$$

The persistence of these herbicides is greatly dependent on soil and environmental conditions with losses of the agents from soil primarily attributed to microbial degradation, volatilization and photodecomposition. Triallate persists for a much longer time than diallate due to both a slightly slower rate of decomposition and a lower volatility (5). Additionally, a residual amount of diallate has been reported to occur on or in treated crops after harvesting (3,5).

Although in terms of usage, the S-chlorallyl thio and dithiocarbamate herbicides are relatively small in comparison to the arsenicals, chlorphenoxy acids, phenylureas, benzoic acids, carbamates and triazines, there are some rather fascinating structure-activity relationships governing their herbicidal and genotoxic activities.

Diallate, triallate and sulfallate are potent mutagens in the Salmonella typhimurium test (Ames assay) with the TA 100 and TA 1535 strains (base-pair substitution mutants) only in the presence of liver microsomal preparations (5-7) indicating that these agents require metabolic activation for their conversion into active mutagens. None of these herbicides caused mutations in strains TA 98 and TA 1535 which are frameshift mutants (7). Diallate was found to be considerably more potent than triallate on TA 100 and TA 1535 showing mutagenic activity of 1 ug/plate (7).

The mutagenic activity of diallate and triallate have been recently examined by a battery of in vitro mammalian

and microbial tests by Sandhu and Waters (8). Diallate
and triallate were mutagenic in the L5178Y mouse lymphoma
thymidine kinase assay and in the Drosophila sex-linked
recessive lethal assay, but both were negative in E. coli
WP2 assay, Saccharomyces cerevisiae D-7 tests for primary
DNA damage and for gene mutation. Both herbicides were
also negative in the mouse micronucleus test. In all
these tests, diallate was much more potent than triallate
and, as noted earlier, in the in vitro tests, the genetic
activity of these herbicides was dependent upon mammalian
activation.

Douglas et al (9) reported that diallate and triallate show dose-related increases without metabolic
activation in S. typhimurium strains TA 1535, TA 100 and
TA 98 indicating that these compounds cause both frameshift and base substitution mutations. The mutagenicity
of both herbicides was greatly enhanced by incubation with
Aroclor 1254 induced rat liver S-9 fraction.

Further studies with these herbicides using mammalian
cells in a number of in vitro tests with Chinese hamster
ovary (CHO) cells combined with metabolic activation were
reported by Douglas et al (9). While diallate and triallate caused dose-related decreases in colony forming
ability, with concomittant dose-related increases in the
frequencies of cells with chromosome damage and in the
number of sister-chromatid exchanges (SCE), only diallate
caused a reduction in DNA molecular weight as determined
by alkaline sucrose gradient (ASG) sedimentation. DNA
damage was negligible even at concentrations of triallate
that reduced colony forming ability to zero. The lesions
in DNA detected by the ASG technique are not necessarily
related to those that produce chromosomal damage. These
data taken together, strongly suggest that both diallate
and triallate are capable of causing mutations in mammals.
Diallate has been reported to be carcinogenic in mice (10)
at a dose of 1000 mg/kg. A more definitive risk evaluation
of these herbicides in terms of inherited disease or
cancer remains to be more rigorously established by
appropriate in vivo methology.

It should be noted that in the study of Douglas et al
(9) where diallate and triallate were found to be mutagenic in S. typhimurium TA 1535, TA 100 and TA 98 without
metabolic activation, that relatively purified agents were
used in contrast to previous studies (5-8) where technical
grade materials were used and the mutagenic activity in TA
100 and TA 1535 required metabolic activation.

Recent elegant studies by Casida, Schupan, Rosen and their co-workers have shed considerable light on the nature of the ultimate mutagen formed on metabolism of the pro-mutagenic herbicides diallate and sulfallate (11-15). For example, the ultimate mutagen formed from diallate is most probably 2-chloroacrolein, resulting from a sequence of sulfoxidation, [2,3]-sigmatropic rearrangement and 1,2 elimination reactions (11-15). A portion of the highly reactive intermediate, diallate sulfoxide (proximate mutagen) is attacked by glutathione in a reaction that competes with its transformation to the ultimate mutagen, 2-chloroacrolein. The rearrangement and elimination reaction sequences to liberate 2-chloroacrolein occur only with the portion of the sulfoxide that is not detoxified by reaction with GSH (to ultimately yield the respective mercapturic and sulfonic acids). The herbicidal activity of the diallate isomers (and triallates) may be due to the sulfoxides acting as carbamoylating agents for critical tissue thiols (12,15-17) or to liberation of 2-chloroacrolein which is similar to potency to the diallate isomers themselves in inhibitive root growth (12,13).

Triallate is somewhat different in terms of its mutagenic activity. For example, triallate sulfoxide (tested immediately after preparation by peracid oxidation) is a potent mutagen (without S-9 activation), whereas its decomposition product, 2-chloroacrylyl chloride, is much less active (13). Since triallate sulfoxide is unstable, it is probably the proximate mutagen. Schuphan et al (13) suggested that it may be necessary for the chloroacrylyl chloride, which is easily hydrolyzed to the non-mutagenic 2-chloroacrylic acid, to be liberated within the bacteria in order to express its mutagenic activity.

The mechanism for the mutagenic activation of sulfallate was recently elaborated by Rosen et al (14). Analogously to diallate, it was found to be metabolized <u>in vitro</u> to 2-chloroacrolein. Although not identified, the proximate mutagen may be formed by hydroxylation of sulfallate or one of its dithio- or thiocarbamate metabolites at the methylene group, adjacent to the sulfur. The mutagenic potencies of sulfallate and 2-chloroacrolein with those of three sulfallate derivatives formed easily on oxidation (11) and two other potential metabolites were compared. Except for 2-chloroacrolein, each compound was inactive or required the S-9 mix for detection or enhanced activity. The mutagenic potency with S-9 mix decrease in the order 2-chloroacrolein >> sulfallate >> sulfallate derivatives and potential metabolites.

In comparison of the mutagenic activities of diallate, triallate and sulfallate, Rosen et al (14) suggested that diallate yields the most potent mutagen(s), possibly because it yields highly active mono- and dichloroacroleins both on the preferred sulfoxidation-rearrangement-elimination sequence and on alpha-hydroxylation. In contrast, the sulfoxidation pathway for triallate yields a weak mutagen, 2-chloroacrylyl chloride, suggesting that alpha-hydroxylation and trichloroacrolein release may be more significant in mutagen formation. 2-Haloacroleins are highly mutagenic compounds (18), whether formed by sulfoxidation-rearrangement reactions involving dechlorination or by alpha-hydroxylation without dechlorination. 2-Chloroacrolein is a potent direct acting mutagen in TA 100 (not dependent on the S-9 mix) when compared with the promutagens cis- and trans-diallate (19). The greater potency of cis- than of trans-diallate may be due to more rapid formation of 2-chloroacrolein in the former case. 2-Chloroacrolein without S-9 metabolic activation and diallate with S-9 activation at 10 to 30 nmoles also gave significant numbers of revertant colonies with S. typhimurium TA 98 for detection of frame-shift mutagen (19).

Haloacroleins (e.g., 2-bromo-, 2,3-dichloro- and 2,3,3-trichloro analogs) in addition to 2-chloroacrolein are much more potent mutagens in S. typhimurium TA 100 assay without metabolic activation than any other aldehydes examined previously. The activity of 2-chloroacrolein is reduced by the S-9 mix possibly due to the oxidation to the acid. Glutathione also reduces the activity, suggesting rapid reaction of 2-chloroacrolein with this nucleophile.

2-Chloroacrolein was found to react rapidly with deoxyadenosine to yield polymeric material as most likely occurring via two distinct types of reaction, e.g., Schiff base formation at the aldehyde carbon and Michael addition at the beta- delta-position. The analogous reaction with DNA and limiting amounts of 2-chloroacrolein was suggested to probably lead to cross-linking (18).

2. **Organophosphorus Insecticides Trichlorofon and Dichlorvos and their Metabolites:** In contrast to the chlorinated hydrocarbons which are being phased out of the pesticide market, organophosphorus pesticides do not pose a serious residue and ecological problem. Thus in recent

years the usage of this class has greatly increased and
these compounds represent a major class of agricultural
pesticides employed today. However, a number of organophosphorus pesticides are chemical alkylating agents and
hence are potential mutagens and/or carcinogens. Thus the
potential genetic hazards presented by the widespread use
of the organophosphorus pesticides warrant scrutiny.

Trichlorofon (O,O-dimethyl-1-hydroxy-2,2,2-trichloro-
ethylphosphonate, Dipterex) is metabolized to dichlorvos
as well as desmethyltrichlorophene, dichloroacetaldehyde
(IV) and 2,2-dichloro-1,1-dihydroxyethanephosphonic acid
methyl ester (V). The genetic effects of trichlorofon are
believed to be due in part to its degradation product
dichloroacetaldehyde and (V) (19). Dichloroacetaldehyde,
a metabolite of the important insecticide dichlorvos
(DDVP) formed by esterase-catalyzed hydrolysis (20-22) (as
well as a metabolite of vinylidene chloride (23-25),
induces dominant lethals in mice (19) and is mutagenic in
S. typhimurium TA 100 with and without metabolic activation (23,26) and in two forward mutation systems in A.
nidulans (23) and S. coelicolor (23). It was negative
when tested in TA 1535. [Trichloroacetaldehyde (chloral)
is an impurity in commercial DDVP and was mutagenic when
tested in Salmonella strain TA 100 with and without
metabolic activation (23)]. Dichloroacetaldehyde has
transient existence and to date has not been detected in
the tissues or excreta of mammals exposed to dichlorvos.

Although the bacterial mutagenicity of dichlorvos has
been largely or entirely explained by DNA methylation
(26), it generally occurs to such a low extent that
methylation of DNA by dichlorvos cannot be considered a
genotoxic hazard to man (20). However, the theoretical
possibility of DNA attack by the dichlorovinyl groups of
dichlorvos has not been adequately studied (20).

It is important that the genotoxic hazard of this
important insecticide be thoroughly evaluated as there is
an apparent extensive exposure to this agent by a large
number of individuals in many countries. Ramel et al (20)
noted that since the metabolic conversion of dichlorvos in
vivo seems to proceed predominantly by esterase-catalyzed
hydrolysis to dimethyl phosphate and dichloracetaldehyde,
both genetic and physiological (including environmental)
effects on esterase activity may have important effects on
the balance between the metabolic pathways for in-vivo
conversion of dichlorvos, perhaps leading to mutagenesis.

References:

1. Sanders, H. J., Chem. Eng. News, Aug. 3 (11981) 20-35
2. Weed Science Society of America, Herbicide Handbook, 3rd ed. (1974)
3. IARC, Monographs on the Evaluation of the Carcinogenic Risk of Chemicals to Man, Some Carbamates, Thiocarbamates and Carbazides, Vol. 12, International Agency for Research on Cancer, Lyon (1976) 69-75
4. Stanford Research Institute, Chemical Economics Handbook - Pesticides, Menlo Park, CA (1976)
5. De Lorenzo, F., Staiano, N., Silengo, L., and Cortese, R., Cancer Res., 38 (1978) 13-15
6. Carere, A., Ortali, V. A., Cardamone, G., and Morpurgo, G., Chem. Biol. Interact., 22 (1978) 297-308
7. Sikka, H. C., and Florczyk, P., J. Agr. Food Chem., 26 (1978) 146-148
8. Sandhu, S. S., and Waters, M. D., 12th Annual Meeting of Environmental Mutagen Society, San Diego, CA, March 5-8 (1981) Abstract Ca-12, p. 123
9. Douglas, G. R., Nestmann, E. R., Grant, C. E., Bell, R. D. L., Wytsma, J. M., and Kowbel, D. J., Mutation Res., 85 (1981) 45-56
10. Innes, J. R. M., Ulland, B. M., Valerio, M. G., Petrucelli, L., Fishbein, L., Hart, E. R., Pallotta, A. J., Bates, R. R., Falk, H. L., Gart, J. J., Klein, M., Mitchell, I., and Peters, J. J., J. Natl. Cancer Inst., 42 (1969) 1101
11. Schuphan, I., and Casida, J. E., Tetrahedron Letters (1979) 841
12. Schuphan, I., and Casida, J. E., J. Agr. Food Chem., 27 (1979) 1060
13. Schuphan, I., Rosen, J. D., and Casida, J. E., Science, 205 (1979) 1013
14. Rosen, J. D., Schuphan, I. Segall, Y., and Casida, J., J. Agr. Food Chem., 28 (1980) 880-881
15. Schuphan, I., Segall, Y., Rosen, J. D., and Casida, J. E., American Chemical Society Symposium Series No. 158, (1981) 65-82
16. Casida, J. E., Gray, R. A., and Tilles, H., Science, 184 (1974) 573
17. Hubbell, J. P., and Casida, J. E., J. Agr. Food Chem., 25 (1977) 404
18. Rosen, J. D., Segall, Y., and Casida, J. E., Mutation Res., 78 (1980) 113-119

19. Fischer, G. W., Schneider, P., and Schueffler, A., Chem. Biol. Interactions, 19 (1977) 205-213
20. Ramel, C., Drake, J., and Sugimura, T., Mutation Res., 76 (1980) 297-309
21. Wright, A. S., Hutson, D. H., and Wooden, M. F., Arch. Toxicol., 42 (1979) 1-18
22. Hutson, D. M., Hoadley, E. C., and Pickering, B. A., Xenobiotica, 1 (1971) 593-611
23. Bignami, M., Conti, G., Conti, L., Crebelli, R., Misuraca, F., Puglia, A. M., Randazzo, R., Sciandrello, G., and Carere, A., Chem. Biol. Interactions, 30 (1980) 9-23
24. Hathaway, D. E., Environ. Hlth. Persp., 2 (1977) 55-59
25. Bronzetti, G., Bauer, C., Corsi, C., et al., Mutation Res., 89 (1981) 179-185
26. Löfroth, G., Z. Naturforsch., 33 (1978) 783-785

The Dependence of Mutagenic Activity on Chemical Structure in Two Series of Related Aromatic Amines

Majdi M. Shahin

L'Oreal Research Laboratories, Aulnay-sous-Bois, France

THE Salmonella/mammalian microsome assay was used to test m-diaminobenzene, four 2,4-diaminoalkylbenzene compounds, and six 2,4-diaminoalkoxybenzene compounds for mutagenicity. Several of the compounds were found to be mutagenic in frameshift tester strains, and the responses were dose-dependent. None of the compounds was mutagenic in the absence of a rat-liver S9 metabolic activation system. In both series of compounds, the size of the substituent groups in the chemical structures had a pronounced effect on mutagenicity. Specifically, m-diaminobenzene and its derivatives with small alkyl or alkoxy substituents were more potent mutagens than the compounds with larger substituent groups, and the compounds with the largest substituent groups were nonmutagenic. The elucidation of relationships between chemical structure and mutagenic activity can lead to an improved ability to predict the toxicologic properties of chemicals. The broad implications of structure-activity relationships for genetic toxicology are discussed.

Through research in chemical mutagenesis, assay systems have been developed for the detection of a variety of genetic endpoints that have implications for human health. Short-term mutagenicity tests are now used in laboratories throughout the world to identify chemicals that are likely to be carcinogens in intact mammals, including man. In this respect, it is important that mutational assays continue to be refined so that they can predict carcinogenesis with high sensitivity and specificity (i.e., with a minimum number of false negatives and false positives).

The most widely used of all mutagenicity tests is the Salmonella/mammalian microsome test (1-2) that was developed by Bruce Ames and his colleagues. In our laboratory, we use this test to evaluate the mutagenicity of series of structurally related chemicals. Our objective is to elucidate relationships between chemical structure and biological activity; the ultimate goal of studies of structure-activity relationships is to acquire the ability to predict the mutagenicity, and thereby the carcinogenicity, of untested compounds.

In this paper we report results of studies of structure-activity relationships among monocyclic aromatic amines. The compounds studied were m-diaminobenzene, a series of four 2,4-diaminoalkylbenzenes (the methyl, ethyl, isopropyl, and n-butyl compounds), and a series of six 2,4-diaminoalkoxybenzenes (the methoxy, ethoxy, isopropoxy, n-propoxy, n-butoxy, and β-hydroxyethoxy compounds). The compounds were tested for their ability to induce mutations in Salmonella typhimurium (3-4). Although none of the compounds was active without metabolic activation, several of the compounds were mutagenic in the presence of an Aroclor 1254-induced S9 metabolic activation system. In the first series of compounds, m-diaminobenzene was the most active mutagen, followed by 2,4-diaminotoluene and 2,4-diaminoethylbenzene; 2,4-diaminoisopropylbenzene and 2,4-diamino-n-butylbenzene were negative, in that neither induced a two-fold increase in the number of revertant colonies in Salmonella. Results obtained in tester strains TA1538 and TA98 with m-diaminobenzene and its 2,4-diaminoalkylbenzene derivatives are presented in Figure 1. The mutagenic compounds produced dose-related increases in the number of revertant colonies, with the slopes of the curves differing for different compounds. The data indicate that larger alkyl groups at the C_1 position confer lower mutagenic activity.

Among the 2,4-diaminoalkoxybenzenes, 2,4-diaminoanisole (the methoxy compound) was the most potent mutagen; the compounds 2,4-diaminoethoxybenzene, 2,4-diaminoisopropoxybenzene, and 2,4-diamino-n-propoxybenzene were markedly less mutagenic than 2,4-diaminoanisole. Figure 2 presents results on the mutagenicity of 2,4-diaminoalkoxybenzene compounds in strains TA1538 and TA98. Like the mutagenicity of 2,4-diaminoalkylbenzenes, the mutagenicity of 2,4-diaminoalkoxybenzenes depends on the size and nature of the alkoxy substituents on the C_1 position of 2,4-diaminobenzene.

Without metabolic activation, none of the 2,4-diaminoalkylbenzenes or 2,4-diaminoalkoxybenzenes that we tested was mutagenic. In addition to the requirement for metabolic activation, the two series of compounds were similar in that mutagenic responses were observed mainly in tester strains that detect frameshift mutations (3-4).

Correlations between structural characteristics of chemicals and mutagenicity in Salmonella, such as those reported here, can be of particular value if they lead to generalizations that aid in predicting carcinogenicity in mammals. A growing body of information suggests that predictions of biological properties on the basis of chemical structure will come to occupy an increasingly prominent place in toxicologic studies. The principal objective of studies of structure-activity relationships in genetic toxicology is to predict mutagenic and carcinogenic effects with a minimum number of false negatives or false positives. Such studies can also contribute to an understanding of mechanisms of mutagenesis, the evaluation and validation of mutational assay systems, and the assessment of risks associated with particular compounds.

(1) Ames, B.N., McCann, J., and Yamasaki, E. Mutation Res. 31, 347-364 (1975)

(2) McCann, J., Spingarn, N.E., Kobori, J., and Ames, B.N. Proc. Natl. Acad. Sci. (U.S.A.) 72, 979-985 (1975)

(3) Shahin, M.M., Bugaut, A., and Kalopissis, G. Mutation Res. 78, 25-31 (1980)

(4) Shahin, M.M., Rouers, D., Bugaut, A., and Kalopissis, G. Mutation Res. 79, 289-306 (1980)

382 M. M. Shahin

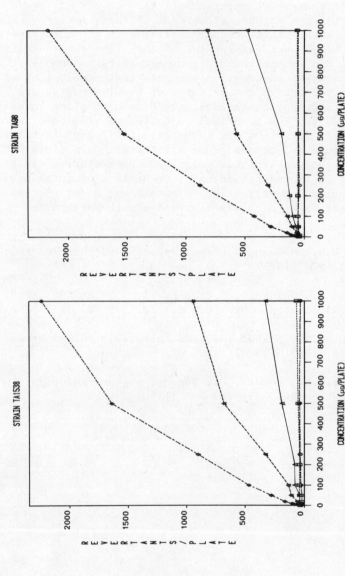

Fig. 1. Dose-response curves for the mutagenicity of m-diaminobenzene and its 2,4-diaminoalkylbenzene derivatives in the presence of an Aroclor 1254-induced rat-liver homogenate. ◇, m-diaminobenzene; ☆, 2,4-diaminotoluene; △, 2,4-diaminoethylbenzene; O, 2,4-diaminoisopropylbenzene; □, 2,4-diamino-n-butylbenzene.

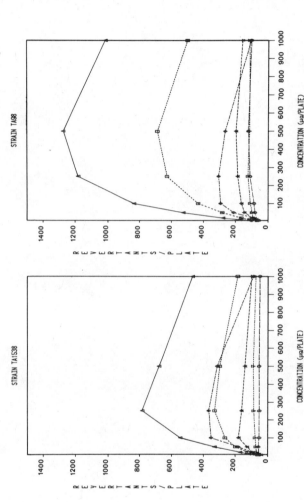

Fig. 2. Dose-response curves for the mutagenicity of 2,4-diaminoalkoxybenzene compounds in the presence of an Aroclor 1254-induced rat-liver homogenate. △, 2,4-diaminoanisole dihydrochloride; ◇, 2,4-diaminoethoxybenzene dihydrochloride; □, 2,4-diaminoisopropoxybenzene dihydrochloride; ☆, 2,4-diamino-n-propoxybenzene dihydrochloride; ▽, 2,4-diamino-n-butoxybenzene dihydrochloride; ○, 2,4-diaminophenoxyethanol dihydrochloride.

a From Shahin et al.(4)

Arylamine-DNA Adduct Formation in Relation to Urinary Bladder Carcinogenesis and *Salmonella typhimurium* Mutagenesis

F. F. Kadlubar, F. A. Beland, D. T. Beranek, K. L. Dooley, R. H. Heflich, and F. E. Evans

National Center for Toxicological Research, Jefferson, Arkansas, U.S.A.

ABSTRACT The metabolic activation of aromatic amines for urinary bladder carcinogenesis is believed to involve sequential hepatic N-hydroxylation and N-glucuronidation. These N-hydroxy arylamine N-glucuronides enter the circulation and are transported to the urinary bladder. Hydrolysis of the conjugates in the acidic environment of the lumen releases free N-hydroxy arylamines which can covalently bind to DNA and are direct-acting mutagens and carcinogens. In in vivo experiments with 2-naphthylamine, 4-aminobiphenyl, 4-nitrobiphenyl, and 2-acetylaminofluorene, the carcinogen-DNA adducts formed in the dog urothelium were shown to be identical to the products obtained by reaction of their N-hydroxy arylamines with DNA in vitro at pH 5. For 2-naphthylamine and 4-aminobiphenyl, C8-substituted deoxyguanosine adducts were found to be major persistent lesions. The same spectrum of adducts was also obtained when the N-hydroxy arylamines were incubated with Salmonella typhimurium TA 1538. By comparing the level of arylamines bound per Salmonella genome with mutation induction, the relative mutagenic efficiency of the adducts were shown to vary over a 5-fold range. These data suggest that the persistence of mutagenic arylamine-DNA adducts may be responsible for the initiation of urinary bladder carcinogenesis.

INTRODUCTION

Aromatic amines were among the first chemicals shown to induce tumors in humans and experimental animals. In industrial workers exposed to these substances and in dogs given the carcinogens in their diet, an appreciable incidence of transitional cell carcinomas of the urinary bladder was observed (reviewed in 1). Present day exposure to aromatic amines still occurs through cigarette smoke, synthetic fuels, and pyrolysis products from cooked foods (2-5; reviewed in 6).

Metabolic studies have shown that arylamine bladder carcinogens are N-hydroxylated in the liver to form highly mutagenic N-hydroxy arylamines. These N-hydroxy derivatives, which may also be formed by the metabolic reduction of nitroaromatics, are carcinogenic at sites of application, and will react spontaneously with DNA, especially at mildly acid pH. Of particular importance to bladder carcinogenesis is the efficient conjugation of these N-hydroxy derivatives by hepatic glucuronyl transferases. The N-glucuronides then enter the circulation and are filtered into the urinary bladder lumen, where under acidic conditions, as is found in human and dog urines, hydrolysis occurs to release the N-hydroxy arylamines. The acidic environment also facilitates formation of electrophilic nitrenium-carbenium ions and the subsequent generation of arylamine-macromolecular adducts (reviewed in 7).

We have previously shown that 2-naphthylamine (2-NA), in vivo, and N-hydroxy-2-NA, in vitro, covalently bind to DNA at the N^6 atom of deoxyadenosine (dA) and at the N^2 and C8 atoms of deoxyguanosine (dG). Of these lesions, the biological persistence of the C8-dG derivative appeared to correlate with target organ specificity (8). In the present report, we compare the arylamine-DNA adducts formed in the dog in vivo from the urinary bladder carcinogens, 4-aminobiphenyl (4-ABP) and 4-nitrobiphenyl (4-NBP) (9), with the adducts obtained by reacting N-hydroxy-4-ABP with DNA at pH 5. In addition, we have investigated the adducts formed in the dog following administration of the hepatic and bladder carcinogen, 2-acetylaminofluorene (refs. 10-11; 2-AAF). Finally, the genetic consequences of these lesions were compared by determining the nature and mutagenic potency of the arylamine-DNA adducts formed in Salmonella typhimurium after exposure to N-hydroxy-2-aminofluorene, N-hydroxy-4-ABP, and N-hydroxy-2-NA.

MATERIALS AND METHODS

Calf thymus DNA (type I), deoxyribonuclease I (type DN-Cl), nuclease P_1, and acid (type I) and alkaline (type III-S) phosphatases were purchased from Sigma Chemical Co. (St. Louis, MO). [2,6,2',6'-(n)^{3}H]4-ABP (385 and 980 mCi/

mmole), [2,6,2',6'-(n)^{3}H]4-NBP (1220 mCi/mmole), [ring-^{3}H]-
N-hydroxy-2-AF (80 mCi/mmole), [2,6,2',6'-(n)^{3}H]N-hydroxy-
4-ABP (33 mCi/mmole), [5,6,7,8-^{3}H]N-hydroxy-2-NA (32 mCi/
mmole), and [ring-^{3}H]2-AAF (746 mCi/mmole) were obtained
from R. Roth, Midwest Research Institute (Kansas City, MO).

The 4-ABP-DNA adducts were prepared in vitro by reacting [^{3}H]N-hydroxy-4-ABP (1.0 mM) with calf thymus DNA (5 mg/ml) at pH 5 in 10 mM potassium citrate buffer (100 ml) for 4 hours under argon. The DNA was isolated and partially hydrolyzed with deoxyribonuclease I as previously described (12). Nuclease P$_1$ (20 µg/ml), acid phosphatase (0.3 units/ml) and alkaline phosphatase (0.5 units/ml) were then added and the incubation was continued for an additional 16 hours at 37º. The adducts were partitioned into water-saturated n-butanol, and the organic phase was washed with 0.5% ammonium sulfate and evaporated under reduced pressure. The residue was redissolved in 50% methanol, and the individual adducts were purified by high pressure liquid chromatography (hplc). 2-NA and 2-aminofluorene (2-AF) adducts were prepared as previously described (12,13).

Male beagle dogs (8-10 kg; Marshall Lab Animals, North Rose, NY) were treated p.o. with the H-carcinogens (60 µmoles/kg) in gelatin capsules and killed 1, 2, or 7 days later by Nembutal injection. Hepatic and urothelial arylamine-modified DNA was isolated as previously described (8). The 4-ABP and 4-NBP modified DNA was hydrolyzed as outlined above; 2-NA- and 2-AF-DNA were hydrolyzed as described (12,13). The individual adducts were separated by hplc and quantified by liquid scintillation counting.

Mutagenicity experiments were conducted in Salmonella typhimurium strain TA 1538 (obtained from B. Ames, Univ. Calif., Berkeley, CA). An overnight culture was concentrated approximately 5-fold by centrifugation (10,000 x g, 10 min.) and resuspension in 0.85% sodium chloride, 50 mM sodium phosphate buffer (PBS), pH 7.2. Immediately preceeding the incubations, the mutagens were dissolved in either ethanol or dimethylsulfoxide. Aliquots (0.5 ml) were added to 19.5 ml of the concentrated bacteria and the mixtures were incubated with shaking for 30 min. at 37º. One ml of the culture was then diluted with 4 ml cold PBS and used to determine reversion frequencies and percent survival (14), while the remainder of the culture was pelleted by centrifugation for arylamine-DNA adduct determination. In order to isolate the bacterial DNA, the lipopolysaccharide cell coat was disrupted by ten liquid nitrogen freeze-thaw cycles, followed by sequential 30-min. incubations at 37° with 10 mg lysozyme and then with 2 ml of 25% sodium dodecyl sulfate. The DNA was further purified by hydroxylapatite chromatography as previously described (15). Enzymatic hydrolysis and hplc were performed in a manner analogous to that used for the dog tissues.

RESULTS

In Vitro Reaction of N-Hydroxy-4-ABP with DNA and Identification of the 4-ABP DNA Adducts. Incubation of calf thymus DNA with [^{3}H]N-hydroxy-4-ABP for 4 hours at pH 5 and subsequent isolation of the modified DNA yielded 15-20 nmoles of bound 4-ABP residues per mg DNA. Upon enzymatic hydrolysis, 90-98% of the radioactivity was recovered by n-butanol extraction and this material was analyzed by hplc. Three u.v.-absorbing, radioactive peaks, A, B, and C, were consistently observed (Fig. 1).

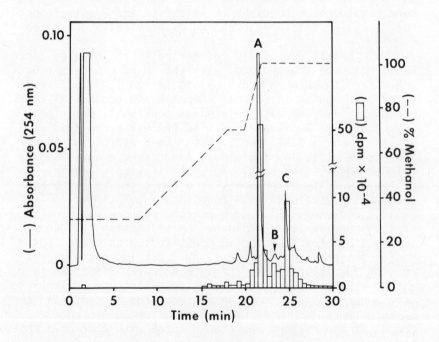

Fig. 1. Hplc profile of the 4-ABP-DNA adducts from calf thymus DNA reacted with [^{3}H]N-hydroxy-4-ABP in vitro at pH 5. Peaks A, B, and C, accounted for 70, 5, and 17% of the total radioactivity, respectively.

Components A and C were treated with bis(trimethylsilyl)trifluoroacetamide: dimethylformamide (1:1) at 37° for 30 min. and analyzed by electron impact in-beam desorption mass spectroscopy (16), which indicated that A was a tetrakis(trimethylsilyl)deoxyguanosin-4-ABP derivative (m^+=722) and C was a tris(trimethylsilyl)deoxyadenosinyl-4-ABP adduct (m^+=634). High resolution ^{1}H-nuclear magnetic resonance (nmr) spectroscopy (Table I) at 270 MHz (component A) and 500 MHz (component C and 4-ABP) were consistent with their identities as N-(deoxyguanosin-8-yl)-4-ABP (A; dG-8-ABP) and N-(deoxyadenosin-8-yl)-4-ABP (C; dA-8-ABP). The latter represents the first example of a carcinogen-DNA adduct formed by substitution at the C8 atom of dA. Additional data on u.v. spectral parameters at pH's 1, 7, and 13 and pH-solvent partitioning

TABLE 1: 270 and 500 MHz ^{1}H-nmr Data[a] for 4-ABP, dG-8-ABP, and dA-8-ABP.

Assignment	4-ABP	dG-8-ABP	dA-8-ABP
NH or NH$_2$ (ABP)	5.19	8.78[b]	9.14[c]
3,5 (ABP)	6.65	7.84[b]	7.97[c]
2,6 (ABP)	7.35	7.59[b]	7.61[c]
3',5' (ABP)	7.37	7.44	7.46
2',6' (ABP)	7.53	7.64	7.65
4' (ABP)	7.21	7.30	7.33
NH (dG)	-	10.9	-
NH$_2$ (dG,dA)	-	6.52	6.80
2 (dA)	-	-	8.03
1' (dG,dA)	-	6.56[d]	6.35[d]
2' (dG,dA)	-	2.74	2.5
2" (dG,dA)	-	2.14	2.03
3' (dG,dA)	-	4.49	4.43
4' (dG,dA)	-	3.99	3.94
5,5" (dG,dA)	-	N.D.	3.77
3'-OH (dG,dA)	-	5.36	N.D.
5'-OH (dG,dA)	-	6.00	N.D.

[a] Chemical shifts are expressed as ppm from TMS; solvent is DMSO-d$_6$. Several chemical shifts are reported to nearest 0.1 ppm or could not be determined (N.D.) due to broad resonances or solvent overlap.
[b,c] Assignments may be reversed.
[d] First order measurements indicated: $J_{1'2'}$= 9.6 Hz, $J_{1'2''}$= 5.9 Hz for dG-8-ABP. $J_{1'2'}$= 9.2 Hz, $J_{1'2''}$= 5.9 Hz for dA-8-ABP. These large J's are consistent with C8-purine substitution (17).

properties (data not shown) were consistent with the assigned structures. The structure of adduct B has not yet been completely established but preliminary nmr and mass spectral data suggest that it is a deoxyadenosinyl-4-ABP derivative (dA-X-ABP).

Identification and Quantitation of Carcinogen-DNA Adducts from the Urinary Bladder and Liver of Dogs Administered 4-ABP, 4-NBP, and 2-AAF.

Administration of the carcinogens 4-ABP, 4-NBP and 2-AAF, to male beagle dogs resulted in the formation of the same major arylamine-DNA adducts that were prepared in vitro by the acidic reaction of their N-hydroxy arylamines with DNA. A typical hplc profile obtained from urothelial DNA after 4-ABP treatment is shown in Fig. 2. In addition to adducts A, B, and C, there were earlier eluting peaks (9,10 min) which co-migrated with alkaline decomposition products of dG-8-ABP. These have been tentatively identified as imidazole ring-opened derivatives (18) and may be an artifact of the isolation procedure.

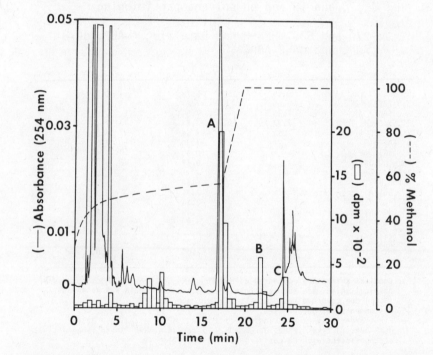

Fig. 2. Hplc profile of the 4-ABP DNA adducts from a dog urothelium at 7 days after dosing. dG-8-ABP (peak A) and its derivative (peaks at 9,10 min); dA-X-ABP (peak B), and dA-8-ABP (peak C) accounted for 72, 13, and 6% of the total radioactivity bound to the DNA.

4-NBP, which is believed to be metabolized to N-hydroxy-4-ABP in vivo (18,19), yielded an adduct profile similar to the 4-ABP-treated urothelium, although the overall level of binding was considerably lower. 2-AAF (data not shown) yielded a single deacetylated adduct in the dog urinary bladder which comigrated with authentic N-(deoxyguanosin-8-yl)-2-AF (cf. ref. 18). Hplc profiles of DNA hydrolysates from the livers of the 4-ABP and 4-NBP dosed dogs were similar to those obtained with the urothelium. In the case of 2-AAF, however, additional peaks were detected which chromatographed with 3-(deoxyguanonsin-N^2-yl)-2-AAF (dG-N^2-AAF) and with N-(deoxyguanosin-8-yl)-2-AAF (dG-8-AAF). This indicated that, like rat liver, metabolically formed N-hydroxy-AAF may be activated by dog hepatic sulfotransferase to an AAF-binding species (20).

A comparative study of the levels of the arylamine-DNA adducts formed by these carcinogens in the dog urothelium and liver is presented in Table 2. The time of sacrifice was 1, 2, or 7 days so that initial and persistent levels of bound adducts could be determined. Urinary excretion of ^{3}H-arylamine metabolites was essentially completed in 24-48 hours.

TABLE 2: Levels of Arylamine-deoxyribonucleoside Adducts in the Urothelium and Liver of Dogs.

Aryl-amine	Tissue(n)[a]	Days After Dose	Arylamine Residues/10^8 Nucleotides				
			dG-C8[b]	dG-N^2	dA-C8	dA-N^6 [c]	Total
2-NA[d]	liver(3)	2	1.7 + 0.3	0.6 + 0.1	--	2.0 + 0.3	4.4 + 0.6
	liver (2)	7	0.7 + 0.1	0.6 + 0.0	--	0.3 + 0.1	1.6 + 0.1
	urothelium (1)	2	8.8	3.1	--	6.6	18.5
	urothelium (1)	7	7.2	2.1	--	2.9	12.2
4-ABP	liver (4)	1	311 + 26	--	29 + 2	80 + 6	420 + 34
	liver (2)	2	607 + 38	--	53 + 9	129 + 6	789 + 53
	liver (4)	7	693 + 83	--	57 + 11	162 + 19	912 + 106
	urothelium (1)	1	306	--	55	67	428
	urothelium (2)	2	551 + 45	--	40 + 12	84 + 20	675 + 77
	urothelium (1)	7	502	--	52	105	659
4-NBP	liver (2)	1	7.2 + 1.9	--	1.1 + 0.3	1.9 + 0.1	10.2 + 2.3
	urothelium (1)	1	15.3	--	3.0	2.9	21.2
2-AAF	liver (4)	2	573 + 81	--	--	--	688 + 115[e]
	urothelium (1)	2	77.3	--	--	--	77.3

[a] Each liver and urothelium was taken from individual dogs; and n represents the number of tissue determinations made. Results are expressed + s.d.
[b] Except for 2-NA which exists completely as an imidazole ring-opened product, the values for other dG-C8 adducts include values obtained for their ring-opened derivatives, which account for 10-20% of the total.
[c] The identification of the ABP-dA adduct is only tentative and is based on preliminary nmr and mass spectral data.
[d] Data taken from ref. 8.
[e] The total includes 55 + 20 dG-N^2-AAF and 60 + 27 dG-8-AAF residues/10^8 nucleotides.

These data show that a C8-substituted deoxyguanosine adduct is a major lesion for both tissues and for all four aromatic amine carcinogens. Furthermore, for 2-NA in the bladder (8) and 4-ABP in the bladder and liver, these adducts appear to be stable, persistent, and non-repairable modifications in the DNA.

Formation of Arylamine-DNA Adducts in Salmonella typhimurium TA 1538 and Mutagenicity of N-Hydroxy-4-ABP, N-Hydroxy-2-AF and N-Hydroxy-2-NA. Incubation of the ^{3}H-labeled N-hydroxy arylamines with Salmonella typhimurium TA 1538 for 30 min. in a liquid suspension assay system resulted in the induction of histidine revertants and in the formation of arylamine-DNA adducts. In Fig. 3, an hplc profile of the arylamine-DNA adducts formed in the bacteria is shown and is remarkably similar to that observed in vitro (Fig. 1) and in the dog bladder in vivo

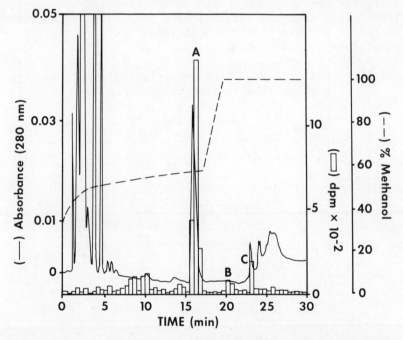

Fig. 3. Hplc profile of 4-ABP DNA adducts from Salmonella typhimurium TA 1538 after treatment with N-hydroxy-4-ABP. dG-8-ABP (Peak A) and its derivatives, dA-X-ABP (Peak B) and dA-8-ABP (Peak C) accounted for 71, 5, and 6% of the total radioactivity bound to the DNA.

(Fig. 2). Similar results were also obtained after incubations with N-hydroxy-2-AF and N-hydroxy-2-NA.

The extent of Salmonella DNA modification by the N-hydroxy arylamines was determined at several mutagen doses and was compared with mutagenicity data obtained concomitantly (see Materials and Methods). The results, shown in Fig. 4, clearly demonstrate that each arylamine, in bind-

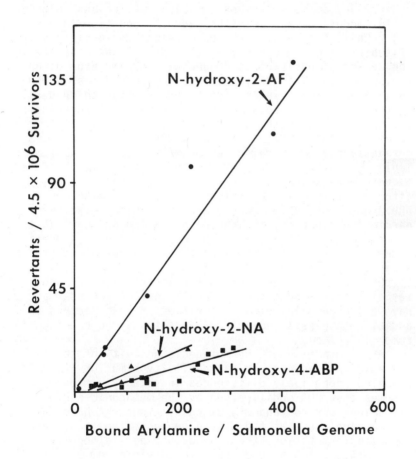

Fig. 4. Mutagenicity of ^{3}H-labeled N-hydroxy-2-AF (●), N-hydroxy-2-NA (▲), and N-hydroxy-4-ABP (■), and their binding to Salmonella typhimurium DNA (4.5 x 10^6 base pairs/genome).

ing to the Salmonella genome, exhibits different mutagenic potency. From these data, a level of 1 mutagen bound/10^6 nucleotides correlates with the induction of 72, 21, and 14 frameshift revertants/10^8 survivors for 2-AF, 2-NA, and 4-ABP, respectively.

DISCUSSION

The in vitro covalent binding of carcinogenic N-hydroxy arylamines with DNA at acidic pH appears to be a useful model for arylamine-DNA adduct formation in vivo (7). While N-hydroxy-2-AF reacts with DNA to yield only a C8-substituted dG adduct (18), N-hydroxy-2-NA reacts with the C8 and N^2 position of dG and with the N^6 atom of dA (12). N-Hydroxy-4-ABP also binds to C8-dG sites in DNA; and, in addition, yields a novel C8-substituted dA derivative (Fig. 1). These same arylamine-DNA adducts are also present in vivo in the dog urothelium and liver after oral doses of 2-AAF, 2-NA, 4-ABP and 4-NBP; and the C8-dG derivatives are the major lesions (ref. 8 and Fig. 2, Table 1). For 2-NA and 4-ABP in the urothelium, the C8-dG adducts were shown to be persistent residues (2-7 days after treatment). In addition, the level of binding, 4-ABP>2-AAF>4-NBP ≃2-NA, correlates well with relative carcinogenic potency for the urinary bladder (9). In the liver, however, the level of binding was 2-AAF≃ 4-ABP>> 4-NBP ≃2NA; and, of these, only 2-AAF has been reported as a hepatocarcinogen in the dog (10). The comparable levels of persistent residues from 4-ABP in the liver, a non-target tissue, indicate that arylamine-DNA adducts may have different biological activities or that factors in addition to persistence of DNA adducts may be required for tumor induction.

An examination of the mutagenic activity of these arylamine-DNA adducts in Salmonella typhimurium TA 1538 revealed appreciable differences in potency. If one assumes that DNA modification by the mutagenic N-hydroxy arylamines occurs randomly among all four DNA bases and that a single hit at the histidine-sensitive locus results in a single mutation, then one adduct per Salmonella genome (4.5 x 10^6 base pairs) should yield one mutation in 4.5 x 10^6 bacteria. By plotting revertants/4.5 x 10^6 survivors vs binding/genome (Fig. 4), the slope x 100 becomes equivalent to mutagenic efficiency and represents the frequency by which the base adduct(s) is expressed as a mutation. For 2-AF, 2-NA, and 4-ABP, these values are 32.4, 9.6 and 6.2%, respectively. These data suggest that

2-AF and 4-ABP adducts can differ by 5-fold in their biological activity (frameshift mutagenicity) and this may account for apparent inconsistencies in correlating in vivo binding with carcinogenicity. Furthermore, since only a dG-C8 adduct was detected in Salmonella upon treatment with N-hydroxy-2-AF, this indicates that dG-C8 arylamine adducts are potentially mutagenic lesions. By comparison, N-hydroxy-4-ABP gave 70% dG-C8 substitution, but was only 20% as efficient as N-hydroxy-2-AF in inducing frameshift mutations. Thus, these data suggest that binding to a specific site by structurally related compounds can result in different mutagenic efficiencies.

ACKNOWLEDGEMENTS

We thank C. Weis, N. Fullerton and G. White for expert technical assistance, P. Freeman for obtaining the mass spectra and L. Amspaugh for preparation this manuscript.

REFERENCES

1. J.L. Radomski, Ann. Rev. Pharmacol. Toxicol. 19:129-157 (1979).
2. C. Patrianakos and D. Hoffmann, J. Anal. Toxicol. 3: 150-154 (1979).
3. M.R. Guerin, C.-H. Ho, T.K. Rao, B.R. Clark, and J.L. Epler, Environ. Res. 23:42-53 (1980).
4. B.W. Wilson, R. Pelroy, and J.T. Cresto, Mutation. Res. 79: 193-202 (1980).
5. T. Sugimura, M. Nagao, T. Kawachi, M. Honda, T.Yahagi, Y. Seino, S. Sato, N. Matsukura, T. Matsushima, A. Shirai, M. Sawamura, and H. Matsumoto, in "Origin of Human Cancer," (H. Hiatt, J.D. Watson, and J.A. Winsten, eds.), p 1561. Cold Spring Harbor Symposium, Cold Spring Harbor, NY, 1977.
6. Y. Yamazoe, K. Ishii, T. Kamataki, and R. Kato, Drug Metab. Disp. 9:292-296 (1981).
7. F.F. Kadlubar, J.A. Miller, and E.C. Miller, Cancer Res. 37:805-814 (1977).
8. F.F. Kadlubar, J.F. Anson, K.L. Dooley,and F.A. Beland, Carcinogenesis 2:467-470 (1981).
9. J.M. Price, in "Benign and Malignant Tumors of the Urinary Bladder" (E. Maltry, Jr., ed.), p. 189, Medical Examination Publ. Co., Flushing, NY, 1971.
10. H.P. Morris and W.H. Eyestone, J. Natl. Cancer Inst. 13:1139-1165 (1953).
11. A.G. Jabara, Cancer Res. 23:921-927 (1963).

12. F.F. Kadlubar, L.E. Unruh, F.A. Beland, K.M. Straub, and F.E. Evans, Carcinogenesis 1:139-150 (1980).
13. F.A. Beland, W.T. Allaben, and F.E. Evans, Cancer Res. 40:834-840 (1980).
14. B.N. Ames, J. McCann, and E. Yamasaki, Mutation Res. 31:347-364 (1975).
15. F.A. Beland, K.L. Dooley, and D.A. Casciano, J. Chromatogr. 174:177-186 (1979).
16. A.P. Bruins, Adv. Mass Spec. 8:246-254 (1980).
17. F.E. Evans, D.W. Miller, and F.A. Beland, Carcinogenesis 1:955-959 (1980).
18. E. Kriek and J.G. Westra, Carcinogenesis 1:459-468 (1980).
19. L.A. Sternson, Experientia 15:260-270 (1975).
20. F.F. Kadlubar, L.E. Unruh, T.J. Flammang, D. Sparks, R.K. Mitchum, and G.J. Mulder, Chem.-Biol. Interactions 33:129-147 (1981).
21. J.H.N. Meerman, F.A. Beland, and G.J. Mulder, Carcinogenesis 2:413-416 (1981).

Mutagenesis in Higher Plants

Mutagenicity Testing of Chemical Mutagens in Higher Plants*

Taro Fujii

Plant Section, National Institute of Genetics, Mishima, Japan

ABSTRACT: Plant systems seem well suited for research of mutagen screening and environmental monitoring. Various plant test systems, such as rice, barley, maize, Tradescantia and soybean are being used for mutagenicity testing of environmental mutagens in Japan. Using graminoceous plants, effects of several alkylating agents, base analogues, some carcinogenic substances or very low dose of radiations were studied, and mutation induction or doubling dose etc. in relation to cell cycle or repair process(es) have been demonstrated. Relationship between mutagenicity and cell cycle in sodium azide, a known potent mutagen was studied with rice and barley. Effects of alkylating agents, air pollutants, as well as small dose of radiations emitted from radioisotopes were examined with mutagen-sensitive stamen hairs of Tradescantia.

Using soybean test system, mutagenic effects of AF-2, L-ethionine and some pesticides were noticed. However, activation of these substances by plant metabolites was not observed in Ames' spot test method. Thus information on plant mutagenesis is accumulating but lacks behind animal or bacterial test systems. More experiments are needed with different test systems in higher plants to establish the mutagenic effects of environmental mutagens or pollutants in order to conserve the stable ecosystem.

* Contribution from the National Institute of Genetics, Japan, No.1389.

INTRODUCTION

The mutagenic action of physical as well as chemical agents in higher plants has been studied extensively with various test systems as reviewed by Ehrenberg[9] and Nilan and Vig[25]. Furthermore, comprehensive review of plant mutagenesis by Nilan[24] has emphasized unique advantages for mutagen screening with plant genetic systems. Because of differences in physiology, metabolism and the insufficient knowledge at molecular level, plants may have limited value for estimating risk of mutagens for man. However, genetic material is DNA throughout prokaryotic to eukaryotic kingdoms, and mutational events merely depend on the structural change of DNA molecules irrespective of the kind of organism. Thus informations obtained from plant material should be evaluated and might be extrapolated to estimate genetic risk in mammals, including man. The comparison of mutagen responses of plants (barley, soybean etc.), mammalian cells <u>in vitro</u> and <u>in vivo</u>, nonmammalian systems (fruits fly) and <u>bacteria</u> has been provided with eight chemicals[5]* The results indicated that plant systems response to these chemicals in a manner quite similar to those of the <u>in vitro</u> mammalian systems.

On the other hand, in my opinion as a plant geneticist, conservation of plant ecosystem without drastic change is indispensable for human life. Pollution of air, land and water by chemical substances may affect plants physiologically and genetically, and accumulation of such damage in plant kingdom will cause disappearance of particular species or varieties, or destruction of stable ecosystem. Consequently, these disturbances will induce unbalance of fauna and flora. Thus mutagenicity testing in higher plants is very important not only for plants themselves but also for human beings.

In the first part of this paper, studies of chemical mutagenesis in plant materials carried out by Japanese scientists will be described referreing to related investigations. Some results obtained recently by present author in soybean and maize will be given in a later part.

*Ethyl methanesulfonate (EMS), methyl methanesulfonate (MMS), ethyleneimine (EI), N-methyl-N^{1}nitrosoguanidine (MNNG), dimethylnitrosamine (DMN), triethylenemelamine (TEM), trenimon (TREN) and mitomycin C (MMC)

Rice and barley

Rice (Oryza sativa) is one of the most widely used material for plant mutagenesis research in Japan, because rice is very familiar and important crop plant for Japanese and Asian peoples. On the other hand, barley (Holdeum vulgare) is used commonly for mutation studies following Stadler's original study. There is an enormous number of mutation experiments with physical or chemical mutagens, and valuable informations such as lethal dose, variation of sensitivity among the strains, mutation spectra, correlation between killing efficiency and mutation frequency, etc were accumulated.

Mutagenic efficiency of 5 alkylating agents was studied by Fujimoto and Yamagata(Table 1)[12]. Among the chemicals used, NEUA showed most potent mutagenicity and the orders of the efficiency may be judged as NEUA>EI>EMS>NMUA>NMUT. Moreover, they observed the differences in mutation spectra between ethylating and methylating agents in that the frequency of chlorophyll deficient mutants is much higher in the case of former chemicals than the latter ones.

Table 1. Frequency of chlorophyll mutations in M_2 generation of seeds treated with alkylating agents[12].

Agent*	Dose mM	Hr	No. of M_2 plants	Mutation frequency (%)
-	-	-	18645	0
EI	20	3	20355	0.45
	60	3	15760	2.85
EMS	50	7	15239	0.08
	80	7	18232	1.60
NEUA	8	5	16722	2.27
	17	5	1o960	4.89
NMUA	2	5	19097	0.21
	5	5	16356	0.20
NMUT	5	5	18753	0.22
	5	5	21245	0.35

* NEUA-N-nitroso-N-ethyl urea, NMUA-N-nitroso-N-methyl urea, NMUT-N-nitroso-N-methyl urethane.

Mutation spectrum induced by several alkylating agents were compared in barley[34,35]. EI induced many sterility and chlorophyll mutants, EI, EMS and NMU (N-nitroso-N-methylurea) induced various kinds of early mutants, but these rates are much lower than those induced by ionizing radiations.

Chromosome breaking activity of 5-bromouracil (BU) has been reported by Kihlman[21]. He suggested that mutations could be induced through transition and/or transversion in the replication of DNA containing incorporated analogue. Effects of base analogues, BU and 2-aminopurine (AP) has been studied in rice by Inoue et al. (Table 2)[19]. Although dose versus effect relationships are not clear, both analogues could induced chlorophyll mutations. Incorporation of these analogues into DNA was confirmed with tritium labelled compounds.

Table 2. Chlorophyll mutations induced by BU and AP[19].

Chemical	Concentration	Treatment (day)	Chlorophyll mutation frequency ($\times 10^{-4}$)
BU	1/2 saturated solution	7	76
		14	47
	saturated solution	7	40
		14	102
AP	1.1×10^{-3}M	7	43
		14	57
	2.2×10^{-3}M	7	7
		14	32
-	-	-	6

Yamagata and Fukui[40] also studied mutagenicity of BU, AP, 5-bromodeoxyuridine (BUdR) and 5-fluorodeoxyuridine (FUdR) using rice and observed much higher mutation induction in BU and AP than latter two chemicals. In barley, Yamaguchi[41] studied effects of diethyl sulfate (dES) and BUdR on soaked seeds of various G_1 and S phases of the first division cycle, and much higher mutation frequency was recorded when these chemicals were applied at S phase. The results indicated that mutagenicity is dependent upon the mitotic stage of the cell at the time of treatment. In his further experiment with barley, post-treatment of caffeine, EDTA and BUdR on gamma-ray exposed seeds Yamaguchi showed an enhancement of chlorophyll mutation frequency[42]. Remarkable increase was observed with EDTA post-treatment, considerably with caffeine, and slight increase was seen with BUdR.

Caffeine increased gamma-ray induced damage,[1] and has a potentiating effect on EMS or NMU induced mutational damage[34]. Thus caffeine enhances damage induced in barley seeds by radiation and alkylating agent treatments, and this has been ascribed to the inhibition of repair process[43]. The action of caffeine is highly specific and depend on the type of DNA damage induced by the mutagens and on the repair mechanism of the damage (Table 3).

Table 3. Effects of caffeine post-treatment (5 mM) on the seedling damage induced by gamma-ray, EMS and NMU[34].

Time of caffeine treatment after seed soaking	Seedling height(cm) 14 days after seed soaking*		
	Gamma-rays (20 kR)**	EMS(0.5%) ***	NMU(0.05%) ***
4 hr 20 min-28 hr 40 min	6.4 (68.8)		
28 40 -52 40	5.7 (128.9)		
52 40 -76 40	8.9 (115.6)		
none	6.7 (98.5)		
6 hr -30 hr		9.9(99.0)	8.9(86.4)
30 -54		7.4(89.1)	9.1(86.6)
54 -78		9.3(89.5)	7.6(69.0)
none		9.8(104.0)	11.1(102.7)

  * Relative values to the respective controls (non-irradiated or non-treated lots) are given in parentheses.
  ** Seeds presoaked for 4 hr were irradiated by gamma-rays and treated with caffeine for 24 hr at various post-irradiation times.
*** Seeds presoaked for 4 hr were treated with EMS, NMU, respectively, for 2 hr and treated with caffeine at various post-treatment times.

It is evident from these data that efficient repaire process(es) on the mutagen induced genetic damage take place in early stage of germination, and that interference with the process(es) of repair synthesis leads to more mutations, suggesting that such repair synthesis is a significant factor in mutation induction[42].

Acridine orange (AO) in the presence of light has been shown to produce structural chromosome abnormalities in plant cells[21]. Yamaguchi's group studied the effect of AO on the recovery of injury induced by alkylating agents and radiations in rice seeds[33]. Post irradiation treatment with AO potentiated the radiation injury on the seedling growth, and EMS, dES or NMU induced damages potentiated also with AO post-treatment. In other words, AO inhibits repaire process as does caffeine.

After discovery of mutagenicity of sodium azide (NaN$_3$) in barley considerable attention has been payed on the mutagenic effects of this agent[27]. It should be noted that azide is one of the first major environmental mutagens whose potent mutagenicity was first recognized in plant materials[26]. Effects of NaN$_3$ was studied under several treatment conditions in dry and pre-soaked seeds of barley by Hasegawa and Inoue[14]. By treating dry seeds, appreciable increase in chlorophyll mutation frequency was noticed with NaN$_3$ treatment of 10^{-4}M for 2 hr. In experiments with pre-soaked seeds, 0 to 24 hr soaked seeds were treated with 10^{-3}M NaN$_3$, and highest mutation (chlorophyll deficient) frequency was observed in 12 hr pre-soaking. Similar findings were also observed with rice plants and support the view that NaN$_3$ is most effective when it is applied during S-phase[15].

Very recently, effect of heavy water (D$_2$O) was examined by Yamaguchi et al.[44]. Differences in seedling growth were compared in barley seeds which were presoaked for various time, and then treated with D$_2$O for 2 hr. They observed suppression of seedling growth with D$_2$O treatment as shown in Table 4.

Table 4. Decrease of the length of first leaves of D$_2$O treated barley seeds[44].

Pre-soaking (hr)	No. of seedlings measured	Mean height (cm)	Variance	F-value
0	38	10.3	0.33	-
10-12	40	9.8	0.81	2.907**
12-14	39	9.0	0.66	8.088**
14-16	39	9.1	0.74	7.179**
16-18	36	8.5	0.46	12.343**
18-20	37	8.2	0.71	12.617**

Maize (Zea mays)

Wx-gene: In estimating mutagenic effect of physical or chemical agents, difficulties arise from two different factors, the extremely large number of individuals needed in experiment to accomplish a reliable screening test, and the relatively high dose which is necessary for obtaining a doubling of the spontaneous mutation frequency. One may be overcome these difficulties is to use waxy-mutation in pollen grains in barley and maize etc.[6, 23] Pollen grains of higher plants have many advantages in genetic analysis. They are haploid germ cells, can produced in large numbers by a single individual, and can be examined in a very large populations with appropriate genetic marker such as wx-gene which would be distinguished quite easily by iodine staining. Induction of wx-mutation by gamma-rays and EMS was compared and their fine structure analysis has been carried out by Amano[2]. The results indicate that ionizing radiations seemed to induced deletion type mutations while EMS predominantly induced point mutations. Mutagenic properties of 32 pesticides were studied with this system[29]. Rosichan et al.[30] proposed improved method for mutagen screening with wx-gene of barley, and demonstrated the high degree of genetic resolution possible with this method.

Yg_2-gene: Yellow-green character controlled by Yg_2-gene is one of the useful character for measuring somatic mutation. Following mutagen treatment of heterozygous Yg_2/yg_2 seeds, mutation results in yellow-green stripes in the leaf of resulting seedlings[8, 23]. In this system, mutations induced with EMS and NaN_3 have been reported[7]. Mutagenicity of AF-2 and L-ethionine has been also reported with this system[10,11].

Tradescantia

Tradescantia genetic system developed by late Dr. A. H. Sparrow is well known and is a unique test system for mutation monitoring. Mutations for color of the highly mutagen-sensitive stamen hairs have been quite widely employed to monitor not only for low doses of physical and chemical mutagens but also for air pollutants[31]. Ichikawa and Takahashi[17] studied mutability of small doses of gamma-rays using stable KU-9 and mutable KU-20 strains having about 5-fold difference in spontaneous mutation frequency. Appreciable increase of mutational events from blue to pink in the stamen hairs higher than spontaneous level could be scored about 3 R exposures and mutation frequency increased linearly with increasing doses up to 50 R.

Tano and Yamaguchi[36] examined mutational effect of radioactive iodine (^{131}I) with clone 02. Zero to 100 nCi of ^{131}I solution was applied at the center of the inflorescence, and mutation frequency increment was observed at very low doses. The doubling dose of ^{131}I was calculated to fall between 0.3 and 1.0 rad[4]. The finding is in good accord with previous results[35], the mutations increased with doses as low as 0.25 rad of X-rays.

Inflorescences of clone 02 were treated with 0-0.5 % EMS solution for 2-16 hr, and clear relationship between mutation frequency and concentration or duration of the treatment was reported[18]. Mutagenic potential of air pollutants, benzo (alpha) pyrene (BP) and N-nitrosomethyamine (NDA) was examined with clone 02 by Tano and Yamaguchi (unpubl.). Mutational events increased with 5 and 10 μg treatments of BP and 1-100 ng treatments of NDA.

Soybean (Glycine max)

Varieties T219 and L65-1237 are characterized by the presence of a gene Y_{11}, which controls plant color. Because of the incomplete dominant nature of Y_{11} gene, different kinds of genic and/or chromosomal abnormalities, such as somatic crossing over, chromosome structural change, forward and back mutations could be detected as leaf spot of different colourations, and therefore, this specific character was used widely for detection of environmental mutagens[38, 39]. Mutagenic effects of furylfuramide (AF-2) and L-ethionine were examined with T-219 strain (Original seeds were kindly supplied by Dr. B. K. Vig). Mutagenicity and carcinogenicity of these chemicals were well investigated in microbial systems, cultured cells and mammalian test systems, but were not analyzed in higher plants. Clear increments in mutation freuqency were seen with concentrations of 0.0001 μg/ml and 0.5 mg/ml of AF-2 and L-ethionine, respectively[10, 11].

Agricultural chemicals which were used commonly in our farming were applied to soybean test system by the present author. No increase of spotting frequency was observed in five chemicals (Antio, Baycid, Diazinon, EPN and Karphos)these they may not have mutagenic effect at concentrations examined (500-2000 times dilution). However, increase in spots was observed by Ekatin [0,0-diethyl S-(2-ethylthioethyl)phosphorodithioate] treatment even at 2000 times dilution which is usual concentration in practical use (Table 5). Results of mutagenicity screening of pesticides by means of the rec-assay with Bacillus subtilis, the reversion assays with Salmonella TA1535, TA98 etc., and Escherichia coli WP2 try and WP2 try hcr strains showed that about 10 % of the 223 pesticides to be positive[20]. Systematic studies of mutagenic efficiency of pesticides in plants are needed

Table 5. Types and frequencies of spots observed on the two simple leaves of $Y_{11}Y_{11}$ soybean plants treated with Ekatin.

Treatment	No. of leaves analyzed	Types and frequencies of spot/leaf*			
		Yl	DG	Db	Total
-(H_2O)	84	1.01	0.95	0.33	2.29
x2000	100	1.22	1.76	0.45	3.43
x1000	94	1.36	2.20	0.45	4.01
x 500	100	1.98	2.73	0.85	5.56

*Yl-yellow, DG-dark green, Db-double spot.

Activation of chemicals by plant metabolism

The activation of chemicals into mutagens by higher plants is a recently discovered hazard[28]. Chemicals not mutagenic in themselves are converted into mutagens through plant metabolism examples can be found in several plant materials, like maize, soybean, Tradescantia etc.[3, 13] Activation of BP by microsomes of Jerusalem artichoke has been reported recently by Higashi et al.[16] Taking these informations into account, activation experiments of L-ethionine were conducted in collaboration of Dr. Tadashi Inoue, with in vivo and in vitro methods. However, we could not detected any difference between L-ethionine treated seed samples and directly applied L-ethionine on the reversion frequency from his^- to his^+, and plant extracts did not activate L-ethionine. These experiments indicated that mutagenicity of L-ethionine observed in soybean might be the direct effect of this chemical itself.

Similar experiments were carried out with 6 agricultural chemicals which were used in the aforementioned mutation experiments with soybean. Again negative results were recorded whereas Ekatin showed mutagenic effect.

CONCLUSION

Studies on genetical as well as ecological effects of environmental mutagens or carcinogens in higher plants are scanty at present. Several plant species, such as maize, barley, soybean, Arabidopsis, Tradescantia etc., offer systems for the analysis of almost all known genic and chromosomal alterations which have been induced in other eucaryotes by a variety of mutagens. On the other hand, majority of higher plants could not escape from their growing site, unlike animals, and therefore, growing plants may more readily be victims of genetic hazard or ecological disturbances directly due to pollution or contamination of their environs. Plants are the only organism to make up organic substances from inorganic elements. Thus mutagenicity testing of environmental mutagens in higher plants is very imprtant and could not be ignored either for fundamental research or for keeping the stable ecosystem in the whole biological kingdom.

ACKNOWLEDGEMENT- The author wishes to express his sincere thanks to Dr. B. K. Vig for his kind supply of soybean seeds and critical reading of the manuscript.

REFERENCES

1. Ahnstrom, G.   Mutation Res. 26, 99-103 (1974)
2. Amano, E.   Gamma Field Symposia 11, 43-59 (1972)
3. Arenaz, P. and Vig, B. K.   Mutation Res. 52, 367-380 (1978)
4. Bingo, K., Tano, S., Numakunai, T., Yoshida, Y. and Yamaguchi, H.   Radiat. Res. 85, 592-595 (1981)
5. Clive, D. and Spector, J. F. S.   Proc. Comparative Chemical Mutagenesis Workshop (de Serres, F. J. ed. NIEHS (1978)
6. Conger, B. V.   Proc. 6th Internatl. Congr. Radiat. Res. Tokyo, 562-567 (1979)
7. Conger, B. V. and Carabia, J. V.   Mutation Res. 46, 285-295 (1977)
8. Constantin, M. J.   Environmental Health Perspectives 27, 69-75 (1978)
9. Ehrenberg, L.   Chemical Mutagens (ed. Hollaender, A) Prenum Press, New York, 2, 365-386 (1971)
10. Fujii, T.   Japan. J. Genet. 55, 241-245 (1980)
11. Fujii, T.   Environmental Exp. Bot. 21, 127-131 (1981)
12. Fujimoto, M. and Yamagata, H.   Japan. J. Breed. (Submitted)
13. Gichner, T., Veleminsky, J. and Underbrink, A. G.   Mutation Res. 78, 381-384 (1980)
14. Hasegawa, H. and Inoue, M.   Japan. J. Breed. 30, 20-26 (1980)
15. Hasegawa, H. and Inoue, M.   Japan. H. Breed. 30, 301-308 (1980)
16. Higashi, K., Nakashima, K., Karasaki, Y., Fukunage, M. and Mizuguchi, Y.   Biochem. International 2, 373-

380 (1981)
17. Ichikawa, S. and Takahashi, C. S.   Mutation Res. 45, 195-204 (1977)
18. Ichikawa, S. and Takahashi, C. S.   Environmental Exp. Bot. 18, 19-25 (1978)
19. Inoue, M., Mori, S. and Hori, S.   Japan. J. Breed. 26, 110-120 (1976)
20. Kada, T., Hirano, K. and Shirasu, Y.   Chemical Mutagens (de Serres, F. J. and Hollaender, A. ed.) Plenum Press, New York, 6, 149-173 (1980)
21. Kihlman, B. A.   Radiat. Bot. 1, 35-42 (1961)
22. Kihlman, B. A.   Hereditas 49, 353-370 (1963)
23. deNettancourt, D., Eriksson, G., Lindgren, D. and Putte, K.   Hereditas 85, 89-100 (1977)
24. Nilan, R. A.   Environmental Health Perspectives 27, 181-196 (1978)
25. Nilan, R. A. and Vig. B. K.   Chemical Mutagenesis (Hollaender, A. ed) Plenum Press, New York, 4, 143-170 (1976)
26. Nilan, R. A., Owais, W. M., Rosichan, J. L. and Kleinhofs, A.   Proc. 6 th Internatl. Congr. Radiat. Res. Tokyo, 568-574 (1979)
27. Nilan, R. A., Sideris, E. G., Kleinhofs, A., Sander, C. and Konzak, C. F.   Mutation Res. 17, 142-144 (1973)
28. Plewa, M. J.   Environmental Health Perspectives 27, 45-50 (1978)
29. Plewa, M. J. and Wagner, E. D.   Environmental Health Perspectives 37, 61-73 (1981)
30. Rosichan, J. L., Arenaz, P., Blake, N., Hodgdon, A., Kleinhofs, A. and Nilan, R. A.   Environmental Mutagenesis 3, 91-93 (1981)
31. Schairer, L. A., Van't Hof, J., Hayes, C. G., Burton, R. M. and deSerres, F. J.   Environmental Health Perspectives 27, 51-60 (1978)
32. Smith, H. H., Bateman, J. I., Quastler, H. and Rossi, H. H.   Biol. Effects of Neutron and Proton Irradiations, IAEA, Vienna, 2, 233-248 (1964)
33. Soomro, A. M. and Yamaguchi, H.   Radioisotopes 26, 38-41 (1977)
34. Soomro, A. M. and Yamaguchi, H.   Japan. J. Breed. 27, 13-38 ((1977)
35. Sparrow, A. H., Underbrink, A. G. and Rossi, H. H. Science 176, 916-918 (1972)
36. Tano, S. and Yamaguchi, H.   Radiat. Res. 80, 549-555 (1979)

37. Ukai, Y. and Yamashita, A.   Japan, J. Breed. 29, 255-267 (1979)
38. Vig, B. K.   Mutat. Res. 31, 49-56 (1975)
39. Vig, B. K.   Environmental Health Perspectives 27, 27-36 (1978)
40. Yamagata, H. and Fukui, K.   Memories of the Coll. Agric., Kyoto Univ. 113, 25-37 (1979)
41. Yamaguchi, H.   Environmental Exp. Bot. 16, 145-149 (1976)
42. Yamaguchi, H.   Proc. 6th Internatl. Congr. Radiat. Res. Tokyo, 575-581 (1979)
43. Yamaguchi, H., Tatara, A. and Naito, T.   Japan. J. Genet. 50, 307-318 (1975)
44. Yamaguchi, H., Tatara, A., Ogawa, N. and Imataki, H. Japan. J. Breed. 31, Suppl. 1, 60-61 (Abst in Jap.)
45. Yamashita, A. and Ukai, Y.   Japan. J. Breed. 29, 101-114 (1979)

Maize as a Monitor for Environmental Mutagens

Michael J. Plewa

Institute for Environmental Studies, University of Illinois, Urbana, Illinois, U.S.A.

ABSTRACT The induction of point mutations in germinal and somatic cells after acute or chronic exposure to chemical mutagens was investigated.

THE HAZARDS OF ENVIRONMENTAL MUTAGENS

During the last one and one-half decades the presence of mutagens in our environment has been recognized as a serious hazard to the public health (1,5,6,19). An environmental mutagen or genotoxin is an agent released into the environment that can alter the genome or its proper functioning. Depending upon the developmental stage of an individual, a genotoxin can exert teratogenic effects, precipitate coronary disease, produce mutations involving germinal cells or cause mutations in somatic cells that may become neoplastic.

THE USE OF GENETIC ASSAYS BASED ON HIGHER PLANTS

Presently plant systems serve as the only higher eukaryote monitors for environmental mutagens in the atmosphere, water, municipal sewage sludge and agricultural pesticides (14). I suggest there are five major advantages that angiosperms can provide for fundamental research in environmental mutagenesis: 1) angiosperms are the highest class in the plant kingdom and the complexity of their genome is similar to mammals, 2) both germinal and somatic cell mutation tests can be conducted with high genetic resolution, 3) plants can be exposed to mutagens under acute or chronic regimens, 4) plants provide the best in situ monitors for environmental mutagens, and 5) plants can activate some chemical promutagens into mutagens (16,17).

Zea mays is one of the pillars of classical genetics. More is known about the genetics of maize than of any other higher plant (3). The use of specific locus mutation tests in maize for studies in environmental mutagenesis has been reviewed by Plewa in a report to the Gene-Tox Program of the U.S. Environmental Protection Agency (15).

My laboratory is constructing a multiple end point assay with a rapidly maturing line of maize, Early-Early Synthetic. We have investigated the induction of point mutations by chemical mutagens in germinal and somatic cells using acute or chronic treatment regimens. We have studied mutation at two loci, waxy (wx) and yellow-green-2 (yg-2) after acute or chronic exposure to the mutagens ethylmethanesulfonate (EMS), N-methyl-N'-nitro-N-nitrosoguanidine (MNNG) and maleic hydrazide (MH).

EARLY-EARLY SYNTHETIC

Early-Early Synthetic is a rapidly maturing variety developed by Dr. D. F. Alexander. I have inbred this variety for several generations. Early-Early Synthetic develops from kernel to tassel emergence in approximately 30 days and attains a height of approximately 50 cm. One plant may be grown in soil in a 10 cm diameter pot or with hydroponic culture (18).

TREATMENT REGIMENS

In our experiments kernels were acutely exposed or plants were chronically exposed to a chemical mutagen. An acute treatment is defined as an exposure for only a fraction of a single ontogenic stage of the test organism to a mutagen. A chronic treatment is defined as an exposure encompassing the major ontogenic stages of the sporophyte.

GENETIC ENDPOINTS

The waxy Locus   The wx characteristic is due to a mutation that alters the composition of starch in the endosperm and in the pollen grain. The starch of individuals carrying a Wx allele is composed of a mixture of amylose and amylopectin while wx/wx individuals synthesize starch consisting of amylopectin only (20).

Iodine combines with amylose and forms a blue-black complex. Kernels or pollen grains that carry the Wx allele contain amylose and stain a blue-black color when subjected to an iodine solution. Since the starch in waxy kernels or pollen grains contains no amylose, iodine does not induce a blue-black complex and the material forms a red or tan color (2,11,12,13). The starch type of a pollen grain is controlled by the genetic constitution of that pollen grain, not by the parental sporophyte. Thus, forward mutation of Wx to wx and reverse mutation of wx to Wx are very easy to detect (18).

The Yellow-Green-2 Locus   Homozygous yg-2 plants appear yellow green in color; they are weak and slow to mature. The dominant allele Yg-2 produces a full green color to the sporophyte. Forward mutation at the yg-2 locus can be detected by scoring heterozygous plants for yellow-green sectors in their leaves. If the dominant allele is mutated or inactivated the recessive allele will be expressed in a sector of leaf cells.

RESULTS AND DISCUSSION

Acute Tests at the wx Locus   A time course experiment using 20 mM EMS was conducted. Five groups of five kernels each that were soaked in aerated water for 72 h were treated; a control group was planted immediately after soaking. A second control group consisted of kernels that were soaked for eight additional hours in distilled aerated water. The treatment groups were exposed to 20 mM EMS for 4, 8 and 16 h. The data indicated a time dependent response up to the 8 h treatment. In an experiment analyzing 2.0×10^6 pollen grains the mutation rates among gametophytes (as opposed to the mutation rate in somatic cells per generation) for control 1, control 2 and the three treatment groups were 9.9, 9.9, 11.7, 286.4 and 12.8×10^{-5}, respectively. Clearly

EMS at millimolar concentrations is a potent mutagen in maize germ-line cells. Acute treatments for 5 h with MH at concentrations of from 10 µM to 10 mM gave positive responses. MH concentrations above 2.5 mM were lethal. Positive responses that were not dose dependent ranged from a 9.4 fold increase over the control mutation rate among gametophytes at the 10 µM treatment to a 5.7 fold increase at the 500 µM MH concentration. Kernels were exposed for 5 h to concentrations of MNNG that ranged from 100 µM to 2.5 mM. Positive, dose dependent increases were observed within a concentration range of from 500 µM to 2.5 mM MNNG. The mutation rates at the wx locus among gametophytes for control and treatment groups of 500 µM, 850 µM, 1.25 mM and 2.5 mM were 9.9, 50.6, 63.8, 234.8 and 10.2×10^{-5}, respectively. A population of 5.1×10^6 pollen grains was analyzed in these experiments.

These experiments indicate that the germ-line cells of the young sporophyte can be mutated after an acute exposure to chemical mutagens. A relatively short treatment period (approximately 1/200 of the sporophytic generation) with EMS, MH or MNNG induces mutation in germ-line cells that results in the production of mutant gametophytes. These data have serious implications for humans who are exposed to mutagens especially prior to their reproductive years.

Chronic Tests at the wx Locus   Forward mutation at the wx locus in gametophytes of Early-Early Synthetic was measured after chronic exposure via the soil to the same mutagens used in the acute tests. An experimental design included 20 plants in individual pots with five plants per treatment group. A control of five plants was included with each experiment.

The number of mutant pollen grains and the estimated number of viable pollen grains analyzed are presented as the frequency of mutant pollen grains. The concentration of an agent in the chronic treatment regimen is expressed in moles per plant. This value was calculated by multiplying the molar concentration of the appropriate mutagen solution by the total volume administered per plant.

For MNNG the concentration range included 4.5×10^{-9} to 2.0×10^{-5} moles/plant. No increase in the frequency of mutant pollen grains over control values was observed in four experiments where 7.9×10^6 pollen grains were analyzed. This equivocal response of maize to MNNG is similar to data reported by Russell in the mouse specific locus test. It has been suggested that mutagens as reactive as MNNG are poor inducers of germ cell mutation in mammals because the agents are detoxified before they reach the target cells. Our data on chronic exposure to MNNG may indicate a similar process occurs in higher plants.

In four experiments the concentration range of EMS extended from 6.5×10^{-7} to 4.5×10^{-3} moles/plant. Plants exposed to concentrations above 6.0×10^{-5} moles/plant during their sporophytic generation produced only aborted pollen grains. Chronic exposure to EMS induced significant increases in the frequency of mutant pollen grains. This frequency was dose dependent at concentrations of EMS below 1×10^{-5} moles/plant. The concentration of EMS in one experiment ranged from 1.1×10^{-6} to 6.8×10^{-6} moles/plant. The frequencies of mutant pollen grains in the control and in the treatment groups of 1.1, 2.2 and 6.8×10^{-6} moles/plant were 5.11, 10.28, 13.65 and 20.50×10^{-5}, respectively. The frequency of mutant pollen grains increased in a linear manner as a function of dose (coefficient of determination, $r^2 = 0.93$). However, there was no correlation between the exposure of EMS and the percentage of aborted pollen grains (coefficient of correlation, $r = 0.22$). Thus, EMS is a potent mutagen to Early-Early Synthetic when chronically administered.

For MH the concentration range in four experiments was from 5×10^{-11} to 2×10^{-4} moles/plant. The concentration of MH in one experiment ranged from 5.0×10^{-9} to 5.0×10^{-7} moles/plant. The frequency of mutant pollen grains ranged from 4.59×10^{-5} for the control group to 5.54×10^{-4} for the plants exposed to 5.0×10^{-7} moles/plant. A statistically significant increase in the frequency of mutant pollen grains was observed in all treatment groups. This increase was dose dependent and linear, $r^2 = 0.99$. The percentage of pollen abortion increased directly with the dose of MH ($r = 0.96$). These data clearly indicate that MH is an exceedingly potent mutagen in maize germinal cells.

This has direct consequences for humans. Although MH is not a mammalian mutagen, it is a strong plant clastogen (21) and it is mutagenic to Tradescantia (J. Veleminsky, personal communication, 1981, Czech. Nat. Acad. Sci.). The difference in response to MH by mammals and higher plants may be due to the activation of MH to a mutagen by the plants (17). Since MH is used as a plant growth regulator, the human population in the United States is exposed to approximately 170 mg of MH per year via the consumption of potatoes, onions and tobacco (21). Since the primary route of human exposure to MH is via plant material, this exposure may pose serious health risks since some mutagen may be present. The possible plant activation of MH into a mutagen that is active in mammalian cells is being investigated in my laboratory.

The linear dose response curves for the EMS and MH treatments are similar to the data on forward mutation at the wx locus in pollen grains after the parental sporophytes were chronically exposed to gamma rays (7,8). I suggest one way to interpret these dose response curves. The kinetics of mutation induction that would result in a linear dose response curve when germ-line cells are chronically exposed to a mutagen may require two functions. The probability that mutation at the wx locus in a germline cell will be induced by a mutagen is constant. When the sporophyte is very young only a few target cells are available, however, if a mutational event occurs a large cluster or sector of mutant pollen grains will result in the tassel. As the plants mature the number of target cells increases, however, the effect of a mutant event on the resulting number of pollen grains will be significantly diminished. The combination of these two functions may result in a dose dependent distribution of the frequency of mutant pollen grains.

Forward Mutation at the yg-2 Locus   Plants heterozygous for the yg-2 locus appear green. Forward mutation at the dominant allele (Yg-2) allows the expression of the recessive allele as a yellow-green sector in the leaves (9,10). After acute treatment of germinating kernels and after sufficient growth in a plant growth chamber the fourth and fifth leaves were analyzed for sectors.

In the dormant embryo, the primordia of leaves four and five consists of approximately 1,500 cells and 250 cells, respectively. Soaking the kernels for 72 h in aerated water at 20 C assures that these cells have undergone mitosis at least once (4). Prior to treatment the number of target cells has increased to 3,000 cells in leaf four and 500 cells in leaf five.

With an exposure of 8 h to 20 mM EMS, the mean number of sectors per leaf four was 14.63 as compared to

0.35 for controls. The data for leaf five was 8.27 sectors per leaf as compared to 0.20 for controls. These data agree with Conger and Carabia (4). Thus, the exposure of a very young sporophyte to a mutagen can cause somatic mutation that is expressed much later in the sporophytic generation.

We have also conducted chronic administration of EMS to heterozygous yg-2 plants growing in hydroponic culture. Treatment solutions from 1 to 1,000 µM EMS induced a dose dependent increase in the frequency of somatic mutation in maize leaf tissues. At leaf four the mean sectors per leaf for the control, 1 µM, 10 µM, 100 µM and 1 mM EMS concentrations were 0.05, 0.17, 0.21, 0.79 and 0.63, respectively. For the same experiment the data for leaf five were 0.15, 0.06, 0.22, 0.47 and 1.06, respectively. Finally we have analyzed the effect of acute treatment of MH at the yg-2 locus. The concentration of the MH solutions ranged from 0.5 µM to 100 mM. Concentrations of MH above 5 mM are lethal. Acute treatment of MH to germinating kernels is much less effective than EMS in inducing somatic mutation, however, at concentrations of 0.5, 1 and 2 mM MH a positive dose dependent increase in the mean yg-2 sectors per leaf was observed. One experiment involving the chronic exposure of yg-2 heterozygotes to low concentrations of MH has recently been conducted. Chronic exposure to MH concentrations of 1 to 10 µM induced a weak dose dependent increase in the mean number of sectors per leaf.

MULTIPLE END POINT ASSAY

The lack of comprehensive genetic assays in higher eukaryotes is a serious problem in the field of environmental mutagenesis. I contend that for a plant genetic assay to be useful the advantages inherent in angiosperms must be exploited. A single higher eukaryote genetic indicator organism that could detect a wide spectrum of genetic damage with high resolution would be of great value not only to environmental monitoring but also to fundamental research in mutagenesis. I plan to genetically construct a multiple end point assay in the inbred Early-Early Synthetic. This genetic indicator organism will be able to detect 1) forward or reverse mutation at two loci in germinal cells, 2) intragenic recombination within the wx cistron in germinal cells, 3) nondisjunction of chromosomes during meiosis, 4) forward mutation at one locus in somatic cells, and 5) sister chromatid exchanges in chromosomes from cultured root-tips.

The fundamental questions I plan to address with these new lines include 1) a comparison of the kinetics of mutation induction in germinal and somatic cells after acute or chronic exposure to chemical mutagens, 2) an investigation of meiotic and premeiotic mutational events in maize by a mathematical and empirical approach, and 3) a determination of the relationships among point mutation, chromosomal nondisjunction and sister chromatid exchanges after acute or chronic exposure to mutagens.

ACKNOWLEDGEMENTS

This research was funded in part by a grant from NIEHS (ES01895) and the efforts and scientific skills of Elizabeth Wagner and Patrick Dowd are gratefully acknowledged.

LITERATURE CITED

(1) Ames, B. N. Science 204, 587-593 (1979)
(2) Brink, R. A., and MacGillivray, J. H. Am. J. Bot. 11, 465-469 (1924)
(3) Coe, E. H., and Neuffer, M. G. Pages 111-223 in G. F. Sprague, ed. Corn and Corn Improvement. Am. Soc. Agron., Inc., Madison, WI (1977)
(4) Conger, B. V., and Carabia, J. V. Mutation Res. 46, 285-296 (1977)
(5) de Serres, F. J. Pages 75-85 in E. C. Hammond and I. J. Selikoff, eds. Annals N.Y. Acad. Sci. vol. 329 (1979)
(6) Drake, J. W., Abrahamson, S., Crow, J. F., Hollaender, A., Lederberg, S., Legator, M. S., Neel, J. V., Shaw, M. W., Sutton, H. E., von Borstel, R. C. and Zimmering, S. Science 187, 503-514 (1975)
(7) Eriksson, G. Hereditas 50, 161-178 (1963)
(8) Eriksson, G. Hereditas 68, 101-114 (1971)
(9) Fujii, T. Japan J. Genetics 55, 241-245 (1980)
(10) Fujii, T. Environ. Experimental Bot. 21, 127-131 (1981)
(11) Nelson, O. E. Science 130, 794-795 (1959)
(12) Nelson, O. E. Genetics 47, 737-742 (1962)

(13) Neuffer, M. G., Jones, L., and Zuber, M. S. The Mutants of Maize. Crop Sci. Soc. Am. Madison WI 74 pages (1968)
(14) Nilan, R. A. Environ. Health Perspectives 27, 181-196 (1978)
(15) Plewa, M. J. Mutation Res. In Press (1981)
(16) Plewa, M. J., and Gentile, J. M. Mutation Res. 3, 287-292 (1976)
(17) Plewa, M. J., and Gentile, J. M. in F. J. de Serres, ed. Chemical Mutagens: Principles and Methods for Their Detection. 7, 401-420 (1981)
(18) Plewa, M. J. and Wagner, E. D. Environ. Health Perspectives 37, 61-73 (1981)
(19) Sorsa, M. J. Tox. Environ. Health 6, 57-62 (1980)
(20) Sprague, G. F., Brimhall, B., and Hixon, R. M. J. Am. Soc. Agron. 35, 817-822 (1943)
(21) Swietlinski, Z., and Zuk, J. Mutation Res. 55, 15-30 (1979)

Cytogenetics

The Role of Cytogenetics in Evaluation of Human Genetic Risk

Nikolay P. Bochkov

Institute of Medical Genetics of the USSR AMS, Moscow, U.S.S.R.

ABSTRACT The significance of cytogenetic methods for evaluation of human genetic risk is considered. Cytogenetic data can be used for direct risk estimation and for extrapolation (from somatic cells to germ cells, from chromosome aberrations to gene mutations). The registration of chromosome aberrations on peripheral blood lymphocytes permits us to estimate mutagenicity of industrial chemicals. Data on spontaneous level of chromosomal anomalies for various types of pregnancy terminations (spontaneous abortion, stillborns, newborns) and in somatic cells are summarized. Sample sizes for detection of increase in mutagenicity process are calculated.

The theme of this report is limited to consideration of registration of chromosome and genome mutations arising under the action of environmental factors, i.e., evaluation of effects in recored by cytogenetic methods. The report will touch only upon questions concerning risk evaluation. The significant part of investigations and sister chromatid exchanges and mechanisms of action of various mutagenic factors will not be analyzed and discussed.

There is no doubt that the use of cytogenetic methods for risk evaluation in man and mammalia is cheaper than the registration of gene mutations. However, this paper will consider only genetic aspects of the problem, which will be illustrated by some examples from human cytogenetics.

The great advantage of cytogenetic methods is the possibility of their use in all variants of risk evaluation. In some cases the same data can be used in the two ways; for direct estimation and for extrapolation of effects on other cell or mutation types. Thus, cytogenetic evaluation of effects in germ cells can be carried out by analysis of pregnancy termination (spontaneous abortions, stillborns, liveborns) for all types of chromosome and genome mutations. At the same time it is a direct estimation of effects on the chromosome level and data for extrapolation in respect to gene mutations as well.

Cytogenetic effects in somatic cells can be studied under the influence of mutagens in vivo and in vitro. That gives an opportunity to make the most justified extrapolation with the least number of steps. Besides, the experiments with human cell cultures are carried out for investigation of general regularities of mutation process, and that is one of the necessary preconditions of correct extrapolation.

The study of chromosome damage in cells exposed to mutagenic factors in organism is a direct estimation of genetic risk and at the same time a possibility for one-step extrapolation on chromosome and genome mutations in germ cells.

The development of tissue culture and chromosomal preparation techniques provides cytogeneticists with appropriate refined methods for revealing chromosomal mutations and sister chromatid exchanges induced by mutagens in human cells. That provides all variants of experiments in vitro for cytogenetic testing on mutagenicity. The human lymphocyte culture is a significant, the only working, system for studying and monitoring of mutational changes caused in man in vivo.

It is extremely important to stress the significance of cytogenetic methods, for estimation of effects directly in human cells is impossible.

The cytogenetic methods are also significant for genetic monitoring in groups of persons exposed to environmental agents (professional damages, air and water pollution, and others). The necessity of monitoring for these groups in undoubtful when it is required to draw a final conclusion about mutagenic effects in human populations.

While considering the role of cytogenetics for genetic risk estimation of mutagenic factors for man,

it is necessary to stress that the spontaneous level of
chromosome aberrations constitutes an essential part
(perhaps the major one) of the human genetic load. Some
types of chromosome aberrations are transferred from
generation to generation, others are characterised by
lethal effects. So there is no doubt of the importance
of cytogenetic monitoring.

There are good prerequisites for estimation of genetic risk in man, based on the data of spontaneous level
chromosome pathology in human populations. Frequencies
of chromosome anomalies were studied for different
pregnancy terminations (spontaneous abortions, stillborns, liveborns) and in various groups of children
after birth. Analysis of the pooled data shows that the
mean frequencies are firmly reproduced under corresponding sample size. That confirms the fitness of cytogenetic risk estimation in man.

Frequencies of chromosome anomalies in different
pregnancy terminations according to the literature are
presented in Table 1.

Table 1
Frequency of chromosome anomalies in
different pregnancy treminations

Pregnancy termination	Number of observations	Anomalies of karyotype (70)	Inherited cases (% of total number)
Spontaneous abortions	4000	40.0	1.0
Perinatally dead	1000	6.0	7.0-10.0
Newborns (Total)	60,000	0.6-0.7	30.0
Newborns with birth defects	1000	6.0-13.0	2.0-5.0

The summarized data presented in Table 1 indicate a high
percentage of karyotype anomalies among spontaneous
abortions (40.0%). These data can be considered a
weighted mean as it is a generalization from investigations of different pregnancy terminations. It is
necessary to note additionally that almost all cases of

chromosomal anomaly are primarily apparent. Structural changes compose only 1.3 %, and not more than half of them are inherited. Thus, not less than 99 % of chromosome anomalies stipulating spontaneous abortions appeared primarily. That makes the using of cytogenetic investigation of spontaneous abortions for genetic risk estimation much easier. As for control for mutational process, the registration of chromosomal anomalies among spontaneous abortions can be equated with the registration of primarily apparent mutants.

The lethal effects of chromosomal anomalies appear in less degree (to 6.0 %) in perinatal death (before birth). The mean frequency is 2 % but fluctuations can be considerable depending on medical service level. In comparison to spontaneous abortions, structural changes are more frequently noted (19 %). Consequently, the inherited cases compose 7-10 %. Taking into account a low perinatal death-rate, difficulties of cytogenetic analysis and the low frequency of chromosomal anomalies, the significance of cytogenetic investigation of perinatally dead for genetic risk estimation in man can be ruled out.

A number of cytogenetic investigations were carried out in groups of newborns. The results of these investigations in different countries are also presented in Table 1.

During investigation of newborns for genetic risk estimation in man one should take into account that structural changes compose a significant part (40.9 % of total number) of chromosome anomalies. Their suitability for genetic monitoring is not large, as 78 % of them are not primary but family cases. But that fact is especially important for genetic risk estimation for future generations. That material gives an opportunity to estimate the proportion of balanced changes and their contribution to chromosomal pathology of following generations.

From the point of view of genetic monitoring, the total group of newborns cannot be considered a "suitable" object. Because of the high frequency of structural changes among chromosomal anomalies, it is necessary to investigate parents in order to determine the proportion of inherited cases, and that enlarges the number of cytogenetic procedures.

As chromosome anomalies are accompanied by serious congenital malformations, children with birth defects can be examined; and on the basis of this "selective" group one can judge the frequency of chromosomal anomalies in the group of newborns in total. The frequency of children with congenital malformations is 3-4 %, a part of which has multiple malformations. Cytogenetic investigation of such children demonstrated that the frequency of chromosomal anomalies among children with multiple developmental defects was nearly 13 %, and in the total group nearly 0.7 % (Table 1).

Statistic analysis of the frequencies of chromosomal anomalies in various groups permits us to suppose that the most exact estimation of cytogenetic effects from environmental factors could be observation of pregnancy terminations and their cytogenetic investigation.

In Table 2 a calculation of sample size at different levels of mutation, using cytogenetic investigation of all pregnancy terminations (spontaneous abortions, stillborns, liveborns) is presented.

Table 2
Sample size for estimation of increase in mutation process after cytogenetic investigation of all pregnancy terminations.

Increase in mutation level (%)	10	20	50	100
Sample size (number of pregnancies)	6,050	1,550	270	85

The analysis shows that under an increase of 10 % in frequency of chromosomal and genome mutation it is sufficient to investigate cytogenetically 6050 pregnancy terminations (nearly 900 spontaneous abortions, 120 stillborns and more than 5000 newborns), and under doubling only 85 pregnancy terminations.

The conditions and approaches for estimation of genetic effects in germ cells have been considered. The cytogenetic methods have even greater advantages for evaluation of effect in somatic cells and for estimation of environmental mutagenic factors on the whole.

The advantages of cytogenetic studies of human cells are as follows: price, short period of time (not more than 2 weeks), possibility of additional biochemical studies, possibility of medium modification, possibility of investigation of cells exposed to mutagens in the whole organism.

Cytogenetic monitoring is becoming a reliable and objective method of estimating changes in human genetic material when it concerns a suitable size population and when the chemical components affecting the population are known. Practically any factory where workers are exposed to chemicals can be the subject of an analysis of frequencies of chromosomal aberration and sister chromatide exchanges in relation to working period. Undoubtedly such monotoring would help to answer the question concerning genetic damage from professional factors.

A positive methodological moment for use of the cytogenetic method is a good reproduction of the spontaneous level of exchange aberration in human cells. Various authors in different years give approximately similar frequencies. They are summarized in Table 3.

Table 3
Spontaneous level of cytogenetic anomalies in human lymphocytes

Type of anomaly	Mean	Variation range	
		Low	Upper
Chromosome aberrations (per 1000 cells)	12.94	3.0	28.8
Exchanges (per 1000 cells)	0.82	0.63	1.16
Sister chromatid exchanges (per 1 cell)	8.41	6.53	14.0

Numerous examples of use of lymphocyte culture for mutagenicity testing of physical and chemical factors are given in the literature.

Chromosomal damage caused by chemicals in humans has been confirmed for a group of industrial combinations (vinyl-chlorid, ethylene-oxide, epychlorhydrine) and for cytostatics for heavy metals (chrome, lead).

Sister chromatid exchanges are not results of chromosomal morphology damage, but further work on estimation of SCE induction, their comparison with mutagenesis and DNA repair process, methodological simplicity and sensitivity of the method —— all these permit us to suppose that SCE can be regarded as a reliable test for mutagenicity, possibly for monitoring of mutagenic effects.

In conclusion, it is necessary to stress that the whole set of cytogenetic methods permit us to speak about a considerable contribution of cytogenetics to genetic risk estimation in man. They permit us to estimate effects in germ cells of experimental animals and man by means of karyotyping. There is an extremely large set of possibilities for studies of somatic cells. All types of chromosomal aberrations and sister chromatid exchanges can be registered in metaphase cells. In anaphase chromosomal damages are revealed as bridges and fragments. At least in interphase cells chromosomal damages can be judged on the basis of their micro-nuclei. Each method has a specific resolving power. One can choose a more adequate approach depending on one's investigation aims, training level of specialists and laboratory conditions.

A number of questions are waiting for consideration: extrapolation of results of cytogenetic investigations from in vitro to in vivo, from somatic cells to germ cells, from chromosomal mutations to gene mutations, from sister chromatid exchanges to gene and chromosomal mutations.

Basis and Biological Significance of Sister Chromatid Exchange

S. A. Latt, R. R. Schreck, E. Sahar, I. J. Paika, and C. Kittrel

Genetics Division, Children's Hospital Medical Center, Boston, and The Spectroscopy Laboratory, M.I.T., Cambridge, Massachussetts, U.S.A.

ABSTRACT Sister chromatid exchange (SCE) formation can be induced by different types of DNA damage. Experiments with a laser light source were used to show that both 8-methoxypsoralen-DNA monoadducts and crosslinks can induce SCEs in CHO cells. SCE induction by a particular agent may depend on a number of factors, including metabolic activation or removal of the test substance and defects in the ability of a cell to respond to DNA damage. SCE formation exhibits properties that can be related to a number of other processes involving DNA interchange, prompting speculations about the mechanism and biological significance of SCE formation.

INTRODUCTION

Sister chromatid exchanges (SCEs) are cytogenetic events which may be related to DNA damage and/or repair. Analysis of SCE formation continues to be explored as a means of monitoring exposure to mutagenic-carcinogens. Also, alterations in SCE formation have been observed in certain hereditary human diseases associated with a predisposition for neoplasia. Hence, there is considerable interest in understanding the molecular basis and biological significance of SCE formation.

TYPES OF DNA DAMAGE INDUCING SCEs

SCEs can be induced by many types of agents; of these, chemicals capable of alkylating DNA are among the most effective and extensively studied(see 29). Agents capable of crosslinking DNA, such as mitomycin C and 8-methoxypsoralen plus light, can also form DNA monoadducts. Comparison of the effects of mitomycin C and two of its monofunctional derivatives on cultured cells (5) led to the suggestion that these monoadducts, rather than crosslinks, were the important lesions stimulating sister chromatid exchange.

Experiments with 8-methoxypsoralen, in which the relative formation of DNA monoadducts and crosslinks was controlled either by varying the duration of light exposure (31) or the number of brief laser pulses incident on cells exposed to 8-methoxypsoralen (42), showed that both monoadducts and crosslinks could induce SCEs. For example, a single (<20 nsec) laser pulse, which induces monoadducts but few if any 8-methoxypsoralen-DNA crosslinks, is sufficient to stimulate SCE formation (Fig. 1), while multiple pulses are needed to produce a significant number of crosslinks. Conversion of monoadducts to crosslinks is accompanied by additional SCE formation (Fig. 2). Per lesion, the DNA crosslinks were somewhat more effective than were monoadducts, while the aggregate number of SCEs induced by monoadducts exceeded that produced by crosslinks. As is the case with other types of DNA damage, only a small fraction of 8-methoxypsoralen-DNA adducts result in SCE induction(7), suggesting that further subdivision of DNA damage according to its efficiency at SCE induction should be possible.

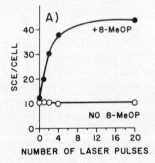 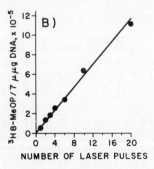

Fig. 1. Induction of SCEs(A) and DNA adducts(B) by exposure of CHO cells or calf thymus DNA to 8-MeOP(0.12 mM) plus brief 15±1 mJ(over a 2 cm^2 area), 355nm, neodymium-YAG laser pulses(42).

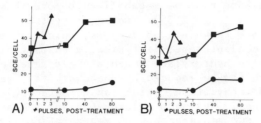

Fig. 2. Induction of SCEs in CHO cells, previously exposed to 8-MeOP and one laser pulse, to subsequent laser pulses, after removal of unreacted 8-MeOP(■). (A) 32 hour culture, (B) 48 hour culture, (●) no initial pulse, (▲), no washout of unreacted 8-MeOP.

VALIDATION OF SCE ANALYSIS AS A SHORT TERM MUTAGEN-CARCINOGEN TEST

Attempts have now been undertaken to standardize evaluation of existing data on SCE formation (29) and to compare SCE analysis with other short term tests (11). Initial results indicate a high concordance between the ability of an agent to induce SCE, especially as measured in vivo, and its activity as a mutagenic carcinogen. A primary limit to this comparison is a paucity of information about the carcinogenicity of different agents. The failure of some known carcinogens to induce SCEs remains a problem that may in part be due to ineffective drug exposure or activation.

FACTORS MODIFYING SCE INDUCTION

The SCE increment caused by a given chemical can depend on: (1) the type of system involved (e.g., in vitro versus in vivo, (2) the cell type exposed, (3) cell age or passage number, and (4) metabolic activation and detoxification of the test compound. In addition, pretreatment of CHO or human cells with multiple low doses of an alkylating agent has been observed to reduce the SCE increment produced by a larger test dose of an alkylating agent, presumably by stimulating repair of DNA damage by that agent (43). Exposure of cells to a second agent e.g., steroid hormones or caffeine, may modify SCE induction by another chemical. However, BrdU, a mutagen which is present in virtually all SCE studies, has an additive but not a synergistic effect on SCEs induced by mitomycin C(10,11).

Many agents must be metabolized before they can react with DNA (e.g. 2,35). Two major systems of drug metabolism are known(e.g. 37). One, which can be induced by 3-methylcholanthrene, is typically characterized by its ability to hydroxylate benzo(alpha)pyrene. The other, which can be induced by phenobarbital, carries out oxidative demethylation, which can be detected flourometrically, using p-chloromethylaniline as the substrate. SCE measurements on regenerating liver from strains of mice which exhibit genetic differences in these enzymes indicate surprisingly little dependence of SCE frequencies on baseline or induced benzo(alpha)pyrene or p-chloromethylaniline demethylase activities (Fig. 3)(48,49). This observation suggests that enzymes other than those commonly assayed, or perhaps steps which remove activated compounds (e.g. reactions catalyzed by epoxide hydrase), may govern the overall response of an organism to a procarcinogen.

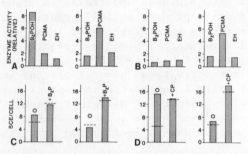

Fig. 3. Comparison of enzyme and SCE induction in mouse liver. C57/Bl(A,C) or DBA/2(B,D) mice were exposed to 3-methylcholanthrene(left hand panels) or phenobarbital(right hand panels) before meassurement of enzyme activity or SCE formation, the latter with or without exposure of the animals to 200 mg/kg benzo(alpha)pyrene (BzP) or 5 mg/kg cyclophosphamide(CP). SCE levels in the absence of any pretreatment are indicated by dashed, horizontal lines(48,49).

SCE FORMATION IN HUMAN HEREDITARY DISEASES PREDISPOSING TO NEOPLASIA.

Cells from patients with Bloom's syndrome(BS) exhibit increased interchange both between sister chromatids (evident as SCEs)(9) and between different chromosomes (evident as quadriradial figures, QRs). The position of QRs in cells from patients with BS, like that of SCEs in general, is typically at the junction of chromosome bands exhibiting bright and dull staining with quinacrine(26). It is tempting to suspect that these junctions contain sequences, perhaps present in multiple copies, that predispose to DNA interchange. The rate of DNA replication fork progression is somewhat reduced in BS (16). However, the magnitude of this retardation is small, and it is not uniformly found in cells cultivated at high density. Induction of SCEs by alkylating agents, which can further retard DNA synthesis, is greater than normal in BS(25). The SCE elevation in BS can be "corrected" by fusing affected cells with cells from non-BS individuals(4).

Exposure of cells from most xeroderma pigmentosum(XP) complementation groups to UV light results in exaggerated induction of SCEs(59) and mutations(34). However, the postreplication repair variant of XP exhibits hypermutability but normal SCE induction when exposed to UV. Hyperinduction of SCEs in XP has also been observed with alkylating agents(59) but this result has been attributed a defect in removal of alkylation damage that is unrelated to XP (17). Overall, studies with XP cells reinforce the impression that SCEs are induced by unremoved DNA damage.

Ataxia telanglectasia (AT) is not associated with an abnormality in SCE formation. Cells from patients with AT are unusually sensitive to agents that can induce double strand DNA breaks. Inhibition of DNA synthesis and stimulation of poly(ADP-ribose) synthesis following exposure of cells to X-rays is reduced in AT(13). The relationship between this observation and the ability of inhibitors of poly(ADP-ribose) synthesis to induce SCEs(40,55) remains unexplained.

Fanconi's Anemia (FA) cells are hypersensitive to agents capable of producing DNA crosslinks(44,46). Baseline SCE frequencies are normal but, depending on experimental details, the increment in SCEs produced by agents capable of crosslinking DNA has been found to be reduced (8,28,30), unchanged (38,45), or greater than normal(3) in FA lymphocytes and unchanged (28) or exaggerated (23)in FA fibroblasts. A general finding in FA cells , especially after exposure to DNA crosslinking agents, is a low mitotic index(15,58). Exaggerated accumulation of cells prior to mitosis can serve as an empirical marker of the response of FA cells to mitomycin C (22).

Toxic effects of agents such as MMC on FA cells are also evident as increased chromosomal breakage. It is of interest that one quarter to one half of the chromatid breaks observed in FA cells with or without exposure to MMC occur at sites of partial SCE formation(12,28), suggesting that failure of successful SCE formation can result in chromosome abberations.

Complementation studies, which have previously been used to characterized BS and XP, are now being applied to the study of FA(60). Mutant rodent cells exhibiting increased sensitivity to DNA crosslinking characteristic of FA have been described(6). Such cells, together with analogous mutant cells exhibiting BS or XP phenotypes(e.g., 54), should be useful in biochemical and cytogenetic studies of phenotypes associated with the respective human diseases.

RELATIONSHIP OF SCEs TO OTHER PROCESSES.

Agents that induce SCEs are typically capable of causing chromosome aberrations. However, the relative increments in SCEs and chromosome aberrations can vary widely, depending on the type of DNA damage involved and the type of cell examined. In contrast, SCEs and QRs show a strong correlation. Experiments comparing the yield of SCE induction with DNA damage at different times in the cell cycle indicate that SCE formation occurs at or very near the DNA replication fork(24,27). It is reasonable to suspect that the somatic recombination underlying QR formation might also be associated with DNA replication.

Meiotic recombination bears superficial resemblance to SCE formation and can, like the latter process, be detected by BrdU-dye techniques (see Allen (1)). However, the two processes may occur at different points of the cell cycle, and mutants of Drosphila that are defective in meiosis show virtually no alteration in SCE frequencies(14). Thus the molecular details of meiotic exchange and SCE formation may differ considerably.

Sister chromatid exchange formation can be considered a special case of mitotic recombination, which often leads to unequal crossing over, due to an interchange out of register between members of a tandemly repeated set of sequences (50,52). Like SCE formation, unequal crossing over can be promoted by clastogens such as UV light (51) or mitomycin C (19). More recently, unequal sister chromatid exchange, mediated by interspersed repeated sequences, has been implicated as the event underlying class switching of immunoglobulin heavy chains (39). If unequal SCE formation is important in the immune response, then it might be possible to utilize the extensive information available about immunoglobulin sequences to characterize SCE formation at a molecular level.

Unequal sister chromatid exchange can promote amplification or diminution of gene copy number. Gene amplification can be detected cytologically by the appearance of relatively large homogeneously staining chromosomal regions or small, acentric entities termed double minutes. These entities have been hypothesized to arise either by unequal SCE formation or by saltatory replication (see 47). Varshavsky(56) has recently postulated that the latter process can be stimulated by tumor promoters. It may prove fruitful to use analysis of amplified sequences within HSRs to investigate the molecular features of SCE formation.

MECHANISTIC CONSIDERATIONS.

Neither the mechanism nor the biological significance of SCE formation is yet known. It has been suggested that SCE formation involves the type of staggered DNA interchange initially described by Holliday(20) for meiotic recombination. However, the experimental evidence for this suggestion rests primarily on density labelling analyses(36) that may be confounded by the presence of small amounts of unusually dense DNA existing in the absence of SCE formation(32,33).

At present, it is not known why SCE formation occurs in so many organisms. One might hypothesize that, in an attempt to minimize misincorporation of a base opposite a DNA damage site, a process might have been developed for a cell first to copy the complementary strand and then to interchange strands, so that the new strand can itself serve as the template at the point of DNA damage. DNA rearrangements expected from such a process have been described in bacteria(41) and have been observed in mammalian cells(18), though the latter were subsequently attributed in large part to events occurring <u>in vitro</u>, during DNA isolation(53). Were a <u>small</u> amount of such replication, capable of bypassing <u>DNA</u> damage, to occur in vivo, it might be associated with exchange of DNA duplexes and evident as SCE formation (Fig. 4).

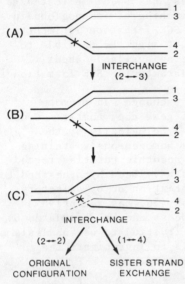

Fig. 4. Schematic diagram of a hypothetical sequence during which replication could bypass DNA damage in a manner predisposing to sister strand exchange. The first interchange, resulting in (B) attaches the new template strand to the nuclear matrix replication apparatus(57). After replication from this strand(C), resolution with or without SCE is hypothesized. No attempt is made to account for steric features of the topological rearrangements necessary for these steps.

Elucidation of the SCE process would be furthered by DNA probes specific for sequences in the vicinity of an SCE. One could then use such probes to search for sequence rearrangement, duplication, or deletion, and to examine the types of sequences that are present in the neighborhood of an exchange. Experimental analysis of SCE formation might than shift from an emphasis on cytological observations to molecular studies facilitated by new developments in recombinant DNA technology.

ACKNOWLEDGMENTS

The research described was supported by grants from the N.I.H.(GM21121 and HD04807) and the American Cancer Society(CD-36F). The expert technical assistance of Ms Lois Juergens is greatly appreciated. Part of this work was performed at the M.I.T. Regional Laser Center, which is a National Science Foundation Regional Instrumentation Facility.

REFERENCES

(1) Allen, J.W. Chromosoma 74, 189-207 (1979)
(2) Ames, B.N., McCann, J., and Yamasaki, E. Mutat.Res. 31, 347-364 (1975)

(3) Berger, R., Bernheim, A., Le Coniat, M., Vecchione, D., and Schaison, G. Cancer Genet Cytogenet 2, 259-267 (1980)
(4) Bryant, E.M., Hoehn, H., Martin, G.M. Nature 279 795-796 (1979)
(5) Carrano, A.V., Thompson, L.H., Stetka, D.G., Minkler, J.L., Mazrimas, J.A., and Fong, S. Mutat. Res. 63 175-188 (1979)
(6) Carver, J.H., Davidson, S.O., Drum, M.A., Hyre, D.A., and Crowley, J.P. Environ Mutagen 3 383 (1981)
(77 Cassel, D.M., and Latt, S.A. Exper Cell Res 128 15-22 (1980)
(8) Cerevenka, J., Arthur, D., and Yasis, C. Pediatrics 67 199-127 (1981)
(9) Chaganti, R.S.K., Schonberg, S., and German, J. Proc Nat Acad Sci US 71 4508-4512 (1974)
(10) Davidson, R.L., Kaufman, E.R., Dougherty, C.P., Ouellette, A.M., DiFolco, C.M., and Latt, S.A. Nature 284 74-76 (1980)
(11) DeSerres, F., and Ashby, J. Evaluation of Short Term Tests for Carcinogens Elsevier, N.Y. (1981)
(12) Dutrillaux, B., Couturier, J., Viegas-Pequignot, E., and Schaison, G. Human Genet 37 65-71 (1977)
(13) Edwards, M.J., and Taylor, A.M.R. Nature 287 745-747 (1980)(17) Fujiwara, Y., Kano, Y., Tatsumi, M., and Paul, P. Mutat REs 71 243-251 (1980)
(14) Gatti, M., Pimpinelli, S., Santini, G., and Olivieri, G. Proc Nat Acad Sci US 77 1575-1579 (1980)
(15) German, J., Caskie, S., and Schonberg, S. J Sup Molec Struct 2 89 (1978)
(16) Hand, R., and German, J. Human Genet 38 297-306 (1977)
(17) Heddle, J.A., and Arlett, C.F. Mutat Res 72 119-125 (1980)
(18) Higgins, N.P., Kato, K., and Strauss, B. J. Mol Biol 101 417-425 (1976)
(19) Hoehn, H., and Martin, G.M. Exper Cell Res 75 275-278 (1972)
(20) Holliday, R. Genet Res 5 282-304 (1964)
(21) Ishii, Y., and Bender, M. Mutat Res 51 411-418 (1978)
(22) Kaiser, T.N., Dougherty, C.P., Lojewski, A.J., and Latt, S.A. Amer J Hum Genet 31 75A (1980)
(23) Kano, Y., and Fujiwara, Y. Mutat Res 81 365-375 (1981)
(24) Kato, H. Cancer Genet Cytogenet 2 69-77 (1980)

(25) Krepinsky, A.B., Rainbow, A.J., and Heddle, J.A. Mutat Res 69 357-368 (1980)
(26) Kuhn, E.M. Chromosoma 66 287-297 (1978)
(27) Latt, S.A. Ann Rev Genet (in press)
(28) Latt, S.A., Stetten, G., Juergens, L.A., Buchanan, G.R., and Gerald, P.S. Proc Nat Acad Sci US 72 4066-4070 (1975)
(29) Latt, S.A., Allen, J.W., Bloom, S.E., Carrano, A.V., Falke, E., Kram, D., Schneider, E., Schreck, R.R., Tice, R., Whitfield, B., and Wolff, S. Mutat Res Revs (in press)
(30) Latt, S.A., Schreck, R.R., Dougherty, C.P., Gustashaw, K.M., Juergens, L.A., and Kaiser, T.N., in Chromosomes and Cancer, J. German ed (in press)
(31) Liu, V.W., Heddle, J.A., Arlett, C., and Broughton, B. (submitted for publication)
(32) Loveday, K.S., and Latt, S.A. Nucl Acids Res 5 4087-4104 (1978)
(33) Loveday, K.S., and Latt, S.A. Amer J Hum Genet 31 103A(1979)
(34) Maher, V.M., Ouellette, L.M., Curren, R.D., and McCormick, J.J. Nature 261 593-595 (1976)
(35) Miller, J.A. Cancer Res 30 559-576 (1970)
(36) Moore, P.D., and Holliday, R. Cell 8 573-579 (1976)
(37) Nebert, D.W., Atlas, S.A., Guenthner, T.M., and Kouri, R.G. pp 345-390 in Polycyclic Hydrocarbons and Cancer H. Gelboin and P.O.P. T'so, eds (1978)
(38) Novotna, B., Goetz, P., and Surkova, N.I. Human Genet 49 41-50 (1979)
(39) Obata, M., Kataoka, T., Nakai, S., Yamagishi, H., Takahashi, N., Yamawaki-Kataoka, Y., Nikaido, T., Shimzu, A., and Honjo, T. Proc Nat Acad Sci US 78 2437-2441(1981)
(40) Oikawa, A., Tohda, H., Kanai, M., Miwa, M., and Sugimura, T. Biochem Biophys Res Commun 97 1311-1316(1980)
(41) Rupp, W.D., Wilde, C.E., Reno, D.L., and Howard-Flanders, P. J Mol Biol 61 25-44 (1971)
(42) Sahar, E., Kittrel, C., Fulghum, S., Feld, M., and Latt, S.A. Mutat Res 83 91-105 (1981)
(43) Samson, L., and Schwartz, J.L. Nature 287 861-863 (1980)
(44) Sasaki, M.S. Nature 257 501-503 (1975)

(45) Sasaki, M.S. pp 285-313 in DNA Repair and Mutagenesis in Eukaryotes eds W.M. Generoso, M. Shelby and F. DeSerres, Academic, N.Y.(1980)
(46) Sasaki, M.S., and Tonomura, A. Cancer Res 33 1829-1835 (1973)
(47) Schimke, R.T., Brown, P.C., Kaufman, R.J., McGrogan, M., and Slate, D.L. Cold Spring Harb Symp Quant Biol 45 785-798 (1980)
(48) Schreck, R.R., and Latt, S.A. Nature 288 407-408 (1980)
(49) Schreck, R.R., Paika, I.J., and Latt, S.A. submitted for publication
(50) Smith, G.P. Science 191 528-535 (1976)
(51) Szostak, J.W., and Wu, R. Nature 282 426-430 (1980)
(52) Tartoff, K.D. Ann Rev Genet 9 355-385 (1975)
(53) Tatsumi, K., and Strauss, B. Nucl Acids Res 5 331-347 (1978)
(54) Thompson, L.H., Carrano, A.V., Brookman, K.W., Minkler, J.L., and Mooney, C.L. Environm Mutagen 3 383-384 (1981)
(55) Utakoji, T., Hosoda, K., Umezawa, K., Sawamura, M., Matsushima, T., Miwa, M., and Sugimura, T. Biochem Biophys Res Commun 90 1147-1152 (1979)
(56) Varshavsky, A. Proc Nat Acad Sci US 78 3673-3677 (1981)
(57) Vogelstein, B., Pardoll, D.M., and Coffey, D.S. Cell 22 79-85 (1980)
(58) Weksberg, R., Buchwald, M., Sargent, P., Thompson, M.W., and Siminovitch, L. J. Cell Physiol 101 311-324 (1979)
(59) Wolff, S., Rodin, B., aand Cleaver, J.E. Nature 265 345-347 (1977)
(60) Yoshida, M.C. Human Genet 55 223-226 (1980)

Virus Induced Cellular Genetic Changes

Warren W. Nichols

Institute for Medical Research, Camden, New Jersey, U.S.A.

ABSTRACT Viruses have been known to interact with host cell genomes since the late 1940's when observations were made in bacterial systems and phage. In the 1960's similar observations were made on infected mammalian cells. The first genome effects seen in virus infected mammalian cells were induced chromosomal changes. This was followed by studies on viral insertion, induced gene mutations and gene transfer. These virus induced events have been shown to be synergistic with chemical mutagens in some systems for both the induction of genetic events and cell transformation.

Reports on the interactions of viruses with host cell genomes started in the late '40's and early 50's in bacterial systems with phage. In mammalian cells observations of these interactions were made in the early '60's. In mammalian systems the primary interest in viral effects on the host genome was in relation to cancer, both from the standpoint of the role and mechanism of virus induction of cancer and as a general model for all types of carcinogenesis. The first reported viral effects on the mammalian cell genome were those of induced chromosomal change, and these were followed by observations on viral insertion, induced gene mutations and gene transfer.

Many viruses have been shown to have cytogenetic effects. All of these will not be enumerated here as there are several extensive reviews available (17,31 39, 40). It should be pointed out, however, that cytogenetic effects are produced by both RNA and DNA viruses; by transforming and tumorigenic viruses; and by non-transforming and non tumorigenic viruses. The types of chromosome abnormalities that are produced include open chromosome breaks that are discontinuities in chromosomes and chromatids that are most likely to result when DNA and/or protein synthesis are inadequate for repair processes. There are also stable and unstable chromosomal rearrangements including chromosome and chromatid intrachanges and interchanges. Premature chromosome condensation (PCC) or chromosome pulverization has also been described. This is usually considered to be an indirect viral effect due to cell fusion between a mitotic cell and an interphase cell. In this case the mitotic cell produces a protein substance that induces condensation in the interphase cell with a characteristic appearance depending on which part of interphase the cell is in (33). The apparent fragmentation seen results when the interphase nucleus is in the S phase of the cell cycle. When the interphase nucleus is in G1 the PCC appearance is that of a long attenuated single-stranded unit, and when in the G2 the PCC appearance is that of an elongated double unit. PCC is also seen however in diploid cells, and in this case the mechanismis not as clear. There are also alterations in chromosome number that are most frequently seen as increases in the number of polyploid cells, but occasionally occur as changes around the diploid number including hypodiploidy. An increased frequency of sister chromatid exchange events has been described in association with SV40 transformation (40,60,61) and with transformation with Rauscher leukemia virus (8).

A generalization can perhaps be made on chromosomal changes in lytic vs. transforming virus infection, in that in the former the effects are usually seen very rapidly, whle in the latter there is often a long latent period between the addition of virus and the first appearance of chromosome defects. In the case of SV40 this can be several weeks, and in EBV transformation of human lymphocytes it can be as long as year after infection before chromosomal abnormalities are seen. In the latter case the chromosome changes may not be virus induced but the result of the increased lifespan that permits the accumulation of spontaneous changes.

As mentioned earlier, it has also been demonstrated that in some infections the viral DNA or a DNA copy of viral RNA are inserted covalently into the host cell DNA. In the case of SV40, Sambrook and his colleagues made this observation in 1968 (47), and with RNA viruses it was made by Temin and his colleagues and Baltimore in 1970 (50,4).

In recent years evidence has also been grown indicating that viruses can induce gene mutations. This has been accomplished using somatic cell genetic techniques. Taylor (48) had supplied precedent for this effect in 1963 with the demonstration that Mu-1 phage produced mutations in bacteria, and also demonstrated that this phage was randomly inserted throughout the bacterial genome rather than at a single specific insertion site as is true with most other bacteriophage. With SV40 virus in a variety of cell systems, gene mutations have been found by Marshak and his group in the USSR (32); T.L. Theile and his colleagues in Germany (51,52,53), and in our own laboratory.

The final type of interaction to mention is gene transfer. This is exemplified by the integration of the herpes simplex virus thymidine kinase gene into TK- cells that are then shown to be able to grow in HAT medium, as was demonstrated by Bacchetti and Graham (2,3), as well as others.

With these as the types of viral host genome interactions that have been described, the remainder of this discussion will be devoted to a group of observations on viral transformation in the context of the somatic mutation theory of cancer.

Evidence for the role of somatic mutation in cancer has been growing steadily over the years by observations such as the demonstration that most, if not all, chemical carcinogens are mutagens (1,34,36), epidemiologic observations (24,26), the increased risk of cancer in chromosome fragility syndromes and abnormalities of DNA repair (6,9 12,15,45), specific chromosomal changes within tumor cells of clinical and experimental malignancies (38,44), and specific prezygotic chromosomal defects that predispose to specific cancers (10,25,43,59).

The epidemiologic work of Knudson (24,26) indicates the probability of a two- or multiple-hit mechanism for cancer. It seems probable that viral transformation is a multi-step process too, on the basis of several observations. In human diploid cells SV40 transformation is seen as a morphologic and kinetic alteration that is subsequently followed by a time when one or a few clones repopulate the culture to derive a permanent transformed line. Girardi (13) observed that crisis occurred at a regular interval after phase 3 or senescent phase-out would be observed in uninfected diploid fibroblasts. This interval was approximately 9 weeks. These observations suggest the possibility that the first stage of transformation that precedes crisis just delays the regularly observed senescence of uninfected diploid cells. This possibility is supported by the observations of Gotoh (14), who carried 91 clones of WI-38 transformed cells, and all of these died out. This also indicates that the rate of survival of crisis is a rather low one.

Another series of observations indicating the probability of a multi-step process in transformation is the work with SV40 deletion mutants. In a series of deletion mutants developed by Paul Berg and his colleagues, mutants of early viral genes were used to infect several animal cell lines and these failed to transform when an assay system based on growth in soft agar was used. The same virus-cell combinations however, did transform when a monolayer assay was used (7,35).

There is no doubt about the importance of the inserted viral genome in transformation. It is known that the inserted viral genetic material is functional in most cases and that turn-off of the viral genome via temperature sensitive viral mutants results in a reversion of many of the acquired phenotype characteristics of transformation back to normal (49). It is not clear however if the induction of transformation or malignancy is produced by the insertion and function of the viral material alone, or if cellular participation or alterations are also necessary. The number of investigators believing that cellular alterations are necessary is steadily growing. Dulbecco (11) has commented that the expression of the viral genome is controlled by cellular genes. He also suggests the possibility that the reason that all cells infected by transforming viruses are not successfully transformed may be because transformation can only be accomplished when the cellular gene balance has been altered in specific ways by rare virus independent genetic events. Hanafusa and colleagues (16) demonstrated defective acute transforming viruses that lacked most of the SrC gene could acquire this information by recombination with cellular genes. Hayward and coworkers (18) demonstrated a similar phenomenon in slow transformation systems by the virus promoter insertion mechanism. Here a virus that inserts into the cell genome has long terminal repeating segments (LTR) at each end of the virus which can promote or amplify cellular gene expression from their insertion.

Croce and his colleagues (29) have recently reported very interesting work on murine teratocarcinoma cells. Stem cells were shown not to support the growth of SV40 or polyoma virus, while when the stem cells were permitted to differentiate they became susceptible. The resistance was not due to blockage at the level of virus adsorption, penetration, uncoating or transport to the nucleus. They went further and used a recombinant plasmid linked to a herpes simplex virus thymidine kinase gene and an SV40 genome and inserted these into the TK-stem cells. The TK gene is expressed in stem cells while the SV40 is not. If differentiation is permitted, then both the TK and SV40 genome are expressed, indicating not only cellular regulation but that it differs with differentiation.

George Klein (22,23), commenting on the EBV transformation of human lymphocytes and comparing this with Burkitt's lymphoma, believes there is an initial step in the EBV transformation that immortalizes the cells, but that for malignant growth in vivo there is a requirement for cytogenetic change. Thus in lymphocytes that are infected with EB virus and grow continually they will not form colonies in agarose or produce tumors in nude mice until cytogenetic changes have taken place. As mentioned earlier, this is usually months or a year after infection. In Burkitt's tumor that is also associated with EB virus, there is a characteristic translocation between chromosomes 14 and 8. While this is associated with this virus infection, it is also seen in Burkitt's tumor, not associated with EBV and with B cell derived ALL, and may be analagous to the mouse lymphoma, below.

Another observation cited to support this concept of a cytogenetic requirement after initiation is the fact that T cell lymphomas in mice are found to be trisomic for chromosome 15, whether they were initiated by chemical, X-ray or virus. Thus, no matter what the initiating agent, the same chromosomal alteration accompanied the development of the tumor.

Ruth Sager (46) has reported that there are high frequencies of rearrangements in mouse cell DNA adjacent to the integrated sites of inserted SV40 genome, and hypothesizes that these may influence the expression of transformation and tumorigenic potential. This may be similar to the observations on RNA viruses made by the Hanafusa and Hayward groups.

Whitehouse (58) has pointed out that the loss of an inhibitor or supressor or suppressor can alter regulation. This could produce results similar to the promoter insertion model or other mechanisms that can produce gene amplification.

It is clear that the cellular genome is important in the susceptibility of cells to virus transformation from studies on various chromosomal abnormalities. However there is lack of clarity on how chromosome abnormalities influence virus transformation. For instance, Fanconi's anemia (27,54), that has an increased level of chromosome breakage and chromosomal trisomies (27,42,55) has been shown to have an increased susceptibility to SV40 transformation, both by the numbers of transformed foci produced at specific multiplicities or infection, and by the percentage of cells that express T antigen 72 hours after infection. Ataxia telangiectasia on the other hand, that also has an increased spontaneous rate of chromosome abnormalities, has been shown to be more susceptible to SV40 induced chromosome breakage (57), but the transformation rate is the same as that found in normal cells (27,57). Bloom's syndrome also has an increased level of chromosome breakage and has been reported in one study to have a decreased transformation rate when infected with SV40 virus, and in another study to have transformation rates similar to normal cells (28,56). Cells from patients with Xeroderma pigmentosum show no increase of transformation frequency with SV40 virus alone or with SV40 virus combined with ultraviolet (20,21,41).

I would like to return to viral induced chromosome

and gene mutations to compare the time sequence of their
appearance and some characteristics of the gene mutational
events. Chromosomal mutations associated with viral
transformation are observed at the time the cells become
T antigen positive when examined sequentially after infection. The frequency of SCE's per cell and anaphase abnormalities are both seen to parallel the increase in T
antigen positive cells. The time when cells become T
antigen positive varies depending on multiplicity of
infection used.

SV40 has also been associated with induced gene
mutations. These seem to be produced earlier than the
chromosomal mutations. In Marshak's (32) experiments
they were observed 24 to 96 hours after infection.
These seem to be related to the viral genome in that
Theile could produce the mutations using extracted SV40
DNA, whle noninfected DNA did not induce any increase
in mutation frequency (52). The mutationsproduced also
seem to be at least partially point mutations rather
than small deletions. In most of the systems examined
mutations to thioguanine or 8-azaguanine resistance
were utlized that could be either true gene mutations
or small deletions, but studies have also been reported
by Luebbe (30) and colleagues that SV40 can induce
revertants of Lesch-Nyhan cells which could not be
explained on the basis of deletion events. The viral
induced mutations are also increased in a synergistic
manner when chemical mutagen treatment is added to the
viral treatment (53). It's interesting that some
chemical mutagens and promoters also increase the rate
of SV40 transformation (19,37,62). Most of the studies
that are reported in viral induced mutations have been
done in animal cell systems or in heteroploid human
cells. It is of interest that in our own laboratory
when examining mutation frequency in diploid cells, an
increased mutation frequency at the hpt locus was seen
in 3 our ot 5 cultures examined. In one culture of
human embryo lung there was no increase noted on six
successive transformation. The reasons for these
differences are not readily apparent at present but
several possibilities such as the timing of the mutational studies after infection will be investigated.

At this point, it can be said there is a probability
of viral transformation being a multiple step event,
that the inserted viral genome into the cell genome is

an important event in transformation, there is the possibility of the cell genome being involved in the transformation process, and transforming viruses have the ability to produce chromosomal and gene mutations. Knudson's two-hit mutational model of carcinogenesis has interesting implications in the light of these observations. His studies indicate that for many or all tumor types there are two subgroups: those that are sporadic and those that have a hereditary predisposition. In the sporadic group, two specific somatic mutations are required, and this is therefore a rare event. In the hereditary subgroup one mutation is already present in a genome from the germline and only a single somatic mutation is necessary for tumor production. This is a much more common event and produces tumors at a younger age that are multifocal and bilateral when applicable. Cell transformation of normal human diploid cells has rarely been achieved with chemical carcinogens or mutagens. With mutation rates of 10^{-5} or 10^{-6}, if two specific mutations are necessary, transformation would be expected only once in 10^{10} or 10^{12} cells per generation. This would indeed be an event so rare that it would be difficult to detect in a cell culture system, and may be why transformation of diploid cells with chemicals is so rarely accomplished. I would like to propose as a hypothesis that viral transformation seen so regularly with human diploid cells is more similar to Knudson's hereditary subgroup of cancers. In the case of viral transforamtion the effect of two mutational events is produced by the insertion and function of the viral genome as one event, and the induction of chromosomal or gene mutation as the second event that leads to the viral transformed state. In this hypothesis the virus would insert randomly in the genome of infected cells, both producing viral messenger RNA and inducing chromosomal and gene mutations. The functioning viral genome would be responsible for the altered cell phenotype and increased growth rate. At the time of crisis most of these cells would degenerate or die, and only those that have an appropriate second mutation produced would survive and propagate with the full malignant phenotype (Figure 1).

It is also conceivable that rarely the complete transformation could occur without viral genome function by induction of two cellular mutations necessary, as in the case of sporadic tumors.

It is likely that either a chromosomal or gene mutation could affect the final stage to full transformation or malignancy much as is seen in hereditary retinoblastoma (25). Here a portion of these cases appears to have a gene mutation predisposing to tumor formation and a second portion has a specific deletion in chromosome 13 leading to tumor formation. In our own laboratory attempt to elucidate this hypothesis by looking at induced chromosomal and gene mutations in relation to transformation by wild type virus and deleted and temperature sensitive mutant viruses are in progress.

In preliminary work with Ts SV40 virus we have made two observations that are of considerable interest and also somewhat unexpected. The first is that human diploid fibroblasts infected at 33°C--the permissive temperature for this mutant--transformed much less readily than is seen at 37°C. This indicates the possibility of a cellular product that facilitates transformation that is not produced as well at the lower temperature. The second observation relates to induced chromosome abnormalities. When foci appeared and cells became T antigen positive there was the appearance of chromosome abnormalities as we had observed with wild type SV40 infection. However, when these cells were put at the restrictive temperature they became negative for T antigen and developed a normal phenotype and the chromosome abnormalities returned to normal levels. These were precrisis cells and the types of abnormality were unstable rearrangements and mostly dicentrics. The possible explanations for this are that it is a selective process, and when cells are put at the restrictive temperature those that are transformed can no longer grown and are overgrown by normal cells that had been suppressed at the permissive temperature. This is contrary to what has been believed however--that the same cells were expressing a different phenotype at the two temperatures.

A second possible explanation is that there are reversible chromosome abnormalities. This too is contrary to previous dogma in that once formed abnormalities were not believed to be able to revert to normal as a general phenomenon. However, there is a category of chromosomal structural change that is reversible at least under some circumstances. These are the cytogenetic alterations that accompany gene amplification. HSR areas can be seen to interchange with double minutes, depending on the selective pressure. A type of change associated with gene amplification that could be reversible should also be investigated here.

(1) Ames, B.N., Sims, P., Grover, P.L. Science 176:47-49 (1972)
(2) Bacchetti, S., Graham, F.L. Proc. Natl. Acad. Sci. USA 74:1590-1594 (1977).
(3) Bacchetti, S., Graham, F.L. IARC Sci. Publ. 24, 1: 495-499 (1978)
(4) Baltimore, D. Nature 226:1209 (1970).
(5) Bishop, J.M., Jackson, H., Quintrell, N., Varmus, H.E. In: Possible Episomes in Eukaryotes, North-Holland, Americal Elsevier, 61-72 (1973).
(6) Bloom, A.D. In: Advances in Human Genetics 3:99-172, H. Harris and K. Hirschhorn, eds. (1972).
(7) Bouck, N., Beales, N., Shenk, T., Berg, P., Di Mayorca, G. Proc. Natl. Acad. Sci. USA 75:2473-2477 (1978)
(8) Brown, R.L., Crossen, P.E. Exptl. Cell Res. 103: 418-420 (1976).
(9) Cleaver, J.E. Nature 218:652-656 (1968).
(10) Cohen, A.J., Li, F.P., Berg, S., Marchetto, D.J., Tsai, S., Jacobs, S.C., Brown, R.S. The New Eng. J. Med.301:592-595 (1979).
(11) Dulbecco, R. J. Gen. Microbiol. 79:7-17 (1973).
(12) Fraumeni, J.F., Miller, R.W. J. Nat. Can. Inst. 38:593-605 (1967).
(13) Girardi, A.J., Jensen, F.C., Koprowski, H. J. Cell. & Comp. Physiol. 65:69-84 (1965).
(14) Gotch, S., Gelb,L., Schlessinger, D. J. Gen. Virol. 42:409-14 (1979).
(15) Hamerton, J.L. In: Human Cytogenetics Clinical Cytogenetics 11:407-441, Academic Press (1971).
(16) Hanafusa, H., Halpern, C.C., Buchhagen, D.L. Kuwai, S. J. Exp. Med. 146:1735-1747 (1977).
(17) Harnden, D.G. In:Chromosomes and Cancer, 151-190, J. German, ed., John Wiley & Sons, N.Y. (1974).
(18) Hayward, W.S., Neel, B.G., Astrin, S.M. Nature 290:475-480 (1981).
(19) Henderson, E.E., Ribecky, R. J. Natl. Can. Inst. 1:33-39 (1980).
(20) Kersey, J.H., Gatti, R.A., Good, R.A. Arronson,

S.A., Todaro, G.J. Proc. Nat. Acad. Sci USA 69: 980-982 (1972).
(21) Key, D.J., Todaro, G.J. J. Invest. Dermatol. 62: 7-10 (1974).
(22) Klein, G. Proc. Nat. Acad. Sci. USA 76:2442-46 (1979)
(23) Klein, G. CIBA Found. Symp. 66:335-358 (1979).
(24) Knudson, A.G. Proc. Nat. Acad. Sci. 68:820-823 (1971).
(25) Knudson, A.G., Meadows, A.T., Nichols, W.W., Hill, R. New Eng. J. Med. 295:1120-1123 (1976).
(26) Knudson, A.G., Strong, L.C. Am. J. Hum. Genet. 24: 513-532 (1972).
(27) Levin, S., Hahn, T. Israel J. Med. Sci. 8:133-136 (1972).
(28) Lin, M.S., Alfi, O.S. Experientia 36:296-97 (1980).
(29) Linnenbach, A., Huebner, K., Croce, C.M. Proc. Nat. Acad. Sci. USA 77:4875-4879 (1980).
(30) Luebbe, L., Strauss, M., Geissler, E. FEBS Proc. Mtg. 56:67 (1979).
(31) Makino, S. Human Chromosomes. Elsevier/North Holland Biomed. Press (1975).
(32) Marshak, M.A., Varshaner, N.B., Shapiro, N.I. Mutation Res. 30:383-396 (1975).
(33) Matsui, W., Weinfeld, H., Sandberg, A.A. J. Nat. Can. Inst. 47:401-411 (1971).
(34) McCann,J., Choi, E., Hamasaki, E., Ames, B.N. Proc. Natl. Acad. Sci. USA 72:5135-5139 (1975).
(35) Mertz, J.E., Carbon, J., Herzberg, M., Davis, R.W., Berg, P. Cold Spg. Har. Symp. Quant. Biol. 39, part 1:69-84 (1975).
(36) Miller, J.A... Cancer Res. 30:559-576 (1970)
(37) Milo, G.E., Blakeslee, J.R., Hart, R., Yohn, D.S. Chem. Biol. Interactions 22:185-197 (1978).
(38) Mitelman, F., Mark, J., Levan, G., Levan, A. Science 176:1340-1341 (1972).
(39) Nichols, W.W. In: The Cell Nucleus, Vol. 2, 437-458, H. Busch, ed., Academic Press, N.Y. (1974).
(40) Nichols, W.W., Bradt, C.I., Toji,L.H., Godley,M., Segawa, M. Cancer Res. 38:960-964 (1978).
(41) Parrington, J.M., Delhanty, J.D.A., Baden, H.P. Ann. Hum. Genet. Lond. 35:149-160 (1971).
(42) Payne, F.E., Schmickel, R.D. Nature New Biol. 230: 190 (1971).

(43) Riccardi, V.M., Sujansky, E., Smith, A.C., Francke, U. Pediatrics 61:604-610 (1978).
(44) Rowley, J.D. Cancer Genet. & Cyto. 2:15-198 (1980).
(45) Ryser,H.J.P. N. Eng. J. Med. 285:721-734 (1971).
(46) Sager, R., Anisowicz, Howell, N. Cold. Spg. Harb. Symp., Mobile Genetic Elements, Vol. 45 (1980).
(47) Sambrook, J., Westphal, H., Srinivassan, T.J., Dulbecco, R. Proc. Natl. Acad. Sci. USA 60:1286-1295 (1968).
(48) Taylor, A.L. Proc. Natl. Acad. Sci. 50:1043-1051 (1963).
(49) Tegtmeyer, P. J. Virol. 15:613-618 (1975).
(50) Temin, H.M., Mizutani, S. Nature 226:1211 (1970).
(51) Theile, M., Krause, H. Studio Biophysica, Berlin, Band 76:45-46 (1979).
(52) Theile, M., Scherneck, S., Geissler, E. Arch. in Virology 65:34, 293-309 (1980).
(53) Theile, M., Strauss, M.,Luebbe, L., Scherneck, S. Krause, H., Geissler, E. Cold Spg. Harb. Symp. Quant. Biol. XLIV, 377-382 (1980).
(54) Todaro, G.J., Green,H., Swift, M.R. Science 153: 1252-1254 (1966).
(55) Todaro, G.J., Martin, G.M. Proc. Soc. Exp. Biol. Med. 124:1232-1236 (1967).
(56) Webb, T., Harding, M. Br. J. Cancer 36:583-591 (1977).
(57) Webb, T., Harnden, D.G., Harding, M. Cancer Res. 37:997-1002.
(58) Whitehouse, H.L.K. In: Chromosomes and Cancer, 42-76, J. German, ed., John Wiley & Sons N.Y.(1974).
(59) Wilson, M.G., Towner, J.W., Fujimoto, A. Am. J. Hum. Genet. 25:57-61 (1973).
(60) Wolff, S., Bodycote, J., Thomas G.H., Cleaver, J. E. Genetics 81:349-355 (1975).
(61) Wolff, S., Rodin, B., Cleaver, J.E. Nature 265: 347-349 (1977).
(62) Yamamoto, N., ZurHausen, H. Nature 280:244-245 (1979).

Environmental Mutagens and Karyotype Evolution in Mammals[1]

Tosihide H. Yosida

National Institute of Genetics, Mishima, Japan

ABSTRACT Relation between the karyotype evolution and environmental mutagens was considered in mammals such as the black rat, the house shrew, the Indian spiny mouse and the Indian muntjac. The karyotype evolution of these species has concentratedly occurred in India, especially southern India. The natural radiation from the cosmic rays and the earth's crust were suggested as causes. The radioactive monazite sands found in Kerala State are a possible source of environmental mutagens. The karyotype evolution and species differentiation seem to be due to chronic irradiation from the above sources for long generations in these animals. Experimental data are cited to prove the above consideration.

What relation is there between karyotype evolution, species differentiation and environmental mutagens? This is an important problem, but is still difficult to discuss with appropriate evidence. The present author (12) has suggested that the karyotype evolution of the black rat, the Indian spiny mouse and Indian muntjac which has occurred in southern India is probably due to the effect of natural radioactive substances. Recently I found another piece of evidence in the house shrew. In the present symposium

[1] Contribution no. 1399 from the National Institute of Genetics, Japan. Supported by Grants-in-Aid for Scientific Research from the Ministry of Education, Science and Culture

I will talk about these four mammals to prove the relation between the karyotype evolution and the environmental mutagens.

1. Karyotype evolution of the black rat

Based on the karyotype characteristics, the black rat, *Rattus rattus*, is classified into the following 4 geographical variants (9): Asian type ($2n=42$), Ceylonese type ($2n=40$), Oceanian type ($2n=38$) and Mauritius type ($2n=42$). Their karyotypes were derived sequentially from each other (Fig.1). According to the opinion of taxonomists black rats occurred originally in the Indo-Malayan region. They should have distributed first in the above area, and then to the other localities of Southeast Asia, East Asia and also Southwest Asia (Asian type). The Ceylonese type black rats were only found in Sri Lanka, while the Oceanian type black rats with 38 chromosomes are distributed widely in the world, except the above areas. The Ceylonese type black rat has developed through the first Robertsonian fusion between pairs 11 and 12 included in the Asian type rat. Such fusion is considered to have occurred in southern India. The Oceanian type black rat was characterized by having the second Robertsonian fusion between chromosome pairs 4 and 7 from the Ceylonese type rat. This fusion is also thought to have occurred in southern India. The Oceanian type black rats might have spread widely over southern India, and then they migrated from southwest Asia to Central Asia and to Europe. From

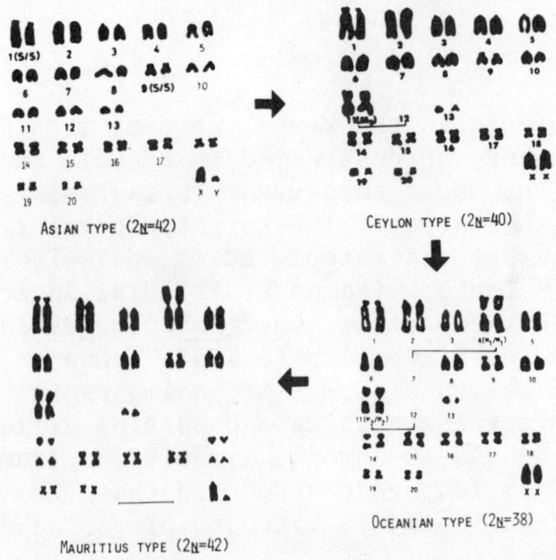

Fig.1. Karyotypes of 4 geographical variants of the black rats (*Rattus rattus*): Asian ($2n=42$), Ceylonese ($2n=40$), Oceanian ($2n=38$) and Mauritius types ($2n=42$). Arrows indicate the direction of the karyotype evolution in the geographical variants.

Europe they spread to other continents of the world. The Mauritius type rat was foundamentally similar to the Oceanian type, but 8 extra small acrocentrics developed by Robertsonian fission were always observed in their karyotype. Noticeable is the fact that the main process of the karyotypes evolution in the black rat occurred in southern India.

2. Karyotype evolution of the house shrew

The second example of the karyotype evolution was found in the house shrew, *Suncus murinus*, distributed in East, Southeast and Southwest Asia. The specimens distributed in East and Southeast Asia and also in northern and Central India showed the 40 chromosomes (Fig. 2A). On the other hand the specimen collected in Sri Lanka was remarkable in having only 32 chromosomes, among which 4 pairs were large metacentrics derived by the Robertsonian fusion of 8 acrocentrics included in the basic karyotype with 40 chromosomes (11) (Fig. 2B). In the specimens distributed in West Malaysia, however, chromosome polymorphism from 40 to 36 chromosomes based on Robertsonian fusion was observed by Yong (7,8). Specimens with 32 or 30 chromosomes have also been observed by Satya-Prakash and Aswathanarayana (4), distributed widely in southern India. Their karyotypes were similar to those found in Sri Lanka, which were derived by the Robertsonian fusion of many acrocentrics. According to some taxonimists, the house shrew was originally distributed in the forest of Central India, and then moved to the other parts of Southwest, Southeast and East Asia (Fig. 3). They should have the basic karyotype with 40 chromosomes. The sequential

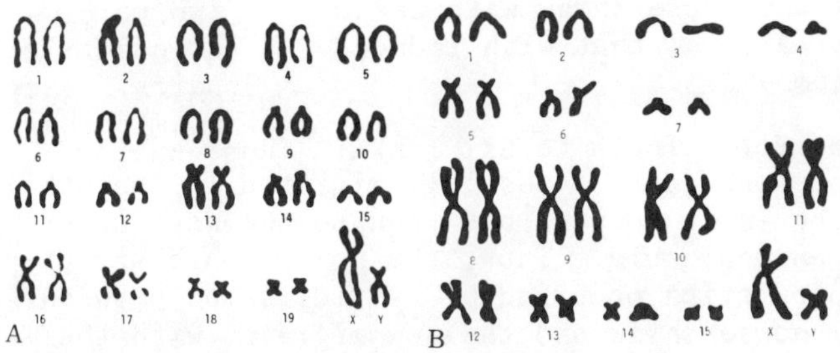

Fig. 2. Karyotypes of 2 geographical variants of the house shrew (*Suncus murinus*). A, basic karyotype with 40 chromosomes. B, karyotype with 32 chromosomes found in Sri Lanka.

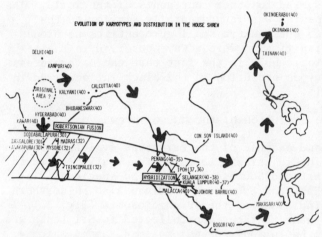

Fig. 3. Distribution map of the house shrew in the world (from Yosida, 1981). Large arrows indicate the direction of migration of the house shrew with 2n=40, and the small arrows, that with reduced chromosome numbers.

Robertsonian fusion should have taken place in the house shrew which moved to southern India. The animals with 32 chromosomes could have developed by the Robertsonian fusion of 8 acrocentric pairs, and those with 30 chromosomes by fusion of two more acrocentric pairs. They seem to have migrated from South India to Sri Lanka. Those having reduced chromosome numbers possibly migrated to the other islands in the Indian Ocean, and then to the west coast of the Malayan peninsula. Thus, the polymorphic karyotype in Malayan population occurred by hybridization between the original house shrew and the new migrants with the reduced chromosome number. An interesting point is that the karyotype evolution by sequential Robertsonian rearrangements in the house shrew occurred also in southern India. This is a very similar phenomenon to the karyotype evolution in the black rats.

3. Karyotype evolution of the Indian spiny mouse

The third example of the karyotype evolution by tandem chromosome fusion was found in the Indian spiny mouse, Mus platythrix, collected from Mysore, South India. The karyotype of this animal was characterized by having 26 acrocentric chromosomes (Fig. 4). Fourteen chromosomes were fewer than in the house mouse with 40 acrocentrics. From the comparison of G-banded karyotypes in the spiny mouse and the house mouse, I obtained evidence that certain chromosome pairs of the spiny mouse could have been derived from the tandem fusion of the house mouse chromosomes (10). Noteworthy is the fact that a strong genetic barrier seems to exist between these two species, because no viable hybrid specimen has been produced so far in reciprocal crosses in the laboratory. This means that these two are distinct species standing at the ultimate stage of speciation. This chromosome evolution seemed also to have occurred in southern India.

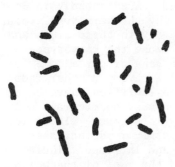

Fig.4. Chromosomes of the Indian spiny mouse (Mus platythrix) (2n=26).

4. Karyotype evolution in the Indian muntjac

The fourth example was the Indian muntjac, Muntiacus muntjak, distributed in the Indian subcontinent. This animal was remarkable for having only 7 chromosomes in the male and 6 in the female (Fig. 5). The Reeve's muntjac, Muntiacus reevesi, distributed widely in Southeast Asia, is quite similar in general morphology to the Indian muntjac, but its diploid number is 46, all elements being acrocentric as in the other deer karyotypes. These two species are known to have been hybridized. A suggestion has been made that the chromosomes of the Indian muntjac could be derived by sequential tandem fusion of several acrocentric pairs included in the original reevesi muntjac, as considered by Wurster and Benirschke (6). It is noticeable that such extreme karyotype evolution in the muntjac occurred also in the Indian subcontinent.

Fig.5. Chromosomes of the Indian muntjac (Muntiacus muntjak) with only 6 in the diploidy (♀).

5. Relation between karyotype evolution and environmental mutagens

The four examples mentioned above are interesting their evolutionary events occurred concentratedly in South India. Then the question arises why the karyotype evolution of these mammals occurred exculsively in these areas and what factor or factors worked to cause the karyotype evolution. It is well known that gene or chromosome mutations are induced by means of X-rays and other radiations. In the natural field, the radiations have been known to arise mainly from two sources: cosmic rays and radioactive materials from the earth's crust. It is well known that the irradiation from cosmic rays increases with an increase of the altitude and latitude. India has many high mountains and is called the roof of the world. There also exist a few areas in the world which embrace a high natural radioactivity. One place typical of these areas is in Kerala State of South India. The carrier of the radioactivity is known to be the black monazite sands, which are known to occur for about 100 miles along the coast of Kerala State. The United Nations (3) reported that the radioactivity of bones of people living in the striped area in Kerala State is about ten or twenty times higher than in the non-striped area. Gruneberg (2) has not found any positive indication on the genetic variation in the skeletal and dental measurements of the rats living in the Quilon area of Kerala State, but Dubinin et al. (1) and Shevchenko (5) found that plants and the mollusk in the high radioactivity area in Siberia had a higher occurrence of chromosome abnormalities. The above discussion on the karyotype evolution makes it possible to state that India, particularly South India, might be the platform for karyotype evolution in mammals, probably by means of natural radiation. The cause of the karyotype evolution of the Mauritius black rat is unknown at the present, although Mauritius island occurred by the submarine eruption.

6. Are chromosomes of the black rat sensitive to radiation?

As already stated, chromosome polymorphism was found often in black rats, but in the Norway rat the variations are seldom found in the natural population. The problem is whether the chromosomal mutation rate is different in those two species. To solve this problem, γ-ray was administrated accutely to male Norway and black rats, and then they were crossed with normal females. Thus, the karyotypes of F_1 offspring were analyzed. The experimental results show that the ratio of the chromosome mutation was not different between the house rat and Norway rat. This suggests that the black rat is not a peculiarly sensitive animal for the production of chromosome mutation. Chromosomal evolution found in several geographical variants, thus, seems to be due to the environmental mutagen's chronic exposure from the natural radiation to the animals for long generations. In other mammals such as the house shrew, the Indian spiny mouse and the Indian muntjac, it could be explained by the same mechanism as in the black rat.

References

(1) Dubinin, N. P., Shevchenko, V. A. and Ponerantseva, M. D. The effect of ionizing radiation on population. In: Radioecology (M. Atomizdat) (1971).
(2) Gruneberg, H. Nature, 222-224 (1964).
(3) Report of the United Nations Scientific Committee on Effects of Atomic Radiation, New York, 1966.
(4) Satya-Prakash, L. K. and Aswathanarayana, N. V. Dunn & Dobzh. Symp. Genet. 265-272 (1976).
(5) Shevchenko, V. A. Personal Information (1978).
(6) Wurster, D. H. and Benirschke, K. Science 168, 1364-1366 (1970).
(7) Yong, H. S. Experientia 27, 589-591 (1971).
(8) Yong, H. S. Experientia 28, 585-596 (1972).
(9) Yosida, T. H. Cytogenetics of the Black Rat - Karyotype Evolution and Species Differentiation. Univ. Tokyo Press, Tokyo and Univ. Park Press, Baltimore (1980).
(10) Yosida, T. H. Cytologia 45, 753-762 (1980).
(11) Yosida, T. H. Jap. Jour. Genet. 56 (1981) (in press).
(12) Yosida, T. H. and Parida, B. B. Proc. Japan Acad. 56(B), 79-84 (1980).

The Induction of Chromosome Aberrations in Germ Cells: A Discussion of Factors That Can Influence Sensitivity to Chemical Mutagens

R. Julian Preston

Biology Division, Oak Ridge National Laboratory,[1] Oak Ridge, Tennessee, U.S.A.

ABSTRACT The intent of this presentation is to discuss the relationships between induced and observed (or recovered) chromosome aberration frequencies, particularly in different germ cell stages following treatment with chemical mutagens. There are many factors that can influence both the induced and observed frequencies, such as the cell cycle stage at the time of treatment and observation, and the extent and relative rates of DNA repair in different cell types. These factors are discussed, often with reference to in vitro cell systems, in an attempt to present a general model of sensitivity to chromosome aberration production by chemical mutagens.

INTRODUCTION

There is a considerable amount of information on the frequencies of chromosome aberrations induced by a wide variety of agents, in a large number of cell types, both in vivo and in vitro (Tabulated in 13). However, much less regard has been paid to the reasons why different cell types appear to have different sensitivities to the same agent. If an attempt is made to estimate genetic hazard it is the differential or

[1]Research sponsored by the Office of Health and Environmental Research, U.S. Department of Energy, under contract W-7405-eng-26 with the Union Carbide Corporation.

relative sensitivity that becomes of great importance. In particular, if extrapolation from laboratory animals to man is attempted, then an understanding of factors that can influence sensitivity to chromosome aberration or mutation induction is essential.

The approach taken here is to consider those factors that appear to or could influence sensitivity, drawing upon available information from a variety of different assays, and to discuss how such factors can influence the sensitivity of germ cells in different stages of gametogenesis to chromosome aberration induction. In some cases, it might be suggested that part of the discussion is like "beating a dead horse," but since even well-documented information is still very often ignored, it seems to require further emphasis.

DISCUSSION

1. Induced vs. observed chromosome aberrations
The majority of microscopically observable chromosome aberrations are cell lethal as a result either of a mechanical inhibition of mitosis (usually by anaphase bridges) or the loss of rather large pieces of chromatin in the form of acentric fragments at division. As a result, cells containing chromosome aberrations will tend to be lost from a cycling population at the first or second division after the induction of the aberrations. In general, chromosome-type inversions and reciprocal translocations, where both chromatids are involved, are non-lethal aberrations in mitotically dividing cells, and their frequency will remain rather constant in a cycling population -- a reduction in frequency resulting from the presence in the same cell of other types of aberrations that are lethal. Chromatid-type translocations can give rise to duplications and deficiencies at the first cell division after induction as a result of particular mitotic segregation of the translocation, and thus their frequency at the 2nd metaphase after treatment will be reduced compared to the induced frequency. Other types of chromatid aberrations are either lethal for the same reasons as chromosome-type aberrations, or give rise to derived chromosome-type aberrations as a result of cell division and a subsequent DNA replication.

Thus, in order to obtain an estimate of the induced frequency of chromosome aberrations, observation has to be made on cells at their first mitosis after treatment. One rider to this statement is that some chemicals appear to produce long-lived lesions whereby aberrations can be induced one or more cell divisions after treatment, and so in order to obtain an overall estimate of induced frequency, 2nd and 3rd mitosis should be analysed. However, the division after treatment at which the cells are being analysed should be known, and it should also be noted that the aberrations induced in cycles subsequent to the treated one will be of the chromatid-type for the majority of chemicals, with accompanying acentric fragments.

The experimental design can usually satisfy the requirement for observations of first division cells with in vitro cell cultures and many in vivo systems, such as bone marrow, spermatogonia, and regenerating liver. However, without proper regard to this, it is quite possible to obtain apparent differences in sensitivity for different cell types or for the same cell type in different species. These considerations have been discussed many times before, but often are ignored.

It is sometimes not possible to analyse cells at their first division after treatment, in which case the interpretation of the results has to be carefully considered. Examples of this are chronic exposures to cycling cell populations and treatment of stem cells where observation has to be made following several cell divisions.

A clear example of the latter situation is where spermatogonial stem cells are the treated cell type, but analysis is made in diakinesis-metaphase I spermatocytes, when reciprocal translocations are the only anticipated aberration type. When mice are treated with chemicals that are known clastogens in spermatogonial cells, no translocations are observed in spermatocytes derived from treated spermatogonial stem cells (10). [In a few instances a very low frequency of translocations was observed, but then only when low doses of the agent were used (6).] The explanation is not that the spermatogonial stem cell is insensitive to the induction of aberrations, but rather that aberrations are lost from the analysed population as a consequence of the cell divisions intervening between treatment and observation. The majority of clastogens induce exclusively chromatid-type aberrations, and the segregation at division of a

symmetrical chromatid exchange results in a probability
of 0.25 of recovering the translocation in a daughter
cell. The presence of other chromatid-type aberrations
that are cell lethal because of loss of acentric chromosomal fragments at division, in the same cell as a
symmetrical translocation, will cause the translocation
to be lost from the analysed spermatocyte population.
The result is that, even if a relatively high frequency
of symmetrical exchanges (reciprocal translocations) are
induced in spermatogonial stem cells, very few if any
will be present in spermatocytes. As mentioned above,
at low doses, translocations are occasionally observed in
spermatocytes following spermatogonial stem cell treatment due mainly to a reduced loss of symmetrical exchanges
as a result of the presence of lethal aberrations in the
same cell - at low aberration frequencies induced by low
doses the probability of this is lower. However, the
low overall aberration frequencies mean that rather few
symmetrical exchanges are induced, and even fewer will be
observed in spermatocytes, due to the segregational loss
discussed above. Therefore, an understanding of the
transmission of aberrations through cell divisions
occurring between treatment and observation is an important component of apparent differences in observed
aberration frequencies in different cell types or in
different species.

2. <u>Cell cycle stage</u>. The stage of the cell cycle
in which cells are present at the time of treatment
influences the sensitivity of a cell to the induction of
chromosome aberrations by radiation or chemical agents.
This is a well-documented observation (9). However, an
understanding of the mechanistic reasons for this
requires some discussion.
 The types of aberrations that can be induced are,
of course, dependent upon cell cycle stage -- chromosome-
type aberrations are induced in G_1 and chromatid-type in
S and G_2. The spectrum of aberration types that can be
induced by a particular agent in general falls into one
of two categories. X-rays and a few chemical agents can
induce a measurable frequency of aberrations in all
stages of the cell cycle, ie. chromosome-type aberrations
in G_1 cells (except when close to S) and chromatid-type
in S and G_2. The majority of chemical agents induce
chromatid-type aberrations exclusively, even for cells

that are treated in G_1. A discussion of the possible reasons for this are given in Section 3 below.

The aberration frequency observed following treatment with chemicals that produce aberrations largely in the S phase, or following passage through the S phase between treatment and observation, will be dependent upon the amount of DNA damage that can result in aberrations remaining in cells at the time of replication. This will be influenced by the duration of the different stages of the cell cycle, and the rate and extent of repair of DNA damage, as well as by the amount of the initial DNA damage. Also the fact that the majority of chemicals induce only chromatid-type aberrations will influence the probability of transmission of aberrations induced by these agents either through several cell divisions or to the F_1 generation, as compared to the probability of transmission of chromosome-type aberrations.

How can these factors influence the sensitivity of different germ cell stages? The fact that most chemicals induce chromatid-type aberrations is, in part, the reason why few if any translocations are observed in spermatocytes following treatment of spermatogonial stem cells or differentiating spermatogonia. This has already been discussed.

It is perhaps worth noting here that similar arguments can explain the fact that following irradiation of maturing oocytes high frequencies of translocations are observed in metaphase I oocytes, whereas very few are recovered in the F_1 (4). The maturing oocyte is a G_2 cell and so the aberrations induced by x-rays will be of the chromatid-type. The segregation of symmetrical chromatid exchanges giving a probability of 0.25 of transmitting the translocation, and the non-transmission of translocations caused by the presence of cell-lethal aberrations in the same cell, will result in very few translocations being transmitted to the F_1 following oocyte irradiation (18, 4).

The other factor of considerable importance in discussions here is that the majority of aberrations are produced during the S phase. This will mean that the time for repair of DNA damage between treatment and replication can greatly influence the frequency of aberrations. This can be clearly seen when a comparison is made of the aberration frequencies observed for example in human lymphocytes treated with methyl methanesulfonate (MMS) or 4-N-nitroquinoline-N-oxide

(4NQO) in G_0 or at different times in G_1, when the time between treatment and the S phase is different. With cells treated in G_0 or early G_1, the frequencies of chromatid aberrations are the same as in untreated controls, whereas with cells treated in late G_1 the frequency of chromatid aberrations is significantly higher than untreated controls (17, 15). The frequencies of chromatid aberrations in cells treated in S (identified as labeled cells with ^{3}H-thymidine) is much higher than for any stage of G_1 (17). Thus the frequency of aberrations is greatly influenced by the interval between treatment and the time of DNA synthesis.

An additional consequence of this is that, particularly following an acute exposure, the proportion of cells, in an asynchronous population, in which there is a reasonably high probability of inducing an aberration will be dependent on the duration of the cell cycle, and the relative lengths of the different stages. For example, a spermatogonial stem cell has a very long cell cycle, with the S phase contributing a rather small fraction of this, and thus there will be only a small fraction of the cells which have a reasonably high probability of producing aberrations. In a large proportion of the cells, the probability of inducing an aberration is negligible. In differentiating spermatogonia, with a much shorter cell cycle, there will be a much larger proportion of the cells in or close to the S phase at the time of an acute exposure, and thus higher aberration frequencies would be expected compared to a stem cell population. A similar argument can apply to chronic exposures, although it will be more complicated.

The effect of duration of the cell cycle, and the relative proportion of cells in the different stages of the cell cycle, can in part account for differences in sensitivity to aberration induction in different species, and in different cell types within a species.

These points are further exemplified by germ cell studies where differences in sensitivities are observed. In experiments of Brewen and Payne (2, 3), superovulated female mice were treated with MMS or triethylenemelamine (TEM) and the oocytes observed at metaphase I, or the female complement was analysed at the first cleavage division, after the treated females were mated to untreated males. No aberrations were observed in metaphase I oocytes following MMS treatment, and a small but significant yield was observed following treatment with TEM.

These observations showed that, as expected, aberrations could not be produced from MMS treatment as there was no DNA synthesis phase between treatment and observation. The aberrations observed after TEM treatment suggested that they could be produced in the absence of DNA synthesis, albeit at a low frequency, but this too is not unexpected in the light of results discussed in the next section, that indicate that very few chemicals can be truly described as S-dependent.

However, when the female complement was analysed at first cleavage division, a high frequency of chromatid aberrations was observed for both MMS and TEM. The fact that the cells passed through the post-fertilization S-phase before observation could account for this. When the interval between treatment and mating was longer than 6-1/2 days for MMS and 10-1/2 days for TEM the frequency of aberrations was greatly reduced compared to earlier mating intervals. Thus the longer the time between treatment and the pronuclear S-phase the more DNA damage that could be repaired before the S-phase, and consequently the lower the probability of inducing an aberration. The fact that high aberration frequencies were seen for longer mating intervals after TEM treatment suggests that the lesions in the DNA produced by TEM are longer-lived. This is indicated by the results of Matter and Jaeger (11) and Burki and Sheridan (5), who found chromatid aberrations at later cleavage divisions following treatment of post-spermatogonial cells. These aberrations, since they were of the chromatid type, must have been produced during the cell cycle at which observations were made.

It is also possible that the influence of time available for repair prior to an S-phase could offer an explanation of the results of Generoso et al (8) that were interpreted as showing that different strains of female mice have different capabilities to repair damage induced by iso-propyl methanesulfonate (IMS) in post-spermatogonia cells. In the experiments of Generoso et al (8) it was shown that the frequency of dominant lethal mutations from treatment of males with IMS was dependent on female stock used, ie. it was highest with T-stock females and lowest with (C3HxC57BL)F_1 females. Similar but less clear cut differences were observed following treatment of males with TEM, benzo(a)pyrene (BzaP) and ethyl methanesulfonate (EMS). The authors concluded that the different female stocks had different repair capabilities with the T-stock being repair deficient and (C3HxC57BL)F_1 being repair proficient. However, there is an alternate explanation or an additional factor to be considered.

The DNA damage induced in male postmeiotic cells is repaired after fertilization, probably by enzymes present in the egg since the cytoplasmic contribution from the spermatozoa is minimal (8). Thus, for an agent that produces the majority of its aberrations during the S-phase, the frequency of aberrations (and dominant lethality) will be influenced by the amount of damage still present in the DNA at the time of replication. This can indeed be influenced by the proficiency of repair, but it can, as discussed above, be influenced by the time available for repair before the cells reach the S-phase. If the different female stocks used in Generoso et al's experiments have different durations of time between fertilization and the S-phase, then the stock showing the highest frequency of dominant lethals, and, incidentally, the highest frequency of chromosome aberrations at the first cleavage metaphase, would have the shorter interval between fertilization and replication. The converse would be true for stocks showing low dominant lethal frequencies.

The fact that EMS, TEM and BzaP produced much smaller differences between stocks could be due to the fact that repair of damage induced by these agents is more rapid than for IMS, and thus would be less influenced by the duration of G_1, since even in the case of a shorter G_1, a fairly high proportion of damage could be repaired before the S phase.

In the case of x-rays, where repair of damage is generally fast, it would be expected that there would be a very small effect of different lengths of G_1 on the frequency of induced aberrations and dominant lethality, provided that the duration of the shorter G_1 is not very short (less than 2 hours). Generoso et al reported that there was no difference in dominant lethality following x-irradiation of the male, whichever female stock was used (8).

It is appreciated that this is a hypothesis, but it does indicate that considerations of the time available for repair of damage between treatment and the S-phase are perhaps also important to take into account.

3. <u>The influence of the rate of DNA repair</u>. In order to produce an interchange it is necessary to have two lesions available to interact. This will require that they be produced close together in time and space (ie. within the rejoining distance). It also seems very probable that the production of an aberration occurs during the repair of DNA damage, and is the result of mis-repair (7). It can also be argued, based on Revell's hypothesis (16, 1), that similar factors apply to all aberration types, ie. if they are indeed the result of an exchange. For the purpose of this discussion, I will consider interchanges.

The frequency of interchanges will be greatly influenced by the rate of repair of DNA damage. If repair is rapid then there will be a greater probability of producing an aberration than when the repair is protracted, since the probability of coincident repairing regions, able to interact is greater. This can be demonstrated by our own results, comparing the aberration frequencies induced in G_1 human lymphocytes by x-rays and 4NQO (12, 15). These experiments utilise the observation that the repair of DNA damage that can be converted into chromosome aberrations can be inhibited by cytosine arabinoside (ara-C). This inhibitory effect is at the resynthesis step of excision repair, thus causing a single strand gap to be produced at a repairing region. The inhibition of repair by ara-C can be reversed by deoxycytidine.

If G_0 and G_1 lymphocytes are x-irradiated with a dose of 200 rads, chromosome-type aberrations are observed. If the cells are incubated with ara-C for 1, 2, or 3 hours after irradiation, and then with deoxycytidine to reverse the inhibition. Greatly increased frequencies of chromosome-type aberrations are observed (12). This is interpreted as being due to the fact that single-strand gaps are accumulated by the inhibition of repair with ara-C and on reversing the inhibition, these gaps interact to produce aberrations. By increasing the probability of having two lesions available to interact, the probability of producing an aberration is increased.

If G_1 lymphocytes (9 hours after stimulation with phytohemagglutinin) are treated with 4NQO (1 x 10^{-6} M) no increase in aberrations compared to untreated controls is observed. However, if the cells are incubated with ara-C for 6 hours after 4NQO treatment, aberrations are induced, and they are of the chromosome-type, indicating that they are produced in G_1 (15). This suggests that the damage induced by 4NQO is repaired rather slowly in G_1 and that the probability of having two lesions available to interact to produce an aberration is low, and hence no aberrations are observed. If this probability is increased by inhibiting repair with ara-C, aberrations can be produced. The fact that they are of the chromosome type shows that they can be produced in the absence of an S-phase. Similar results were obtained following MMS treatment of G_1 lymphocytes. If the repair of damage is very slow or non-existent during G_1, it would be expected that, even in the presence of ara-C, chromatid-type aberrations would be induced as a consequence of replication of damaged DNA, and the subsequent conversion into an aberration. This seems to be what happens following treatment of G_1 cells with mitomycin C (Preston, unpublished results).

Thus the rate of repair of the damage that can be converted into chromosome aberrations can influence both the frequency and spectrum of aberrations. In the case where repair is rapid, as for x-rays, chromosome-type aberrations can be produced in G_1 and chromatid-type aberrations in S and G_2. Furthermore the aberration frequency can be affected by the **rate** of repair such that if the rate of repair of x-ray-induced damage, that is converted into aberrations, is increased then the aberration frequency would be expected to be increased. This is the explanation given for the increase in x-ray-induced aberrations in lymphocytes from Down syndrome individuals compared to that in normal lymphocytes (13).

If the rate of repair is slower, as for 4NQO- and MMS-induced damage, the probability of inducing aberrations in G_1 and G_2 is very low. Furthermore, the aberrations observed will be of the chromatid-type since they will usually only be produced when the cell is in S at the time of treatment or passes through an S phase between treatment and observation. The probability of producing an aberration appears to be increased at the time of replication, or shortly after replication, as a consequence of replicating damaged DNA.

Thus the effectiveness of different chemicals to produce chromosome aberrations in the same germ cell stage can be influenced both by the nature of the initial damage and the rate of its repair. The rate of repair determines the probability of producing an aberration prior to replication (either in G_1 or in G_2 prior to the next replication) and also the proportion of damage remaining in the DNA at the time of replication, and hence the probability of inducing an aberration in the S-phase. The differential response to the induction of aberrations in different germ cell stages of the same species, or the same germ cell stage in different species, can also be influenced by the relative rates of repair of the damage resulting in aberrations.

The three sections above discuss factors that can influence the aberration frequency observed in cells exposed to physical or chemical agents. However, these factors cannot be considered in isolation since they are clearly interactive. There are also other possible causes of variations in aberration frequency, such as the consequences of the specific products of the interaction of an agent with DNA, and the amount and relative proportions of these different products with different agents or in different cell types. Factors such as these are inherent to the discussion presented, and will not be considered separately here.

This paper clearly presents an abbreviated account of the factors that can give rise to real or apparent differences in response of different germ cell stages to the induction of chromosome aberrations. It is hoped that this abbreviation does not detract too much from the points discussed.

REFERENCES

(1) Brewen, J. G. and Brock, R. D. *Mutation Res.* 6, 245-255 (1968).
(2) Brewen, J. G. and Payne, H.S. *Mutation Res.* 37, 77-82 (1976).
(3) Brewen, J. G. and Payne, H. S. *Mutation Res.* 50, 85-92 (1978).
(4) Brewen, J. G., Payne, H. S., and Preston, R. J. *Mutation Res.* 35, 111-120 (1976).
(5) Burki, K. and Sheridan, W. *Mutation Res.* 52, 107-115 (1978).
(6) Cattanach, B. M. and Williams, C. E. *Mutation Res.* 13, 371-375 (1971).

(7) Evans, H. J. In "Progress in Genetic Toxicology" (Eds. Scott, D., Bridges, B. A., and Sobels, F. H.) Elsevier/North Holland, Amsterdam pp 57-74 (1977).

(8) Generoso, W. M., Cain, K. T., Krishna, M., and Huff S. W. *Proc. Nat. Acad Sci (U.S.A.)* 76, 435-437 (1979)

(9) Kihlman, B. A. "Actions of Chemicals on Dividing Cells" Prentice-Hall, Englewood (1966).

(10) Leonard, A., Jeknudt, G., and Linden, G. *Mutation Res.* 13, 89-92 (1971).

(11) Matter, B. E. and Jaeger, I. *Mutation Res.* 33, 251-260 (1975).

(12) Preston, R. J. *Mutation Res.* 69, 71-79 (1980).

(13) Preston, R. J. *Environmental Mut.* 3, 85-89 (1981)

(14) Preston, R. J. Au, W., Bender, M. A., Brewen, J. G., Carrano, A. V., Heddle, J. A., McFee, A. F., Wolff, S. and Wassom, J. S. *Mutation Res.* (in press).

(15) Preston, R. J. and Gooch, P. C. *Mutation Res.* (in press).

(16) Revell, S. H. Advances in Radiation Biology 4, 367-416 (1974).

(17) Sasaki, M. S. *Mutation Res.* 20, 291-293 (1973).

(18) Searle, A. G. and Beechey, C. V. *Mutation Res.* 171-186 (1974).

Genetically Determined Chromosomal Instability in Man

Masao S. Sasaki

Radiation Biology Center, Kyoto University, Kyoto, Japan

ABSTRACT Chromosome aberrations can arise as a consequence of DNA-mutagen interaction. To make an insight into the genetic factors involved in chromosome aberration formation and thier relevance to the etiology of cancer, the chromosomal instability has been studied in several human monogenic syndromes characterized by high cancer propensity. Some recessive diseases are unequivocally associated with abnormalities in DNA metabolisms to cope with DNA-mutagen interaction. Some dominant diseases possess a function to promote chromosome rearrangements while they manifest as deletion type mutations in initiating cancer. The observations strongly point to the involvement of modification of gene expression by deletion, genome rearrangement or DNA insertion element in the pathway to cancer.

INTRODUCTION

Chromosome aberration is, in itself, a major heritable genetic damage, and its genesis in germ line cells constitutes one of the most important topics in the research on genetic risk evaluation from human exposure to environmental mutagens. In addition, a large body of literature exists, pointing to the possible involvement of chromosome aberration formation in mutation and maturation toward cancer (15), and provokes the cytogenetic basis of the somatic mutation theory of carcinogenesis.

There are many human monogenic syndromes with high cancer propensity which show inherent tendency toward chromosomal instability (9). The affected persons display an increased amount of chromosome breakage and rearrangements as well as abnormal susceptibility to chromosome breakage by chemical and physical mutagens. The investigations of the chromosomal instability in such procancer syndromes provide an important key to the understanding of origin and nature of chromosome aberrations and their relevance to the environmental carcinogenesis.

GENETICALLY DETERMINED CHROMOSOMAL INSTABILITY

Genetic diseases associated with an increased chromosomal instability are listed in Table 1. In the affected persons, high frequency of spontaneous chromosome aberrations have been demonstrated in their bone marrow cells, peripheral blood lymphocytes or cultured skin fibroblasts. Some (asterisks in the table) are called "chromosome breakage syndrome" (8) and the frequency of spontaneous chromosome aberrations provides one of the important diagnostic criteria.

Of significant feature is that these disease conditions are all characterized by a high cancer propensity, and the affected persons develop leukemia and other cancers with unusually high frequency. In these disease conditions, chromosome aberrations are primarily of the chromatid type and indicate that the chromosome aberrations in these diseases arise in connection with DNA replication, and

Table 1
Genetic Diseases Associated with Chromosomal Instability

Disorders	Heredity
Fanconi's anemia*	a.rec.
Ataxia telangiectasia*	a.rec.
Bloom's syndrome*	a.rec.
Werner's syndrome	a.rec.
Chediak-Higashi syndrome	a.rec.
Kostmann's agranulocytosis	a.rec.
Glutathione reductase deficiency anemia	a.rec.
Blackfan-Diamond syndrome	a.rec.
Friedreich Ataxia	a.rec.?
Dyskeratosis congenita	X,rec.
Incontinentia pigmenti*	X,dom.
Scleroderma	Multifactorial
Pernicious anemia (hereditary)	?

strongly point to the involvement of some inherent metabolic defects which directly or indirectly relate with the DNA metabolisms. In Fanconi's anemia (FA) and ataxia telangiectasia (AT), DNA repair deficiencies (7, 13, 19, 20) or defective replication response to DNA damage (6, 17) have been suggested. However, their relevance to the increased spontaneous chromosome aberrations is not unequivocally understood. Bloom's syndrome (BS) is unique in that it shows manifold increase in sister chromatid exchange (SCE) (3). Abnormality in DNA replication has been suggested to be an integral component of the high SCE rate and increased chromosomal instability of this syndrome (10).

CHROMOSOME MUTABILITY

In addition to the spontaneous mutability, the susceptibility to mutagens and carcinogens are particularly interesting in view of the somatic mutation theory of carcinogenesis. Table 2 compares the susceptibility to chromosome aberration formation by some mutagens and mutagenic carcinogens as tested in cultured lymphocytes from patients. The chromosomal susceptibilities were studied after treatment with γ-ray, UV or such chemicals as 4-nitroquinoline 1-oxide (4NQO), decarbamoyl mitomycin C (DCMMC), mitomycin C (MMC), methylmethanesulfonate (MMS) or N-methyl-N'-nitro-N-nitrosoguanidine (MNNG), and expressed in the table as relative values as compared with the sensitivity of normal cells. Xeroderma pigmentosum (XP) is highly sensitive to UV, 4NQO and DCMMC, while FA is speci-

Table 2

Chromosomal Susceptibility to Mutagens and Carcinogens

	Relative susceptibility to chromosome breakage by:						
	γ-ray	UV	4NQO	DCMMC	MMC	MMS	MNNG
Xeroderma pigmentosum							
Classical	0.9	16.6	8.7	15.6	0.9	1.0	1.2
Variant	1.0	1.1	1.0	1.2	1.0	1.1	---
Fanconi's anemia	1.1	2.8	1.1	2.9	32.3	1.1	1.0
Ataxia telangiectasia	2.2	---	1.1	---	1.0	1.0	---
Werner's syndrome	0.8	1.0	1.0	---	0.9	0.8	---
Progeria	1.1	---	1.1	---	1.3	0.9	---
Incontinentia pigmenti	---	---	3.2	---	0.7	3.7	---
Dyskeratosis congenita	1.1	1.0	1.0	---	1.0	1.0	---
Control	1	1	1	1	1	1	1

fically sensitive to MMC, and moreover AT is highly radiosensitive. The mutagen specific sensitivity suggests that some inherent defects to cope with specific type of DNA damage are responsible for the hypersensitivity to these mutagens. In XP, deficiency in endonuclease-mediated nucleotide excision repair has been unequivocally demonstrated (4). The deficiency in the repair of DNA interstrand crosslinks has been suggested in FA (7, 9), and defective repair of radiation induced DNA damage (13, 19) and defective replication response to radiation damage (6, 17) have been observed in AT. Spontaneous chromosome aberrations in XP are within limits of normal range. The hypersensitivity to chromosome breakage by UV and UV-mimetic chemicals means that XP is a conditional chromosome breakage syndrome. In recessively transmitted diseases, homozygotes do not significantly contribute to the cancer incidence in the population as a whole, but the chromosome instability in these procancer diseases indicates the importance of chromosome aberrations in the development of cancer as well as the importance of the genetic control mechanisms of DNA-mutagen interaction and its biological consequences.

PROBLEMS ASSOCIATED WITH HETEROZYGOTES

The heterozygotes for recessive genes are relatively common, and thus in view of the ecogenetics in environmental carcinogenesis, a possible contribution of heterozygotes to cancer incidence in general population is of particular interest. Swift (25) has proposed that the FA, AT and BS heterozygotes may contribute to familial aggregation of cancer. Actually, there are some indications that heterozygotes differ from normal cells in their response to mutagens. Cells from some AT heterozygotes show moderate hypersensitivity to X-rays (18), and FA heterozygotes were hypersensitive to the induction of SCE by nitrogen mustard (1). It should be also noted that many mutagens can induce somatic recombination by isolocus chromatid exchange between homologous chromosomes (24), or by plyploidization-segregation parasexual cycle (21), which will give rise to the cancer-prone new homozygous recombinant cells in the soma of heterozygotes. However, problems associated with heterozygotes are largely unexplored and need further researches.

NATURE OF DOMINANTLY EXPRESSED PROCANCER MUTATIONS

Dominant mutations are particularly interesting in that a single inherent mutation is expressed to initiate cancer in particular type of tissue at particular stage of the development. None of them have ever been identified as an alteration in biochemical property. According to Knudsn's "two mutation model" for the induction of cancer (14), in dominant disease, there is one mutation already inherited in the germ lines and a second somatic mutation then results in the tumor.

As to the nature of germinal mutation, association of certain inborn chromosomal abnormalities to specific type of embryonal tumors is interesting. In a series of cytogenetic survey of embryonal tumors, we found three patients with retinoblastoma associated with a deletion of a band q14 of chromosome 13 (13q-) and one associated with 13/X translocation. We also found a case with aniridia-associated Wilms' tumor due to deletion of a band p13 of chromosome 11. These observations indicate that the structural (interstitial deletion) or functional (inactivation) deletion at particular chromosome loci leads to a specific type of tumors.

In the hereditary form of these diseases and other dominant diseases, no such constitutional chromosome abnormalities were detected. However, attention has recently been given to the emergence of karyotypically aberrant cells in cultured skin fibroblasts from unaffected skin of patients or carriers of some dominant diseases of high cancer risks (23). Such aberrant cells are non-consti-

Table 3

Autosomal Dominant Diseases Exhibiting Non-constitutional Chromosome Rearrangements in Cultured Skin Fibroblasts

Disorders	References
Familial polyposis coli	27
Gardner's syndrome	27
Peutz-Jeghers' syndrome	27
Nevoid basal cell carcinoma syndrome	11
Prokeratosis of Mibelli	26
Retinoblastoma	22
Medullary thyroid cancer syndrome (Type 2B)	23
Hereditary colorectal carcinoma	22
Familial childhood leukemia	23

tutional mutants and usually invole translocations and inversions. This is unique in that, unlike those in recessively transmitted diseases, they emerge without any indication of an elevated level of spontaneous chromosome breakage and that in some disease conditions, the chromosomal sites involved in the rearrangements tend to cluster to a particular part of a particular chromosome. The presence of such mutant karyotypes in cultured skin fibroblasts has been found in various disease conditions of dominant mode of inheritance (Table 3). The observations indicate that some dominant mutations promote chromosome rearrangements in the somatic cells.

Table 4 shows the results of observations on the frequency of non-constitutional mutant karyotypes in the cultured skin fibroblasts from patients with retinoblastoma. Although the appearance of mutant karyotypes in sporadic cases not associated with 13q- anomaly were variable, but it is interesting that such mutant karyotypes were not found in patients associated with 13q- while they are present in patients with hereditary retinoblastoma. This

Table 4

Frequencies of Cells with Non-constitutional Mutant Karyotypes in Cultured Skin Fibroblasts from Patients with Retinoblastoma (B: bilateral; U: unilateral)

Heredity	Patients	Tumor	Constitutional karyotypes	Mutant karyotypes
Sporadic	GM1142	B	46,XX,del(13)	0 (%)
	NaYo	B	46,XX,del(13)	0
	NiSa	B	46,XY,del(13)	0
	YaMa	B	46,XY/46,XY,del(13)	0
	OhNa	B	46,X,t(13;X)	0
	MaMa	B	46,XX	38.0
	TaRi	B	46,XX	10.0
	YoHi	B	46,XY	4.0
	IuRi	B	46,XX	0
	TaTo	B	46,XY	0
	NaNo	U	46,XY	70.0
	TaMi	U	46,XX	0
	YaMe	U	46,XX	0
Hereditary	Ito-Mo	U	46,XX	8.3
	ItoSi	B	46,XY	8.1
	As-Mo	U	46,XX	18.0
	AsKe	B	46,XY	2.0

indicates that germinal mutation responsible for hereditary retinoblastoma has essentially the same function (possibly deletion effect) as 13q- in initiating tumors, but differs in its ability to promote chromosome rearrangements. Such mutator characteristics of dominantly expressed procancer mutant genes readily lead us to the speculation that they might be analogous to the DNA insertion elements or gene control elements.

In bacteria, yeast, maize and Drosophila, the insertion elements have been demonstrated to act as agents for large scale rearrangements of DNA which is often site specific, and also act as control elements for the expression of host genes (2, 5, 12, 16). The insertion elements will interfere with the expression of the downstream portion of the host genome by promoter-related flip-flop mechanisms or transcriptional competition, and hence result in turining off (functional deletion) of the gene expression (16). Such promoter mechanism of modification of gene expression may also be possible by rearrangements of DNA, and can explain the involvement of chromosome rearrangements in the causation of cancer.

All these evidence obtained from research on chromosomal instability are pointing to the importance of the modification of gene expression afforded by deletion, genome rearrangements or DNA insertion mutations in the pathway to cancer. Such genomic changes may result from DNA-mutagen interaction as well as an interaction with viral genome. There is ample evidence for that such interactions are also under the genetic control, and constitute one of the most significant topics in the research on ecogenetics of environmental carcinogenesis.

ACKNOWLDGMENTS

I am grateful to Drs. A. Kaneko, H. Tanooka, Y. Ejima, J. Utsunomiya, Y. Tsunematsu, Y. Sato, G. Akaishi, T. Sekine, I. Tsukimoto and S. Takai for their cooperation, and to Drs. H. Takebe and S. Kondo for discussion. This work was supported by Grants-in-Aid for Cancer Research, and for Environmental Sciences, from the Ministry of Education, Science and Culture, and by the Subsidy for Cancer Research from the Ministry of Health and Welfare.

(1) Berger, R., Bernheim, A., le Coniat, M., Vecchione, D., and Schaison, G. Cancer Genet. Cytogenet. 2, 259-267 (1980).
(2) Calos, M.P., and Miller, J.M. Cell 20, 579-595 (1980).
(3) Chaganti, R.S.K., Schonberg, S., and German, J. Proc. Nat. Acad. Sci. USA 7, 4508-4512 (1974).
(4) Cleaver, J. Proc. Nat. Acad. Sci. USA 63, 428-435 (1969).
(5) Cohen, S.N. Nature 263, 731-738 (1976).
(6) De Wit, J., Jaspers, N.G.J., and Bootsma, D. Mutation Res. 80, 221-226 (1981).
(7) Fujiwara, Y., Tatsumi, M., and Sasaki, M.S. J. Mol. Biol. 113, 635-649 (1977).
(8) German, J. Birth Def. Orig. Art. Ser. 5, 117-131 (1969).
(9) German, J. Prog. in Medical Genet. 3, 61-101 (1972).
(10) Hand, R., and German, J. Proc. Nat. Acad. Sci. USA 72, 758-762 (1975).
(11) Hepple, R., and Hoehn, H. Clin. Genet. 4, 17-24 (1973).
(12) Howe, M.M., and Bade, E.G. Science 190, 624-632 (1975).
(13) Inoue, T., Hirano, K., Yokoiyama, A., Kada, T. and Kato, H. Biochim. Biophys. Acta 479, 497-500 (1977).
(14) Knudson, A.G. Proc.Nat.Acad.Sci.USA 68, 820-823 (1971).
(15) Levan, G. and Mitelman, F. Chromosomes Today 6, 363-371 (1977).
(16) Nevers, P. and Saedler, H. Nature 268, 109-115 (1977).
(17) Painter, R.B., and Young, B.R. Proc. Nat. Acad. Sci. USA 77, 7315-7317 (1980).
(18) Paterson, M.C. In "DNA Repair Mechansisms", Hanawalt, P.C. et al. (eds.), Academic Press, New York, pp. 207-218 (1978).
(19) Paterson, M.C., Smith, B.P., Lohman, P.H.M., Anderson, A.K., and Fishman, L. Nature 260, 444-447 (1976).
(20) Sasaki, M.S. Nature 257, 501-503 (1975).
(21) Sasaki, M.S. Mutation Res. 54, 224-225 (1978).
(22) Sasaki, M.S., and Ejima, Y. Gann Monograph on Cancer Research (1981), in press.
(23) Sasaki, M.S., Tsunematsu, Y., Utsunomiya, J., and Utsumi, J. Cancer Res. 40, 4796-4803 (1980).
(24) Shaw, M.W., and Cohen, M.D. Genetics 51, 181-190 (1965).
(25) Swift, M. Birth Def. Orig. Art. Ser. 12, 133-144 (1976).
(26) Taylor, A.M.R., Harnden, D.G., and Fairburn, E.A. J. Nat. Cancer Inst. 51, 371-378 (1973).
(27) Utsunomiya, J. Pathology of Carcinogenesis in Digestive Organs, Farber, E., et al. (eds.), Univ. Park Press, Baltimore, pp. 305-321 (1977).

Formation of Chromatid-Type Aberrations from Chemically Induced DNA Lesions: Information Obtained with Hydroxyurea and Caffeine

B. A. Kihlman, H. C. Andersson, K. Hansson, B. Hartley-Asp,[1] and F. Palitti[2]

Department of General Genetics, University of Uppsala, Uppsala, Sweden

ABSTRACT Root-tip cells of *Vicia faba* and human lymphocytes were exposed to mutagenic agents with S-dependent and S-independent effects and post-treated with hydroxyurea (HU) and/or caffeine. Most of the observations reported here were obtained in experiments with S-dependent agents: the herbicide maleic hydrazide in *Vicia* and the trifunctional alkylating agent thiotepa in lymphocytes. The HU treatments were given during the last 3 hours before fixation/harvesting in both materials. The caffeine treatments were given during S in *Vicia* and during the last 3 hours before harvesting in lymphocytes. In both materials the frequency of the induced chromatid aberrations was greatly enhanced by the post-treatments. The results suggest that a large proportion of the chromatid aberrations are formed during G_2-prophase from DNA lesions induced before replication of the chromosomal sites involved.

[1] Present address: Research Laboratories, AB Leo, S-251 00 Helsingborg, Sweden

[2] Present address: Centro di Genetica Evoluzionistica, Istituto di Genetica, Università di Roma, I-00185 Rome, Italy

INTRODUCTION: PRODUCTION OF CHROMATID-TYPE ABERRATIONS
BY S-DEPENDENT AND S-INDEPENDENT MECHANISMS

It is customary to distinguish between two main types of mechanism for the production of chromatid aberrations. When induced by ultraviolet light, alkylating agents and many other chemical substances, aberrations arise by an S-dependent mechanism. This means that the DNA lesions induced by these agents do not give rise to chromatid aberrations before they have become replicated. Thus, DNA lesions induced by alkylating agents in G_2 do not result in chromatid aberrations until the second cell cycle after their induction. Chromatid aberrations may, however, be observed in the first mitosis after treatment when the lesions have been induced during G_1 or S in unreplicated chromosomes or chromosome sites. This S-dependence could mean that aberrations are formed as mistakes when chromosome segments with lesions are replicated. Evans and Scott (4) have coined the term "misreplication" for this mechanism of forming aberrations. Another possibility is that the lesions must be replicated before they can give rise to chromatid aberrations. The aberration would not then be formed until some time after replication. The aberrations induced by an S-dependent mechanism are always of the chromatid type.

In contrast to UV and alkylating agents, ionizing radiation and certain antibiotics, such as those belonging to the bleomycin group, induce aberrations by an S--independent mechanism. In the first mitosis after treatments with these agents, chromatid-type aberrations are found in cells exposed during S and G_2. When G_1 cells are exposed to these agents, chromosome-type aberrations are normally formed. Apparent sub-chromatid-type aberrations are found in cells exposed during prophase. Aberrations induced by S-independent agents are believed to be formed when the processes for repair of DNA damage fail or make mistakes ("misrepair" according to Evans (3)).

This paper deals with the question whether there is any tendency for chromatid-type aberrations to be formed at a particular stage of the cell cycle from DNA lesions induced by S-dependent agents. For these agents we know that the induced DNA lesions are long-lived in unreplicated chromosomes and may persist from one cell cycle to another. But what happens when the lesions are replicated? Are the aberrations formed in connection with the replication process or later during the cell cycle?

Our approach to this problem has been to use agents with the capacity of modifying the induced aberration frequency when given as post-treatments. Agents capable of reducing the induced aberration frequency would be ideal for our purpose since chromatid aberrations, particularly exchanges, are irreversible events. Once formed, no post-treatment can make them revert to the normal configuration. However, although several types of post-treatments have the capacity to potentiate the frequency of chromatid aberrations, none are known to reduce the induced aberration frequency.

In the absence of methods that could provide the final solution to the problem, we have to be content with the information that can be obtained with potentiating agents. By using this type of agent, we should be able to find out how late during the cell cycle it is possible to influence the formation of chromatid-type aberrations from the induced DNA lesions. We should further be able to tell whether there is any stage of the cell cycle during which the process of aberration formation is particularly sensitive to the effect of the potentiating agents. In our studies of these questions, we have been using hydroxyurea (HU) and caffeine as potentiating agents. Most of the observations reported here concern the effect of S-dependent agents.

RESULTS AND DISCUSSION

Experiments with hydroxyurea Hydroxyurea (HU) represents a group of potentiating agents which have in common the ability to inhibit the synthesis of immediate DNA precursors. It is an efficient inhibitor of the enzyme ribonucleotide reductase which catalyzes the reduction of riboside diphosphates to the corresponding deoxyriboside diphosphates (12). Thus, in the presence of HU, the formation of the immediate precursors of DNA is strongly suppressed. Since this seems to be the most characteristic if not the only action of HU at the molecular level, it is a reasonable assumption that the inhibitory action of HU on the synthesis of the immediate precursors of DNA is responsible for its action at the chromosomal level.

We have found that post-treatments with HU strongly potentiate the frequency of chromatid-type aberrations induced by mutagenic agents (7, 8, 10, 11, 12). The HU effect has been demonstrated in root tips of *Vicia faba*, in Chinese hamster cells (Cl-1 and CHO lines) and in human lymphocytes. A potentiating effect of HU post-treatments has been observed both for chromosome damage induced by agents with S-independent effects and for chromosome damage induced by agents with S-dependent effects. In addition to an increase of the total frequency of aberrations, the HU post-treatment causes a change in the proportion of the various types of aberration. The frequency of certain types of chromatid aberration such as gaps and breaks is dramatically increased, whereas the frequency of exchanges is only slightly affected. In order to be effective, the HU post-treatment has to be given late in interphase. In root tips of *Vicia faba* the strongest potentiation is obtained 2-5 hours before metaphase, that is, in the second half of G_2.

These findings show that lesions induced in S by agents with S-independent effects and in G_1 and S by agents with S-dependent effects may not be transformed into chromosomal aberrations until late G_2. From this it follows that aberrations produced by agents with S-dependent effects must not necessarily be formed by misreplication during S. The data further show that the aberrations obtained by the combined mutagen and HU treatment are formed late in interphase, possibly at a time when the process of chromosome condensation has already begun. As alredy mentioned, agents that produce the HU-type of potentiating effect on induced chromosome damage have in common the ability to inhibit deoxyribonucleotide synthesis. We believe, therefore, that these agents potentiate the frequency of induced chromosome damage because they suppress some form of DNA synthesis that takes place late in G_2 when chromosomal DNA has been damaged by mutagenic agents. It seems likely that the DNA synthesis affected is required for the repair of the induced DNA damage and that chromatid-type aberrations arise when this repair process fails or makes mistakes.

Experiments with caffeine A potentiating effect of the same magnitude or even larger than that produced by HU can be obtained by post-treatments with caffeine. However, at the chromosomal level the caffeine effect differs in many respects from that of HU. At the molecular level the mode of action of caffeine is not well understood, but it is clear that it does not produce its effect by the same mechanism as HU.

In comparison with the HU effect, the caffeine effect is more variable. It is frequently different in different materials and also different for agents with S-independent and S-dependent effects. In contrast to HU, caffeine potentiates all types of chromatid aberrations, breaks as well as exchanges, that is, the caffeine effect is mainly quantitative.

Most striking is the caffeine effect for S-dependent chromosome damage. When given as a post-treatment to agents producing this type of effect, caffeine causes a 3- to 9-fold increase in the frequency of aberrations. Pre-treatments are ineffective. The enhancement of the induced chromosome damage is paralleled by a similar effect of caffeine on induced cell killing. However, these effects of caffeine are not observed in all types of cell. A strong potentiation by caffeine has been found in all plant cells studied and in cultured mammalian cells of rodent origin. In contrast, in normal human fibroblasts the lethal and chromosome-damaging effects of S-dependent agents are only slightly, if at all, affected by post--treatments with caffeine (9 and 20).

Studies on the enhancement by caffeine of induced cell killing in rodent cells and of induced chromosome damage in root-tip cells have indicated that in order to be effective, caffeine must be present during the first S-phase after mutagen treatment (19, 24). At the molecular level it has been demonstrated by many workers that the DNA newly synthesized after mutagen treatment is of lower molecular weight than that newly synthesized in untreated cells (for references, see 9 and 20). Apparently gaps are formed in the newly-synthesized DNA strands opposite base damage in the parental (template) strands. Caffeine inhibits the filling-in of the gaps in some but not all types of cell (for references, see 14 and 20). The inhibitory effect of caffeine on the filling-in of the post-replicative gaps in the newly--synthesized DNA was found to correlate rather well with the potentiating effect of caffeine on induced chromosome damage and cell killing. It was, therefore, suggested that the inhibitory effect of caffeine on gap filling is responsible for the potentiating effect of caffeine on induced chromosome damage and cell killing (13, 21).

However, it has been demonstrated both in plant and

in animal cells that caffeine potentiation of X-ray-induced chormosome damage and cell killing is most marked for cells irradiated late in the cell cycle (1, 13). A potentiating effect of caffeine post-treatments on damage induced in G_2 cannot easily be explained as a result of an inhibitory effect of caffeine on the filling-in of post-replicative gaps in newly-synthesized DNA. And quite recently we have obtained results in experiments with human lymphocytes which have forced us to revise considerably our earlier conception of the caffeine effect (12).

We had been comparing the potentiating effects of caffeine and HU on chromosome damage induced in G_2 by S-independent agents (X-rays and Streptonigrin) and found these effects to be remarkably similar. Both the caffeine and the HU post-treatments caused about a 3--fold increase in the frequency of aberrations. The aberrations consisted mainly of chromatid breaks. The post--treatments were given together with colcemid during the last 3 hours before harvesting the lymphocyte cultures. In order to be able to distinguish between the effects of caffeine and HU, we gave the same type of post-treatments to cells that had been exposed to the trifunctional alkylating agent thiotepa during G_1, 6 hours after stimulation of the cultures. On the basis of previous results obtained by us and other workers, we did not expect any effect of caffeine in this case for two reasons: 1) in order to enhance chromosome damage induced by agents with S-dependent effects, caffeine should be present during S and 2), the chromosome damage induced in normal human cells by agents with S-dependent effects was supposed to be comparatively insensitive to post--treatments with caffeine. However, to our great surprise we found that the frequency of thiotepa-induced chromatid aberrations was increased by post-treatments with caffeine during the last 3 hours before harvesting to the same extent as by a similar post-treatment with HU.

To make sure that caffeine exerted its effect in G_2 cells, autoradiographs were prepared from cultures that had been exposed to tritiated thymidine simultaneously with caffeine. Of the cells that had been exposed to both thiotepa and caffeine, 6 % were labelled at metaphase in late replicating regions of the chromosomes. The frequency of aberrations was 0.6 per labelled and 0.7 per unlabelled metaphase. After exposure to thiotepa alone, no labelled metaphases were found and the frequency of aberrations per unlabelled metaphase was 0.1. Thus, there can be no doubt that caffeine exerted its potentiating effect in G_2 cells. Since this caffeine potentiation could hardly be explained as a result of an inhibitory effect of caffeine on the filling--in of post-replicative gaps, it was necessary to look for another explanation.

There have been several reports in the literature that caffeine reduces the inhibitory effects of chemical and physical agents on DNA synthesis (2, 6, 15, 17, 20, 21, 25). It has further been found by many authors that the radiation-induced G_2-delay is counteracted by caffeine (16, 22, 23, 25, 26). On the basis of these findings, Painter suggested that caffeine enhances cell killing and chromosome damage because it prevents cells from going through delays that would allow them to repair DNA damage before it can be expressed (17, 18). The idea that caffeine enhances cell killing because it counteracts an inhibitory effect on DNA synthesis is supported by results recently reported by Gascoigne et al. (5). In the experiments of these authors, Baby Hamster Kidney (BHK 21/C13) cells were exposed first to N-methyl-N-nitrosoguanidine (MNNG) and then post-treated with caffeine. The caffeine post-treatments enhanced the MNNG-induced cell killing and counteracted the MNNG-induced inhibition of DNA synthesis but had no effect on the joining of newly-synthezised DNA fragments.

Since we found Painter's hypothesis attractive and worth considering, we have studied the effects of post--treatments with caffeine on the reduction of mitotic activity induced by X-rays and thiotepa in human lymphocytes. After exposure to both agents, post-treatments with caffeine almost completely reversed the induced mitotic delay. In contrast to caffeine, HU enhanced the X-ray- and thiotepa-induced reduction of mitotic activity.

Since our results are in good agreement with Painter's hypothesis, we are now inclined to accept it. We suggest that there is a mechanism for repair of DNA damage late in G_2 that in human lymphocytes is affected by both caffeine and hydroxyurea but in different ways. Whereas caffeine shortens the time available for repair, HU, as an inhibitor of deoxyribonucleotide synthesis, reduces the supply of material required for repair. Both effects would result in an increased frequency of unrepaired and misrepaired chromosome damage.

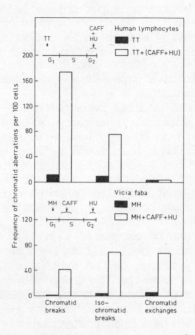

Fig. 1. The effects of 2.5×10^{-3}M caffeine (caff) and 5×10^{-3}M hydroxyurea (HU) on the frequencies of chromatid aberrations obtained in human lymphocytes at 50 hours after a 2-hour exposure to 5×10^{-5}M thiotepa (TT) and in *Vicia* root tips at 23 hours after a 2-hour exposure to 5×10^{-5}M maleic hydrazide (MH).

Experiments with caffeine + hydroxyurea It remains to say a few words about experiments in which post-treatments with caffeine have been combined with post-treatments with HU. First some results of conventional experiments in which a caffeine post-treatment during S has been combined with an HU post-treatment during late G_2--prophase. These experiments were carried out with root--tip cells of *Vicia faba* that had been exposed to the herbicide maleic hydrazide (MH) while in G_1-S. The 5--hour treatment with 2.5×10^{-3}M caffeine was given immediately or shortly after the exposure to MH and the post-treatment with 5×10^{-3}M HU during the last 2½-3 hours before fixation. The caffeine post-treatment caused alone a 5-fold increase in the total frequency of aberrations, whereas the HU post-treatment alone resulted in a doubling of the MH-induced aberration frequency. When MH-treated roots were exposed first to caffeine and then to HU, the frequency of aberrations was 10 times higher than that produced by MH alone. Thus, the caffeine post-treatment did not affect the degree of potentiation caused by HU (Fig. 1).

As pointed out earlier, the fact that the production of chromatid aberrations is S-dependent could mean either that the aberrations can be formed only during S or that the primary lesion is altered by the replication process in such a way that it is capable of giving rise to an aberration at a later stage of the cell cycle. Let us assume that the chromatid aberrations induced by MH normally are formed during S and that caffeine enhances this process. The fact that the HU potentiation of MH--induced chromosome damage is of the same relative magnitude in the presence and absence of a caffeine post--treatment must then mean that caffeine not only enhances the formation of chromatid-type aberrations during S, but also the formation of lesions that in the presence of HU give rise to aberrations during G_2-prophase. This is quite possible, of course, but a simpler explanation would be that caffeine has only one effect during S and that is to increase the frequency of lesions that are modified by replication in such a way that at a later stage of the cell cycle they may give rise to aberrations. From this interpretation it follows that a majority, if not all, of the MH-induced lesions do not give rise to aberrations until G_2.

In plant cells caffeine seems to be capable of affecting S-dependent chromosome damage only during S. Following Painter's hypothesis we suggest that caffeine enhances S-dependent chromosome damage in plant cells by promoting the replication of unrepaired DNA lesions.

In human lymphocytes, on the other hand, caffeine is active also during G_2. At this stage the effect of caffeine would be to prevent a G_2-delay that would allow the cell to repair damage to chromosomal DNA before the condensation process sets in. When caffeine is given together with HU, the potentiation obtained of the induced chromosome damage is usually larger than the sum of the potentiation produced by the two agents alone. A 10-fold increase in the frequency of aberrations is quite frequently observed and this applies to chromosome damage induced by both S-dependent (Fig. 1) and S-independent agents. Thus, in these cases, it is quite clear

that 90 % of the aberrations observed are formed during
the last 3 hours before harvesting from lesions induced
at a much earlier stage of interphase. It is tempting
to conclude that all chromatid-type aberrations are formed at late interphase-early prophase, but, admittedly
we have no evidence that this actually is the case.

ACKNOWLEDGEMENTS

This work was supported by a grant from the Swedish
Natural Science Research Council.

REFERENCES

(1) Busse, P.M., Bose, S.K., Jones, R.W., and Tolmach, L.J. *Radiat. Res. 71*, 666-677 (1977)
(2) Byfield, J.E., Murnane, J., Ward, J.F., Calabro-Jones, P., Lynch, M., and Kulhanian, F. *Br. J. Cancer 43*, 669-683 (1981)
(3) Evans, H.J. In "Genetical Aspects of Radiosensitivity: Mechanisms of Repair". Int. Atomic Energy Agency, Vienna, pp. 31-48 (1966)
(4) Evans, H.J. and Scott, D. *Proc. R. Soc. (London), Ser. B 173*, 491-512 (1969)
(5) Gascoigne, E.W., Robinson, A.C., and Harris, W.J. *Chem.-Biol. Interactions 36*, 107-116 (1981)
(6) Griffiths, T.D., Carpenter, J.G. and Dahle, D.B. *Int. J. Radiat. Biol. 33*, 493-505 (1978)
(7) Hansson, K., and Hartley-Asp, B. *Hereditas 94*, 21--27 (1981)
(8) Hartley-Asp, B., Andersson, H.C., Sturelid, S., and Kihlman, B.A. *Environ. Exp. Bot. 20*, 119-129 (1980)
(9) Kihlman, B.A. "Caffeine and Chromosomes", Elsevier, Amsterdam, 504 pp. (1977)
(10) Kihlman, B.A., and Andersson, H.C. *Environ. Exp. Bot. 20*, 271-286 (1980)
(11) Kihlman, B.A., and Hartley-Asp, B. *Hereditas 59*, 439-463 (1968)
(12) Kihlman, B.A., Hansson, K., Palitti, F., Andersson, H.C., and Hartley-Asp, B. In *Progr. Mutat. Res. 4*, "DNA-Repair, Chromosome Alterations and Chromatin Structure" (A.T. Natarajan, G. Obe and H. Altmann, eds.), Elsevier, Amsterdam (in press)

(13) Kihlman, B.A., Sturelid, S., Hartley-Asp, B., and Nilsson, K. *Mutat. Res. 26*, 105-122 (1974)
(14) Lehmann, A.R. *Life Sci. 15*, 2005-2016 (1974)
(15) Murnane, J.P., Byfield, J.E., Ward, J.F., and Calabro-Jones, P. *Nature 285*, 326-329 (1980)
(16) Oleinick, N.L., Brewer, E.N., and Rustad, R.C. *Int. J. Radiat. Biol. 33*, 69-73 (1978)
(17) Painter, R.B. *J. Mol. Biol. 143*, 289-301 (1980)
(18) Painter, R.B., and Young, B.R. *Proc. Natl. Acad. Sci. U.S.A. 77*, 7315-7317 (1980)
(19) Rauth, A.M. *Radiat. Res. 31*, 121-138 (1967)
(20) Roberts, J.J. *Adv. Radiat. Biol. 7*, 211-436 (1978)
(21) Roberts, J.J., Sturrock, J.E., and Ward, K.N. *Mutat. Res. 26*, 129-143 (1974)
(22) Scaife, J.F. *Int. J. Radiat. Biol. 19*, 191-195 (1971)
(23) Snyder, H.M., Kimler, B.F., and Leeper, D.B. *Int. J. Radiat. Biol. 32*, 281-284 (1977)
(24) Swietlinska, Z., Zaborowska, D., and Zuk, J. *Mutat. Res. 17*, 199-205 (1973)
(25) Tolmach, L.J., Jones, R.W., and Busse, P.M. *Radiat. Res. 71*, 653-665 (1977)
(26) Tomasovic, S.P. and Dewey, W.C. *Radiat. Res. 74*, 112-128 (1978)
(27) Walters, R.A., Gurley, L.R., and Tobey, R.A. *Biophys. J. 14*, 99-118 (1974)

Metabolism of Mutagens and Carcinogens

Metabolic Rationale for the Weak Mutagenic and Carcinogenic Activity of Benzo(e)pyrene

James K. Selkirk, Michael C. MacLeod, Betty K. Mansfield, and Anne Nikbakht

Biology Division, Oak Ridge National Laboratory, Oak Ridge, Tennessee, U.S.A.

ABSTRACT Benzo(e)pyrene is a weak carcinogen and mutagen in contrast to theoretical calculations that predict substantial carcinogenic potential for this positional isomer of benzo(a)pyrene. Earlier short-term studies have shown metabolism in intact cells strongly favors attack of B(e)P by the mixed function oxidases at the K-region and distal to the bay-region. Current studies have extended incubation times to four days to determine if long-term induction would result in formation of bay-region metabolites and could account for the weak positive carcinogenic and mutagenic activity reported in the literature.

INTRODUCTION

Polyaromatic hydrocarbons are metabolized by all species and tissues to a number of ring substituted hydrogenated derivatives as detoxification products (6). It has been fully established that this class of compounds, which are activated by the microsomal mixed function oxidases, form reactive electrophiles which covalently modify cellular macromolecules and exert profound biological effects including cytotoxicity, mutagenicity, and malignant transformation. While a relatively large number of oxygenated products are formed, it would appear that the principal carcinogenic intermediate for benzo(a)pyrene is the 7,8-diol-9,10-epoxide which is produced as a secondary oxidation step on the primary 7,8-diol metabolite by the same mixed function oxidases (Fig. 1). This assump-

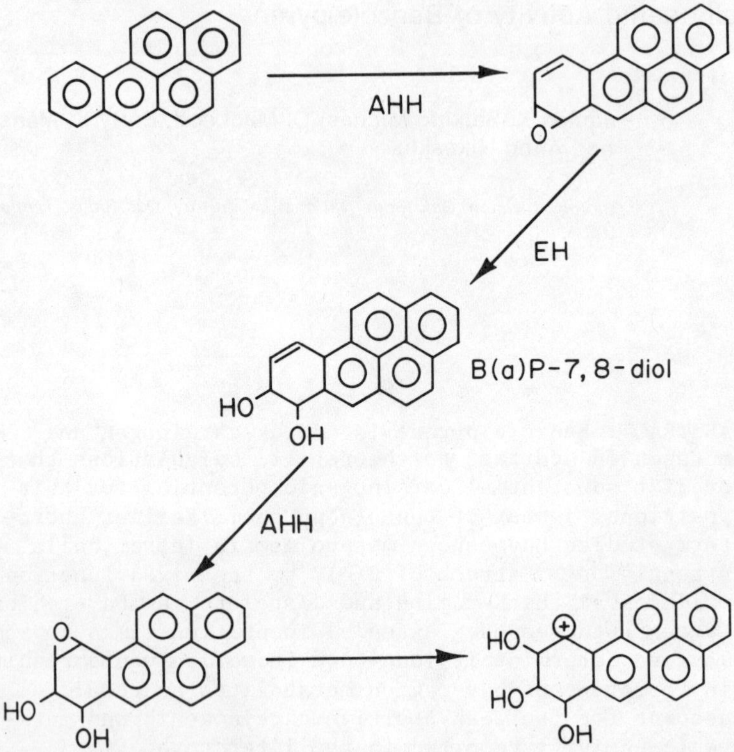

Figure 1. Primary and Secondary Metabolism of Benzo(a)-pyrene to Form the Putative Ultimate Carcinogenic Electrophile.

tion is supported by the far greater tumorigenic and mutagenic activity of the 7,8-diol, both in *in vivo* (4) and *in vitro* systems (3) and also by the fact that the 7,8-diol-9,10-epoxide appears to be the major form of benzo(a)pyrene that binds to DNA (7). In addition, there is the possibility that nuclear mixed function oxidases may play an additional role by activating the 7,8-diol within the nucleus to the reactive diol-epoxide since the 7,8-diol has been shown to pass readily through cell membranes (1). In general, metabolism of benzo(a)pyrene by cell-free systems presents a significantly different profile of metabolites when compared to intact cell systems which is primarily due to the presence of enzymes in whole cell incubations which catalyze conjugate formation to glucuronic acid, glutathione, and sulfates. Since all eukaryotic species, both susceptible and resistant to malignant transformation, possess the same microsomal mixed function oxidase activity product, and, therefore, the same qualitative profile of metabolites for benzo(a)pyrene, it is evident that they are all capable of forming the highly reactive carcinogenic intermediate diol-epoxide. The fact that there is great disparity in species susceptibility to this carcinogen suggest a greater degree of complexity in the cellular processing of benzo(a)pyrene between species and tissues than is currently understood.

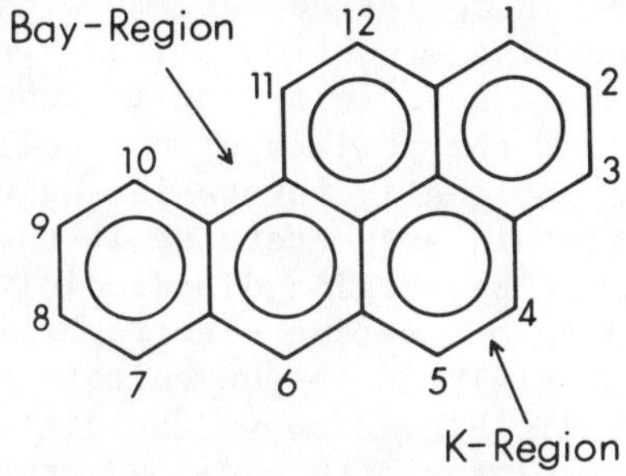

Benzo(a)pyrene

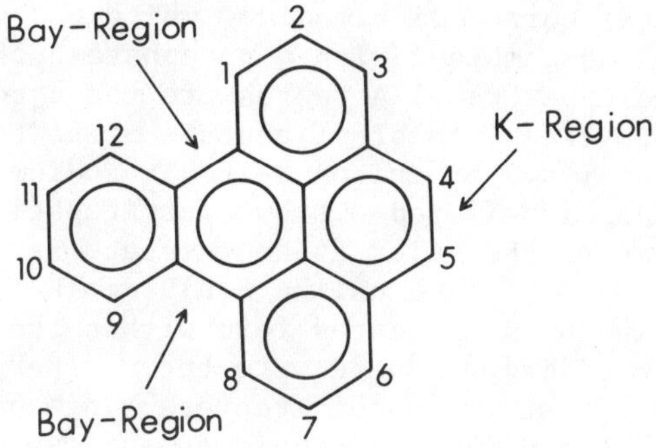

Benzo(e)pyrene

Figure 2. Isomeric Benzo(a)pyrene and Benzo(e)pyrene Showing Both the Reactive K-Region and Bay-Region.

In contrast to benzo(a)pyrene, the geometric isomer, benzo(e)pyrene (Fig. 2), displays extremely low carcinogenic potency (6). As previously shown, incubations with hamster embryonic cells produce little or no "bay-region" intermediates, but prefer to attack the molecule in the K-region vicinity (2). Therefore, it may be postulated that benzo(e)pyrene, although possessing two equivalent bay-regions, and, therefore, holds a high carcinogenic potential, displays very weak carcinogenic potency, because the cellular biochemistry prefers to metabolize the molecule away from the bay-region area. This is reinforced by the fact that synthetic B(e)P-9,10-dihydrodiol does not favor formation of its analogous epoxide to any significant degree in cell-free systems without previous induction of the mixed-function oxidases (8). We have, therefore, utilized benzo(e)pyrene in long-term culture to further probe the ability of the cell to recognize the most metabolically favorable regions of the benzo(e)pyrene molecule and determine if there is significant formation of bay-region dihydrodiol that would presumably lead to a reactive electrophilic diol-epoxide by maintaining B(e)P in tissue culture over a 4-day incubation period. Fig. 3 shows the distribution of benzo(e)pyrene metabolites after 24-hour incubation, (1 µg/ml) in the intracellular and extracellular compartments.

Panel B represents the hydroxylated intermediates found in the surrounding medium, which was decanted from the cells and combined with a phosphate buffered saline wash. The upper Panel A represents the cytoplasmic metabolites extracted into ethyl acetate after the cell membranes were lysed by triton X-100 in sodium deoxycholate and the nuclei pelleted out by centrifugation at 1,000 x g. In both cases, the major 24-hour metabolite was the K-region-4,5-diol, which unlike B(a)P diols, was observed to exist in an unconjugated form within the cytoplasmic space. In addition, there were three other discernable peaks, the first presumably representing an oxidation product of 4,5-diol to its quinone and two more polar peaks (P1 and P2) which were presumed to be phenolic. As has been reported for benzo(a)pyrene, the extracellular space did not contain free phenolic metabolites (2). Also, the toxic phenols are rapidly conjugated and ejected from the cells into the surrounding medium.

It was apparent in this series of experiments that the bay-region 9,10-diol (D_1) was not observable under these conditions and quite possibly the reason for the lack of carcinogenic and mutagenic potency of this molecule. Since *in vitro* mutagenesis assays often require longer treatment periods, it was important to know if there could be an induction of this putative reactive intermediate with time as the various microsomal and cytoplasmic enzyme systems were induced by continued presence of B(e)P.

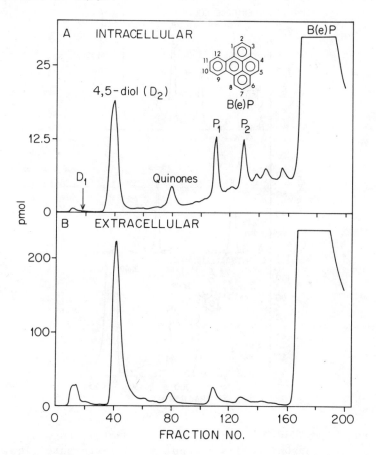

Figure 3. High-Pressure Liquid Chromatographic Profile of Benzo(e)pyrene Metabolites Formed During 24-hour Incubation with Hamster Embryo Cells.

When incubation periods were extended to four days, formation of ethyl-acetate-soluble metabolites reached a plateau between 24 and 48 hours of incubation and were qualitatively identical to the 24-hour profile, except for the ratio of peaks Q and P_1 (Fig. 4, Panel A). However, it was interesting to observe that these compounds were produced at sufficient concentration within the cell so they were now able to pass through the cell membrane

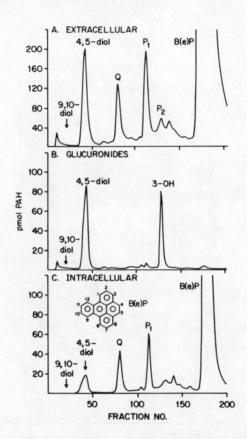

Figure 4. Four-Day Incubation of Benzo(e)pyrene by Hamster Embryonic Cells. Panel A, High-pressure liquid chromatographic assay; Panel B, Water-soluble glucuronide conjugates showing P_2 as the new metabolite 3-OH-B(a)P; Panel C, Hydroxylated metabolites ejected from the cell into the surrounding medium.

into the surrounding medium (Fig. 4, Panel C). However, when the non-ethyl-acetate extractable radioactivity, remaining behind in the water-soluble fraction, was treated with β-glucuronidase (Fig. 4, Panel B), it was discovered that in addition to the 4,5-diol glucuronide conjugate, P2 was also forming a glucuronide. This metabolite was identified by co-chromatography and fluorescence spectra against an authentic sample of 3-hydroxy-B(e)P. Indicated by an arrow at the lefthand side of the chromatograph is the position where B(e)P-9,10-diol would chromatograph under these conditions. In addition, finding 3-hydroxy-B(e)P as a major product of B(e)P metabolism, shows that other non-bay-region positions on the B(e)P molecule are also favored as sites for enzyme attack which yield additional metabolic evidence for the probable lack of carcinogenic or mutagenic activity by this theoretically reactive molecule. Since we have previously shown that there are low levels of binding of benzo(e)pyrene to nuclear protein and DNA (2), in spite of a low rate of turnover of benzo-(e)pyrene in cells, it may be suggested that the longer half-life of benzo(e)pyrene in cells in culture may allow accumulation over a longer period of time than is possible for benzo(a)pyrene. Previous *in vivo* experiments lend some support to this assumption (5).

Another possible rationale derives from the probability that 3-hydroxy-B(e)P is a non-enzymatic ring-opening derivative of 2,3-benzo(e)pyrene epoxide which when opened to form the 3-hydroxy derivative would also yield a reactive electrophilic intermediate that may be the species of B(e)P interacting with cellular macromolecules.

It is apparent that the biochemical pathway followed by any potential carcinogenic molecule is dictated by its chemical and physical characteristics once it is incorporated into the cell. While physical-chemical calculations can predict the probability that a given molecule will be mutagenic or carcinogenic, a multitude of parameters, including the relative induction of mixed function oxidases and cytoplasmic transferases must be taken into account as well as the relative attractiveness of a given metabolic intermediate for a reactive site on the enzyme. In addition, relative enzyme activity may be both a function of species and tissues, in which case all these physical, chemical and biochemical parameters will combine to ultimately dictate how much a potentially carcinogenic molecule will follow the pathway toward a biologically reactive nucleophile and how much of the chemical will be rapidly detoxified and rendered harmless to the cell.

REFERENCES

(1) Bresnick, E., Chuang, A. H. L., and Bornstein, W. A. *Chem.-Biol. Interactions* 24, 111-115 (1979)
(2) MacLeod, M. C., Cohen, G. M., and Selkirk, J. K. *Cancer Research* 39, 3463-3470 (1979)
(3) Mager, R., Huberman, E., Yang, S. K., Gelboin, H. V., and Sachs, L. *Inter. J. Cancer* 19, 814-817 (1977)
(4) *Public Health Service Department* 149, 566-624 (1978)
(5) Scribner, J. C. *J. Natl. Cancer Inst.* 50, 1717-1719 (1973)
(6) Selkirk, J. K. *Modifiers of Chemical Carcinogenesis* pp. 1-31 (1980).
(7) Sims, P., Grover, P. L., Swaisland, A., Pal, K., and Hewer, A. *Nature* 252, 326-328 (1974).
(8) Wood, A. W., Levin, W., Thakker, D. R., Yagi, H., Chang, R. L., Ryan, D. E., Thomas, P. E., Dansette, P. M., Whittaker, S., Turujmah, Lehr, R. E., Kurma, S., Jerina, D. M., and Conney, A. H. *J. Biol. Chem.* 254, 4408-4413 (1979).

This research was supported by the National Cancer Institute under Interagency Agreement 40-5-63 and the Office of Health and Environmental Research, U.S. Dept. of Energy, under contract W-7405-eng-26 with the Union Carbide Corporation.

N-Hydroxylation of Mutagenic Heterocyclic Aromatic Amines in Protein Pyrolysates by Cytochrome P-450

Ryuichi Kato, Yasushi Yamazoe, Kenji Ishii, and Tetsuya Kamataki

Department of Pharmacology, School of Medicine, Keio University, Tokyo, Japan

ABSTRACT The enzymatic mechanism by which heterocyclic aromatic amines in amino acid and protein pyrolysates are N-hydroxylated and activated to mutagens was studied. The activities of N-hydroxylation and mutagenic activation of Trp-P-2 were found in nuclei as well as in microsomes of liver cells. The activities to form N-hydroxylated products and mutagens from Trp-P-2, Glu-P-1 and Glu-P-2 could be reconstituted with cytochrome P-450 purified from rat liver microsomes. In these experiments, the formation of N-hydroxy-Trp-P-2 correlated well with the number of revertants. Based on these and other results, we confirmed that N-hydroxy-Trp-P-2 was a common intermediate for the induction of revertants in Salmonella mutation test. The proposed mechanism for the covalent binding to DNA of N-hydroxy-Trp-P-2 has also been shown.

INTRODUCTION

Sugimura and his coworkers (4) have paid much efforts on the evaluation of environmental mutagens, especially of those in foods. Thus, they have isolated more than ten potent promutagens in pyrolysates of amino acids and proteins (5). Some of the promutagens isolated so far have been proven to be carcinogens in experimental animals. Trp-P-2 was isolated from pyrolysates of tryptophan and proteins, and Glu-P-1 and Glu-P-2 were from glutamic acid. The mutagenicity of Trp-P-2, Glu-P-1 and Glu-P-2 is reportedly about 17, 8 and 0.3 times as great as that of aflatoxin

B_1, respectively, on the basis of μg of these compounds (2). In those studies the promutagens induced mutation of Salmonella typhimurium only when liver 9,000 x g supernatant (S-9) fraction of PCB-treated rats was included in the incubation medium. We report herein on the N-hydroxylation and mutagenic activation of Trp-P-2, Glu-P-1 and Glu-P-2 by an enzyme in nuclei as well as microsomes of rat hepatic cells and by purified cytochrome P-450 enzyme system.

RESULTS AND DISCUSSION

Enzyme systems responsible for the activation of mutagenic compounds are located not only in microsomes but in nuclear membranes. Though the activities in nuclear membranes are much lower than those in microsomes, the activation in nuclear membranes are recognized as an important process for carcinogenesis of compounds since nuclear enzymes are located closer to DNA. Thus, the N-hydroxylation and mutagenic activation of Trp-P-2 by liver nuclei were examined (Table 1). Liver nuclei isolated from untreated rats were capable of N-hydroxylating Trp-P-2. The activity was about 13 % that of microsomes on mg of protein basis. The activity in the mutagenic activation was also detected. The rate of mutagenic activation by nuclei was about 11 % that by microsomes. Treatment of rats with PCB or 3-methylcholanthrene resulted in great enhancements in the nuclear N-hydroxylation and in mutagen production activities as seen with microsomes (5,6). Phenobarbital showed minimal effects on both nuclear and microsomal activities. The microsomal mutagen-producing activity was inhibited by introduction of carbon monoxide in the atmosphare of incubation medium, indicating that the mutagenic

Table 1. Effects of inducers on nuclear and microsomal N-hydroxylation and mutagenic activation of Trp-P-2

Pretreatment	Nuclei		Microsomes	
	N-Hydroxylation of Trp-P-2[a]	Number of revertants[b]	N-Hydroxylation of Trp-P-2[a]	Number of revertants[b]
None	0.06	46	0.46	438
PCB	0.95	1978	12.30	20850
3-Methylcholanthrene	0.68	1450	4.59	6858
Phenobarbital	0.09	179	0.41	2341

a: nmol N-OH-Trp-P-2/mg protein/15 min
b: The number of revertants of S. typhimurium TA98 resulted from 10 nmoles of Trp-P-2 activated by nuclei or microsomes (50 μg protein)

activation of Trp-P-2 was catalyzed by cytochrome P-450 in microsomes. Among inhibitors of cytochrome P-450 known, 7,8-benzoflavone is a specific inhibitor of a particular form(s) of cytochrome P-450 which is inducible by 3-methylcholanthrene. The mutagenic activation of Trp-P-2 by nuclei was inhibited by addition of 7,8-benzoflavone at a concentration of 10 µM. n-Octylamine and SKF525-A inhibited the activities but to much lesser extents than did 7,8-benzoflavone. The extents of the decrease in the number of revertants seen with nuclei were quite similar to those with microsomes. These results probably suggest that a form of nuclear cytochrome P-450 responsible for the mutagenic activation of Trp-P-2 is the same as that of microsomes.

To confirm the involvement of cytochrome P-450 in the N-hydroxylation of Trp-P-2, Glu-P-1 and Glu-P-2, a reconstituted system containing purified cytochrome P-450 was utilized. Both Glu-P-1 and Glu-P-2 were metabolized by cytochrome P-450 to three products as judged by the elution profiles from HPLC (Fig. 1). The formation of these metabolites depended on the forms of cytochrome P-450. All metabolites were formed in greater amounts when forms of cytochrome P-450, PCB P-448 and MC P-448, purified from

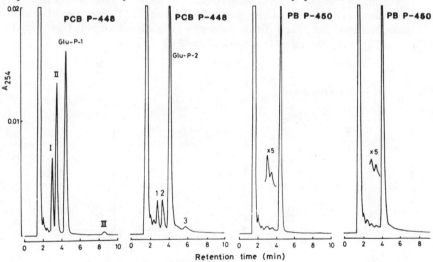

Fig. 1. Formation of metabolites of Glu-P-1 and Glu-P-2 by cytochrome P-450 purified from microsomes of PCB- or phenobarbital-treated rats

microsomes of PCB- and MC-treated rats, respectively, which had absorption maxima at about 447 nm in carbon monoxide difference spectra, were used. PB P-450 purified from phenobarbital-treated rats showed poor activity in the N-hydroxylation of these substrates. The fractions designated as peak II and peak 2 on the charts contained N-hydroxylated products which were mutagenic metabolites of Glu-P-1 and Glu-P-2, respectively, as judged by the Salmonella mutation assay in which no enzyme system was added. In accordance with these results, PCB P-448, but not PB P-450, was an efficient catalyst for the mutagenic activation of Glu-P-1 and Glu-P-2. Other forms of cytochrome P-450, including MC P-450 and PCB P-450 purified from 3-methylcholanthrene- and PCB-treated rats, respectively, also catalyzed the mutagenic activation but to much lesser extents (Fig. 2) Similar experiments conducted with Trp-P-2 showed that PCB P-448 and MC P-448 had higher capabilities in the N-hydroxylation and mutagen production. The N-hydroxy-Trp-P-2 eluted from a HPLC column induced revertrants in the Salmonella assay without aids of any enzyme systems.

N-Hydroxy-Trp-P-2 bound covalently to externally added calf thymus DNA (3). The amount bound to DNA was decreased by inclusion of dithiothreitol. However, the addition of dithiothreitol together with L-serine, ATP and seryl-tRNA synthetase purified from Baker's yeast enhanced the covalent binding to DNA of N-hydroxy-Trp-P-2 (7).

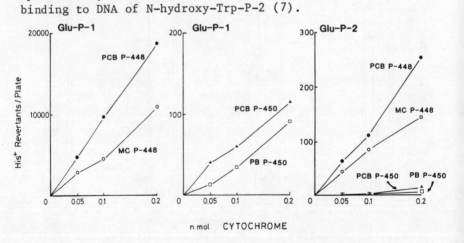

Fig. 2. Mutagenic activation of Glu-P-1 and Glu-P-2 by cytochrome P-450 purified from microsomes of rats treated with different inducers

Based on these results, we propose the mechanism of activation of Trp-P-2 as summarized below. First, Trp-P-2 is hydroxylated at 3N-position by cytochrome P-450. Secondly, N-hydroxy-Trp-P-2 is a mutagen. Thirdly, N-hydroxy-Trp-P-2 itself is capable of binding to DNA via two pathways. Although we do not have direct evidince, N-hydroxy-Trp-P-2 may spontaneously form a nitroxide radical to bind to DNA. Also, in the presence of seryl-tRNA synthetase and dithiothreitol which probably acts as a stabilizer of both N-hydroxy-Trp-P-2 and seryl-tRNA synthetase, the seryl-conjugate of N-hydroxy-Trp-P-2 may bind to DNA via a nitrenium ion. Other heterocyclic aromatic amines involving Glu-P-1 and Glu-P-2 are also assumed to be activated in a manner similar to Trp-P-2. The exact chemical mechanisms for the activation of these compounds are under investigation in our labolatory.

ACKNOWLEDGEMENTS

These studies were supported by a Grant-in-Aid for Cancer Research from the Ministry of Education, Science and Culture of Japan.

REFERENCES

1) Ishii, K., Yamazoe, Y., Kamataki, T., and Kato, R. Cancer Res. 40, 2596-2600 (1980)
2) Kawachi, T., Nagao, M., Yahagi, T., Takahashi, Y. Sugimura, T., Takayama, S., Kosuge, T., and Shudo, T. in "Advances in Midical Oncology, Research and Education. Vol. 1 Carcinogenesis" (Margison, G. P. ed) Pergamon Press, New York (1979)
3) Mita, S., Ishii, K., Yamazoe, Y., Kamataki, T., Kato, R. and Sugimura, T. Cancer Res. 41 3610-3614 (1981)
4) Sugimura, T., and Nagao, M. CRC Crit. Rev. Toxicol. 6 189-209 (1979)
5) Sugimura, T., Wakabayashi, K., Yamada, M., Nagao, M., and Fujii, T. in "Mechanisms of Toxicity and Hazard Evaluation" (Holmstedt, B., Lauwerys, R., Mercier, M., and Roberfroid, M. ed.) pp 205-217 Elsevier/North Holland Biomedical Press, Amsterdam (1980)
6) Yamazoe, Y., Ishii, K., Kamataki, T., Kato, R. and Sugimura, T. Chem.-Biol. Interns. 30, 125-138 (1980)
7) Yamazoe, Y., Tada, M., Kamataki, T., and Kato, R. Biochem. Biophys.Res. Commun. in press (1981)

Microsomal Metabolic Activation, Mutagenicity and Antimutagenicity

Marcel Roberfroid,[1] Chehab Razzouk,[1] François Hervers,[2] and Heinz Günter Viehe[2]

[1]Unit of Biochemical Toxicology and Cancerology, and [2]Unit of Organic Chemistry, Université Catholique de Louvain, Brussels, Belgium

ABSTRACT

The complex metabolism of chemical promutagens controls the concentration of the molecular species at the target site. Metabolic interactions between the various pathways, chemical induction and/or modification of specific enzymes, species or even individual variations are all factors which may influencethe net results of the metabolic balance. With 2-acetylaminofluorene and 2-aminofluorene as substrate the effects of those factors are discussed. Since free radicals could be involved in the molecular mechanism of genotoxicity, radicophiles which trap those intermediates have been tested as potential antimutagenic agents. Preliminary data indicate that they indeed inhibit mutagnicity in vitro.

INTRODUCTION

Enzymes catalyzing the metabolism of promutagens are among the most important endogenous factors which may influence their genotoxic effect(s). Such a metabolism is generally very complex including reactions of phase 1 (functionalization reactions) and reactions of phase 2, which are essentially conjugation processes (1). The reactions of phase 1, which usually preceed those of phase 2, are mainly catalyzed by endoplasmic reticulum bound enzymes. Among those enzymes, cytochrome P-450-dependent mixed function, oxydases play a major role in initiating the metabolic transformations of those agents.

With most chemical mutagens, the activity of the metabolizing enzymes directly controls the concentration of the molecular species at the target site(s) (5,11). The balance between the various metabolic pathways (phase 1 and 2) influences either the structure or the amount of the active intermediate(s) and consequently the intensity and the nature of this interaction(s) with the essential macromolecules (nucleic acids and/or proteins). The enzymes which metabolize chemical mutagens are thus key elements which determine both the importance and the nature of their potential genotoxic effects.

When testing chemicals for potential genotoxicity in vitro, the nature, activity and "quality" of the enzymatically active tissue preparation used are essential factors which can influence the results and consequently their interpretations. An incomplete or even a false figure of mutagenicity and thus of potential carcinogenicity may be obtained by using, in vitro, inadequate enzymatic systems or by underestimating their limits.

Since the final aim of mutagenicity testing is to predict potential genotoxicity and carcinogenicity for humans, an ideal "in vitro" test should include a complete metabolic system which mimics precisely the metabolism of the exposed individuals including the balance between activation and inactivation pathways. Both endogenous and exogenous factors may however influence that balance particularly in humans. An ideal metabolic system thus does not exist and each enzymatic preparation is more or less imperfect. The correct interpretation of the results of mutagenicity testing therfore requires a clear understanding of those limits.

Among those limitations, the purposes of this presentation are : to discuss the role of metabolic interactions and the importance of enzymatic induction in controling the net production of the active intermediate(s) and to emphasize the importance of species or even individual variations in term of metabolic balance between activation and inactivation in metabolic processes.

Mutagenic metabolites have often electrophilic properties (7). Some of them are however free radicals (14,2). Moreover, by one electron exchange, the reaction between an electrophile and a nucleophile could imply free radicals intermediates (fig. 1).

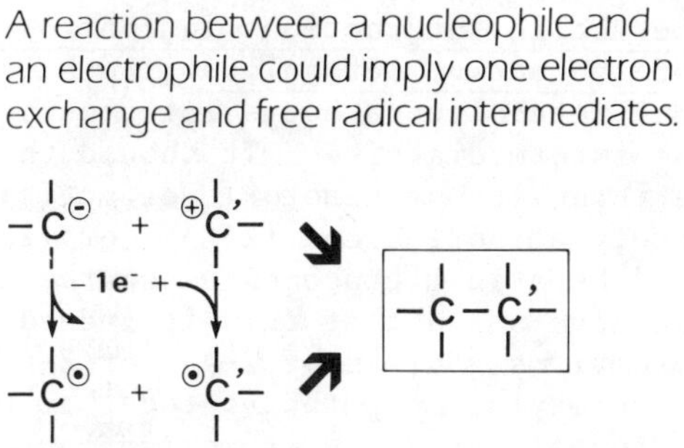

A reaction between a nucleophile and an electrophile could imply one electron exchange and free radical intermediates.

Fig. 1

Beyond their metabolic activation, the genotoxicity of chemicals could thus be modified or even inhibited if compounds do exist which trap radical intermediates. The new chemical concept (15,16) of radicophily claims that captodative susbstituted radicals are stabilized (fig.2) and that capto-dative substituted olefins are efficient radical traps (fig.3). The present report will show preliminary evidences for a possible role of radicophiles as antimutagenic agents.

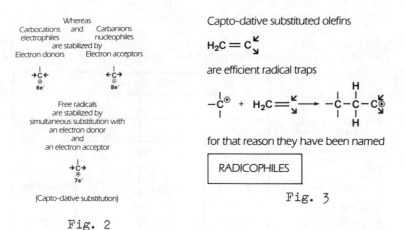

Fig. 2   Fig. 3

MICROSOMAL METABOLISM AND MUTAGENICITY

Role of metabolic interactions in controling the production of the active metabolite. The balance between the various metabolic pathways influences the amount of the active intermediate(s). It should thus also influences the intensity of genotoxic effect(s).
Using 2 acetylaminofluorene (2-AAF) or (2AF) as a promutegen and hepatic microsomes as source of metabolizing enzymes we have shown that this is indeed the case. The microsomal metabolism of 2-AAF or 2-AF includes both N- and C-hydroxylation pathways together with reduction of the hydroxamic acid and isomerization (N— C transoxygenation). In some species, a microsomal amidase hydrolyzes the N-acetyl bond (12). Experimental evidences have recently accumulated (13) which demonstrate the importance of that last step in the enzyme mediated mutagenicity of 2-AAF. The metabolic interaction between the N- and C- hydroxylation pathways, which controls the net production of the primary active metabolite (N-hydroxy 2-AAF), however influences the intensity of the genotoxic effect of that chemical.
As shown in table 1, the N-hydroxyla-e activity of control rat liver microsomes is weak as compared to the monooxygenase activities which catalyze the C_5 and C_7 hydroxylations. The apparent K_M's of those three enzymes are of the same order of magnitude.

Table 1 Effects of 3 MC pretreatment of rat and hamster on the mixed function oxydases dependent metabolism and on the microsomal mediated mutagenicity of 2AAF. For the ames test, the concentration of 2AAF was 10 µg/plate.

2-AAF metabolism and mutagenicity	Liver microsomes			
	Control		3-MC induced	
	Rat	Hamster	Rat	Hamster
N-hydroxylase				
$K_M \times 10^{-6}$ M	0.53 ± 0.02	0.93 ± 0.15	1.00 ± 0.10	0.23 ± 0.03
Vm pmols/min/mg prot.	13.2 ± 0.4	139.6 ± 7.2	432 ± 7	1664 ± 56
C_5-hydroxylase				
$K_M \times 10^{-6}$ M	0.42 ± 0.08	0.14 ± 0.02	0.13 ± 0.03	0.16 ± 0.05
Vm pmols/min/mg prot.	30.5 ± 1.5	352 ± 11	212 ± 14	160 ± 13
C_7-hydroxylase				
$K_M \times 10^{-6}$ M	2.8 ± 0.3	0.25 ± 0.02	0.20 ± 0.02	0.34 ± 0.13
Vm pmols/min/mg prot.	112 ± 5	511 ± 12	396 ± 8	501 ± 80
Ames test (TA 1538)				
Nber His+ rev/plate	700 ± 80	1600 ± 150	300 ± 35	2520 ± 200

The induction of mixed function oxidases by 3-methyl-cholanthrene (3-MC) increases both N- and C- hydroxylase activities. Since the N-hydroxylation is the first step in the activation of 2-AAF one would expect such an induction to be accompained by an increase in the intensity of its microsomal mediated mutagenicity.
Surpringly however this is not the case and, in our experimental conditions, thè genotoxic effect decreases when 2-AAF is activated by microsomes isolated from 3-MC treated rats as compared to control animals.
In hamster livers however the effect of 3-MC pretreatment is, as expected, an increased capacity of their microsomes to activate 2-AAF as a mutagen in the Ames test even though the induction of the N-hydroxylase is less marked than in the rat hepatic tissue.
We have previously suggested (9) an explanation for such a peculiar effect of 3-MC in the rat which explains also its inhibition of the hepatocarcinogenicity by 2-AAF. The inducing effects of 3-MC in that particular species (rat) are accompanied by a significant modification of the apparent K_M value for the arylamide N-hydroxylase. That increase in K_M value is an indirect effect which expresses the fact that some C-hydroxy metabolites of 2-AAF are competitive inhibitors of the N-hydroxylase. There is thus a metabolic interaction between the N- and C- hydroxylations pathways. After 3-MC, the overproduced phenols retro- inhibit the N-hydroxylase which even though it is induced, produces no or less hydroxamic acid. Since in the hamster liver the inducing effect of the polycyclic aromatic hydrocarbon is restricted to the N-hydroxylase, the net result is an over production of the primary active metabolite and an increased genotoxicity. There is no overproduction of phenols and the net amount of N-hydroxy 2-AAF increases.

Metabolic interaction exists in chemical metabolism. It can be markedly modified by induction and it may be species specific. Since, in some instances, it influences the genotoxic effects of a given chemical, it is a factor which must be considered when interpreting results of mutagenicity testing in vitro.

<u>Species specificity and specific metabolic pathways</u>:
Beyond the consequences of qualitative differences in chemical metabolism, there are also important species variations in biological effects of chemicals which

can be explained by the existence of specific metabolic pathways. It has, for example, long been known that guinea-pig is resistant to the hepatocarcinogenic effect of 2-AAF (4) even though its liver S9 efficiently activates that chemical to a mutagen (6).
Razzouk et al.(10) have recently clearly demonstrated that guinea-pig liver microsomes do indeed contain arylamide N-hydroxylase which is as active as that of the rat. However when 2-AAF is incubated in the presence of guinea-pig liver microsomes supplemented with NADPH, the hydroxamic acid accumulates only during the first three to five minutes. It then rapidly disappears. The same is true for the number of his $^+$ revertants per plate induced in the Ames test with liquid preincubation of micromolar concentration of 2-AF in the presence of guinea-pig liver microsomes. That number increases during the first five to ten minutes of incubation and then declines rapidly (table 2).

Table 2 Effect of incubation time on the guinea pig liver microsomal mediated N-hydroxylation and mutagenicity of 2-AF (0.25 µM). The spontaneous reversion rate of TA 1538 strain was 15 ± 3.

Time of incubation min.	Microsomal metabolism pmols N-OH2AF/mg prot.	Microsomal mediated mutagenicity rev His$^+$/plate
1	100 ± 10	–
2	200 ± 25	–
5	–	65 ± 5
10	300 ± 37	80 ± 10
15	250 ± 20	73 ± 8
20	175 ± 20	60 ± 6
30	80 ± 6	45 ± 3

By incubating micromolar concentration of N-hydroxy 2-AAF it was demonstrated, by the same authors (10), that guinea-pig liver microsomes catalyze a very efficient metabolic transformation of that product to an inactive C_7 phenolic metabolite. Such a peculiar metabolic process seems to be present also in human liver microsomes. The resistance of guinea-pig to the hepatocarcinogenic effect of 2-AAF is thus not due to the absence but rather to the presence of an enzyme activity. The absence of N-hydroxylase would not have been in agreement with the high capacity of guinea-pig liver S9 to activate

that arylamide to a mutagenic intermediate in the Ames test.
Whereas the presence of a specific detoxification pathway is perfectly compatible with that genotoxwc effect.
The very high concentration of 2-AAF which are usually applied in the Ames test may indeed rapidly saturate that enzyme.
Important interspecies variations do exist in the metabolism of chemical mutagens. Such differences are both qualitative and quantitative. The detection of a mutagenic effect will thus strongly be influenced by the origin of the metabolic system used. The extrapolation of such an effect to an other species is thus always hazardous.

The metabolism of a chemical is a key event in its mutagenicity : Most chemical mutagens need to be activated. Such an activation represents generally only one or a few of the the many xenobiotic pathways. The mutagenic effect(s) of a chemical result(s) from a balance between the various pathways most of which prevent that effect.
The balance between the various pathways which controls the genotoxic effect(s) may vary enormously from one species to the other. Human is the animal species for which the knowledges concerning the metabolism of chemicals are the most scarcely. From the point of view of a biochemist involved in research on xenobiotic metabolism, the extrapolation of mutagenicity testing to human carcinogenicity risks is a difficult or even, in many cases, an impossible task.

FREE RADICALS, RADICOPHILY AND MUTAGENICITY

As indicated in the introduction, free radicals could play a major, yet unknown, role in the mechanism of mutagenic activity of many chemicals.
Endogenous or exogenous compounds active as radical traps could thus be important agent in preventing or controlling such a genotoxic effect.
Since capto-dative substituted olefins are efficient radical traps leading to the formation of stabilized radical adduct which may easily dimerize (fig. 4).
they have been tested in the Ames test in the presence of various direct or indirect acting mutagens.

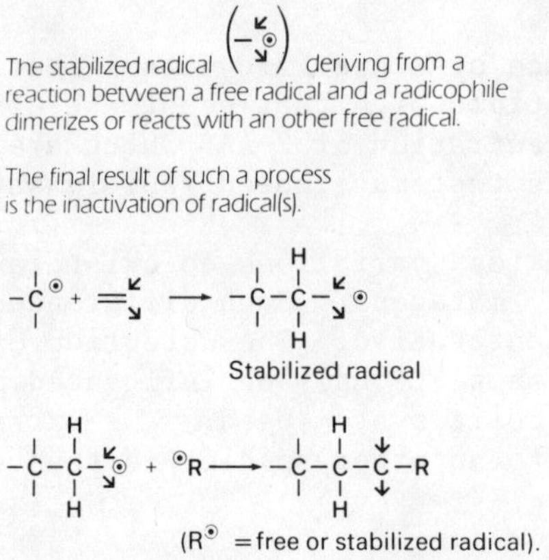

The stabilized radical $\left(-\overset{K}{\underset{Y}{C}}\odot\right)$ deriving from a
reaction between a free radical and a radicophile
dimerizes or reacts with an other free radical.

The final result of such a process
is the inactivation of radical(s).

$$-\overset{|}{\underset{|}{C}}\odot + \overset{K}{\underset{Y}{=}} \longrightarrow -\overset{|}{\underset{|}{C}}-\overset{H}{\underset{H}{\overset{|}{C}}}-\overset{K}{\underset{Y}{\odot}}$$

Stabilized radical

$$-\overset{H}{\underset{H}{\overset{|}{\underset{|}{C}}}}-\overset{|}{\underset{|}{C}}-\overset{K}{\underset{Y}{\odot}} + {}^\odot R \longrightarrow -\overset{H}{\underset{H}{\overset{|}{\underset{|}{C}}}}-\overset{|}{\underset{|}{C}}-\overset{\downarrow}{\underset{\downarrow}{C}}-R$$

Fig. 4

($R^\odot$ = free or stabilized radical).

As shown in figures 5 and 6 one of such olefins :
1-,morpholino 1-cyanoethylene is indeed an inhibitor
of the in vitro mutagenicity of 2-nitrofluorene, N-methyl
nitro nitrosoguanidine,2 aminoanthracene, benzo(a)pyrene,
nitrosoguanidine or 2 acetylaminofluorene. When necessary, liver S_9 prepared from arochlor 1254 treated rats
was added to the incubation mixture.

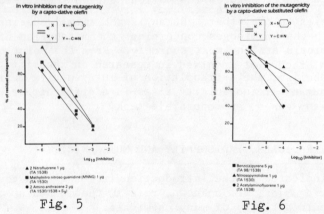

Fig. 5   Fig. 6

Inhibition of both carcinogenicity and mutagenicity
(8 -17) is a well known phenomenon which has already been
analyzed. Most of the identified chemicals act
however by inhibiting the metabolic activation step.
The application of the concept of radicophily opens
a new way in that research by allowing the synthesis of
(pre)mutagens trapping agents.

CONCLUSION

Chemical mutagenicity is a very complex process which involves, in most cases, the biotransformation of the promutagenic agent and the high reactivity of metabolic intermediates. The understanding of the molecular mechanisms of such a genotoxic effect requires the comlete knowledge of both the qualitative and the quantitative aspects of the metabolic part. Due to our actual ignorance of those processes, the interpretation, in term of potential risks for humans, of the results of mutagenicity testing must be prudent.
Except for a very few examples, the chemical nature of the reactive intermediate(s) involved in mutagenicity is unknown. The very preliminary data presented here seems to indicate that a better knowledge of the chemistry of the mechanism of this deleterious biological effect could lead to the discovery of potential antidote(s). Toxicology which has limited itself, to much, to the identification and the description of effects could thus become an essential part of health prevention.

REFERENCES

1. Bartsch,H.,Kuroki,T.,Roberfroid,M. and Malaveille,C. *Chemical Mutagens 7,* Hollander,A. and de Serres,F. eds. Plenum Press,New York -London in press(1981)

2. Bartsch,H.,Dworkin,C.,Miller,E.C. and Miller,J.A. *Biochem.Biophys*.Acta *304,*42-55 (1973)

3. Batzinger,R.R.,Suh-Yun,L.O.and Bueding,E.*Cancer Res.* 38,4478-4485 (1978)

4. Gutman,H.R.,and Bells,P.*Biochem,Biophys*.Acta,*498,* 229-243 (1977)

5. Jenner,P.,and Testa,B;,*Xenobiotica,8,*1-25 (1978)

6. Mc Gregor,D.,*Muta. Res.*,*30,*305-316 (1975)

7. Miller,E.C.,*Cancer Res.*,*38,*1471-1496 (1978)

8. Rahimtula,A.D.,Zacharias,P.K. and O'Brien,P. *Biochem,J.*,*164,* 473-475 (1977)

9. Razzouk,C.,Mercier,M.,And Roberfroid,M.B.,*Cancer Res*, *40*, 3540-3546 (1980)

10. Razzouk,C.,Mercier,M. and Roberfroid,M.B.,*Cancer Lett*. *9*, 123-131 (1980)

11. Roberfroid,M.B., *Arch.Toxicol.* 46,181-193 (1980)

12. Roberfroid,M.B.,*Excepta Medica,Congress series*, in press.

13. Schut,H.A.J.,Wirth,P.J.,and Thorgeirsson,SS., *Molec. Pharmacol.,14,* 682-692 (1978)

14. Selkirk,J.K.,*J.Natl.Cancer Inst.* 64, 771-_74 (1980)

15. Stella,L.,Janousek,Z.,Merényi,R. and Viehe,H.G. Angew.Chem.Int.edit.*18*, 917-932 (1979)

16. Viehe,H.G.,Merényi,R.,Stella,L. and Janousek,Z. Angew.Chem.Int.sdit. *18*, 917-932 (1979)

17. Wattenberg,L.W., *Cancer Res.* 35,3326-3331 (1975)

Modulation of Mutagenesis by Biological Substances

Hikoya Hayatsu

Faculty of Pharmaceutical Sciences, Okayama University, Okayama, Japan

ABSTRACT Modulation of the activity of environmental mutagens by biological substances of small molecular weight was investigated by use of the Salmonella/microsome assay. Hemin was found to strongly inhibit the mutagenicity of polycyclic compounds, e.g., the amino acid pyrolysis products Trp-P-1, Glu-P-1, IQ, etc, and benzo(a)pyrene and aflatoxin B_1. It was found that hemin can form a complex with Trp-P-1 and -2. This complex formation is probably the reason for the lack of S9-mediated metabolic activation of Trp-P-2 in the presence of hemin. Metabolically activated form of Trp-P-2 can also form a complex with hemin, and, as a result, can no longer cause mutation. A diethyl ether extract of human feces was found to contain inhibitors for a range of mutagens. The inhibitors were identified as oleic and linoleic acids. Since these fatty acids are commonly present in fats, natural samples containing fats are expected to show little mutagenicity. This was shown to be the case for grilled ground beef samples. Possible physiological significance of these modulations is discussed.

INTRODUCTION

The salmonella/microsome assay of mutagenicity (1) is widely used for detecting environmental mutagens. It is known that the addition of certain reagents to this assay system can cause modulation of the activity of mutagens. For example, cysteine suppresses the mutagenicity of MNNG[1] (13) and it enhances the activity of Trp-P-1 (11). Search for such modulators is important in two ways. In one way, the presence of such a modulator in the test sample may give "false negative" or "false positive" results. In another, there is the possibility that such modulators can act also in vivo. With regard particularly to the latter possibility, a screening of biological substances for their mutagenesis-modulating effect is obviously desirable. In our effort to that direction, we found that hemin and its analogs are inhibitors for mutagenesis inducible by compounds bearing polycyclic structures (2,3), and that oleic and linoleic acids are also inhibitors for many mutagens (7,8).

INHIBITORY EFFECT OF HEMIN ON MUTAGENESIS

When hemin was present in the preincubation mixture (15) of the Ames test, the mutagenicity of the tryptophan-pyrolyzate product Trp-P-1 was strongly inhibited. Thus, the number of His$^+$ revertant colonies decreased to less than 5 % of original value when 75 nmole hemin was present in the preincubation mixture containing 1.8 nmole Trp-P-1, Salmonella typhimurium TA98 and S9 mix. No killing of the bacteria was detected in this treatment, indicating that the observed decrease in the revertant colonies was not a reflection of killing. Similarly, other amino acid- and protein-pyrolyzate mutagens were also subject to the inhibition by hemin (ref.3 and unpublished results) (see Table 1). The mutagenesis induced by polycyclic aromatic hydrocarbons was also strongly inhibited by hemin (2): Consistently, 1-2 equivalents of hemin added to this class of mutagen exerted a 50 % inhibition. There are mutagens that cannot be inhibited by hemin. This suggests

[1]Abbreviations: MNNG, N-methyl-N'-nitro-N-nitroso-guanidine; Trp-P-1, 3-amino-1,4-dimethyl-5H-pyrido-[4,3-b]indole; Trp-P-2, 3-amino-1-methyl-5H-pyrido-[4,3-b]indole; Glu-P-1, 2-amino-6-methyldipyrido-[1,2-a:3',2'-d]imidazole; Glu-P-2, 2-amino-dipyrido-[1,2-a:3',2'-d]imidazole; Amino-α-carboline, 2-amino-9H-pyrido[2,3-b]indole; Aminomethyl-α-carboline, 2-amino-3-methyl-9H-pyrido[2,3-b]indole; IQ, 2-amino-3-methylimidazo[4,5-f]quinoline; MeIQ, 2-amino-3,4-dimethylimidazo[4,5-f]quinoline; MeIQ$_x$, 2-amino-3,8-dimethylimidazo[4,5-f]quinoxaline; AF-2, 2-(2-furyl)-3-(5-nitro-2-furyl)acrylamide

Table 1. Mutagens examined for inhibition by hemin

Mutagens inhibited	Mutagens *not* inhibited
Trp-P-1, Trp-P-2, Glu-P-1, Glu-P-2, Amino-α-carboline, Aminomethyl-α-carboline, IQ, MeIQ, MeIQ$_x$, Activated Trp-P-1, Activated Trp-P-2, Activated Glu-P-1, Benzo(a)pyrene, Benzo(a)pyrene 4,5-epoxide, 3-Methylcholanthrene, 7,10-Dimethylbenz(a)anthracene, Chrysene, Aflatoxin B$_1$, 2-Acetylaminofluorene, 2-Nitrofluorene	AF-2, MNNG, 4-Nitroquinoline 1-oxide, Nitromin, N-Methyl-N-nitrosourea, N-Nitrosodi-n-butylamine, Quinoxaline 1,4-dioxide, Carbadox

that the effect of hemin is not the one on the bacteria but on the mutagens themselves. The mutagens which were inhibited by hemin generally have polycyclic structures, i.e. the structures consist of three or more rings. A possible mechanism of the inhibition is an interaction between hemin and the mutagen, both having planar structures. In fact, formation of a complex between hemin and Trp-P-1 can be demonstrated by spectrophotometric studies and by precipitating the complex from aqueous solutions (unpublished results). Furthermore, the direct mutagen Trp-P-2*, obtained by treatment of Trp-P-2 with S9 and by subsequent purification on high performance liquid chromatography, can also form a complex with hemin, and the mutagenesis inducible by the Trp-P-2* was strongly inhibited by hemin. An observation consistent with this inhibition was that the inactivation of a B. subtilis transforming DNA by trp-P-2* was abolished when severalfold excess of hemin was added to the reaction mixture (unpublished results).

In a separate experiment, it was shown that hemin can inhibit the S9-mediated metabolic activation of Trp-P-2. Therefore, the inhibition of the Trp-P-2 mutagenesis by hemin takes place at two stages, i.e. at the metabolic activation of Trp-P-2 and at the attack of the activated Trp-P-2 molecule on DNA.

Several biological pyrrole pigments were tested for their inhibitory effects. Protoporphyrin was inhibitory to Trp-P-1, Trp-P-2 and Aminomethyl-α-carboline, although the effects were weaker than those observed for hemin (3).

Similarly, inhibitory effects weaker than those of hemin were found for biliverdin, bilirubin and chlorophyllin in the mutagenesis induced by benzo(a)pyrene (2). Chlorophyllin and biliverdin also showed inhibitions to Trp-P-1, Trp-P-2, Glu-P-1, Glu-P-2, Amino-α-carboline and Amino-methyl-α-carboline (3).

INHIBITORY EFFECT OF OLEIC AND LINOLEIC ACIDS

Since there are several reports from laboratories of foreign countries on fecal mutagens (5,9,14), we ourselves were interested in the search for mutagens in feces of Japanese. We prepared an ether-extract of human feces and explored the possible presence of mutagens in the sample. However, the results of the Ames tests on the sample were all negative. When we mixed this sample with Trp-P-1 in the preincubation mix, we found that the mutagenicity of Trp-P-1 was strongly suppressed. This indicated the presence of mutagenesis-inhibitory factor(s) in the feces extract. We fractionated the inhibitory principles by thin layer chromatography on silica gel, and found that they were oleic and linoleic acids. The presence of 0.07 mg oleic acid/plate inhibited almost completely the mutagenesis of Salmonella TA98 inducible with Trp-P-1 (1.5 nmole) in the presence of S9. 0.05 mg Linoleic acid/plate gave similar results. Oleic and linoleic acids showed inhibitory effects on the mutagens Trp-P-1, Glu-P-1, Amino-α-carboline, IQ, benzo(a)pyrene and aflatoxin B_1 (7). These effects were found by using the Salmonella TA98/S9 system. Since oleic acid is cytotoxic to Salmonella TA100, it was necessary to use other test systems to investigate the effect of oleic acid on base pair substitution mutagens. The system of E. coli WP2,try,hcr/S9 (10) was suitable for this study, and oleic acid was found to inhibit the mutagenicity of N-nitrosodimethylamine (+ S9)(12). In this system, oleic acid did not affect the mutagenicities of 4-nitroquinoline 1-oxide (- S9), MNNG (- S9) or N-methyl-N-nitrosourea (- S9).

Oleic acid inhibited the mutagenesis inducible by the activated Trp-P-1. This suggested a direct interaction between oleic acid and the mutagen. In the case of the inhibition for N-nitrosodimethylamine, oleic acid was found to block the metabolic activation of the mutagen, hence the effect is an indirect one.

The fatty acids are commonly present in fats.

Therefore, samples containing fats may give negative results in mutagenesis assays. In fact, the mutagenicity of the basic fraction of cooked ground beef (4,6) became negative when the acidic fraction of the beef, which contained oleic acid, was mixed with it (8). This provides a warning for those who do the screening of mutagens by the Ames assay: Fatty acids must be removed from the samples to avoid false negatives.

CONCLUDING REMARKS

The porphyrin pigments and the fatty acids are the constituents of organisms. The possibility that they serve as real protectors of higher organisms against the threat of mutagens merits further evaluation. In addition, the presence of these inhibitors in foods raises the possibility that they may change the fate of mutagens in digestive tracts by chemically binding with the mutagens.

ACKNOWLEDGMENTS

I thank Drs. M. Nagao, T. Sugimura, T. Kawachi and S. Nishimura of the National Cancer Center Research Institute for the gift of mutagens and for stimulating discussions. This work was supported by a Grant for Cancer Research from the Ministry of Education, Science and Culture, a Grant-in-Aid for Cancer Research from the Ministry of Health and Welfare, and a grant from the Naito Science Foundation.

REFERENCES

(1) Ames, B. N., McCann, J., and Yamasaki, E. *Mutation Res.*, *31*, 347-364 (1975)
(2) Arimoto, S., Negishi, T., and Hayatsu, H. *Cancer Letters*, *11*, 29-33 (1980)
(3) Arimoto, S., Ohara, Y., Namba, T., Negishi, T., and Hayatsu, H. *Biochem. Biophys. Res. Comm.*, *92*, 662-668 (1980)
(4) Commoner, B., Vithayathil, A. J., Dolara, P., Nair, S., Madyastha, P., and Cuca, G. C. *Science*, *201*, 913-916 (1978)

(5) Ehrich, M., Aswell, J. E., Van Tassell, R. L., and Wilkins, T. D. *Mutation Res., 64,* 231-240 (1979)
(6) Felton, J. S., Healy, S., Stuermer, D., Berry, C., Timourian, H., Hatch, F. T., Morris, M., and Bjeldanes, L. F. *Mutation Res., 88,* 33-44 (1980)
(7) Hayatsu, H., Arimoto, S., Togawa, K., and Makita, M. *Mutation Res., 81,* 287-293 (1980)
(8) Hayatsu, H., Inoue, K., Ohta, H., Namba, T., Togawa, K., Hayatsu, T., Makita, M., and Wataya, Y. *Mutation Res., 91,* 437-442 (1981)
(9) Kuhnlein, U., Bergstrom, D., and Kuhnlein, H. *Muation Res., 85,* 1-12 (1981)
(10) Mochizuki, M., Suzuki, E., Anjo, T., Wakabayashi, Y., and Okada, M. *Gann, 70,* 663-670 (1979)
(11) Negishi, T., and Hayatsu, H. *Biochem. Biophys. Res. Comm., 88,* 97-102 (1979)
(12) Negishi, T., Ohara, Y. and Hayatsu, H. *in Seventh International Meeting on N-Nitroso Compounds: Occurrence and Biological Effects, Bartsch, H., O'Neill, I. K., Castegnaro, M., and Courcier, M. M., eds, International Agency for Research on Cancer, Lyon,* in press (1981)
(13) Rosin, M. P., and Stich, H. F. *Mutation Res., 54,* 73-81 (1978)
(14) Wang, T., Kakizoe, T., Dion, P., Furrer, R., Varghese, A. J., and Bruce, W. R. *Nature, 276,* 280-281 (1978)
(15) Yahagi, T., Nagao, M., Seino, Y., Matsushima, T., Sugimura, T., and Okada, M. *Mutation Res., 48,* 121-130 (1977)

Biotransformation and Interaction of Chemicals as Modulators of Mutagenicity and Carcinogenicity

Silvio De Flora

Institute of Hygiene, University of Genoa, Genoa, Italy

ABSTRACT The influence of some biotransformation and interaction phenomena on the genotoxic properties of chemicals and on their potential effects *in vivo* was investigated in the *Salmonella*/microsome test and in repair-deficient *E. coli*. Over 30 metabolising systems, including human preparations, were used to detect the selective activation or deactivation of some mutagens. In many cases, relationships could be detected between chemical features and metabolic trends, and a number of noncarcinogenic mutagens were found to undergo metabolic deactivation *in vitro*. The effect of human gastric juice on the activity of some mutagens, the formation of mutagenic nitrosoderivatives and other interactions were investigated also with the aim of improving the overlapping of mutagenicity and carcinogenicity data. Exposure of some chemicals either to UV light or to sunlight had various effects, including the conversion of known procarcinogens into direct-acting mutagens.

INTRODUCTION

Bacterial test systems have provided an important tool not only as screening methods but also as experimental models for investigating, under controlled conditions, a variety of problems associated with the initiation of cancer. I report here the results of some studies aiming at elucidating the influence of metabolic and/or interaction mechanisms on the genotoxic properties of chemicals and on their potential effects *in vivo*.

GENOTOXICITY OF CHEMICALS IN BACTERIAL TEST SYSTEMS

An analysis of mutagenicity patterns in the *Salmonella*/microsome (Ames) reversion test has been so far completed in this laboratory for 121 organic or inorganic chemicals (Table 1) and 24 complex mixtures (Table 2). The experimental protocol intentionally included a number of noncarcinogenic mutagens - e.g. 19 out of 29 reported in four literature papers (1,20,34,35), i.e. **A6-8,B2-4,C4,C8,D2,E1,H1-2,L8,M2,N13, N18-19,O4-5** - as well as nonmutagenic carcinogens, e.g. **C2, C9,F3,F10,L9-10,N1,N16**. The genotoxic activity of some compounds (**F8-9,F12-17,H2,H4,I2-3,L7-8,N5,N12,N14-15,O4,P1,P6**) or mixtures (see later) was also evaluated by checking their selective lethality towards five trp^- *E.coli* strains, including WP2 (repair-proficient), WP2uvrA ($uvrA^-$), WP67 ($polA^-$), CM871 ($uvrA^-,recA^-,lexA^-$) and TM1080 ($polA^-,lexA^-$,plasmid R391). With the exceptions of **N5** (positive) and **O4** (negative), the results were in agreement with those of the Ames test and showed the convenience of using a battery of strains lacking a variety of DNA repair mechanisms.

INFLUENCE OF CHEMICAL FEATURES ON MUTAGENICITY PATTERNS

In vitro metabolic studies do not take into account pharmacodynamic processes and some host mechanisms, and additionally they are subject to the same variability factors also affecting *in vivo* evaluations. Moreover, the quantitative ratio of test chemicals to metabolising systems may be quite different *in vitro* or in the organism. For these reasons, extrapolation of *in vitro* findings to animals or to humans must be very careful. Nevertheless, the assessment of qualitative metabolic trends in the presence of S9 fractions or of other biological preparations may provide useful data to improve the evaluation of potential health hazards.

Such a view is validated by some evident associations between chemical features of test compounds and the three mutagenicity parameters investigated in the Ames test, i.e. (1) the molecular mechanism of reversion, (2) the mutagenic potency (also influencing the spectrum of reverted strains and (3) the metabolic trends. For instance, the mutagenic compounds of Table 1 falling in the classes of inorganics (**F1,O4-5**), aliphatics (**F3-4,N7,N11-13,N15**), hydrazides (**F11, F13**) and epoxides as **M2,N18-20** (plus mat. 2 of Table 2) are all direct-acting base-substituting agents of medium-low potency, whose mutagenicity is often decreased by S9 mix. On the contrary, mutagenic polycycles (**A2-6,C4-8**), heterocycles (**I1**), benzidines (**D1-2**) and azo compounds (**E1-2**) are all frameshift agents of medium-high potency requiring metabolic activation by S9 mix. These patterns can be modified by introducing some radicals in the molecule. For instance, NO_2 groups clearly favour the mutagenic response (1) by broadening the spectrum of sensitive strains (e.g. **B1** *vs* other PAH's, **N12** *vs* other aliphatics, **G1** *vs* **G2**), (2) by enhancing the mutagenic potency (e.g. **F7** *vs* **F5**) and (3) by afford-

Table 1. Study of 121 compounds in the Ames reversion test

Compound	Source	TA1535	TA1537	TA1538	TA98	TA100	Range of activity (nmoles/plate)	Potency (revs/nmol)	Metabolic effect
A. POLYCYCLIC AROMATIC HYDROCARBONS									
1) 3-Hydroxybenzo(a)pyrene	D	+	+	+	+	+	0.7–75	38	MI
2) Benzo(a)pyrene (BaP)	h	–	+	+	+	+	0.4–200	185	AC
3) Benz(a)anthracene	m	–	+	+	+	+	20–170	12	AC
4) 3-Methylcolanthrene	D	–	+	+	+	+	1.9–298	24	AC
5) 9,10-Dimethylanthracene	a	–	w	+	+	+	$1-24 \times 10^2$	2.1	AC
6) Dibenz(de,kl)anthracene (Perylene)	v	–	+	+	+	±	10–80	1.9	AC
7) Dibenz(a,h)anthracene-5,6-oxide	D	–	w	w	w	+	2–17	23	SD
8) 1-Naphthol	b	–	–	–	–	–	-10^4		
B. NITRO AROMATICS & HETEROCYCLES									
1) 2-Nitronaphthalene	h	+	+	+	+	+	57–578	4.4	NC
2) Nitrofurantoin	u	–	–	–	w	+	0.2–3.4	775	NC
3) 5-nitro-2-furamidoxime	B	–	–	–	w	+	0.1–1.9	740	D
4) 5-Nitro-2-furoic acid	h	–	–	–	±	+	100–700	4.3	D
5) 2,4-Dinitrophenol	b	–	–	–	–	–	-2.7×10^3		
C. AROMATIC & HETEROCYCLIC AMINES & DERIVATIVES									
1) Aniline	d	–	–	–	–	–	-2.2×10^5		
2) Aminoantipyrine (AAP)	d	–	–	–	–	–	-1.2×10^5		
3) Aminopyrine (AP, Piramidone)	d	–	–	–	–	–	-2.2×10^3		
4) 1-Naphtylamine	v	–	–	+	+	+	$0.2-3 \times 10^3$	0.71	AC
5) 1-Aminoanthracene	h	–	+	+	+	+	13–830	27	AC
6) 2-Aminoanthracene	v	+	+	+	+	+	1–415	560	AC
7) 2-Aminofluorene	h	–	w	+	+	+	3.5–110	180	AC
8) 4-Acetylaminofluorene	C	–	–	+	+	+	220–1800	0.14	AC
9) Para-rosaniline	n	–	–	–	–	–	-6.6×10^3		
D. BENZIDINE DERIVATIVES									
1) Benzidine dichloride	p	–	–	+	+	–	97–3110	0.25	AC
2) 3,3'-Diaminobenzidine	v	–	w	+	+	–	20–1900	4.2	AC
E. AZO COMPOUNDS									
1) N-Hydroxy-4-aminoazobenzene·HCl	A	–	–	+	+	+	50–3200	0.62	I
2) N,N-Dimethyl-4-aminoazobenzene	h	–	–	w	+	+	$1-4 \times 10^3$	0.11	AC
F. HYDRAZINE DERIVATIVES									
1) Hydrazine hydrate	i	+	–	–	–	–	$1-10 \times 10^3$	0.007	D
2) 1,1-Dimethylhydrazine (1,1-DMH)	i	–	–	–	±	–	$2-16 \times 10^4$	0.0008	SI
3) 1,2-Dimethylhydrazine (1,2-DMH)	n	–	–	–	–	+	$1-4 \times 10^3$	0.009	NC
4) Carbamylhydrazine	m	+	–	–	–	–	$8-70 \times 10^3$	0.0009	D
5) Phenylhydrazine	d	–	+	–	–	–	$5-50 \times 10^2$	0.033	SD
6) 1-Carbamyl-2-phenylhydrazine	d	–	–	–	–	–	-1.3×10^5		
7) 2,4-Dinitrophenylhydrazine	a	–	+	+	+	+	$2-50 \times 10^2$	0.96	D
8) Mebanazine oxalate	l	–	–	w	w	–	$1-5 \times 10^3$	0.07	NC
9) Phenelzine·H$_2$SO$_4$	z	+	–	–	–	+	$5-80 \times 10^2$	0.19	D
10) Procarbazine·HCl (Natulan)	k	–	–	w	w	–	$2-16 \times 10^4$	0.0005	SI
11) Isonicotinic acid hydr. (Isoniazid)	y	w	–	–	–	±	$7.5-25 \times 10^4$	0.0004	D
12) Iproniazid	k	–	–	–	–	–	-1.7×10^5		
13) Nialamide	q	+	–	–	–	+	$0.5-40 \times 10^3$	0.62	D
14) Isocarboxazid	k	–	–	–	–	–	-56×10^3		
15) Hydralazine·HCl	f	+	–	–	+	+	$5-50 \times 10^3$	0.05	NC

Table 1 (continued)

Compound	Source	TA1535	TA1537	TA1538	TA98	TA100	Range of activity (nmoles/plate)	Potency (revs/nmol)	Metabolic effect
16) Dihydralazine·H_2SO_4	f	−	+	+	+	+	$15-30\times10^3$	0.03	NC
17) Endralazine mesilate	g	−	−	+	+	−	$3-22\times10^3$	0.003	NC
G. 4-SUBSTITUTED QUINOLINO-N-OXIDES									
1) 4-Nitroquinolino-N-oxide (4NQO)	D	+	+	+	+	+	$0.3-6.2$	2880	MD
2) 4-Hydroxylaminoquinolino-N-oxide	D	−	−	+	+	+	$3-182$	56	MD
3) 4-Aminoquinolino-N-oxide (4AQO)	B	w	w	+	+	+	$0.2-31\times10^3$	0.38	I
H. ACRIDINE DERIVATIVES									
1) Acriflavine·HCl	b	−	+	+	+	+	$50-3690$	0.43	I
2) ICR 191·2HCl	r	−	+	+	+	+	$1-200$	480	MD
3) ICR 191-OH·2HCl	x	−	+	−	−	−	$50-250$	14.5	MD
4) ICR 170·2HCl	r	−	+	+	+	+	$10-200$	270	D
5) ICR 170-OH·2HCl	x	−	+	−	−	−	$100-200$	3.3	D
I. MISCELLANEOUS HETEROCYCLES									
1) Aflatoxin B1 (AFB1)	v	−	−	+	+	+	$0.2-5$	1950	AC
2) Cimetidine	w	−	−	−	−	−	-8×10^4		
3) Ranitidine·HCl	j	−	−	−	−	−	-2.8×10^3		
4) Riboflavine	v	−	−	−	−	−	-1×10^4		
5) Chloracizine·HCl	c	−	−	−	−	−	-2.5×10^3		
6) N-Demethyldiazepam	s	−	−	−	−	−	-1.5×10^4		
L. MISCELLANEOUS ORGANICS									
1) Cholesterol	v	−	−	−	−	−	-1×10^4		
2) Prostaglandin E2 (PGE2)	v	−	−	−	−	−	-1×10^2		
3) L-Thyroxine Na salt	v	−	−	−	−	−	-2×10^4		
4) Saccharin Na salt	v	−	−	−	−	−	-5×10^5		
5) Streptomycin	v	−	−	−	−	−	-1×10^3		
6) Benzetonium chloride	h	−	−	−	−	−	-1×10^2		
7) Captan	u	+	+	w	+	+	$3-33$	26	D
8) Folpet (Phaltan)	e	+	+	w	+	+	$16-133$	8	NC
9) Dieldrin	u	−	−	−	−	−	-2.6×10^4		
10) DDE	h	−	−	−	−	−	-3.2×10^4		
M. STYRENE & DERIVATIVES									
1) Styrene	b	−	−	−	−	−	-5×10^3		
2) Styrene epoxide	m	+	−	−	−	+	$2.4-24\times10^3$	0.06	D
3) Styrene glycol	m	−	−	−	−	−	-1.4×10^5		
N. MISCELLANEOUS ALIPHATICS & EPOXIDES									
1) Carbon tetrachloride	n	−	−	−	−	−	-6.5×10^4		
2) Methanol	d	−	−	−	−	−	-2.5×10^6		
3) Ethanol	d	−	−	−	−	−	-1.7×10^6		
4) Ethyl ether	d	−	−	−	−	−	-1×10^6		
5) Formaldehyde	b	−	−	−	−	−	-1.3×10^3		
6) Acetone	d	−	−	−	−	−	-1.3×10^6		
7) β-Propiolactone	v	+	−	−	−	+	$3-55\times10^2$	1.2	NC
8) Ascorbic acid	p	−	−	−	−	−	-1.4×10^4		
9) Monosodium glutamate	b	−	−	−	−	−	-1.2×10^5		
10) Ethylenediaminetetracetate·2Na	d	−	−	−	−	−	-1.2×10^3		
11) Dimethylnitrosamine (DMN)	h				−	+	$1.3-2.7\times10^5$	0.014	AC

Table 1 (continued)

Compound	Source	TA1535	TA1537	TA1538	TA98	TA100	Range of activity (nmoles/plate)	Potency (revs/nmol)	Metabolic effect
12) N-Methyl-N-nitroso-N'-nitroguanidine (MNNG)	D	+	+	+	+	+	0.8-68	530	NC
13) Trimethylphosphate	b	-	-	-	-	+	$6-11 \times 10^5$	0.0003	SI
14) Dimethylsulfoxide (DMSO)	b	-	-	-	-	-	-1.4×10^6		
15) Methylmethane sulfonate (MMS)	n	+	-	-	-	+	85-2727	1.5	NC
16) Thioacetamide	d	-	-	-	-	-	-6.7×10^4		
17) Sodium diethyldithiocarbamate	t	-	-	-	-	-	-9.4×10^3		
18) 1,2-Epoxybutane	m	+	-	-	-	+	$7-56 \times 10^4$	0.003	NC
19) 2,3-Epoxy-1-propanol (Glycidol)	m	+	-	-	-	+	$1.7-27 \times 10^3$	0.06	NC
20) 1,1,1-Trichloro-2,3-propenoxide	a	+	-	-	-	+	$0.6-25 \times 10^2$	1.2	SD
O. MISCELLANEOUS INORGANICS									
1) Sodium chloride	d	-	-	-	-	-	-1.6×10^3		
2) Sodium hydroxide	d	-	-	-	-	-	-5×10^4		
3) Anhydrous sodium bicarbonate	ß	-	-	-	-	-	-5×10^5		
4) Sodium nitrite	b	+	-	-	-	+	$4-14 \times 10^4$	0.005	SD
5) Sodium azide	n	+	-	-	-	+	15-150	27	D
6) Sodium sulfite	n	-	-	-	-	-	-8×10^3		
7) Potassium cyanide	n	+	-	-	-	-	-3×10^3		
8) Potassium permanganate	n	-	-	-	-	-	-3.2×10^3		
9) Cadmium chloride	b	-	-	-	-	-	-2.7×10^3		
10) Sulfamic acid	n	-	-	-	-	-	-5.1×10^4		
P. CHROMIUM COMPOUNDS									
1) Sodium dichromate	n	-	w	w	w	+	50-250	4.4	MD
2) Potassium chromate	b	-	w	w	w	+	80-410	2.9	MD
3) Calcium chromate	b	-	w	w	w	+	60-290	3.2	MD
4) Strontium chromate	d	-	w	w	w	+	60-250	2.8	MD
5) Ammonium chromate	d	-	w	w	w	+	50-320	3.7	MD
6) Lead chromate	d	-	w	w	w	+	65-380	2.3	MD
7) Zinc yellow	o	-	w	w	w	+	90-590	1.4	MD
8) Chromium trioxide	n	-	w	w	w	+	40-220	5.1	MD
9) Chromium orange	o	-	w	w	w	+			MD
10) Molybdenum orange	o	-	w	w	w	+			MD
11) Chromium yellow	o	-	w	w	w	+			MD
12) Chromyl chloride: liquid phase	b	-	w	w	w	+	100-430	1.8	MD
" " : vapour "						+			MD
13) Chromium carbonyl	b	-	-	-	-	-	-2.3×10^3		
14) Chromic chloride	b	-	-	-	-	-	-3×10^4		
15) Chromic nitrate	t	-	-	-	-	-	-2×10^4		
16) Chromic potassium sulfate	b	-	-	-	-	-	-1.6×10^4		
17) Chromic acetate	b	-	-	-	-	-	-7×10^4		
18) Neochromium	o	-	-	-	-	-	-5×10^4		
19) Chromium alum	o	-	-	-	-	-	-4.3×10^4		
20) Chromite (contaminated with Cr^{6+})	o	-	-	-	-	+	2 mg (spot test)		

Footnotes to Table 1:

Compounds: see ref. *6* for the formula of pure chemicals, for the composition of industrial products (P7,9-11,18-20) and for other remarks concerning 106 out of the 121 compounds listed in this table.

Source of compounds: a=J.T.Baker Chem., b=BDH, ß=British Chrome & Chem., c=Bruco, d=Carlo Erba, e=Chem.Manufacture Lab., f=Ciba-Geigy, g=DGI-Sandoz, h=Ega-Chemie, i=Fluka, j=Glaxo, k=Hoffmann-La Roche, l=ICI, m=ICN-K&K Lab., n=Merck, o=Montedison, p=Noury-Baker, q=Pfizer, r=Polysciences, s=Ravizza, t=Riedel-de Haen, u=Serva Feinbiochemica, v=Sigma, w= Smith,Kline & French, x=Terochem, y=Vecchi & C.Piam, z=W.R.Warner.
Gifts: A=Dr.Y.Hashimoto, B=Dr.M.Nagao, C=Dr.E.Yamasaki, D=IIT Res.Inst.

Reverted strains: + =more than 3-fold,dose-related and reproducible increase of revertants; w =2 to 3-fold increase; ± =reproducible but less than 2-fold increase of revertants. The range of spontaneous revertants of the 5 tester strains is reported in Ref. 6.

Range of activity: min-max levels of activity of mutagenic chemicals or max dose tested of negative chemicals, corresponding in most cases either to 100 ul of liquid compounds, or to the limit of solubility of solid compounds, or to toxic levels.

Potency: calculated by dividing the number of net revertants at the top level of the linear part of dose-response curves, in the most sensitive strain and under the most favourable metabolic condition (i.e. with or without S-9 mix), by the corresponding amount of compound.

Metabolic effect: the effect of S-9 mix containing liver S-9 fractions from Aroclor-treated rats on the mutagenic response was evaluated by varying the amounts both of compound and of S-9 fraction. Since the metabolic effects vary in different points of dose-response curves, they are not expressed in numerical terms, but rather provide a somewhat arbitrary indication of in vitro metabolic trends. Key: NC =no change, AC = activation, MI =marked increase, I =increase, SI =slight increase, MD = marked decrease, D =decrease, SD =slight decrease.

Table 2. Study of complex mixtures or of contaminated materials in the Ames test. See footnote to Table 1 for the key.

Test material	TA1535	TA1537	TA1538	TA98	TA100	Metabolic effect	Reference
1) An Agaricus bisporus extract containing agaritine	−	±				NC	9
2) Plastic plates sterilized with ethylene oxide	+	−	−	−	+	SD	6
3-5) Three lab reagents containing 0.1% azide	+	−	−	−	+	D	10
6-9) Four vaccines containing formalin	−	−	−	−	−		6
10-16) Seven oil dispersants	−	−	−	−	−		30
17) A sample of paraffin	−	−	−	−	−		30
18) A sample of Arabian crude oil & its DMSO extract	−	−	−	−	−		30
19) An industrial effluent	−	−	−	−	−		28
20-23) Two pairs of fly ash samples from a dual-fired (coal and fuel oil) power plant:							
−downstream electrostatic precipitator (June 78)				w	−	NC	
−electrostatic precipitator hoppers (" ")				w	−	SI	
−electrostatic precipitator hoppers (July 78)				−	−		
−upstream electrostatic precipitator (" ")				−	−		
24) Unfractioned cigarette smoke condensate	−	−	+	+	±	AC	

ing a direct mutagenicity to cyclic compounds (**B2-4**). The NH$_2$ group appears to be important for the mutagenicity of some activable compounds (e.g. **C4-C8**), as also inferred from the comparison with structurally related chemicals lacking this group (e.g. **D2** vs **D1**, **C4** vs **A8**). On the contrary, geminal CH$_3$ groups determine a decrease of mutagenicity, regardless of the atoms bearing them, being present either in negative compounds (**C3,F12,N14**), or in borderline mutagens whose activity is slightly enhanced by S9 mix (**F2,F10,N13**), or in weak mutagens requiring liquid preincubation with S9 mix and bacteria (**E2,N11**). Other structure-activity relationships can be inferred from the observation of Table 1. Of particular interest is the class of hydrazine derivatives (**F1-17**), which apparently is the only one showing a poor homogeneity of results. Actually, as described more in detail in another paper (*14*), both the structural features of these compounds (e.g. the more or less pronounced planarity of molecules) and the addition of a variety of radicals to the H$_2$N-NH$_2$ group justify the broad variations of mutagenicity patterns among hydrazines and hydrazides.

METABOLIC DEACTIVATION OF NONCARCINOGENIC MUTAGENS

A number of mutagens negative in animal assays show a decreased activity in the presence of liver S9 fractions or of other metabolising systems. This means that these compounds should not be classified as 'false positives', but rather as true mutagens possibly undergoing metabolic deactivation. Some examples, as reported in previous papers (*4-6, 10*), include Na nitrite (**O4**) and azide (**O5**), 5-nitro-2-furamidoxime (**B3**) and 5-nitro-2-furoic acid (**B4**), the pesticide captan (**L7**), dibenz(a,h)anthracene-5,6-oxide (**A7**) and styrene oxide (**M2**), while the possible detoxification of two other noncarcinogenic epoxides, i.e. 1,2-epoxybutane (**N18**) and 2,3-epoxy-1-propanol (**N19**), could not be demonstrated with the S9 fractions used. Of particular interest is the case of the antitumor ICR compounds. In fact, ICR191 (**H2**) (noncarcinogenic) was deactivated more efficiently and rapidly than ICR170 (**H4**) (carcinogenic) by mouse (liver > lung) or rat (liver > testis > kidney > lung > striated muscle > spleen) S9 fractions. The frameshift activity of HPLC-pure ICR191-OH (**H3**), also negative in the same animal system (*25*), was sharply decreased by liver S9 preparations as well (*6*).

COMPARATIVE EFFECT OF VARIOUS METABOLISING SYSTEMS ON MUTAGENS

Some mutagens have been comparatively tested in the presence of over 30 metabolising systems from different animal species, including human preparations, also under the influce of diseases, drugs, inducers or peculiar diets (Fig.1). The assessment of metabolic behaviour may even be useful to understand the target of activable or deactivable mutagens, especially when the ultimate DNA-binding metabolites are formed within target cells. This may be the case for chromium compounds. In fact, although Cr(III) is the putative ultimate

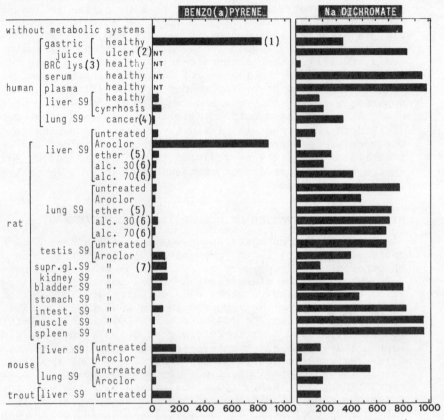

NT= not tested. (1) Preincubated with gastric juice and subsequently with Aroclor rat liver S9. (2°) Ulcer patients treated with an antisecretory drug. (3) Blood red cell lysates (see ref.26). (4) Peripheral tissue surrounding the cancer. (5) Daily narcosis with ether for 7 days. (6) Diet low in proteins and lipoproteic factors plus 20% alcohol instead of drinking water (ref. 3), administered for 30 or 70 days. (7) Suprarenal glands.

Fig.1. Mutagenicity of benzo(a)pyrene (1 µg) and dichromate (30 µg) in the absence or in the presence of fluids, cell lysates or S9 mix (0.5 ml) containing 100 µl S9 fractions from human, rat (Sprague-Dawley), mouse (Swiss albino) or trout (*Salmo gairdnerii*) tissues.

mutagen, only Cr(VI) is able to pass cell membranes and is thus the hazardous form of this metal outside cells. On this ground, reduction of Cr(VI) to Cr(III) by gastric juice (*11*) or inside erythrocytes (*26*) is consistent with its probable lack of carcinogenicity by the oral route or at distance from the administration site (*18*). As shown in Fig.1, the mutagenicity of Cr(VI) was efficiently decreased - through NADPH-requiring and Aroclor-inducible enzymic pathways - by human (liver > lung), mouse (liver > lung), trout (liver) and rat (liver > suprarenal glands > kidney > testis > stomach > lung) S9 fractions, while no reduction was afforded by other rat S9 fractions (bladder, intestine, striated muscle and spleen) (*4,26,31*). On the other hand, conversion of Cr(III) to Cr(VI) could be demonstrated *in vitro* only in the presence of strongly oxidising agents and not of any of the metabolising systems tested (*27*). The reduction of Cr(VI) by S9 mix is presumably the expression of an intracellular detoxification, since the Cr(III) produced in the endoplasmic reticulum is likely to be trapped by cytoplasmic ligands. Conversely, its lack of reduction - e.g. in striated muscles of locally injected animals - suggests that Cr(VI) can reach the cell nucleus, where such a conversion can occur close to the DNA target. The demonstration that also human lung preparations have some activity in reducing Cr(VI) mutagenicity is of peculiar relevance. In fact, this finding suggests that the initiation of lung cancer by chromium in humans (*18*) may depend on a quantitative balance between the inhaled metal deposited in the lower respiratory tract and cell defense mechanisms, which might explain the conflicting nature of epidemiologic studies.

MUTAGENS AND CARCINOGENS IN THE GASTRIC ENVIRONMENT

Preincubation of mutagens with human gastric juice is compatible with the subsequent addition of liver S9 fractions (*11*), thus reproducing *in vitro* two consecutive (gastric-hepatic) metabolic steps which are crucial for the biotransformation of foreign chemicals introduced by the oral route. The mutagenicity of some compounds was found to be affected in such an environment, either in the sense of decrease (e.g. dichromate (**P1**) and azide (**O5**)), or of stabilization (e.g. captan (**L7**)) or even of potentiation (e.g. ICR170 (**H4**)). These changes were generally pH-dependent - which has been demonstrated also for other mutagens , e.g. nitrosamines (*24*) - but in the case of dichromate also a chemical reduction was involved (*11*). The lack of carcinogenicity of azide - as assessed by the oral route or by gastric intubation (*36*) - can be also justified by the loss of mutagenicity afforded both by gastric juice and by liver S9 fractions (*7,10,11*).

The most typical and largely investigated interaction in

the organism, and chiefly in the gastric environment, is the
formation of mutagenic and carcinogenic N-nitrosoderivatives.
We have been investigating this phenomenon by using human
gastric juice as the reaction medium and by checking variability factors (e.g. doses of precursor compounds, pH, temperature, preincubation time, inhibitors) in order to mimic an
in vivo situation under controlled experimental conditions.
By this way, we have found that also primary amines readily
combine with nitrite to form direct-acting mutagenic derivatives (the corresponding diazonium salts), which are stable in
the case of AAP (**C2**), a nonmutagenic carcinogen (*20,21*), while the benzenediazonium salt formed from aniline (**C1**) is then
converted into toxic phenols (*2*). The tertiary amine AP (**C3**)
reacts with nitrite to form DMN (**N11**), whose detection requires a further preincubation step with bacteria and S9 mix (*6*).

The sensitivity and flexibility of the *Salmonella*/micro-
some test is demonstrated by the examples of ICR170 (**H4**)(tertiary amine on the side chain) and, more efficiently, of ICR
191 (**H2**) (secondary amine), which are both S9-deactivable fra̅-
meshift agents mainly reverting the *hisC3076* mutation. In
fact, their nitrosoderivatives are mutagens requiring metabolic activation and showing a distinctive frameshift specificity (*hisD3052*) (*13,15*). The most suggestive case of nitrosation was provided by the widely used anti-ulcer drug cimetidine
(**I2**), since a mutagenic response was detected even by adding
nitrite to the gastric juice of patients receiving this drug.
Ascorbic acid efficiently prevented such a reaction (*12*). Another potent inhibitor of H_2-histamine receptors, i.e. ranit-
idine (**I3**) though with somewhat different patterns, also forms
a nitrosoderivative inducing base-pair substitutions in both
his^- *S.typhimurium* and trp^- *E.coli* and is responsible for a
higher lethality in repair-deficient *E.coli*.

INTERACTION BETWEEN CHEMICALS IN THE AMES TEST

Some interactions investigated in the progress of our *in
vitro* studies are shortly summarized in the following lines:
(1) BaP (**A2**) + TCPO (**N20**) (inhibitor of epoxide hydrolases) =
slight increase of BaP mutagenicity. (2) BaP + seawater or
(3) BaP + each of 7 oil dispersants (materials 10-16 of Table
2) = mutagenicity of BaP unaffected (*29,30*). (4) BaP + Arab-
ian crude oil or its DMSO-extract or (5) BaP + crude oil +
oil dispersants = mutagenicity of BaP eliminated (*29,30*).
(6) BaP + mais oil = decrease of BaP mutagenicity (*30*).
(7) PGE2 (**L2**) + BaP (**A2**) or 2AF (**C7**) or 4HAQO (**G2**) or dichromate (**P1**) = no effect on mutagenicity nor on metabolic activation (BaP,2AF) or deactivation (4HAQO,dichromate). (8) Na
nitrite (**O4**) + sulfamic acid (**O10**) = mutagenicity of nitrite
efficiently removed at equimolar doses (*12*). (9) Thioacetami-

de (**N16**) + thyroxine (**L3**) = no effect. (10) Cr(VI)+ reducing chemicals (ascorbic acid (**N8**), Na sulfite (**O6**)) or metabolites (NADH,NADPH,GSH) = loss of Cr(VI) mutagenicity (*26*). (11) Cr(III)+ K permanganate (**O8**) = oxidation to mutagenic Cr(VI) (*27*). (12) Na dichromate (**P1**)+ BaP or (13) dichromate + cigarette smoke condensate (CSC) (**24**,Table 2) = in the absence of S9 mix, mutagenicity of dichromate unaffected by BaP and lowered by CSC, due to reducing agents in CSC. In the presence of S9 mix, less than additive response, presumably due to inhibitory or competitive effects of Cr(VI) on BaP- or CSC-activating metabolic pathways. Although the above findings may add some useful information, it is clear that other possible factors acting *in vivo* should be also taken into account for a more complete evaluation.

ACTIVATION OF PROMUTAGENS BY ULTRAVIOLET AND SUNLIGHT

The epidemiologic importance of sunlight in the etiology of human cancer is well recognized (*19*), and the effects and mechanisms of action of UV light have been widely investigated, together with some possible interactions with chemicals (*37*). However, with few important exceptions (e.g. refs. *17,22,23*), in my knowledge the circumstance that UV and/or sunlight can also determine an activation of promutagens in the extracellular environment has been scarcely explored, especially if compared with the very rich literature on the metabolic activation of the same compounds.

We have been investigating this phenomenon by exposing thin layers of DMSO solutions of organic compounds either to sunlight or to UV light, at the center of a source composed of four 254 nm 4-Watt lamps, arranged in a vertical position at a distance of 100 mm each from the other. As inferred from experiments with strain TA98 (Table 3) and TA100 (data not shown) of *S.typhimurium*, both these physical agents variously

Table 3. Effects of UV light (30 min treatments) and of sunlight (30 min exposures, i.e. 21 to 34 Cal/sq cm) on the mutagenicity of some compounds dissolved in DMSO.

Compound (amount per plate)	Number of revertants ($m \pm 2s$ of triplicates)					
	untreated		UV light		sunlight	
	S9-	S9+	S9-	S9+	S9-	S9+
DMSO (100 ul)	21±4	28± 7	23± 6	29± 2	24± 5	27±6
Benzo(a)pyrene (10 ug)	20±2	267±29	65±11	72±18	87±21	171±23
3-Hydroxybenzo(a)pyrene (2 ug)	54±8	481±36	38± 3	33±11	45± 8	56±5
3-Methylcolanthrene (10 ug)	24±5	228±19	98±23	83± 8	89±25	197±26
Perylene (20 ug)	25±9	93±14	143±11	138±28	121±31	112±19
2-Aminofluorene (5 ug)	25±3	924±92	114±19	107±24	188±36	616±59
3,3'-Diaminobenzidine (200 ug)	19±6	980±96	35±8	1566±85	38±5	1385±98
p,p'-DDE (5 mg)	22±5	26± 6	44±8	53± 6	35±2	39±5
Aflatoxin B1 (2 ug)	27±3	573±62	25±6	816±79	29±8	694±48
Cigarette smoke cond.(0.2 cig.)	24±7	438±51	23±4	247±33	26±6	209±18

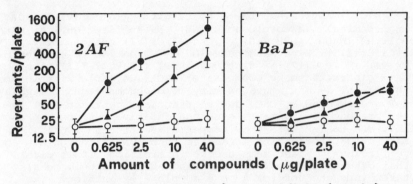

o **untreated**   ● **sunlight (30 min)**   ▲ **UV light (30 min)**

Fig.2. Number of revertants (m±2s of tripicates) induced by varying amounts of 2AF and BaP in *S. typhimurium* TA98, in the absence of S9 mix.

affected the mutagenicity of all test substances and determined the appearance of a direct mutagenicity in some compounds, i.e. BaP,3MC,perylene,2AF and, borderline,DAB and p,p'-DDE. Of peculiar interest is the case of the chlorinated hydrocarbon p,p'-DDE (L10), a carcinogenic pesticide negative in the Ames test (20,21) which, more convincingly following UV irradiation, acquired a weak yet reproducible mutagenicity towards both TA98 and TA100, either with and without S9 mix. On the contrary, both UV and sunlight decreased the direct (S9-deactivable) mutagenicity of the pesticide captan (L7) in TA 100 (data not shown).

The UV- and sunlight derivatives of 2AF and BaP yielded positive results when tested, without S9 mix, in repair-deficient *E.coli*. The effects of UV light could be prevented by shielding chemical solutions with glass slides, while sunlight was equally activge even when test solutions were kept in sealed glass tubes.

Activation of promutagens was related to the doses of irradiated compounds (Fig.2). Time-course experiments demonstrated that (a) activation can be detected even after a very

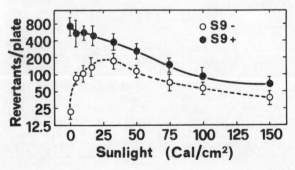

Fig.3. Number of revertants (m±2s of triplicates) induced by 2AF (5 ug) in *S. typhimurium* TA98 following exposure to sunlight in a sealed glass tube.

short exposure period (e.g. 1 min of UV treatment), (b) a
long-lasting exposure may result in a decrease of direct mutagenicity (Fig.3) and (c) the optimal doses of radiations
vary from compound to compound. The activated mutagens appear
to be fairly stable. In fact, only a slight decline of direct
mutagenicity could be detected after 30 days of conservation
in the dark at 24-28°C of both BaP and 2AF photoderivatives.

No attempt has been so far made to identify the directacting products. UV irradiation is known to generate ozone
from molecular oxygen, while the energy of visible light is
sufficient to elevate oxygen to singlet oxygen, which is very
reactive (16). Experiments with photosensitizers (riboflavine
and eosin) and quenchers (azide) of such a reaction are now
in progress.

Independently of the putative mechanisms, the findings
herein reported appear to involve some practical implications. From the technical point of view, the possible photoactivation of promutagens, even in glass tubes, introduces
a new element of uncertainty in the performance and interpretation of both *in vitro* and *in vivo* assays. Another important implication is that, due to the ubiquity of PAH's and
their derivatives, we must infer that large amounts of direct-acting frameshift mutagens are presumably disseminated in
our environment. Such a conclusion, together with the demonstrated reaction of PAH's with ozone (33) or with nitrogen oxides (32), may justify the findings that environmentally detected mutagens - e.g. in the atmosphere or in waters or resulting from combustion processes - are generally frameshift
agents which do not require metabolic activation. Finally,
the possibility that UV-treatment may activate some chemicals
following either direct human exposure or its use as a disinfectant agent (e.g. in the treatment of chemically contaminated drinking water) seems to deserve attention.

ACKNOWLEDGEMENTS

I thank Professors F.L. Petrilli and V. Boido for useful suggestions
and Drs. C. Bennicelli, P.Zanacchi and A. Camoirano for their valuable
assistance. The data on the intensity of solar irradiation were kindly
supplied by Professor G. Flocchini (Institute of Geophysics, University of
Genoa, Italy).
The studies reported in this paper were mainly supported by Italian
C.N.R. (Progetto Finalizzato Controllo della Crescita Neoplastica). Preliminary investigations with human lung biopsies were carried ou in the
framework of a study supported by C.N.R. (Gruppo Cardiorespiratorio).

REFERENCES

(1) Bartsch,H., Malaveille,C., Camus,A-M., Martel-Planche,G.,
    Brun, G., Hautefeuille,A., Sabadie,N., Barbin,A., Kuroki,
    T., Drevon,C., Piccoli,C. and Montesano,R. In *Molecular
    and Cellular Aspects of Carcinogen Screening Tests*

(Montesano,R. *et al* Eds.), IARC, Lyon, 179-241 (1980).

(2) Boido,V., Bennicelli,C., Zanacchi,P. and De Flora,S. *Toxicol. Letters* **6**, 379-383 (1980).

(3) Chandrasekharan,N. *Austr.J.Exp.Biol.Med.Sci.* **49**, 383-395 (1971).

(4) De Flora,S. *Nature (London)* **271**, 455-456 (1978).

(5) De Flora,S. *Ital.J.Biochem.* **28**, 81-103 (1979).

(6) De Flora,S. *Carcinogenesis* **2**, 283-298 (1981).

(7) De Flora,S. *Mutation Res.* **85**, 185-186 (1981).

(8) De Flora,S., Zanacchi,P., Bennicelli,C., Picciotto,A. and Boido,V. *11th Ann.Meet.E.E.M.S.*,Budapest,July 6-11 (1981).

(9) De Flora,S., Cajelli,E. and Brambilla,G. *IRCS Med.Science* **7**, 185 (1979).

(10) De Flora,S., Coppola,R., Zanacchi,P. and Bennicelli,C. *Mutation Res.* **61**, 389-394 (1979).

(11) De Flora,S. and Boido,V. *Mutation Res.***77**,307-315 (1980).

(12) De Flora,S. and Picciotto,A. *Carcinogenesis* **1**,925-930 (1980).

(13) De Flora,S. and De Flora,A. *Cancer Lett.***11**,185-189(1981).

(14) De Flora,S. and Mugnoli,A. *Cancer Lett.***12**,279-285 (1981).

(15) De Flora,S., Zanacchi,P., Camoirano,A. and Bennicelli,C. *Mutation Res.Lett.*,in press (1981).

(16) Foote,C.S. *Science* **162**, 963-970 (1968).

(17) Gibso,T.L., Smart,V.B. and Smith,L.L. *Mutation Res.* **49**, 153-161 (1978).

(18) IARC Monographs on the Evaluation of the Carcinogenic Risk of Chemicals to Humans,IARC,Lyon,**23**,205-323 (1980).

(19) Maugh II,T.S. *Science* **205**,1363-1366 (1979).

(20) McCann,J., Choi,E., Yamasaki,E. and Ames,B.N.*Proc.Natl. Sci. USA* **72**,5135-5139 (1975).

(21) McCann,J. and Ames,B.N. *Proc.Natl.Acad.Sci.USA* **73**,950-954 (1976).

(22) McCoy,E.C., Hyman,J. and Rosenkranz,H.S. *Biochem.Biophys.Res.Comm.***89**,729-734 (1979).

(23) McCoy,E.C. and Rosenkranz,H.S.*Cancer Lett.***9**,35-42 (1980).

(24) Negishi,T. and Hayatsu,H. *Mutation Res.***79**,223-230 (1980).

(25) Peck,R.M., Tan,T.K. and Peck,E.B. *Cancer Res.* **36**,2423-2427 (1976).

(26) Petrilli,F.L. and De Flora,S. *Mutation Res.***54**,139-147 (1978).

(27) Petrilli,F.L. and De Flora,S.*Mutation Res.***58**,167-173 (1978).

(28) Petrilli,F.L., De Renzi,G.P., Palmerini Morelli,R. and De Flora,S. *Water Res.* **13**,895-904 (1979).

(29) Petrilli,F.L., De Renzi,G.P. and De Flora,S. *3rd Intern. Congr.Industr.Waste Water and Wastes*,Stockholm,Febr.6-8 (1980).
(30) Petrilli,F.L., De Renzi,G.P. and De Flora,S. *Carcinogenesis 1*,51-56 (1980).
(31) Petrilli,F.L. and De Flora,S. *Chromates Symposium 80 (ORC/IHF)*,Rockville,Maryland,September 16-18,1980.
(32) Pitts,J.N.Jr., Van Caunwenberghe,K.A., Grosjean,D., Schmid,J.P., Fitz,D.R., Belser,W.L.Jr.,Knudson,G.B. and Hynds,P.M. *Science 202*,515-519 (1978)
(33) Pitts,J.N.Jr., Lokensgard,D.M., Ripley,P.S., Van Cauwenberghe,K.A., Van Vaeck,L., Shaffer,S.D., Thill,A.J. and Belser,W.L.Jr. *Science 210*,1347-1349 (1980).
(34) Purchase,I.F.H., Longstaff,E., Ashby,J., Styles,A., Anderson,D.,Lefevre,P.A. and Westwood,F.R. *Br.J.Cancer 37*,873-959 (1978).
(35) Sugimura,T., Sato,S., Nagao,M., Yahagi,T., Matsushima,T., Seino,Y., Takeuchi,M. and Kawachi,T. In *Fundamentals in Cancer Prevention* (Magee,P.N. *et al* Eds.),Univ.of Tokyo Press, 191-215 (1976).
(36) Ulland,B., Weisburger,E.K. and Weisburger,J.H. *Toxicol. Appl.Pharmac.25*,446 (1973).
(37) Ullrich,R.L. In *Carcinogenesis-A Comprehensive Survey* (Slaga,T.J.Ed.),Raven Press,New York,*5*,167-184 (1980).

Mutagens and Carcinogens in the Digestive Tract

Polynuclear Aromatic Hydrocarbons and Catechol Derivatives as Potential Factors in Digestive Tract Carcinogenesis

Stephen S. Hecht, Steven Carmella, Keizo Furuya, and Edmond J. LaVoie

Naylor Dana Institute for Disease Prevention, American Health Foundation, Valhalla, N.Y., U.S.A.

ABSTRACT Analyses of human feces and urine were carried out to determine levels of benzo[a]pyrene (B[a]P) and catechols, respectively. Eight volunteers consumed charcoal broiled meat containing B[a]P; the dose was approximately 9 µg/person. B[a]P was not detected in the feces (<0.1 µg/person). In contrast, rats excreted unchanged B[a]P (11% of the dose) in feces after consuming charcoal broiled meat. Catechol and 4-methylcatechol were detected in human urine as conjugates. Individuals on non-restricted diets excreted 10 ± 7.3 mg catechol/24 hr and 3.4 ± 2.3 mg 4-methylcatechol/24 hr. Studies with restricted diets indicated that diet was the major factor in determining levels of urinary catechol. Results of a dose-response study on the cocarcinogenicity of catechol and B[a]P on mouse skin are also reported.

INTRODUCTION

Polynuclear aromatic hydrocarbons (PAH) are formed in the incomplete combustion of organic matter and are therefore widely distributed in the environment. Extensive studies on cigarette smoke and its carcinogenic effects have indicated that PAH, together with promoters and cocarcinogens, are important factors in respiratory carcinogenesis (14,25). However, the potential roles of PAH and cocarcinogens in digestive tract carcinogenesis are not well defined despite their common occurrence in the diet. For example, only limited data are available on carcinogenesis by PAH in the large intestine. 3-Methylcholanthrene, which is not an environmental PAH, induces tumors of the large intestine in randomly bred and in two inbred lines of Syrian golden hamsters (7,15). Benzo[a]pyrene B[a]P causes forestomach tumors in mice (24).

The proximate carcinogens of PAH on mouse skin and in mouse lung are frequently dihydrodiols which can form "bay region" dihydrodiol epoxides (6). For B[a]P, the proximate and ultimate carcinogens are thought to be (-)-trans-7,8-dihydro-7,8-dihydroxyB[a]P and (+)-trans-7β,8α-dihydroxy-9α,10α-epoxy-7,8,9,10-tetrahydroB[a]P, respectively. Metabolism of B[a]P by cultured human colon results in formation of these compounds and the corresponding DNA adducts (1). Human fecal bacteria can hydrolyze conjugates of 7,8-dihydro-7,8-dihydroxyB[a]P and of other B[a]P metabolites which are excreted in the bile (20). Thus, either free B[a]P or its metabolites which reach the colon could act as carcinogens at this site. Effects of PAH on the digestive tract may be enhanced by dietary cocarcinogens. For example, catechol is known to be a powerful cocarcinogen with B[a]P on mouse skin (26). Compounds related to catechol occur widely in the diet, especially as components of fruits and vegetables. In this report, we will review our studies on the analysis of human feces for B[a]P after consumption of charcoal broiled meat. In addition, we have examined the dose-response relationship of catechol and B[a]P as cocarcinogens on mouse skin and have assessed the levels of catechol and 4-methylcatechol in human urine. We hope to establish a scientific foundation for further studies on the possible role of PAH and cocarcinogens in digestive tract cancer.

<u>Analysis of Human Feces for B[a]P after Consumption of Charcoal Broiled Beef</u>   We first carried out model studies with rats in order to develop methology to analyze feces for B[a]P and its metabolites (12). Groups of ten rats were gavaged with [^{14}C]-BaP at levels of 4.0, 0.4, and 0.04 μmol/rat. Excretion of label in feces averaged 77%, in agreement with earlier studies (13,17). Unchanged B[a]P in feces was determined by high pressure liquid chromatography and ranged from 5.6-13% depending on the dose. Lower levels of un-

changed B[a]P have been detected by other investigators in the feces of mice, rats, and Syrian golden hamsters (3,5,17,19). Several B[a]P metabolites were also detected in rat feces. These included 3-hydroxyB[a]P, 9-hydroxyB[a]P, B[a]P-3,6-dione, B[a]P-1,6-dione, 4,5-dihydro-4,5-dihydroxyB[a]P, and 7,8-dihydro-7,8-dihydroxyB[a]P. The detection of these metabolites is in agreement with in vitro studies of B[a]P metabolism and with investigations of biliary metabolites of B[a]P (10,27). A considerable portion of the excreted radioactivity was not identified and at least 40% consisted of material which could not be extracted from the feces.

When charcoal broiled beef containing 52.7 µg/kg B[a]P was administered to rats, approximately 11% of the B[a]P was excreted unchanged in the feces. These results agreed with the model studies described above. To extend these studies to humans, eight volunteers consumed an average of 360 g charcoal broiled beef containing 24 µg/kg B[a]P. Thus, each volunteer consumed about 9 µg B[a]P and based on the model studies, we would have expected excretion of about 1 µg B[a]P in feces. However, we were unable to detect B[a]P even though our detection limit was 0.1 µg B[a]P/person. In this study, we did not analyze for metabolites of B[a]P in feces because the available methods were not sensitive enough.

Further studies are clearly necessary to assess the distribution of B[a]P and other PAH in man. The study outlined above suggests that there may be significant differences between humans and rats in the excretion of B[a]P and its metabolites. It is possible that less unchanged B[a]P is excreted in human feces compared to rat feces. A recent study with Syrian golden hamsters showed that less unchanged B[a]P was excreted in the feces compared to rats and that the amount excreted was dependent on diet (19). Differences in metabolism by fecal microflora in rats and man are also likely to have a major effect on excretion of B[a]P and its metabolites.

<u>Cocarcinogenesis by Catechol and B[a]P on Mouse Skin: Dose Response.</u> The cocarcinogenicity of catechol with B[a]P on mouse skin was first demonstrated by Van Duuren and Goldschmidt (26). They applied 2 mg catechol and 5 µg B[a]P three times weekly. In a study of

the cocarcinogenic fractions of tobacco smoke, we found significant activity in several fractions, one of which was predominantly catechol. In that study 0.25 mg of the catechol fraction was applied to mouse skin together with 3 μg B[a]P (11). In the present dose-response study, the high dose was 0.25 mg catechol and 3 μg B[a]P, applied 5 times weekly for 59 weeks. In the other groups, the amounts of catechol applied with 3 μg B[a]P were 0.1 mg, 0.01 mg, and 0.001 mg. The results are summarized in Table 1. The data demonstrate that the 0.1 mg dose of catechol was cocarcinogenic but the 0.01 mg dose was not. Thus, catechol is probably a somewhat weaker cocarcinogen with B[a]P than are benzo[e]pyrene, pyrene, and fluoranthene (26).

Table 1. Dose-Response Relationship in Cocarcinogenicity of Catechol and B[a]P on Mouse Skin[a]

Experimental Group	% of Animals w/skin tumors	Number of Skin Tumors	
		Squamous Cell Papillomas	Squamous Cell Carcinomas
Catechol(0.25mg)+BaP(3μg)	72	30	56
Catechol(0.1mg)+BaP(3μg)	66	7	24
Catechol(0.01mg)+BaP(3μg)	18	1	7
Catechol(0.001mg)+BaP(3μg)	24	5	5
Catechol(0.25mg)	21	2	4
BaP (3μg)	11	1	2
Vehicle control	0	0	0

[a]Each group consisted of 30 female CD-1 mice. Solutions were applied 5 times weekly in 0.1 ml acetone to the shaved backs of mice. The experiment was terminated after 59 weeks of treatment

Analysis of Human Urine for Catechol and 4-Methylcatechol. The presence of catechol in human urine has been reported previously but no quantitative data are available (8,9,18). We developed a simple gas chromatographic method for the analysis of catechols in human urine. Urine samples were treated with β-glucuronidase and sulfatase, acidified, and extracted with ether. The ether extracts were silylated and analyzed by glass capillary gas chromatography. A typical gas chromatogram of the catechol fraction of human urine is illustrated in Figure 1. Peaks were identified by combined gas chromatography-mass spectrometry. For the quantitative analysis of catechol and 4-methylcatechol, U-^{14}C-catechol was employed as internal standard.

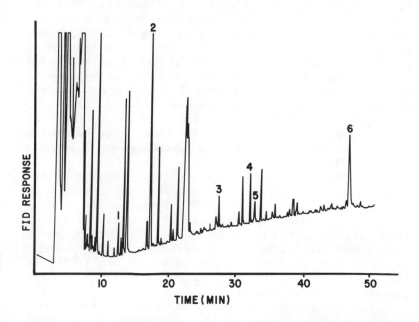

Figure 1. Gas chromatogram of the catechol fraction of human urine. Peak identities, as determined by combined GLC-MS; 1, phenol; 2, p-cresol; 3, catechol; 4,4-methylcatechol; 5,hydroquinone; 6,p-hydroxyphenylacetic acid.

Levels of catechol and 4-methylcatechol, which occurred predominantly as conjugates, in the 24 hr urine of 11 non-smoking donors on non-restricted diets are presented in Table 2a. Table 2b summarizes the urinary catechol levels of non-smoking donors who consumed a restricted uniform diet. The levels of catechols in the urine of smokers on the same restricted diet are summarized in Table 3. The data from Tables 2 and 3 are presented graphically in Figure 2.

The results of this study on a limited number of donors indicate that diet is the major factor contributing to catechol levels in human urine. Catechol levels in the urine of donors who consumed a uniform restricted diet were lower than those of donors whose diet was not restricted and the variation between subjects was markedly reduced. The restricted diet was comprised primarily of meat, poultry and carbohydrates. Coffee and tea were not allowed and consumption of fruit and vegetable products was extremely limited. Thus, these products would appear to contain precursors to urinary catechol. This is in agreement with a previous study which demonstrated that excretion of catechols in rat urine was reduced or eliminated when the rats consumed diets free of plant derived material (2). Several compounds which occur in plants, including quinic acid, shikimic acid, vanillin, and various phenolic acids have been shown to be metabolic precursors to catechol in experimental animals (4,16,21,22).

The levels of urinary catechol among smokers on the restricted diet appeared to increase in proportion to the amount of catechol inhaled. However, the amounts of urinary catechol which can be attributed to smoking are insignificant when compared to those found among non-smokers on normal non-restricted diets. Thus, catechol derived from tobacco smoke is not likely to play a major role as a causative factor in bladder cancer induced by smoking.

The influence of diet on urinary levels of 4-methylcatechol appeared to be opposite to that observed for catechol. Dietary restriction resulted in an increase in urinary 4-methylcatechol. Apparently, catechol and 4-methylcatechol have different dietary precursors.

Table 2. Catechols in the Urine of Non-smokers on Non-restriced and Restriced Diets.

Donor Number	Urine Volume (1/24h)	Urinary Catechols (mg/24h)[a,b]	
		Catechol	4-Methylcatechol
a) Non-restricted diet			
1	1.80	12	2.7
2	1.25	12	1.5
3	0.74	1.2	N.D.[c]
4	2.23	4.5	N.De[d]
5	0.88	10	3.6
6	2.62	6.3	N.De
7	1.48	2.6	N.D.
8	0.71	9.7	4.1
9	1.02	13	5.8
10	0.93	30	7.2
11	0.53	9.6	4.5
Mean ± S.D.		10±7.3	3.4±2.3
b) Restricted diet			
1	0.96	5.9	9.8
2	1.46	4.4	11
3	0.74	2.5	8.0
8	0.67	5.0	6.6
9	1.09	3.6	7.3
10	1.01	5.8	7.7
12	0.53	3.6	6.6
Mean ± S.D.		4.4±1.2	8.1±1.7

a. calculated by the isotope dilution method
b. predominantly as conjugates
c. N.D.= not detected.
d. N.De= not determined due to interfering peak.

Table 3. Catechols In The Urine Of Smokers On Restricted Diets

Donor Number	Number of Cigarettes[a]/24h	Catechols Inhaled[b]		Urine volume l/24h	Urinary Catechols[c],[d] (mg/24hr)	
		Catechol	4-Methylcatechol		Catechol	4-methylcatechol
13	6	1.6	0.1	1.64	4.4	4.7
14	7	1.9	0.1	1.15	4.1	2.1
15	15	4.0	0.3	1.25	5.1	8.0
16	27	7.3	0.6	1.15	11	8.2
17	30	8.1	0.6	1.44	9.4	7.5
Mean ± S.D.					6.8 ± 3.0	6.1 ± 2.6

a. A typical 85 mm non-filter U.S. cigarette, containing 272 µg/cig catechol and 21 µg/cig 4-methylcatechol in its mainstream smoke.
b. Based on standard machine smoking conditions and assuming uniform inhalation among donors.
c. Calculated by the isotope dilution method.
d. Predominantly as conjugates.

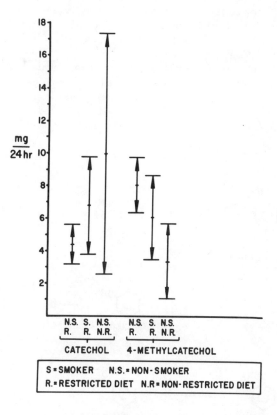

Figure 2. Effects of diet and smoking on urinary levels of conjugated catechol and 4-methylcatechol.

Prospects   Carcinogenic PAH such as B[a]P are known to occur in broiled and smoked foods, as well as in other dietary constituents. Relatively potent cocarcinogens such as catechol and 4-methylcatechol are either present in the diet or are formed metabolically and are excreted as their conjugates to the extent of 1-30 mg/day in human urine. Studies on the disposition of PAH in humans have so far been rather inconclusive as have bioassays of PAH for carcinogenicity in the digestive tract. The distribution of catechol and related compounds in man has not been investigated. Unconjugated catechol may be present in the stomach or colon. For example, catechol can be formed from various dietary components such as vanillin upon incubation with rat caecal contents or in vivo (4,16,21,22, see Figure 3) and oral or rectal adminstration of tannic acid to humans resulted in excretion of catecholic acids in urine (23). The cocarcinogenicity of catechol and its conjugates in target organs other than mouse skin and with carcinogens other than B[a]P also need to be determined. These studies are necessary in order to realistically evaluate the role of PAH and cocarcinogens in digestive tract carcinogenesis.

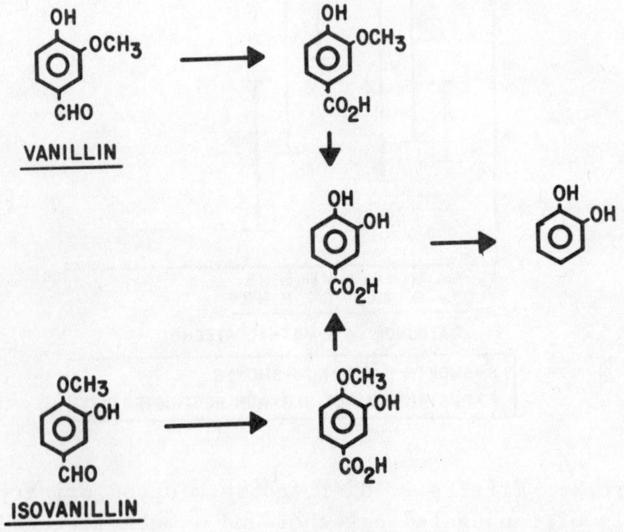

Figure 3. Metabolism of vanillin and isovanillin to catechol (Ref 22)

References

(1) Autrup, H., Harris, C.C., Trump, B.F., and Jeffrey, A.M. Cancer Res. 38, 3689-3696 (1978)
(2) Bakke, O.M. J. Nutr 98, 209-216 (1969)
(3) Berenblum, I. and Schoental, R. Biochem J. 36, 92-97 (1942)
(4) Booth, A.N., Robbins, D.J., Masri, M.S. and DeEds, F. Nature 187, 691 (1960)
(5) Chalmers, J.G. and Kirby, A.H.M. Biochem J. 34,

1191-1195 (1940)
(6) Conney, A.H., Levin, W., Wood, A.W., Yagi, H., Lehr, R.E. and Jerina, D. In "Advances in Pharmacology and Toxicology, Vol 9" (Y. Cohen, ed) Pergamon Press, Oxford, 41-52 (1979)
(7) Della Porta, G. Cancer Res 21, 575-579 (1961)
(8) Dirmikis, S.M. and Darbre, A. J. Chromatog 94, 169-187 (1974)
(9) Euler, von U.S., Lishayko, H. Nature 183, 1123 (1959)
(10) Falk, H.L., Kotin, P., Lee, S.S. and Nathan, A. J. Natl Cancer Inst 28, 699-724 (1962)
(11) Hecht, S.S., Carmella, S. Mori, H., and Hoffmann, D. J. Natl Cancer Inst 66, 163-169 (1981)
(12) Hecht, S.S., Grabowski, W., and Groth, K. Fd. Cosmet Toxicol 17, 223-227 (1979)
(13) Heidelberger, C. and Weiss, S.M. Cancer Res 11, 885-891 (1951)
(14) Hoffmann, D., Schmeltz, I., Hecht, S.S. and Wynder, E.L. In "Hydrocarbons and Cancer, Vol 1" (H.V. Gelboin and P.O.P. T'so, eds) Academic Press, N.Y., 85-117 (1978)
(15) Homburger, F., Hsueh, S.S., Kerr, C.S. and Russfield, A.B. Cancer Res 32, 360-366 (1972)
(16) Indahl, S.R. and Scheline, R.R. Xenobiotica 3, 549-556 (1973)
(17) Kotin, P., Falk, H.L. and Busser, R. J. Natl Cancer Inst 23, 541-555 (1959)
(18) Lewis, S., Kenyon, C.N., Meili, J., and Burlingame, A.L. Anal. Chem. 51, 1275-1285 (1979)
(19) Mirvish, S.S., Ghadirian, P., Wallcave, L., Raha, C., Bronczyk, S., and Sams, J.P. Cancer Res 41, 2289-2293 (1981)
(20) Renwick, A.G. and Drasar, B.S. Nature 263, 234-235 (1976)
(21) Scheline, R.R. Acta Pharmacol et Toxicol 26, 189-205 (1968)
(22) Strand, L.P. and Scheline, R.R. Xenobiotica 5, 49-63 (1975)
(23) Tompsett, S.L. J. Pharm. Pharmacol. 10, 157-161 (1958)
(24) Triolo, A.J., Aponte, G.E., and Herr, D.L. Cancer Res 37, 3018-3021 (1977)
(25) U.S. Dept of Health and Human Services. "Smoking and Health, a Report of The Surgeon General" DHEW Publiction No (PHS) 79-50066. U.S. Government Printing

Office, Washington, D.C., 1979
(26) VanDuuren, B.L. and Goldschmidt, B.M. J. Natl Cancer Inst 56, 1237-1242 (1976)
(27) Yang, S.K., Deutsch, J., and Gelboin, H.V. In "Polycyclic Hydrocarbons and Cancer, Vol 1" (H.V. Gelboin and P.O.P. T'so, eds) Academic Press, N.Y. 1978, 205-231

This study was supported by Helen J. Britt Memorial Grant for Cancer Research from the American Cancer Society, BC-56.

Approaches to the Identification of Human Colon Carcinogens

Michael C. Archer, Mark T. Goldberg[1], and W. Robert Bruce[1]

Department of Medical Biophysics, University of Toronto, Ontario Cancer Institute, Toronto, Canada

ABSTRACT Three approaches used to identify carcinogens responsible for colon cancer are summarized: chemical assays of known colon carcinogens, short-term <u>in vitro</u> tests for putative carcinogens present in the colon and site-specific <u>in vivo</u> assays for genotoxic agents. They are compared for specificity, sensitivity and relevance and a suggested priority for their use is given.

In order to prevent disease one must first know something of its cause. The more clearly the causative factors can be defined, the more specific the preventive measure that can be prescribed.

Our present understanding of the cause of colon cancer is limited. Epidemiological evidence suggests that the causative factors relate to diet, perhaps to an excess of fat or meat or to a deficiency of fibre or vegetables (1 ,9). Preventive measures based on this limited understanding necessarily lead to crude and stringent preventive measures that are likely to be costly and disruptive. We need to know the causative agents more exactly. We need to know them in chemical terms. With a more complete understanding of the nature of the chemical carcinogens and related compounds that are responsible for colon cancer we should be able

[1]Present address: Ludwig Institute for Cancer Research, Toronto Branch, 9 Earl Street, Toronto, Canada, M4Y 1M4

to develop more appropriate and specific changes to our food sources and food processing in a manner that is least disruptive to our society.

In this paper we summarize 3 approaches we are using to identify carcinogens that may be responsible for colon cancer. We will briefly describe the methods, the early results obtained with each and then consider the relative advantages of each approach.

1) <u>Chemical assays of known colon carcinogens</u>. Our first approach to investigate colon carcinogenesis and its prevention, has been to determine whether the lower gastrointestinal tract contains any chemicals or their precursors known to induce tumours of the colon in experimental animals. If such compounds were present, we could devise ways of reducing their levels. We have concentrated our efforts so far, on analysis of the N-nitroso compounds and their precursors, nitrate and nitrite in fresh human feces. Several N-nitroso compounds such as nitrosobis(2-hydropropyl)amine, nitrosobis(2-oxopropyl)amine, nitrosomethyl(acetoxylmethyl)amine and nitrosomethylurea, are known colon carcinogens.

Analyses of human fecal samples by an analytical method that is sensitive and free from interferences indicated that nitrate and nitrite levels in the intestine range from zero to 12 μmol/kg (0-0.9 ppm) and 5 to 19 μmol/kg (0.3-0.9 ppm) respectively (16). We also showed that when deliberately added to feces samples, nitrate and nitrite were destroyed during a 2-hour incubation period in a reaction that depended on the presence of microorganisms. Our results indicate that nitrate and nitrite levels in the intestine are lower than previously proposed (17), and suggest that conditions in the lower gastrointestinal tract favour denitrification.

Using a method for nitrosamine analysis that gives high recovery values and that is free from artefactual synthesis of nitrosamines, we have shown that human feces do not contain volatile nitrosamines (detection limit 0.1 - 0.5 μg/kg) (13). This result may be compared with that of Wang et al. (18) who reported that most normal human feces contained measurable levels of nitrosodimethylamine and nitrodiethylamine. We have repeated the analytical method used by Wang et al., and have shown that it is subject to artefactual formation of nitrosamines, though the exact problem has not been identified. We also showed that nitrosation reactions

are not catalyzed by fecal organisms. Following a 2-day
anaerobic incubation of feces with either a secondary
amine (dimethylamine, dipropylamine or morpholine) or
nitrite, no nitrosamine was formed. When the amine
and nitrite were added together, nitrosamine was formed,
but at a level of 2-20% of that formed in autoclaved
feces under the same conditions. Nitrosamines were stable
following anaerobic incubation with feces for up to 4
days. These results suggest that fecal organisms inhibit
the chemical formation of nitrosamines instead of
catalyzing it. When morpholine and nitrate were added
together, nitrosomorpholine was formed. In this case,
morpholine nitrosates so rapidly that it intercepts
nitrite formed by the action of nitrate reductase before
the nitrite can be further reduced. However, very high
concentrations of morpholine and nitrate (g/kg), which
are far from the conditions in normal feces, were required
to form measurable nitrosomorpholine.

While Asatoor and Simenhoff (2) and Johnson (11)
have reported that amines can be formed in the intestine,
the concentration of these amines must be low, since no
nitrosamines could be detected in our experiments, even
after adding high concentrations of exogenous nitrite.
It is possible that nitrosamines could be made at low
levels close to the intestinal wall where the microenvironment may be aerobic. It is also possible that
nitrosamines might be formed in the upper gastrointestinal
tract and transported unchanged into the colon. In view
of the stability of the nitrosamines in feces, and their
complete absence from fresh feces, however, these
possibilities seem most unlikely. While volatile nitrosoamines do not appear to be present in the lower gastro-
intestinal tract, other known colon carcinogens such as
1,2-dimethylhydrazine or its derivatives, 4-aminobiphenyl,
3-methylcholanthrene or 3'-methyl-2-naphthylamine might
be present. Sensitive and selective analytical methods
are required for their analysis.

2) Short-term in vitro tests for putative carcinogens
present in the colon. The second approach has been to
examine the contents of the colon for presumptive
carcinogens with the use of short-term mutagen assays.
We have concentrated our efforts, so far, on the contents
of the feces that are soluble in organic solvents and on
chemicals that produce mutations on the Salmonella tester
system. We have found that about 30% of our donors in

Toronto excrete easily measured quantities of a mutagen that is directly active on TA-100 and TA-98 (4,7). Separation studies suggest that this material is due to one group of compounds with similar chromatographic properties and with a similar u.v. spectrum. These compounds appear to be relatively stable in feces but become extremely sensitive to u.v. light, heat, oxygen and acid as they are purified. Their structure is not yet established although it seems likely that they are pentaenes with an allylic hydroxyl group (5).

Diet studies have shown that the levels of fecal mutagens are reduced in individuals receiving supplementary ascorbic acid and α-tocopherol (6), or these vitamins combined with bran fibre (7). It seems unlikely that these dietary factors are responsible for the marked individual-to-individual differences in the levels of these compounds or could explain the marked differences in levels in different populations. Other factors possibly relating to other components in the diet, physiological factors or bacterial flora may be important.

The relation of this mutagen to the causation of colon cancer is not established. Although measurable levels of the mutagen are more frequent in populations at high risk for colon cancer than in those at low risk (8), the levels in patients with diagnosed colon cancer are similar to those in patients with other lower GI problems (3).

3) <u>Site-specific in vivo assays for genotoxic agents</u>. The third approach to investigation of colon carcinogenesis in this laboratory has been to attempt to detect genetic damage in the target tissue and then to examine which variables in the diet affect this damage. With this goal in mind, we have developed a colonic micronucleus assay.

Micronuclei arise as a result of chromosome breakage prior to mitosis. At mitosis, acentric fragments of chromosomes remain in the cytoplasm of daughter cells and when stained with chromatin-specific stain, appear as "micronuclei". Thus, the micronucleus frequency is an indication of the rate of chrosomal aberration. The micronucleus assay is usually associated with the bone marrow (10), but the method appears to work equally well for the cells of the colonic epithelium.

The intestinal epithelium is regenerated by stem cells populating the bottom of crypts (12). Cells derived from these stem cells migrate up the crypt until, after several days, they reach the intestinal lumen and are sloughed off. We have collected epithelial cells from the upper third of crypts in the colon using the chelation method of Reodiger and Truelove (15) followed by a brief trypsinization. We have obtained single cell preparations which, when fixed in acid alcohol, dropped onto slides and stained with 4% Geisma can be scored for micronuclei. Duplicate slides were coded and scored independently by two individuals. Each scored 500 cells per slide for the incidence of micronuclei, and their scores were pooled to yield a micronucleus incidence per 1000 cells.

To test the sensitivity of this assay, C57BL/J6 mice were injected 1,2-dimethyldrazine (DMH), a colon carcinogen which is not detected as a mutagen by the Salmonella tester system (14). One to 4 days after a single i.p. injection, the mice were sacrificed by cervical dislocation, their colons were removed and everted, and single cell preparations were made as described above. A dose-response curve for DMH shows that a single dose of this carcinogen produces a significant increase in the spontaneous micronucleus rate which peaks 3 days after injection. Presumably the delay of 3 days from injection to the peak effect reflects the time necessary for migration of coloncytes from the proliferating stem cells at the base of the crypts to a site near the colonic lumen where they are harvested.

The background incidence in the colon of untreated animals is higher than in other mammalian tissues such as polychromatic erythrocytes, hepatocytes and spermatids. There are two possible explanations for this high background rate. The first is that the colonic epithelial cell line tolerates a higher error rate than other mammalian cells. This is possible since the epithelial cells are sloughed off into the fecal stream after a few days. The other and more likely explanation is that mutagens may exist in murine feces just as they do in human feces and may be responsible for the high background rate of genetic damage. With this latter possibility in mind, we attempted to lower the background incidence of micronuclei in mouse colon through dietary intervention. Mouse chow was supplemented with (a) 1% of the bile acid binding agent cholestyramine (Questran®), (b) a combination of 1.5% Vitamin E as the free tocopherol and 4 ppm selenium as sodium selenite in the drinking water or (c) 1% each of ferulic and caffeic acids, phenolics frequently present in plant material.

While the first two supplements produced no effects, the third supplement, ferulic and caffeic acids, was effective in lowering the background micronuclei in mouse colonic epithelial cells by 34%, a significant decrease ($p < 0.05$).

It is instructive to compare the three approaches we have used for attempting to identify colon carcinogens for specificity, sensitivity and relevancy. In considering these we should also consider which method or which combination of methods is most likely to lead to a definition of the causative factors for colon cancer.

The first approach, the chemical assay of known carcinogens, offers the greatest possibility for specificity and sensitivity. The problem is, however, relevance. Analytical techniques allow the determination of chemical carcinogens at levels that may be well below the level where biological effects can be observed. Furthermore we do not know in advance that cancer of the colon in man is due to any of the known colon carcinogens in experimental animals. It is possible that none of the known colon carcinogens are present in the human colon, but if they are present this would certainly be of immediate concern and possible relevance.

The second approach, <u>in vitro</u> mutagen assay, is specific only in the biological sense; it may be reasonably sensitive and it is relevant in that agents that produce genetic damage are usually carcinogenic. However, the mutagen assay as we have used it, could fail to detect the important carcinogen. First it is possible that the assay might fail to detect a carcinogenic compound in the feces. This problem could be reduced of course, by the application of several short-term assays. A second, more difficult failure, would result if the relevant exposure of colonic cells is not from the bowel contents at all but from the blood. The detection of mutagens here might pose a more difficult problem. Nevertheless it seems clear that the presence of genotoxic chemicals in the lumen of the bowel does pose a possible hazard and studies of the nature of these compounds and their relation to cancer are relevant.

The third approach, the demonstration of genetic lesions in the target cells in the colon, again has little specificity except in the biological sense. It is not clear what the sensitivity of the approach is likely to be, though animal studies and limited studies with patients appear to demonstrate levels of aberrations in the colon that are substantially greater than the background in other cell populations in the body. Diet manipulations that lead to a reduction in the frequency of mutations in the colonic cells of man would thus appear to be of great relevance though they would not in themselves give information on the specific nature of the genotoxic chemical.

In retrospect it would seem that the most effective use of the three approaches might well be in the reverse order of that described. First it might be possible to determine what factors in the Western diet lead to genetic damage in the human gastrointestinal epithelium. In vitro short-term tests could then be used to isolate the agents that might be important. Finally, when the nature of the compounds is established, specific analytical methodology could be developed to measure with high sensitivity chemicals of known relevance.

ACKNOWLEDGEMENTS

Supported by the Ontario Cancer Research and Treatment Foundation, the National Cancer Institute of Canada, and the Ludwig Institute for Cancer Research.

REFERENCES

(1) Armstrong, B., and Doll, R. Int. J. Cancer 15, 617-631 (1975)
(2) Asatoor, A.M., and Simenhoff, M.L. Biochim. Biophys. Acta 111, 384-392 (1965)
(3) Bruce, W.R. and Dion P.W. Am. J. Clinical Nutr. 33, 2511-2512 (1980)
(4) Bruce, W.R., Varghese, A.J., Furrer, R., and Land, P.C. Cold Spring Harbor Conference Cell Proliferation 3, 1641 (1977)
(5) Bruce, W.R., Varghese, A.J., Land, P.C. and Krepinsky, J.J.F. Banbury Report 7, 227-233 (1981)
(6) Dion P.W., Bright-See, E., Smith, C.C. and Bruce, W.R. (In preparation)
(7) Dion, P.W. and Bruce, W.R. (In preparation)
(8) Ehrich, M., Aswall, J.E., Van Tassell, R.L. and Wilkins, T.R. Mutation Research 64, 231 (1979)

(9) Graham, S., Dayal, S., Swanson, M., Mittleman, E.A., and Wilkinson, G. <u>Natl Cancer Inst. 61</u>, 709 (1978)
(10) Heddle, J.A. <u>Mutation Research 18</u>, 187-190 (1973)
(11) Johnson, K.A. <u>Med. Lab. Sci. 34</u>, 131-143 (1977)
(12) Leblond, C.P. and Cheng, H. In: Stem Cell of Renewing Cell Population. <u>Academic Press,</u> London. 7-31 (1976)
(13) Lee, L., Archer, M.C., and Bruce, W.R. <u>Cancer Research</u> (In press)
(14) McCann, J., choi, E., Yamasaki, B. and Ames, B.N. <u>Procedings National Academy of Science 72</u>, 5135-5139 (1975)
(15) Roediger, W.E.W. and Truelove, S.C. <u>Gut 20</u>. 484-488 (1979)
(16) Saul, R.L., Kabir, S.H., Cohen, Z., Bruce, W.R. and Archer, M.C. <u>Cancer Res., 41</u>, 2280-2283 (1981)
(17) Tannembaum, S.R., Fett, D., Young, V.R., Land, P.C. and Bruce, W.R. <u>Science 200</u>, 1487-1489 (1978)
(18) Wang, T., Kakizoe, T., Dion, P., Furrer, R., Varghese, A.J. and Bruce, W.R. <u>Nature 276</u>, 280-281 (1978)

Endogenous Formation of N-Nitroso Compounds and Gastric Cancer

S. R. Tannenbaum, P. Correa, P. M. Newberne, and J. G. Fox

Department of Nutrition and Food Science, Massachusetts Institute of Technology, Cambridge, Massachusetts, U.S.A.

ABSTRACT Endogenous formation of N-nitroso compounds in the gastric environment contributes a major proportion of man's exposure to these mutagens and carcinogens. We have been studying the factors which control the formation of nitrite and its reactivity with amines and amides. The studies have been carried out both *in vitro* in model systems simulating gastric contents and *in vivo* in dogs with permanent gastric fistulas. Of major importance is the consideration of blocking of carcinogen formation by natural inhibitors including ascorbic acid and α-tocopherol. The significance of these reactions is explained relative to a proposed etiologic model for gastric cancer.

INTRODUCTION

Nitrate has been identified in several locales as a component of the environment possibly associated with greater than ordinary risk of certain cancers. Nitrate is a stable chemical species, but is readily converted to nitrite in biological systems.

A carcinogenic agent may result from the reaction of nitrite with a nitrogen-containing compound to form an N-nitroso compound. Exposure to these N-nitroso compounds might occur in several different ways. The first possibility is through preformed compounds in the environment, particularly in food and water. A second possibility is through formation of these compounds in the body by the reaction of ingested nitrite with an appropriate nitrogen compound. A third possibility is through the reaction of endogenously formed nitrite with a nitrogen compound.

The conversion of nitrate to nitrite occurs normally in the saliva, leading to possible formation of N-nitroso compounds in the mouth, or more probably in the acid environment of the stomach. Other possible sites for nitrite formation and reaction include the intestinal tract, the hypochlorhydric stomach, and the infected bladder. Many other factors are important in catalyzing or blocking the ultimate effect of nitrite, but it is particularly important to mention ascorbic acid and α-tocopherol which may play an important role in inhibiting endogenous nitrosation.

An Etiologic Model. Epidemiological studies around the world have demonstrated an association between exposure to nitrate and gastric cancer. We have proposed a model for the disease in which chronic gastritis leads to conditions which favor carcinogen formation (1). We have recently described this model as outlined below (2). The essential characteristic of this model is that it proposes a chain of events which lead from the normal stomach to gastric cancer, and that interruption of any part of the sequence leads to a reduced risk for gastric cancer. In other words, the model provides a scientific framework for risk factors identified through epidemiological investigations, and in addition suggests possible modes for intervention in high-risk populations.

In Phase 1, an injury occurs to the gastric mucosa which leads to elevation of the pH of the stomach (Phase 2). The nature of the injury does not appear to be important as long as there is a permanent loss of parietal glands. Some examples of such injuries which have an elevated risk include the following:

1. Antrectomy followed by surgical restablishment of GI tract continuity (e.g. Bilroth I and II). In these cases the loss of parietal glands is accompanied by subsequent reflux of bile into the stomach causing irritation of the mucosa and elevation of gastric pH. The remaining gastric stump is often found to have focal intestinal metaplasia.

2. Pernicious anemia is associated with autoimmune loss of parietal glands, decreased secretion of hydrochloric acid, etc.
3. Dietary factors are suggested primarily by the results of epidemiological studies. Injury to the gastric mucosa could be caused by excessive intake of salty foods, consumption of very hard foods (e.g. coarsely ground corn) or by other unknown factors that lead to loss of surface mucous cells and exposure of gastric pits. This is a largely untapped area and could be the subject of more extensive research.

In Phase 2 of the model, elevation of gastric pH above 5 leads to rapid growth of bacteria, many of which have nitrate reductase activity. The source of the bacteria is mainly swallowed saliva and/or duodenal reflux. Due to the rich nature of alkalized gastric juice as a microbiological growth medium, coupled with a large inoculum and an aerobic environment, growth of bacteria is very rapid. Therefore, cycling of pH above and below pH 5 will be followed by cycling of bacterial growth, in phase with pH.

The reduction of nitrate to nitrite in Phase 3 will be determined by the presence of a suitable microflora and the availability of nitrate. The extent of accumulation of nitrite will be determined by the concentrations of nitrate-reducing bacteria and of nitrate, and by the absence or presence of factors which lead to nitrite destruction. The reaction of nitrite with nitrogen compounds to form carcinogenic N-nitroso compounds is the final Phase. For gastric cancer this would most likely be a nitrosamide, since these types of compounds and not nitrosamines have been shown to experimentally cause gastric cancer in rats and guinea pigs. Thus, each phase is essential in order for continuity to the next to be possible. Nitrate *per se* cannot be of etiologic significance to gastric cancer unless there is a substantial nitrate-reducing population in the stomach. Even if there was the potential for nitrite formation, it would be theoretically possible to block N-nitrosation, e.g., through the use of blocking agents such as ascorbic acid or α-tocopherol (6).

Intragastric Nitrosation. In order to develop a better understanding of how nitrosation proceeds in the stomach, and how the reaction can be blocked, we have carried out experiments with dogs which have been implanted with permanent gastric fistulas (3).

The kinetics of intragastric nitrosation and its quantitative relation to carcinogenesis has been discussed by Mirvish (4). Earlier work from our laboratory on the formation of N-nitrosopyrrolidine in the dog stomach (5) raised several questions not answered by earlier and subsequent experiments in *in vitro* or rodent model systems. In particular, the rate of nitrosamine formation in the catheterized stomach was subject to pronounced catalytic effects and rapid disappearance of the nitrosamine, possibly through direct absorption via the gastric mucosa. Since the intragastric nitrosation of nitrogen compounds is subject to a large number of variables, only some could be controlled in our experiments. An equation to describe the rate of intragastric nitrosation should minimally include concentrations of major and competing reactants, pH, catalysts, volumes, and terms for absorption and gastric emptying. We have been only partially successful in reaching that goal through the use of a nonabsorbable marker, and the measurement of nitrite and pH as a function of time. In the fasting animal dimethylnitrosamine (NDMA) forms within minutes after addition of the reactants if the pH is below 5. The presence of food in the stomach slows the reaction, and the additional presence of ascorbic acid greatly depresses the amount of NDMA formed. The concentration of nitrite in gastric juice declines rapidly after its introduction, but the concentration of NDMA declines even more rapidly than nitrite. This suggests that NDMA is probably rapidly absorbed directly from the dog stomach, in contrast to earlier results in rodent experiments.

Several important conclusions can be drawn from our experiments. The major controlling factors in the rate of formation appear to be nitrite stability and intragastric pH. When the pH is greater than 5 nitrite stability increases and the rate of nitrosation is very low, comparable to the situation in man (7). When stimulation of acid production drops the pH to below 5 nitrosation occurs with a concomitant decrease in nitrite stability, and therefore the loss of nitrite from the stomach.

The effect of ascorbic acid on inhibition of nitrosation is qualitatively similar to that found *in vitro*, however a failure to achieve complete inhibition could be due to incomplete mixing, or partitioning of reactants.

In another paper in this Symposium (Ohshima and Bartsch) an approach to measuring endogenous nitrosation is presented which utilizes the reaction of nitrite with proline to form N-nitrosoproline. Since nitrosoproline is not metabolized, the integral amount of nitrosation can be determined from the quantity appearing in the urine. We believe that a combination of the nitrosoproline method and studies on intragastric nitrosation will ultimately lead to a better understanding of the mechanisms and extent of endogenous nitrosamine formation.

ACKNOWLEDGEMENT

This work was supported by the National Cancer Institute Grant No. 1-P01-CA26731-02.

REFERENCES

(1) Correa, P., Haenszel, W., Cuello, C., Tannenbaum, S., and Archer, M. *Lancet* ii, 58-60 (1975)
(2) Correa, P., Haenszel, W., and Tannenbaum, S.R. *NCI Monograph* (in press)
(3) Lintas, C., Fox, J., Tannenbaum, S.R., and Newberne, P.M. *Carcinogenesis* (in press)
(4) Mirvish, S.S. *Toxicol. Appl. Pharmacol.* 31, 325-351 (1975)
(5) Mysliwy, T.S., Wick, E.L., Archer, M.C., Shank, R.C., and Newberne, P.M. *Br. J. Cancer* 30, 279-283 (1974)
(6) Tannenbaum, S.R., and Mergens, W. *Annals of New York Academy of Sciences* 355, 267-277 (1980)
(7) Tannenbaum, S.R., Moran, D., Falchuk, K.R., Correa, P., and Cuello, C. *Cancer Lett.* 14, 131-136 (1981)

Ingestion and Excretion of Nitrate in Humans and Experimental Animals

Hajimu Ishiwata and Akio Tanimura

National Institute of Hygienic Sciences, Tokyo, Japan

ABSTRACT   The ingestion, distribution, secretion, metabolism and excretion of nitrate and nitrite in humans and experimental animals were described. The ingestion amounts of nitrate and nitrite in Japanese were higher than the ADI recomended by WHO. The main metabolic pathway of nitrate was as follows. Food-borne nitrate was absorbed from the upper small intestine and nitrite was absorbed from the stomach and upper small intestine. A portion of nitrate was secreted into saliva and then reduced to nitrite by nitrate-reducing bacteria in the oral cavity. Blood nitrate was also secreted into the lower digestive tract, and nitrate secreted was reduced to nitrite by the intestinal flora. Most of nitrate ingested was excreted in urine. The amount of nitrate excreted in urine was less than that ingested. But it is known that nitrate excretion exceed when little nitrate was ingested.

INTRODUCTION

The biochemical and pharmacological investigations of nitrate in humans and animals have expanded because nitrate is reduced to nitrite, a precursor of carcinogenic N-nitroso compounds, in the body. Main metabolic pathways of nitrate had been elucidated in 1970's, and further detailed studies are developing. In this paper, the ingestion, distribution, secretion, metabolism and excretion of nitrate and nitrite in humans and experimental animals are described.

INGESTION OF NITRATE AND NITRITE

Daily intake of dietary nitrate was 218-408 mg in Japanese and 75-131 mg in Western people, and was higher than the acceptable daily intake of nitrate (corresponds to 219 mg nitrate/60kg body weight) recommended by WHO(1) in Japanese. Daily intake of dietary nitrite in both Japanese and Western people was less than ADI (8 mg nitrite/60 kg body weight). In Japanese, saliva supplied with 93.2% of total intake of nitrite, and vegetables supplied with 89.9% of nitrate. In this case, salivary nitrate was not included on total.

NITRATE AND NITRITE IN SALIVA

The average concentrations of salivary nitrate and nitrite in Japanese (n=368) were 73.0±77.8 ppm and 16.5± 20.8 ppm (mean±S.D.), respectively. When the daily secretion volume of saliva was assumed as 1 L, the amount of salivary nitrite was higher than ADI. Salivary nitrate and nitrite concentrations were closely related with the ingestion of nitrate. The precursor of nitrite in whole saliva was nitrate in ductal saliva in which no nitrite was contained, i.e. nitrate secreted into saliva from salivary glands was reduced to nitrite by nitrate-reducing bacteria in the oral cavity. Figure 1 shows the effect of

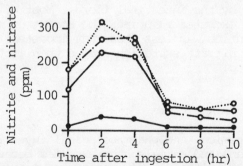

Fig. 1. Effect of sodium nitrate (200mg) intake on the concentrations of nitrate and nitrite in human saliva ●—●, nitrite in whole saliva; ○—○, nitrate in whole saliva; ○----○, nitrate in Stenon's ductal saliva; ○—○, nitrate in Wharton's ductal saliva. No nitrite was detected in any ductal saliva.

nitrate intake on the concentrations of nitrate and nitrite in whole and ductal saliva(2).

ABSORPTION AND DISTRIBUTION IN THE GASTRO-INTESTINAL TRACT

In guinea-pigs, nitrite was absorbed from the ligated stomach and upper small intestine, and nitrate was absorbed from the upper small intestine but not from the stomach.
Nitrate and nitrite concentrations in the gastric juice obtained from healthy humans (n=12) were 43.4±42.0 and 0.9±1.4 ppm, respectively. According to Saul et al. (3), human feces and ileostomy fluid contained nitrate and nitrite at the level less than 1 ppm. It is not yet known whether food-borne nitrate and nitrite reach to the lower digestive tract or not. The distribution of nitrate and nitrite in the rat gastro-intestinal tract is shown in Table 1 (4).

Table 1. Distribution of nitrite and nitrate in the rat digestive tract (PPM)

		STOMACH	SMALL INTESTINE				CAECUM	LARGE INTESTINE	SERUM
			1	2	3	4			
LOW-NITRATE DIET*	NO_2^-	0.1±0.1	N.D.	N.D.	N.D.	N.D.	N.D.	N.D.	0.8±0.1
	NO_3^-	4±2	4±2	2±1	1±1	N.D.	N.D.	N.D.	N.D.
HIGH-NITRATE DIET**	NO_2^-	6.6±8.2	4.4±1.1	1.6±1.0	4.6±2.8	0.8±1.2	N.D.	N.D.	1.0±0.3
	NO_3^-	1232±208	N.D.	3±3	3±3	1±3	N.D.	3±1	72±7

*NITRITE: 0.1 PPM, NITRATE, 15 PPM
**5000PPM OF NITRATE WAS ADDED TO THE LOW-NITRATE DIET.
RATS WERE FED ON EACH DIET FOR A WEEK, AND SACRIFICED IN THE MORNING.

SECRETION OF NITRATE INTO THE LOWER DIGESTIVE TRACT

The rat lower small intestine, caecum and large intestine were ligated avoiding occlusion of blood vessels, a small hole was made into the ligated intestines, the intestinal contents were washed out, and the hole was ligated. Five milligrams of nitrate was injected into the abdominal vein. Nitrate injected intravenously was secreted into the ligated lower small intestine, caecum and large intestine(4) (Fig.2). Neither nitrate nor nitrite was detected in any ligated intestines in which contents had not been washed out. Nitrate spiked in the intestinal contents disappeared rapidly and a little accumulation of nitrite was observed. These results agreed with the report by Witter et al.(5). N-Nitrosodimethylamine was absorbed from the rat lower small intestine, caecum and large intestine(4). Sukegawa et at.(6) reported that human milk (n=202) contained 9.7 ppm of nitrate and less than

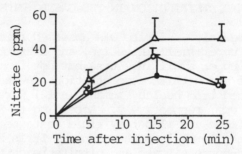

Fig. 2. Secretion of blood nitrate into the ligated lower digestive tract in rats

●, lower small intestine; ○, caecum; △, large intestine
Five milligrams of nitrate was injected into the abdominal vein. The contents of the ligated intestines were previously replaced with saline.

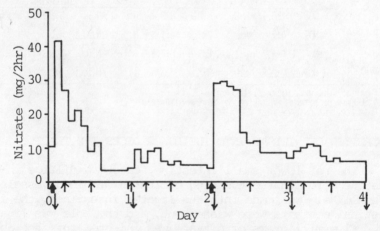

Fig. 3. Effect of sodium nitrate (200mg) intake on the urinary excretion of nitrate in a subject

↑, ingestion of a low-nitrate diet; ⬆, ingestion of 200 mg of sodium nitrate with the low-nitrate diet; ↓, excretion of feces. The amount of urinary nitrate during the sleeping time was shown as 2-hours equivalent.

0.3 ppm of nitrite.

URINARY EXCRETION OF NITRATE

The average concentration of urinary nitrate (n=33) was 74.4±42.5 ppm, and no nitrite was detected in any samples. Nitrite appeared by the urinary tract infection. The urinary excretion of nitrate was closely ralated with the ingestion of nitrate (Fig. 3)(7). The excess excretion of nitrate into urine was observed in humans(8) and rats(9,10) when little amount of nitrate (less than 10 mg /day in human) was ingested.

The metabolic fate of nitrate and nitrite are summarized in Scheme 1.

REFERENCES
(9) Green,L.C., Tannenbaum,S.R., and Goldman,P. *Science* *212*, 56-58 (1981)
(7) Ishiwata, H., Mizushiro,H., Tanimura,A. and Murata,T. *J. Food Hyg. Soc. Japan 19*, 318-322 (1978)
(4) Ishiwata,H., Sakai,A., and Tanimura,A. *ibid.(in press)*
(2) Ishiwata,H., Tanimura,A., and Ishidate,M. *ibid. 16*, 89-92 (]975)
(3) Saul,R.L., Kabir,S.H., Cohen,Z., Bruce,W.R., and Archer,M. *Cancer Res. 41*, 2280-2283 (1981)
(6) Sukegawa,K., Ariga,H., Nishimura,K., Ryu,F., and Hayashi,Y. *J. Japanese Soc. Food Nutr. 31*, 215-219 (1978)
(8) Tannenbaum,S.R., Young,V.R., Green,L., and Ruiz de Luzuriaga,K. *IARC Scientific Publications No. 31* pp. 281-287 (1980)
(1) WHO *World Health Organization Technical Report Series No. 539* p. 17 (1974)
(5) Witter,J.P., Balish,E., and Gatley,S.J. *Appl. Environ. Microbiol. 38*, 870-878 (1979)
(10) Witter,J.P., Gatley,S.J., and Balish,E. *Science 213*, 449-450 (1981)

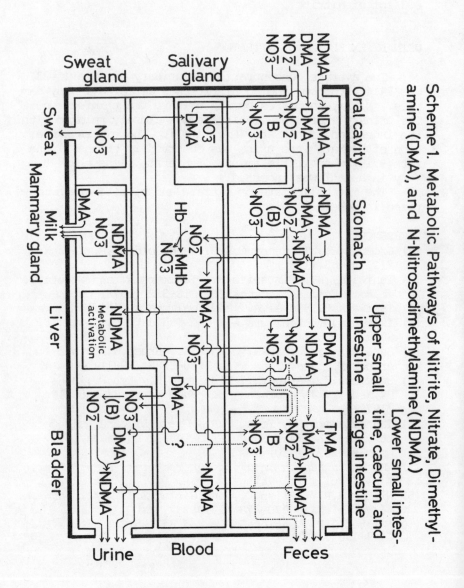

Scheme 1. Metabolic Pathways of Nitrite, Nitrate, Dimethylamine (DMA), and N-Nitrosodimethylamine (NDMA)

Quantitative Estimation of Endogenous Nitrosation in Humans by Measuring Excretion of N-Nitrosoproline in the Urine

Hiroshi Ohshima and Helmut Bartsch

Programme of Environmental Carcinogens and Host Factors, International Agency for Research on Cancer, Lyon, France

ABSTRACT Endogenous formation of N-nitrosoproline (NPRO) was demonstrated in a human subject by monitoring its excretion in the urine after ingestion of vegetable juice or pickled vegetables (as a source of nitrate) and L-proline. NPRO, when ingested in food or formed *in vivo*, was rapidly and almost completely eliminated from the human body. NPRO has been reported to be non-carcinogenic and non-mutagenic. Thus, monitoring of urinary NPRO after ingestion of the precursors may provide a quantitative index for endogenous nitrosation. Based on these observations, an experimental protocol is described to estimate the extent of endogenous nitrosation in individuals or groups of subjects at high cancer risk in which endogenously formed N-nitroso compounds are suspected to be associated with an increased incidence of cancer, such as that of the stomach and the esophagus.

INTRODUCTION

There is strong evidence to support that in the human body carcinogenic N-nitroso compounds can be formed by the interaction of nitrosatable amino compounds and nitrosating agents (4,8,14,16,17,20). However, in spite of the fact that most N-nitroso compounds are carcinogenic in wide range of animal species (7,10), evidence implicating these compounds in the etiology of human cancer has not been unequivocally presented. The question that now requires clarification is whether the amounts of N-nitroso compounds formed endogenously correlate with an increased incidence of specific types of human cancer. However, the lack of suitable methods to estimate endogenous nitrosation in humans has hindered attempts to obtain such data. For those reasons, we have recently developed a simple and sensitive procedure for monitoring the extent of nitrosation in experimental animals and humans (14,15). In the following, we present our recent findings on the formation of N-nitroso compounds in humans; based on these observations, we also describe an experimental approach which could be useful to estimate the extent of nitrosation *in vivo* in high risk populations or human individuals, in which endogenously formed N-nitroso compounds are being suspected to be linked with an increased incidence of cancer, such as that of the stomach and esophagus.

ENDOGENOUS FORMATION OF NPRO IN HUMAN SUBJECTS AFTER INGESTION OF NITRATE AND PROLINE

On the basis of animal experiments (2,6,15), we have concluded that monitoring of urinary levels of N-nitrosoproline (NPRO) could be a suitable procedure to estimate endogenous nitrosation in humans. The rationale for applying this procedure in human subjects is based on the following observations: i) NPRO has been reported not to be carcinogenic and mutagenic (10,12), ii) after gavage of rats with [^{14}C]-NPRO, the $^{14}CO_2$ production and DNA alkylation were negligible (2), but urinary excretion of NPRO (as the unchanged compound) was rapid and almost complete (2,6,15), iii) urinary levels of NPRO in rats gavaged with proline and nitrite reflected well endogenous nitrosation of proline (15), iv) in humans, preformed NPRO ingested in the aqueous extract of broiled dried-squids, was also rapidly and almost quantitatively eliminated in the urine within 24 hr after its ingestion, as shown in Fig. 1 (15). Thus, the amount of NPRO excreted in the 24-hr urine was an indicator of daily endogenous nitrosation (14).

Based on the absence of adverse biological effects of NPRO, one male volunteer ingested beetroot juice as a source of nitrate, and then 30 min later, L-proline, a naturally occurring amino acid in foodstuff and in the body. Simultaneous ingestion of these precursors resulted in a significant increase in urinary levels of NPRO (Fig. 1). The level of NPRO excreted in the 24-hr urine

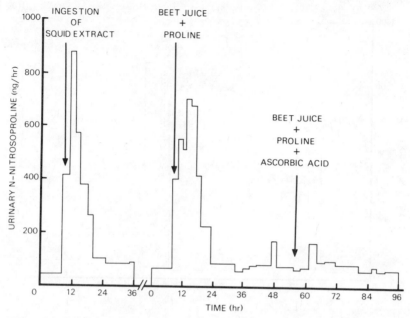

Fig. 1 Urinary excretion of NPRO in a human subject after ingestion of preformed NPRO and after ingestion of beetroot juice and proline.

increased exponentially with the dose of nitrate ingested, and appeared to increase proportionally with the dose of proline ingested. At the highest dose of nitrate (325 mg NO_3^-) and proline (500 mg) investigated, excretion of NPRO ranged from 16.6 to 30.0 (mean 23.3) μg per 24 hr per person. The simultaneous intake of ascorbic acid (or α-tocopherol) was found to inhibit nitrosation of proline in the human body (Fig. 1) (14).

Endogenous formation of NPRO was also observed to occur in a human subject who ingested 100 g of pickled vegetables (Chinese cabbage) as a source of nitrate and nitrite, followed, 1 hr later, by 100 mg proline (Fig. 2). The pickled vegetable ingested (1 sample) contained about 750 ppm (mg/kg) sodium nitrate and 4.5 ppm sodium nitrite. Preformed NPRO was also detected in the pickled vegetable at a concentration of 35 ppb (μg/kg). Therefore, intake of the vegetable alone led to an increase in urinary NPRO (Fig. 2). When 100 mg proline was ingested 1 hr after intake of 100 g of the vegetable, urinary levels of NPRO were further increased (Fig. 2). These results also

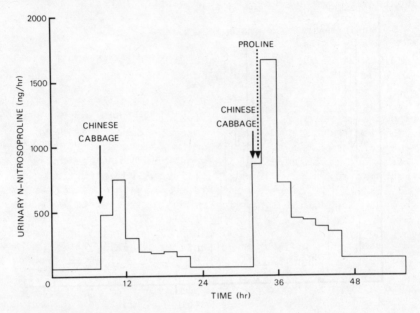

Fig. 2 Formation of NPRO *in vivo* in a human subject after ingestion of pickled vegetable (Chinese cabbage) and then, 1 hr later, *L*-proline.

support the notion that *N*-nitroso compounds can be formed in the body after simultaneous ingestion of nitrate/nitrite containing vegetables and nitrosatable precursors, which can be present in foodstuff or in the stomach.

A possible mechanism for the formation of NPRO in the human body may be as follows: Nitrate ingested in vegetable juice or the pickled vegetable is resecreted in saliva. The resecreted nitrate is then reduced to nitrite by the oral microflora (11,18,21). Proline, when ingested 30 min after intake of nitrate, probably reacts with salivary nitrite to give NPRO under the acidic conditions of the stomach. When proline is taken simultaneously with ascorbic acid, the latter compound competes with proline for nitrite, resulting in inhibition of NPRO formation in the body.

Thus, NPRO, when formed *in vivo*, appeared to be excreted unchanged in the urine without significant loss caused by resorption or metabolism. Using a sensitive and selective chemiluminescence method (14), NPRO excreted in the 24-hr urine after ingestion of its precursors can be measured as a quantitative index for nitrosation *in vivo*.

MONITORING OF ENDOGENOUS NITROSATION IN HUMAN SUBJECTS AT HIGH RISK FOR STOMACH CANCER

Endogenous formation of N-nitroso compounds has been associated with an increased risk of stomach cancer. For example, an elevated risk for stomach cancer has been observed in patients with chronic atrophic gastritis or with pernicious anemia, and in patients who have undergone gastric surgery such as Billroth-II gastrectomy (5, 13,16,19). It has been postulated that the achlorhydric stomach found in such patients may provide a suitable milieu for intragastric formation of N-nitroso compounds by the presence of a large number of bacteria, which may be involved in the conversion of nitrate to nitrite and subsequent nitrosation *in vivo* (4,16,20). In fact, higher levels of N-nitroso compounds were more frequently found in fasting gastric juice samples of such patients than in those of normal subjects (17). However, the extent of endogenous nitrosation occurring in these individuals at high risk for stomach cancer has not been determined.

In collaboration with the Turku University Hospital (Turku, Finland), the Regina Elena Institute (Rome, Italy), the Surgical Service of Fjordungssjukrahusid (Akureyri, Iceland) and the Division of Epidemiology and Biostatistics (DEB) of IARC (Dr N. Muñoz), studies are now underway aiming to estimate the extent of nitrosation in these high risk patients described above. The experimental protocol includes the ingestion of vegetable juice (as a source of nitrate), followed by the amino acid L-proline. Specimens of 24-hr urine are being collected and analyzed for NPRO as an index for endogenous nitrosation. The results obtained will be correlated with diagnostic findings on the presence of precancerous lesions and data on the gastric microenvironment such as the pH and bacterial flora.

EVALUATION OF HUMAN EXPOSURE TO N-NITROSO COMPOUNDS AND NITRATE IN HIGH RISK AREAS OF ESOPHAGUS CANCER

As described above, after ingestion of a single dose of nitrate and proline, NPRO can be formed in the human body. However, humans are more likely to be continuously

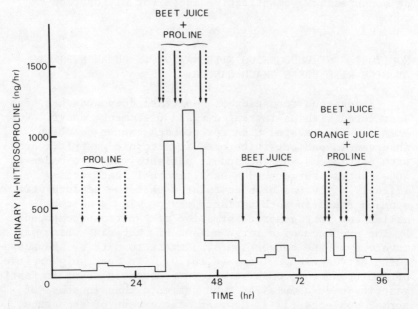

Fig. 3 Formation of NPRO *in vivo* in a human subject after ingestion of beetroot juice and proline.

exposed to nitrate/nitrite, which may be present in foodstuff and drinking water, or may be formed endogenously in the body. In order to investigate experimentally such a situation, where chronic exposure is involved, the following 4 series of experiments were conducted in one male volunteer (Fig. 3): i) 3 times a day, 100 mg proline was taken 1 hr after each meal (consisting of a low nitrate diet), hereby, only trace amounts of NPRO were excreted in the urine (1.7 µg per 24 hr per person), ii) nitrate-rich meals (low-nitrate diet plus 100 ml beetroot juice containing 130 mg NO_3^-) were taken 3 times a day together with 3 doses of proline; as a result, an increased amount of NPRO was excreted in the urine (13.4 µg/24 hr), iii) the intake of 3 nitrate-rich meals without proline did not significantly increase the urinary NPRO (2.9 µg/24 hr), iv) when orange juice (a possible source of vitamin C) was taken at the end of each nitrate-rich meal, followed by proline, urinary excretion of NPRO was markedly suppressed (4.0 µg/ 24 hr *versus* 13.4 µg in exp. ii   The results obtained from these experiments indicate that endogenous nitrosation may occur continuously in the body following chronic nitrate exposure. However, even when the amount of ingested nitrate was high enough to give rise the formation of N-nitroso compounds in the human body, the presence of dietary vitamin C effectively inhibited nitrosation (Fig. 3). These results suggest that nitrosation *in vivo* may be modified by many factors such as the presence of catalysts or inhibitors in foods. Thus, in addition to data on individual exposure to nitrate, assessing of the extent of endogenous nitrosation may be also required to establish a possible causal relationship between the amounts of N-nitroso compounds formed and an increased incidence of specific types of human cancers.

Many epidemiological studies have associated an increased risk of stomach cancer and esophageal cancer with an elevated exposure to nitrate (1,3,5,9,22). For example, in some provinces in Northern China, esophageal cancer is a prevalent disease, and a positive correlation has been reported between the cancer risk and the levels of nitrate/ nitrite in the drinking water, or the intakes of pickled vegetable and mouldy foods (3,22). In collaboration with the Cancer Institute of the Chinese Academy of Medical Sciences (Beijing, People's Republic of China) and Dr N. Muñoz (DEB, IARC), 24-hr urine specimens were collected in Linxian and Jiaoxian countries, a high and low incidence area for the esophageal cancer, respectively. These samples are being analyzed for nitrate/nitrite contents and for N-nitroso compounds excreted in the urine. As the subjects living in the high incidence area are assumed to be chronically exposed to nitrate/nitrite through ingestion of pickled vegetables or drinking water, 24-hr urine specimens will also be collected in subjects following multiple ingestion of proline (1 hr after each meal, 100 mg 3 times a day). The levels of urinary NPRO will be compared with those in urine specimens collected from subjects without intake of proline. The extent of nitrosation occurring in individuals and their exposure to nitrate will be evaluated from the excreted amounts of NPRO and nitrate in the urine. The results will be compared with the presence or absence of pre-cancerous lesions of esophageal cancer.

In conclusion, based on the findings on endogenous formation of NPRO in one human subject, a protocol is described, allowing the estimation of the extent of endogenous nitrosation, which is suspected to occur in individuals or groups of human subjects at high risk for stomach and esophageal cancer. By applying this procedure, results of on-going studies may reveal whether endogenously formed N-nitroso compounds are etiological factors involved in the disease.

ACKNOWLEDGMENT

We wish to thank Miss Y. Granjard for secretarial help.

REFERENCES

(1) Armijo, R. and Coulson, A.H. *Int. J. Epidem.*, *4*, 301-309 (1975)
(2) Chu, C., and Magee, P.N. *Cancer Res.*, *41*, 3653-3657 (1981)
(3) Coordination Group for Research on Etiology of Esophageal Cancer in North China *Chinese Med. J.*, *1*, 167-183 (1975)
(4) Correa, P., Haenszel, W., Cuello, C., Tannenbaum, S. and Archer, M. *Lancet ii*, 58-59 (1975)
(5) Cuell, C., Correa, P., Haenszel, W., Gordillo, G., Brown, C., Archer, M. and Tannenbaum, S. *J. Natl. Cancer Inst.* *57*, 1015-1020 (1976)
(6) Dailey, R.E., Braunberg, R.C. and Blaschka, A.M. *Toxicol.*, *3*, 23-28 (1975)
(7) Druckrey, H., Preussmann, R., Ivankovic, S. and Schmähl, D. *Z. Krebsforsch.* *69*, 103-201 (1967)
(8) Fine, D.H., Ross, R., Rounbehler, D.P., Silvergleid, A. and Song, L. *Nature* *265*, 753-755 (1977)
(9) Fraser, P., Chilvers, C., Beral, V. and Hill, M.J. *Int. J. Epidemiol.* *9*, 3-11 (1980)
(10) IARC Monographs on the Evaluation of the Carcinogenic Risk of Chemicals to Humans *17*, Some N-Nitroso Compounds (1978)
(11) Ishiwata, H., Boriboon, P., Nakamura, Y., Harada, M., Tanimura, A. and Ishidate, M. *J. Food Hyg. Soc. Jpn* *16*, 19-24 (1975)
(12) Mirvish, S.S., Bulay, O., Runge, R.G. and Patil, K. *J. Natl. Inst.* *64*, 1435-1442 (1980)
(13) Mosbech, J. and Videbaek, A. *Br. Med. J. ii*, 390-394 (1950)
(14) Ohshima, H. and Bartsch, H. *Cancer Res.* *41*, 3658-3662 (1981)

(15) Ohshima, H., Béréziat, J.C. and Bartsch, H. Presented at the 7th International meeting of N-nitroso compounds - Occurrence and Biological Effects - Tokyo (1981)
(16) Ruddell, W.S.J., Bone, E.S., Hill, M.J., Blendis, L.M. and Walters, C.L. *Lancet ii*, 1037-1039 (1976)
(17) Schlag, P., Bockler, R., Ulrich, H., Peter, M., Merkle, P. and Herfarth, C. *Lancet i*, 727-729 (1980)
(18) Spiegelhalder, B., Eisenbrand, G. and Preussmann, R. *Food Cosmet. Toxicol. 14*, 545-548 (1976)
(19) Stalsberg, H. and Taksdal, S. *Lancet ii*, 1175-1177 (1971)
(20) Tannenbaum, S.R., Moran, D., Rand, W., Cuello, C. and Correa, P. *J. Natl. Cancer Inst. 62*, 9-12 (1979)
(21) Tannenbaum, S.R., Weisman, M. and Fett, D. *Food Cosmet. Toxicol., 14*, 549-552 (1976)
(22) Young, C.S. *Cancer Res. 40*, 2633-2644 (1980)

Genetic Factors and Epidemiology

Genetic Factors in the Response to Environmental Mutagens

H. J. Evans

MRC Clinical and Population Cytogenetics Unit, Western General Hospital, Edinburgh, U.K.

ABSTRACT Inherent human diversity in the response to environmental mutagens may be expressed in terms of variation in cancer susceptibility and mutagen induced chromosomal damage. Constitutional chromosomal changes and single gene defects that predispose to specific cancers, and mutations that confer hypersensitivity to mutagen-induced chromosome aberrations and sister chromatid exchanges are discussed, and evidence for chromosomal instability and mutagen hypersensitivity presented for Friedrich's ataxia and for a dominant form of ataxia. The contribution of such mutations to mutagen sensitivity and/or cancer predisposition within populations is also considered.

It had been evident that the human species encompasses a wealth of genetic diversity long before the many recent and rapid advances in our understanding of the genetics of man. This diversity is reflected in a considerable degree of genetic variation in man's responses to his environment and, *a priori*, we would expect that differences between individuals in their responses to the actions of environmental mutagens might be the rule rather than the exception. Interest in our varied responses to environmental mutagens follows from the important damaging effects of these mutagens both on the individuals who are exposed to them and on any subsequent offspring of those so exposed. The fact that the majority of known mutagens are also carcinogens and the known association between mutagenesis and carcinogenesis focusses our interest on genetic variations that result in an enhanced, or diminished, sensitivity of individuals in a given pedigree to the induction of somatic mutations and/or to the associated early development of specific or non-specific malignancies. In our studies on such individuals we are concerned with three aspects: First, in defining the nature of the response with reference to a variety of different mutagens with different, and at least partly understood, modes of action.

Second, with utilising the different mutations that exist to unravel the processes involved in mutagenesis and carcinogenesis and to understand host defence mechanisms that may operate, particularly at the DNA and chromosomal levels.

Third, with defining the risks to individuals and families.

Now all types of cancer are to be found in virtually all populations, but the fact that the predominant type, or types, may differ between populations and that migrant populations tend to take on the patterns of cancer that are characteristic of their new environment, strongly reinforces the argument that the bulk of human cancers are environmentally induced (17). Thus, the major risk factors are certainly environmental and maybe summarised as involving the places in which we live, the food we eat and drink, our social habits and our occupations. However, if we look in detail within any one population it is evident that, within any group embracing a given set of environmental conditions, the distribution of cancer incidence is not uniform and that there are so-called 'cancer families' in which the incidence of cancer in blood relatives, but not in unrelated spouses, is very significantly in excess of the over-all group average incidence. Such a family is for example illustrated in Figure 1.

In some of these cancer families the excess cancer risk may appear to be inherited as a single dominant gene and here there are many examples of families for example, with breast cancer, medullary carcinoma of the thyroid, or polyposis coli. Excess cancer risks associated with the inheritance of a recessive gene include a variety of heritable conditions such as Bloom's syndrome, xeroderma pigmentosum, ataxia telangiectasia, Fanconi's anaemia and others and I shall return to these later. What I should mention here, however, is that the cancers in many

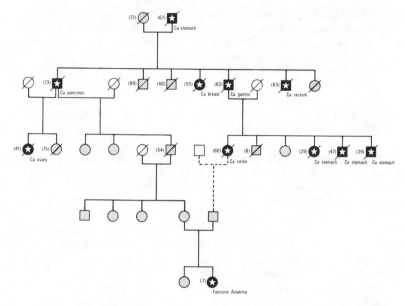

Fig. 1. Pedigree of a family in which 10 of the 17 deaths of blood relatives were due to malignant disease. The propositus was a case of Fanconi's anaemia. Types of cancer and age are indicated.

of these cancer families are relatively specific, as for example skin cancer with xeroderma and colon cancer with polyposis coli, etc., and this specificity tends to highlight the heritability in these cases. In other instances the cancers that may arise may be more diverse and unless there is an associated disease state that is clearly inherited, or unless we are specifically interested in cancer heritability, then such families may go undetected since cancer is a common disease.

The facts that, with few notable exceptions, all known carcinogens are mutagens, together with the considerable evidence that mutation is involved in one or more of the two or four step processes that are believed to be necessary for transformation to a malignant state, have been taken to imply either (i) that the inherited mutational change that is associated with some heritable cancer predispositions is itself one of the necessary steps, or (ii) that the mutational change sensitises ("hypersensitivity") the individual to further mutational changes that may result from an inability to deal effectively with normal cellular metabolites or with acquired environmental mutagens. Hypersensitivity will be discussed in some detail by others, but let us consider some of the general principles that are emerging by reference to a few examples.

(A) CONSTITUTIONAL CHROMOSOMAL ANOMALIES ASSOCIATED WITH CANCER PREDISPOSITION

Aside from an increased risk of breast cancer in 47,XXY Klinefelter males having an additional X chromosome (15) and an increased leukaemia risk in Down's syndrome patients with trisomy 21 (16), there are now at least three well-known inherited constitutional chromosome structural changes that are associated with the early development of specific malignancies. These are: retinoblastoma associated with a specific deletion of part of the long arm of chromosome 13 (del 13q14) (36); Wilms' tumour with aniridia associated with a specific deletion of part of the short arm of chromosome 11 (11p13) (24); and familial renal cell carcinoma associated with a translocation involving chromosomes 3 and 8 (t(3:8)(p21; q24) (6). Recently, some suggestive evidence has also appeared indicating a specific association between a constitutional deletion of the short arm of chromosome 20 (20p12.2) and medullary carcinoma of the thyroid (32).

We might consider that since these cancers are organ-specific and, at least in the case of retinoblastoma and Wilms' tumour, are childhood cancers, the chromosomal changes associated with them might represent important genetic changes affecting the morphogenesis of the particular organs involved, i.e. the retina and the kidney, and reflect one of the steps in the neoplastic process. We might also ask whether these mutations themselves result in an increased sensitivity to the effects of external mutagens and, perhaps surprisingly, the answer here appears to be a qualified and limited yes. Limited because published information relates only to retinoblastoma and qualified since the evidence for increased sensitivity to X-ray induced cell killing in retinoblastoma is equivocal, in that cells from some patients with a deletion of 13q14 are clearly X-ray sensitive (33), but cells from other similar patients are not (22). It may well be that there are two separate loci at 13q14, one involved with retinoblastoma and the other with X-ray sensitivity and that the actions and consequences of one locus does not depend upon the other.

(B) CHROMOSOMAL INSTABILITY SYNDROMES ASSOCIATED WITH CANCER PREDISPOSITION

The three classical chromosomal instability syndromes are Bloom's syndrome (BS) (12), Fanconi's anaemia (FA) (26) and ataxia telangiectasia (AT) (14). All are inherited as autosomal recessives, all show a predisposition to the early development of various malignancies and all are associated with spontaneous chromosome aberration frequencies in blood lymphocytes or skin fibroblasts that are significantly in excess of those observed from cells of normal healthy individuals.

The causes of these instabilities are not at all clear; all show increased frequencies of chromatid and indeed chromosome-type aberrations, but BS in unique in showing an extraordinarily high incidence of sister chromatid exchange (4). There is no evidence that any of these conditions are associated with virus infections and although it has been tempting to speculate (8) that the enhanced spontaneous aberration yields may reflect the presence of transposable DNA elements, or jumping genes, this seems unlikely since, in BS (3) and FA (35) at least, the instability is corrected when the cells are hybridised with those from normal individuals. There is evidence that lymphocytes from AT (27) and BS (10) patients secrete factors into the plasma which can cause aberrations in cells from normal individuals cultured in that plasma. It has also recently been shown that the chromosome breakage abilities of the factors secreted by these cells can be nullified in the presence of superoxide dismutase and that the clastogenic activity of these extracts is probably mediated via O_2^- radicals (10). In this regard it is of interest to note that Joenje et al. (20) have recently shown that in the case of FA cells the chromosomal instability may well be largely a consequence of the inability of the cells to protect themselves against the genetic toxicity of oxygen (Figure 2).

It would seem, then, that cells from these three chromosomal instability syndromes are probably unable to neutralise substances which may be normal constituents of

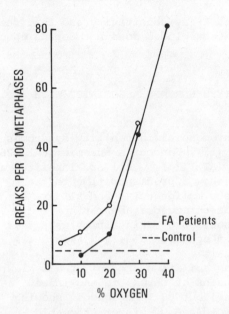

Fig. 2. Chromosome aberrations in blood lymphocytes of Fanconi's anaemia patients (—) and controls (---) cultured at different oxygen tensions (redrawn from Joenje et al. (20))

metabolising cells which, if not inactivated, would result in producing chromosome aberrations. It is perhaps not surprising therefore to find that when cells from these patients are exposed to various mutagens they show a much enhanced response as compared with normal cells. There is much evidence to show that BS cells are sensitive to ultra-violet light, AT cells to X-rays and bleomycin, and FA cells to cross-linking agents such as mitomycin C. There is also some evidence that AT cells may have some anomaly in their systems for identifying and dealing with DNA damage (23) and that FA cells may be defective in repairing DNA cross-links (25) and have reduced nuclear levels of topoisomerase (34).

I shall return later to DNA repair defective syndromes, but should ask here whether there are other genetically determined conditions, or syndromes, in man that are associated with chromosomal instabilities? The answer is yes, although in some the elevated aberration yields may be secondary consequences, as is probably the case in many of the collagen diseases (11). In our own laboratory we have discovered that Friedrich's ataxia (FRA) - a condition similar to but distinguishable from AT - is a chromosomal instability syndrome. Moreover,

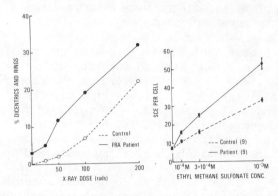

Figs. 3 and 4. Elevated spontaneous and X-ray-induced aberrations and enhanced response of Friedrich's ataxia lymphocytes to SCE induction by ethyl methane sulfonate.

cells from FRA show an ultrasensitivity to chromosome damage induced by X-rays (Figure 3) and to the induction of sister chromatid exchange by certain chemical mutagens (Figure 4). We have also discovered a similar ultrasensitivity to X-ray (Figure 5) and chemical mutagen induced chromosome damage (Figure 6) in cells from patients with a dominant form of ataxia, but this latter condition does not show an elevated spontaneous frequency of aberrations and cannot be considered as a chromosomal instability syndrome in the usual sense.

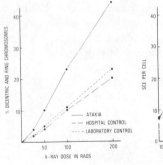

Figs. 5 and 6. Enhanced response to X-ray-induced chromosome aberrations and EMS-induced SCEs in lymphocytes of a patient with dominant ataxia.

(C) INHERITED CONDITIONS ASSOCIATED WITH DEFECTIVE DNA REPAIR AND CANCER PREDISPOSITION

Xeroderma pigmentosum (XP) is the classical DNA repair defective syndrome in man and although much is known concerning its genetics, and indeed the nature of the repair defects (5), the molecular biology of the defective DNA repair is by no means well understood. XP cells are, nevertheless, not chromosomally unstable, but they are very sensitive to the induction of chromosome damage by ultra-violet light and "UV-like" chemical mutagens and in the majority of cases are not sensitive to X-rays. The ultrasensitivity of cells of AT and FA to various mutagens, and referred to above, has also been considered to reflect defects in DNA repair, but the evidence is not unequivocal. Similarly, there are at least nine other syndromes in man (9), which include Cockayne's syndrome, Gardner's syndrome, various progerias etc., which at one time or another have been claimed to involve defective DNA repair. The claims in all these cases are questionable, but what is not in question is the fact that in each case cells from affected individuals show an ultra-sensitivity to cell killing, the induction of chromosome aberrations or the induction of sister chromatid exchanges following exposure to various mutagens. At the present time, then, we can identify at least a dozen or so heritable states, some of which may involve defective DNA repair, and all of which confer enhanced mutagen sensitivity on the part of the cells of individuals with these conditions. In the majority of these cases the enhanced mutagen sensitivity is governed by the inheritance of a pair of recessive alleles and although each condition may in its own right be of relatively rare prevalence, the question naturally arises as to whether heterozygote carriers for these genes also have some measure of sensitivity?

(D) HETEROZYGOTES

Swift and his colleagues have, for some years, been interested in the cancer incidence in heterozygote carriers of the genes responsible for FA (28), AT (30) and XP (29) and have provided some persuasive evidence that, at least in some cases, the incidence of cancer is enhanced. Parallelling the observations of an increased cancer incidence in heterozygotes for these conditions, are the findings that heterozygotes in at least some families with Fanconi's anaemia (Evans, unpublished) and heterozygotes in some families with AT (7) show higher spontaneous aberration frequencies than normal controls. Moreover, it would appear that heterozygotes may show a similar, but less dramatic, increased sensitivity to chromosome damage on exposure to certain mutagens (1).

Now each of the three conditions referred to are themselves relatively rare (see Table 1), but the calculated frequency of heterozygotes yields substantial numbers. For example, AT may have a prevalence of around 1 in 40,000 livebirths and thus, if one allele is involved, some 1% of the population will carry the gene. FA is less common with something like 0.3% of the population being carriers, but FRA in the UK is, if anything, slightly more common than AT and the frequency of the FRA gene has again been shown to be of the order of around 1 in 100 (13).

Table 1. Estimates of the heterozygote frequency for some of the autosomal recessive syndromes associated with mutagen hypersensitivity and/or cancer predisposition

Syndromes	Incidence (approx.)	Estimated heterozygote frequency (%)
Ataxia telangiectasia	1 in 40,000	1
Friedrich's ataxia	1 in 48,000	0.9
Bloom's	In Ashkenazi Jews	0.5 (?)
Fanconi's anaemia	1 in 360,000	0.3
Werner	1-22 in 10^6	0.2-0.9
Xeroderma pigmentosum	1-4 in 10^6	0.2-0.4

Considering the known half dozen or more different syndromes that are associated with an enhanced response to mutagen-induced chromosome damage, and a known or implied DNA repair defect, and if we expect that there may well be around 100 different genes that might be associated with DNA repair and DNA recombination processes, it must follow that many of us are heterozygous at one or more loci involved in these processes. We must therefore expect a range in efficiency for DNA repair and mutagen sensitivity throughout the population. It would seem very probable that many of the mutations involving DNA-housekeeping genes would be homozygous lethals, so that the half dozen or so known DNA repair defect syndromes in man must represent only a small fraction of individuals with a diminished DNA repair capacity and an increased mutagen sensitivity.

(E) OTHER HERITABLE FACTORS THAT MAY INFLUENCE THE RESPONSE TO ENVIRONMENTAL MUTAGENS

In our discussion we have been especially interested in genetic variations that influence mutagen sensitivity and cancer predisposition at the DNA and chromosomal levels, and have specifically ignored variations in processes which are important in the metabolism and inactivation of direct acting, or activation of indirect acting, mutagens/carcinogens. In terms of environmental chemical mutagenesis, it is probable that genetic variations in metabolising systems, such as the cytochrome P-450 system (21), play a larger role than variations in DNA repair capacities. We are all of us familiar with differences between populations or individuals in enzymes such as catalase, superoxide dismutase and alcohol dehydrogenase, to name but three of a large number of enzymes that are important in the metabolism of indirect acting carcinogens. However, this whole area is not part of our remit, although I think it is relevant to briefly allude to one suggestive and rather intriguing example that is emerging from our own work.

My colleagues and I (18) have shown that cigarette tobacco tar is a potent inducer of sister chromatid exchanges in human lymphocyte chromosomes, but it turns out that when we expose lymphocytes from cigarette smoking lung cancer patients to tobacco tar, they show greater sensitivity to the induction of SCE than cells of younger healthy smokers, who in turn show a higher response than healthy smokers of advanced age (19). There is evidence for genetic factors predisposing to lung cancer (31) and the enhanced sensitivity of lymphocyte chromosomes from lung cancer patients to SCE induction by tobacco tar may well reflect a heritable enhanced sensitivity.

Finally, before concluding I think we should not

forget that just as there are alleles that confer mutagen sensitivity and cancer predisposition, there will be alleles that confer resistance to the effects of mutagens. It does not necessarily follow, however, that resistance to mutagens will result in a lower cancer incidence and this reminds me of a recent intriguing report that describes a cancer family in which a variety of cancers showed a pattern of inheritance consistent with an autosomal dominant gene with a pleiotropic effect and incomplete penetrance (2). The interesting finding here was that skin cells from four of five cancer patients in the family showed a very marked increased resistance to cell killing by γ-irradiation relative to cells from spouses and other controls. There are various speculative interpretations that may be advanced to account for this finding, but an increased radio-resistance allied to an increased cancer risk is something of an exception and, to paraphrase Bateson, we would do well "to treasure our exceptions".

REFERENCES

(1) Auerbach, A.D. and Wolman, S.R. *Nature 271*, 69-71 (1978).
(2) Bech-Hansen, N.T., Blattner, W.A., Sell, B.M., McKeen, E.A., Lampkin, B.C., Fraumeni, J.F. and Paterson, M.C. *Lancet 1*, 1335-1337 (1981).
(3) Bryant, E.M., Hoehn, H. and Martin, G.M. *Nature 279*, 795-796 (1979).
(4) Chaganti, R.S.K., Schonberg, S. and German, J. *Proc. Nat. Acad. Sci. USA 71*, 4508-4512 (1974).
(5) Cleaver, J.E. and Bootsma, D. *Ann. Rev. Genet. 9*, 19-38 (1975).
(6) Cohen, A.J., Li, F.P., Berg, S., Marchetto, D.J., Tsai, S., Jacobs, S.C. and Brown, R.S. *N. Engl. J. Med. 301*, 592-595 (1979).
(7) Cohen, M.M., Shaham, M., Dagan, J., Shmueli, E. and Kohn, G. *Cytogenet. Cell Genet. 15*, 338-356 (1975).
(8) Evans, H.J. In *Genes, Chromosomes, and Neoplasia* (F.E. Arrighi, P.N. Rao and E. Stubblefield, eds.), pp. 511-527, Raven Press, N.Y. (1981).
(9) Evans, H.J. In *Sister Chromatid Exchange* (S. Wolff, ed.), pp. 297-374, John Wiley, N.Y. (1981) (in press).

(10) Emerit, I. In *DNA Repair, Chromosome Alterations and Chromosome Structure* (A.T. Natarajan, H. Altmann and G. Obe, eds.), North-Holland/Elsevier, Amsterdam (1981) (in press).
(11) Emerit, I. and Michelson, A.M. In *Abstracts of VI Int. Cong. Radiat. Res., Tokyo,* p. 82, JARR, Tokyo (1979).
(12) German, J. In *Chromosomes and Cancer* (J. German, ed.), pp. 601-617, John Wiley & Sons, N.Y. (1974).
(13) Harding, A.E. and Zilkha, K.J. *J. Med. Genet. 18*, 285-287 (1981).
(14) Harnden, D.G. In *Chromosomes and Cancer* (J. German, ed.), pp. 619-636, John Wiley & Sons, N.Y. (1974).
(15) Harnden, D.G., MacLean, N. and Langlands, A.O. *J. Med. Genet. 8*, 460-461 (1971).
(16) Harnden, D.G. and O'Riordan, M.L. *Lancet 1*, 260-261 (1973).
(17) Higginson, J. In *Environmental Carcinogenesis: Occurrence, Risk Evaluation and Mechanisms* (P. Emmelot and E. Kriek, eds.), pp. 9-24, Elsevier/North-Holland, Amsterdam (1979).
(18) Hopkin, J.M. and Evans, H.J. *Nature 279*, 241-242 (1979).
(19) Hopkin, J.M. and Evans, H.J. *Nature 283*, 388-390 (1980).
(20) Joenje, H., Arwert, F., Eriksson, A.W., de Koning, H. and Oostra, A.B. *Nature 290*, 142-143 (1981).
(21) Nebert, D.W. In *Enzymatic Basis of Detoxication Vol. 1* (W.B. Jakoby, ed.), pp. 25-68, Academic Press, N.Y. (1980).
(22) Nove, J., Little, J.B., Weichselbaum, R.R., Nichols, W.W. and Hoffman, E. *Cytogenet. Cell Genet. 24*, 176-184 (1979).
(23) Paterson, M.C. In *DNA Repair, Chromosome Alterations and Chromosome Structure* (A.T. Natarajan, H. Altmann and G. Obe, eds.), North-Holland/Elsevier, Amsterdam (1981) (in press).
(24) Riccardi, V.M., Sujansky, E., Smith, A.C. and Francke, U. *Pediatrics 61*, 604-610 (1978).
(25) Sasaki, M.S. In *DNA Repair Mechanisms* (P.C. Hanawalt, E.C. Freidberg and C.F. Fox, eds.), pp. 675-684, Academic Press, N.Y. (1978).
(26) Schroeder, T.M. and Kurth, R. *Blood 87*, 96-112 (1971).
(27) Shaham, M., Becker, Y. and Cohen, M.M. *Cytogenet. Cell Genet. 27*, 155-161 (1980).

(28) Swift, M. *Nature 230*, 370-373 (1971).
(29) Swift, M. and Chase, C. *J. Natl. Cancer Inst. 62*, 1415-1421 (1979).
(30) Swift, M., Sholman, L., Perry, M. and Chase, C. *Cancer Res. 36*, 209-215 (1976).
(31) Tokuhata, G.K. and Lilienfield, A.M. *J. Natl. Cancer Inst. 30*, 289-312 (1963).
(32) Van Dyke, D.L., Jackson, C.E. and Babu, V.R. In *Human Gene Mapping 6*, Birth Defects: Original Article Series, The National Foundation, March of Dimes (1981) (in press).
(33) Weichselbaum, R.R., Nove, J. and Little, J.B. *Proc. Natl. Acad. Sci. USA 75*, 3962-3964 (1978).
(34) Wunder, E., Burghardt, U., Lang, B. and Hamilton, L. *Hum. Genet. 58*, 149-155 (1981).
(35) Yoshida, M.C. *Hum. Genet. 55*, 223-226 (1980).
(36) Yunis, J.J. and Ramsay, N. *Am. J. Dis. Child. 132*, 161-163 (1978).

Diseases of Environmental-Genetic Interaction: Preliminary Report on a Retrospective Survey of Neoplasia in 268 Xeroderma Pigmentosum Patients

Kenneth H. Kraemer[1], Myung M. Lee[1], and Joseph Scotto[2]

[1]Laboratory of Molecular Carcinogenesis and [2]Biometry Branch,
National Cancer Institute, Bethesda, Maryland, U.S.A.

ABSTRACT: Xeroderma pigmentosum (XP) is the prototype disease of environmental genetic interaction with hypersensitivity to the neoplastic effects of ultraviolet (UV) exposure of the skin and eyes. Life-long protection from UV in a few XP patients has resulted in virtual absence of these abnormalities. Cultured cells from XP patients are hypersensitive to killing and mutagenesis induced by UV and also to certain chemical mutagens and carcinogens which may impinge upon internal organs. In a preliminary review of the English language literature we have identified 268 XP patients in papers published during the past 100 years. 92% of the patients were less than 25 years old and their survival probability was markedly reduced. Half of the XP patients had developed cutaneous neoplasms with median age of onset of 10 years. A substantial number of XP patients had oral cavity lesions, all before 20 years of age. Two XP patients had primary brain tumors and 2 had leukemia. Other internal neoplasms (testicle, lung, breast, uterus) were reported.

INTRODUCTION

This report will deal with the prototype disease of environmental-genetic interaction, xeroderma pigmentosum (XP) [38]. Some examples of how cellular studies of XP have given us insights into the molecular basis of mutagen hypersensitivity will be presented. In addition, an illustration will be given of how clinical studies of XP may yield valuable information concerning the effects of environmental mutagens on us all.

XP is a rare, autosomal recessive, progressive degenerative desease with sun sensitivity, increased susceptability to cutaneous and ocular neoplasms and defective DNA repair [37,39,58]. Approximately half of the XP patients do not give a history of acute sun sensitivity. They tan normally but do develop excessive freckling at an early age. With repeated sun exposure XP patients usually develop numerous cancers of the skin. However, pigmentary abnormalities and neoplasms generally are not found on sun protected portions of the skin such as the buttocks. In patients with XP, the eye also shows abnormalities. However, these are usually confined to the anterior portion which is the only portion to receive ultraviolet. The fundus of the eye (uveal tract and retina), which is shielded from UV radiation by the cornea and lens, is usually spared. Early diagnosis of XP and rigorous protection from sunlight has been shown to prevent most of the serious sequelae in several XP patients [39].

Corresponding to the clinical UV hypersensitivity is an in vitro cellular hypersensitivity to ultraviolet radiation. In 1968, Cleaver reported that cells from patients with XP had defective DNA repair [10]. Subsequent investigations by numerous scientists around the world during the past 13 years have led to a better understanding of different DNA repair pathways present in humans. DNA repair defects have been reported for cells from different XP patients in each of the major DNA repair pathways: nucleotide excision repair, base excision repair, post-replication repair, and photoreactivation repair [39]. By implication, this provides information about the process of DNA repair in people who do not have XP.

Cultured cells from XP patients have been shown to be hypermutable to ultraviolet radiation [39,43]. This observation, in conjunction with the clinical hypersensitivity to ultraviolet induced neoplasia in XP patients, provides strong support for the theory that somatic mutations are important in the induction of ultraviolet-induced cancers.

XP cells have a similar hypersensitivity to cell killing or mutagenesis following treatment with certain environmental mutagens, carcinogens and drugs such as benzo[a]pyrene derivatives, nitroquinoline oxide, psoralen plus UV radiation and nitrofurantoin [39,43]. The relationship between ultraviolet radiation hypersensitivity and clinical disease seems clear in XP. By analogy, studies of clinical disease in XP may provide insights into the possible clinical importance of other environmental mutagens to which their cells are hypersensitive.

The remainder of this paper will provide a preliminary report of a retrospective literature survey we are conducting in an attempt to obtain both qualitative and quantitative information concerning clinical disease in XP. We hope to be able to record the types of abnormalities found in XP patients and to assess their prevalence. In this manner we may gain clues as to possible environmental mutagen exposures of significance. Our goal is to survey all of the English language literature describing clinical features of individual patients with XP. The minimum criteria for entry into the study is the diagnosis of XP and an age of observation. The unit of the study is thus the individual patient.

Articles are abstracted using a standard form with questions covering clinical signs and symptoms believed to be relevant to XP. Those include age of onset and type of clinical symptoms, ethnicity, cutaneous, ocular and neurological abnormalities, neoplasms, family history, treatment, and other possibly related abnormalities such as immunologic deficits. A separate form is prepared for each patient indicating the literature citation and the clinical and laboratory features. Some patients have extensive information while others have only their diagnosis and an age of observation. The forms are reviewed and the data entered into a computer.

This paper will present data from a preliminary analysis of 268 patients with XP reported in 76 articles [1-9,11-16, 18-36,39-42,44-50,52-66,68-83]. This represents approximately half of the patients of whom we have collected information and may be about one third of our eventual total.

TABLE 1   Year of Publication

	Frequency	Cum Freq	Percent	Cum Percent
1870-1879	5	5	1.866	1.866
1910-1919	1	6	0.373	2.239
1920-1929	15	21	5.597	7.836
1930-1939	23	44	8.582	16.418
1940-1949	16	60	5.970	22.388
1950-1959	22	82	8.209	30.597
1960-1969	47	129	17.537	48.134
1970-1979	124	253	46.269	94.403
1980-present	15	268	5.597	100.000

Table 1 shows a summary of the journal articles from which the information was abstracted. The 77 papers were published over a period of a little more than 100 years. There was information on a total of 268 patients. The first

XP patients were described by Kaposi in 1874 [27]. Reports on half of the patients included were published after 1970.

Information on 72 patients came from papers published by American authors. The next most frequent group was from Japan (53), followed by United Kingdom (25), Turkey (17), India (14), France (14), Egypt (13), Germany (11), and 11 other countries with less than 10 patients each. We report on the country of origin of the author of the paper because this information is available for many more patients than is the country of birth of the patients. Though there may be significant exceptions, in general, the country of origin of the article is a good indicator of the country of origin of the patients. However, these figures may not be representative of the relative frequencies of XP in these countries. They can be taken to indicate at a minimum that some XP patients were identified in these countries.

One hundred thirty five of the XP patients were male and 125 were female. This is close to the equal ratio expected with an autosomal recessive disease. In 8 patients the sex was not indicated in the reference. This points up one of the major limitations of the study -- incomplete information. In general, we have limited this investigation to information published in the literature and have not attempted to gather additional information on published patients from other sources.

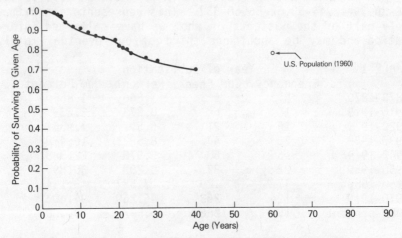

Figure 1 Survival of xeroderma pigmentosum patients.

The group of XP patients was rather young, in that 60% were less than 15 years old and more than 92% were less than 25 years old at time of last observation. However, 8 patients were older than 54 years. In two of the patients XP was diagnosed prenatally and a therapeutic abortion was performed. At the time of last observation 223 patients were alive and 39 or 14% were dead. Assuming that the data from the various countries, races, and time periods can be pooled, and that the mortality experience as reported in the literature is representative of the actual mortality, Figure 1 shows the survival probability calculated from this data. This graph plots the probability of surviving to a given age as a function of age. The survival probability was markedly reduced for this group of XP patients. For comparison, the probability of surviving to age 60 years for the U.S. population in 1960 is indicated. We can see that this same survival probability was attained by the XP patients in their third to fourth decade.

Information was sought concerning actinic keratoses. Most references did not give this information. However, 62 patients or 23% of the total population had at least one actinic keratosis according to the authors of the papers. This pre-malignant lesion is very common in the general population but is usually associated with advanced age and many years of sun exposure. 70% of the XP patients with a history of actinic keratoses developed these lessions before 20 years of age.

Table 2 shows the number of patients with malignant skin neoplasms. 142 patients, that is more than half of the XP patients, were reported to have had malignant skin neoplasms. We have information as to whether the skin neoplasms were basal cell carcinomas or squamous cell carcinomas in 50 patients. 16 of the patients had at least one basal cell carcinoma, 31 had at least one squamous cell carcinoma and 6 had at least one basal cell carcinoma as well as at least one squamous cell carcinoma. This is in contrast to the situation in the white population of the U.S. where basal cell carcinomas are more frequent than squamous cell carcinomas by a factor of at least 3 to 1. It suggests that though the frequency of both is increased, normal DNA repair may be more important in preventing squamous cell carcinomas than basal cell carcinomas.

TABLE 2 Number of XP patients with malignant skin neoplasms

	Frequency	Percent
Not Stated	79	29.5
Neoplasm Present	142	53.0
Neoplasm Absent	47	17.5
Totals	268	100.0

Analysis of skin neoplasm prevalence by age group showed that 30% of the patients less than 15 years old and 70% of these in the 16-24 year group were reported to have cutaneous neoplasms. Reports of 84 XP patients listed the age of onset of the cutaneous neoplasms. Figure 2 shows the cumulative percent distribution of the age of skin cancer diagnosis in XP patients with cutaneous neoplasms in comparison to that of people in the U.S. population with basal cell or squamous cell carcinomas. 50% of these patients developed their cutaneous neoplasms before age 10. This is in marked contrast the age distribution of almost 30,000 people in the U.S. who developed basal cell or squamous cell carcinomas during a one year survey period beginning June, 1977 [67]. 50% of these people who developed neoplasms were over age 60. Thus there is a striking decrease in the median age in the distribution of neoplasms in the patients with XP.

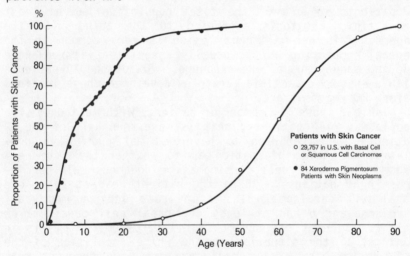

Figure 2 Age distribution of patients with skin cancer.

Eye tumors were reported in 24 of the patients. The eye is also exposed to ultraviolet radiation and ocular tissues have been shown to have defective DNA repair [39,58]. Thus the increased prevalence of ocular neoplasms may be explained on the basis of ultraviolet exposure.

Table 3 lists non-skin, non-eye neoplasms in XP. These are taken from the 268 patients' histories listed previously plus additional cases not yet entered into the computer [39]. We see a substantial number of oral cavity lesions including squamous cell carcinomas of the tongue, gum and palate. These oral neoplasms all were reported in XP patients less than 20 years old. This age distribution contrasts markedly with that of the U.S. population as assessed by the Third National Cancer Survey 1969-1971 [17]. In this survey, less than 1% of 6,531 people with buccal cavity and pharyngeal cancer were younger than 20 years old. Thus, here again there is a marked change in the age distribution of neoplasms. The XP patients developed oral cavity neoplasms at a much younger age than in the general population.

TABLE 3 Non-skin, Non-eye Neoplasms in XP

Site	Neoplasm	Age Range	Number of Patients
Oral Cavity (Tongue, Gum, Palate)	SCC	3-18 years	17 (9 biopsied)
Brain	Medulloblastoma	14 yr	1 (1 biopsy)
	Glioblastoma	15 yr	1 (1 biopsy)
	Astrocytoma	21 yr	1 (0 biopsy)
Leukemia	Acute lymphatic	3 yr	1 (1 histology)
	Myelogenous	32 yr	1 (1 histology)
Breast	Carcinoma	38 yr	1 (1 biopsy)
Lung	Bronchogenic Ca	62 yr	1 (1 biopsy)
Uterus cervix	Adenocarcinoma	51 yr	1 (1 autopsy)
Testicle	Sarcoma	12 yr	1 (death certif)
Eye (non-UV)	Melanoma	33 yr	1 (1 biopsy)

Table 3 also indicates 3 patients with brain tumors of which 2 had histologic confirmation, 2 patients with leukemia and 1 each with breast, lung, uterus and testicular neoplasms. The ocular melanoma listed was called to my attention by our symposium chairman, Dr. Takebe. This Japanese XP patient developed a melanoma in the uveal tract, a portion of the eye that is shielded from UV radiation. Histologic sections of this tumor were reviewed by Dr. Zimmerman, Chief of Ocular Pathology of the Armed Forces Institute of Pathology, Washington, D.C. This case is highly unusual in that melanomas are much less common in Japanese than in Caucasians. Table 3 thus indicates a likely increase in oral cavity neoplasms in the XP patients and suggests that there may be an increase in other internal neoplasms such as brain tumors and leukemia.

This data, particularly that of the oral cavity neoplasms, raises the question as to what factors might be responsible for these neoplasms. In considering the oral cavity neoplasms, it seems unlikely that the tongue receives significant sun exposure. In people standing upright, the tongue is well shielded from sunlight. Further, there is no evidence of actinic damage on the tongue adjacent to squamous cell carcinomas. The most likely oral cavity carcinogens in children have not been determined. They are not likely to have been chewing tobacco or betel nuts. However, pyrolysis products from charbroiled foods, nitrosamines or flavenoids are mutagens present in ingested foods. Cells from one XP patient have been reported to have reduced DNA repair after damage by mutagenic tryptophan pyrolysis products [51].

In conclusion, I would like to point out again some of the limitations of a retrospective literature study such as this. We are limited by the information presented in a given paper and usually do not have the opportunity for independent verification or for obtaining additional information. Nevertheless, the study affords a baseline for comparison of a prospective study. The Xeroderma Pigmentosum National Registry is being established by Alan Andrews, James German, Clark Lambert and Kenneth Kraemer in the U.S. This Registry will be attempting to answer many of these questions prospectively.

ACKNOWLEDGMENTS:

We wish to thank James Larson of NCI and Meg Seldin and her staff at ORI for invaluable assistance in computer analysis of this data; and Dr. Robert Tarone of the Biometry Branch, NCI for statistical advice.

REFERENCES:
1. Andrews GC. Arch Dermatol Syphil 13:588-589, 1926.
2. Bartholdson L, et al. Scand J Plastic Reconst Surg 11:173-177, 1977.
3. Becker SW, et al. Arch Dermatol Syphil 25:915, 1932.
4. Bell ET, Rothnem TP. Am J Cancer 30:574-576, 1937.
5. Berkel AI, Kiran O. Turkish J Dermatol 16:43-52, 1974.
6. Berlin C, Tager A. Dermatologica 116:27-42, 1958.
7. Brockington WS, Postlethwait RW. Am Surgeon 18:50-57, 1952.
8. Canright CM. Arch Dermatol Syphil 29:668-670, 1934.
9. Chance B. Am J Ophthalmol 9:689, 1926.

10. Cleaver JE. Nature 218:652-656, 1968.
11. Cleaver JE, et al. J Invest Dermatol 77:96-101, 1981.
12. Cole HN, Driver JR. Arch Dermatol Syphil 31:721, 1935.
13. Copeland MM, Martin HE. Am J Cancer 16:1337-1357, 1932.
14. Corlett WT. J Cutan Dis 33:164-170, 1915.
15. Corson EF, Knowles FC. Arch Dermatol Syphil 18:284-289, 1928.
16. Cripps DJ, et al. J Invest Dermatol 56: 281-286, 1971.
17. Cutler SJ, Young JL (eds) Third National Cancer Survey: Incidence Data. National Cancer Inst Monograph 41, Bethesda, US Dept Health Edu Welfare, 1975, Table 19a, p100.
18. Dennie CC. Arch Dermatol Syphil 28:426, 1933.
19. Dupuy JM, Lafforet D. Clin Immunol Immunopath 3:52-58, 1974.
20. El-Hafnawi H, et al. Br. J Dermatol 74: 201-213, 1962.
21. Goldman L, et al. Arch Dermatol 33:272-278, 1961.
22. Gray AMH. Br J Dermatol Syphil 50:103-105, 1938.
23. Guerrier CJ, et al. Dermatologica 146:211-221, 1973.
24. Guy J. Arch Dermatol Syphil 26:576, 1932.
25. Handa J, et al. J Comput Asst Tomogr 2:456-459, 1978.
26. Harris LC, Keet MP. J Pediatrics 57:759-768, 1960.
27. Hebra F, Kaposi M. On Diseases of the Skin Including the Exantomata. Tay W (trans). London, The New Sydenham Society, 1874, vol 3, pp 252-258
28. Jensen OA. Acta Ophthalmoligica 40: 96-103, 1962.
29. Jung EG, Bantle K. Birth Defects 7:125-128, 1971.
30. Kalk F, Jenkins T. S Afr J Surg 16:260-261, 1978.
31. Keddie FM. Arch Dermatol Syphil 53:302-303, 1946.
32. Keeler CE. NCI Monograph 10:349-359, 1963.
33. Keijzer W, et al. Mutat Res 62:183-190, 1979.
34. Khanna VN, et al. J Indian Med Assn 45:550-553, 1965.
35. King H, Hamilton CM. Arch Dermatol Syphil 42:570-575, 1940.
36. Koller PC. Br J Cancer 2:149-156, 1948.
37 Kraemer KH. In Nichols WW, Murphy DG (eds) DNA Repair Processes. Miami, Symposium Specialists, 1977, pp 37-71.
38. Kraemer KH. Arch Dermatol 116:541-542, 1980.
39. Kraemer KH. Xeroderma Pigmentosum In Demis DJ, Dobson RL, McGuire J (eds) Clinical Dermatology. Hagerstown, Harper & Row, 1980, Unit 19-7 pp 1-33.
40. Loewenthal LTA, Trowell HC. Br J Dermatol 50:66-71, 1938.
41. Lynch FW. Arch Dermatol Syphil 29:858-873, 1934.
42. Lynch HT, et al. Arch Dermatol 96:626-635, 1967.
43. McCormick JJ, Maher VM. In Friedberg EG, Hanawalt PC (eds) DNA Repair. NY, Marcel Dekker,1981, vol 1, pp501-521.
44. Mathur SP. Ophthalmologica 140:33-36, 1960.

45. Moore C, Iverson PC. Cancer 7: 377-382, 1954.
46. Moragas JM, Gimenez JM. Dermatologica 140:65-74, 1970.
47. Moss HV. Arch Dermatol 92:638-644, 1965.
48. Nicholas J, et al. Arch Dermatol Syphil 16:764, 1927.
49. Nurnberger. Internatl J Dermatol 12:169-172, 1973.
50. Oettle AG. NCI Monograph 10:197-214, 1963.
51. Okui T, Fujiwara Y. J Rad Res (in press).
52. Pawsey SA, et al. Quart J Med 48:179-210, 1979.
53. Rauschkolb RR, et al. Arch Dermatol 89:482-483, 1964.
54. Reed WB, et al. Arch Dermatol 91:224-226, 1965.
55. Reed WB, et al. J Am Med Assn 207:2073-2079, 1969.
56. Reed WB, et al. Arch Dermatol 113:1561-1563, 1977.
57. Reese AB, Wilber IE. Am J Ophthal 26:901-911, 1943.
58. Robbins JH et al. Ann Intern Med 80:221-248, 1974.
59. Ronchese F. Arch Dermatol 94:739-741, 1966.
60. Saebo J. Br J Ophthal 32:398-411, 1948.
61. Salamon T, et al. Arch Derm Forsch 251:277-280, 1975.
62. Saleem TA, et al. Lancet 2:264-265, 1980.
63. Sambasivan TS et al. Indian J Dermatol 11:79-81, 1966.
64. Sanctis C de, Cacchione A. Riv Sper Freniat 56:269-292, 1932.
65. Schamberg JF. Arch Dermatol Syphil 3:204, 1921.
66. Schamberg JF. Arch Dermatol Syphil 14:604-605, 1926.
67. Scotto J, et al. Incidence of Non-melanoma Skin Cancer in the United States, 1977-1978: Preliminary Report. Bethesda, Md, DHEW Publ No (NIH) 80-2154, 1980.
68. Seale. Arch Dermatol Syphil 16:666, 1927.
69. Sevel D. Br J Ophthal 47:687-689, 1963.
70. Short J, et al. Cancer 31:449-454, 1973.
71. Siegelman MH, Sutow WW. J Pediat 67:625-630, 1965.
72. Silberstein AG. Am J Dis Child 55:784-791, 1938.
73. Smithers DW, Wood JH. Lancet 1:945-946, 1952.
74. Stefanini M, et al. Human Genet 634:1-6, 1980.
75. Stevanovic DV. Arch Dermatol 84:53-54, 1961.
76. Stonham FV, Nagpaul DC. Indian Med Gazette 76:75-77, 1941.
77. Takebe H, et al. Cancer Res 37:490-495, 1977.
78. Thomas J, et al. Indian Genet 16:327-329, 1979.
79. Withers SM, Coleman WG. Arch Dermatol Syphil 2:27-34, 1920.
80. Wolf HF de. Arch Dermatol Syphil 29:139, 1934.
81. Wolff-Rouendaal D de. Ophthalmologica 173:290-291, 1976.
82. Wysenbeek AJ, et al. Clin Oncology 6:361-365, 1980.
83. Yamamoto M, Kujino J. Nippon Ika Daigaku Zasshi 37:259-263, 1970.

Surveillance of Human Populations for Germinal Cytogenetic Mutations

Ernest B. Hook[1] and Philip K. Cross[2]

[1]Birth Defects Institute, Division of Laboratories and Research, New York State Department of Health, Albany, New York and [2]Department of Pediatrics, Albany Medical College, Albany, New York, U.S.A.

ABSTRACT The term "surveillance" is used to denote ongoing review of background rates in a population; "monitoring" to denote continuing analysis of rates in populations at presumptive high risk because of exposure to some hazard. Concern about a hazard may warrant systematic studies of cytogenetic abnormalities in embryos, fetuses, or livebirths that are unlikely to be undertaken for surveillance alone. For the latter, indirect and opportunistic approaches may be all that are possible. Several such methods are discussed, including i) use of birth certificate reports for following Down's syndrome phenotype, ii) use of laboratory records for following interchange trisomy 21 and 13, iii) use of birth defect registry and surveillance system data on Down's syndrome in livebirths, iv) phenotypic surveillance of livebirths followed by selective cytogenetic examination, v) analyses of results of prenatal cytogenetic diagnoses as reported to a registry.

MONITORING VS SURVEILLANCE

The terms *monitoring* and *surveillance* as applied to disease outcomes imply similar activities, although operationally they are occasionally used with somewhat different reference. We suggest that the terms be distinguished as follows: *Surveillance* of a population for human mutations or other evidence of human morbidity and mortality should be understood as the systematic collection and regular review of data on fluctuations in baseline or background rates of events — events usually occurring in a large civil jurisdiction. An analogue from another field of public health is the ongoing review of reports of infectious diseases. *Monitoring* of a population should be understood as the long-range observation of individuals who are at presumptive high risk for adverse outcomes because of specific life events. An analogue is the long-range observation of individuals who are or have been in high-risk occupations, such as uranium or asbestos mining, for development of malignancies and other chronic diseases.

Insofar as human mutations are concerned these distinctions have important logistic implications. Public health authorities and other government agencies are quite likely to be interested in searching prospectively for the occurrence of possible mutations in a population after some catastrophic event — i.e. in monitoring — as part of the complete evaluation of morbidity and mortality. (Even if the catastrophe is not recognized until long after the event retrospective study may not be adequate to document the nature and long-range consequences of such an episode.) Directed, ad hoc studies of mutations, although expensive, may well appear justified in order to uncover or exclude mutagenic effects of putative or known toxic substances.

Surveillance of mutations in human populations, on the other hand, has a much lower priority. Investigations such as large-scale cytogenetic study of consecutive livebirths or spontaneous embryonic and fetal deaths are much less likely to be funded if they are solely for the purpose of determining background rates.

For this reason it is worth considering in detail some opportunistic methods for surveillance of populations for mutations — using, for example, existing data sources — that are less costly than ad hoc screening. (Further discussions of the costs and yields of ad hoc studies of cytogenetic and specific locus mutations appear in references 7, 13 and 14.) Surveillance of somatic cytogenetic mutations is difficult, if not impossible to accomplish by opportunistic methods and so will not be considered here.

USE OF VITAL RECORDS

For many years Down's syndrome, among other defects, has been reported and coded on birth certificates in New York State. About 95 to 97% of all Down's syndrome cases are associated with 47,+21. The rest are associated with interchange trisomies, i.e. 46 chromosomes with an unbalanced Robertsonian translocation. Therefore surveillance of the phenotype of Down's syndrome is one crude method of surveillance for 47,+21.

For Upstate New York the rate of Down's cases reported on birth certificates dropped from 0.52 per 1,000 livebirths in 1963 to 0.37 in 1978. Some of this change is attributable to the declining proportion of older mothers among women having livebirths and the well-known fact that rates of 47,+21 are much higher in older than in younger mothers. (See discussion in reference 10.) This change can be adjusted for by analyzing changes in the annual (direct) standardized rate of Down's syndrome. This is the rate that would occur in any year in a reference population of fixed maternal-age composition if the maternal-age-specific rates in this reference population were the same as those in the population under study. (See e.g. reference 3.) Standardization of rates thus takes account of maternal-age differences between populations and of changes over time. Despite the drop in crude rates cited above, the standardized rate per 1,000 livebirths was almost constant: 0.46 in 1963 and between 0.46 and 0.49 in 1978, depending on adjustments for selecive abortion. (See below.) (The reference population is all Upstate New York white livebirths in 1963-1974.)

In Manitoba a marked increase in Down's syndrome has been reported in livebirths born to mothers 35 and over in 1970 to 1974, compared to 1965 to 1969. (2) Based on data subsequently made available to us, the calculated increase for this age group was 54% in the crude rate and 65% in the standardized rate, but for those under 35 the crude rate diminished slightly (table 1). Stimulated by this finding, we reexamined the New York State data, stratifying by maternal age for these years.(11) Our data showed evidence for a similar temporal trend in older mothers, albeit of lesser magnitude; the standardized rate increased by about 30% (table 2).

The Manitoba data had been based upon a virtually complete community survey of all instances of Down's syndrome, but our own survey had been based upon birth certificate reports only. One difficulty in relying on birth certificates is that they are notoriously incomplete. Most studies have shown completeness of reporting for Down's syndrome of only about 30 to 50%. Since the change over time in our own data may have been attributable to a change in reporting, an investigation of the completeness of reporting in these two intervals, stratified by maternal age was necessary. From data on cytogenetically confirmed cases, the completeness estimates of birth certificate reports were as follows: 1965-1969: maternal age <35, $31/89 = 34.8\%$; maternal age $\geqslant 35$, $25/63 = 39.7\%$; all ages, $56/152 = 36.8\%$; 1970-1974: maternal age $<35, 61/199 = 30.7\%$; maternal age $\geqslant 35$, $26/98 = 26.5\%$; all ages, $87/297 = 29.3\%$. Thus the completeness of reporting of Down's syndrome had actually declined and the increase in reported rates had occurred despite, not because of, the change in reporting on birth certificates. (11)

Four general types of explanations are possible for such a change: i) an increase in the gametic mutation rate for chromosome abnormalities associated with Down's syndrome in women 35 and over; ii) an increase in survival of Down's syndrome embryos and fetuses; iii) statistical fluctuation; and iv) some further undetected reporting artifact peculiar to cases born to older mothers. While the last two possibilities appear implausible, they cannot be excluded.

Evidence for changes in jurisdictions other than New York and Manitoba would also be of interest. There is suggestive (but not statistically significant) evidence for such change in Montreal (11) and in Safford County Borough, United Kingdom, (4) but not in data available from Israel (11) or Australia or in recent analyses from British Columbia (5, R. Ward personal communication). Aside from birth certificates reports we know of no other readily available, relevant data sources in the U.S.A. The Center for Disease Control in Atlanta, for example, only began collecting data on Down's syndrome in 1969, and so there are relatively few data for the first five-year interval (1965-1969) for comparison with 1970-1974.

Another issue is whether the trend persisted beyond 1974. Interpretation of data for 1975 and later years is complicated by the extensive use of prenatal diagnosis and selective abortion of Down's syndrome fetuses, particularly by women aged 35 and over. These data must be adjusted for the 70% to 80% likelihood that a Down's syndrome fetus 16 to 20 weeks would, if not electively terminated, survive to livebirth (8) and for the likelihood that if it did survive, it would be reported on a birth certificate. These adjusted rates for 1974-1978 are shown in table 3. For maternal ages 35 and older, in all years after 1974, the adjusted standardized rates are greater than the standardized rate of 2.16 per 1,000 for the years 1965-1969 (table 3).

Our comparison of 1970-1974 with 1965-1969 was based on the grouping of years by Evans et al. (2) However, the New York State data by single year suggest that there were actually two periods in which the rates appeared to change for maternal ages 35 and older, once between 1967 and 1968, the other between 1972 and 1973. True, there had been a change in Upstate New York in 1968 in the coding of livebirths with two or more malformations reported on birth certificates, but we had taken this factor into consideration by hand-tabulating all livebirths with multiple malformations coded in earlier years. Nevertheless, it is conceivable that some unadjusted change in coding or reporting on birth certificates influenced the apparent change in rate in 1968, but we are unaware of any, nor of any reason why there should have been any artifactual change in 1973.

Another problem with use of birth certificates is their imperfect accuracy. Anyone undertaking case-control studies of individuals with phenotypic diagnoses of Down's syndrome will encounter occasional individuals suspected of having Down's syndrome — and so identified in some record, such as the birth certificate — who on subsequent cytogenetic study are found to have normal chromosomes. Unpublished observations suggest that the proportion of such cases is of the order of 5 to 10%. Such individuals may, however, be trisomic at least in part for the 21st chromosome. There are well-authenticated instances of cryptic mosaicism and translocations missed on more superficial study. (See extensive discussion in reference 10.)

This example indicates some of the frustrations and difficulties in using birth certificates and other vital records for surveillance. We strongly recommend that anyone relying on such sources devote particular attention to fluctuations in methods of reporting, coding (especially of livebirths with two or more malformations), completeness, and, if possible, accuracy.

INTERCHANGE TRISOMY 21 AND CYTOGENETIC LABORATORY EXPERIENCE

Another type of opportunistic method for surveillance revealed indirect evidence for a change in the rate of interchange trisomic Down's syndrome in livebirths in New York State in 1973 and later years. In the experience of cytogenetic laboratories reporting to the New York State Chromosome Registry the proportion of Down's syndrome attributable to interchange trisomy (i.e. unbalanced Robertsonian translocations) rose abruptly in 1973, compared to earlier years for which data are available, and stayed high in 1974 and 1975. (9,12a,12b) Thereafter there has been considerable fluctuation, but in general the rates have been higher than in 1968-1972. (See table 4.)

Such a change in the proportion of interchange trisomic Down's syndrome could be attributable to either a real two-to-threefold increase in the rate of such translocations or a drop to one-half to one-third in the rate of 47,+21. The latter, however, would result in a marked fall in the total phenotype, which is ruled out by other evidence, such as that cited above from birth certificates. Therefore the best explanation for the trend is a true increase in the livebirth rate of interchange Down's sydrome. (Similar but less substantial evidence exists for a change in interchange trisomic Patau's syndrome around – but not exactly at – the same time. Patau's syndrome is also produced by unbalanced Robertsonian translocations.)

An obvious question is the relationship of the change in interchange trisomy in 1973, as reported to the Chromosome Registry, to the change in phenotypic Down's syndrome, as reported on birth certificates in the same year. These trends appear to be unrelated. The rise in interchange Down's syndrome is almost exclusively in individuals born to mothers under 30 who account for 80% of livebirths. The increase in phenotypic Down's syndrome, while it occurs in both older and younger mothers, is particularly prominent in older mothers. Comparison is somewhat obscured because of the different age boundaries used in the two analyses. Nevertheless it is clear that although the change in translocations may account for some of the rise in phenotypic Down's syndrome in the under-35 group in 1973, it did not contribute to the increase in those 35 and over.

A case-control study of the interchange trisomies, investigating possible environmental causes of the increase in translocations found no apparent explanation for the change. If an undetected artifact or statistical fluctuation is not responsible for the change, it could be attributable to: i) a true increase in the mutation rate for Robertsonian translocations associated with interchange trisomy; and/or ii) a change in segregation ratios of translocation chromosomes, i.e. less alternate but more adjacent segregation, and thus a corresponding drop in de novo balanced rearrangements; and/or iii) better differential survival of interchange trisomic fetuses.

Proportional surveillance of interchange trisomics is of course limited to a few cytogenetic outcomes but does not depend upon the existence of a chromosome registry. Pertinent data can be obtained relatively easily from cooperative local diagnostic cytogenetic laboratories.

BIRTH DEFECT REGISTRIES AND SURVEILLANCE SYSTEMS

In some jurisdictions birth defect registries or surveillance systems for birth defects are in place. These are likely to provide more complete data on Down's syndrome than birth certificate reports. Nevertheless, the problem of changes in sources of ascertainment must be carefully addressed. For example, data from the British Columbia Registry (15) have been widely cited as indicating a rise in maternal-age-specific rates of Down's syndrome in 1964-1967 compared to 1952-1963. For 1952-1955, 1956-1959, and 1960-1963 the age-adjusted rates per 1,000 livebirths were 1.24, 1.13, and 1.14 respectively. They then rose to 1.39 in 1964-1967, to 1.50 in 1968-1971, and to 1.51 in 1972-1973. But new sources of ascertainment were introduced in 1964 and 1966, so it is not at all surprising to find an increase in the rate of cases reported to the registry.

Some surveillance systems, e.g. the U.S. Birth Defect Monitoring Program (1), use only newborn hospital discharge records. These are likely to be more complete than birth certificate reports. For example, in metropolitan Atlanta where the Center for Disease Control collects data from all possible sources on birth defects diagnosed through the first 12 months of life, in 1968-1976 215/229 = 93.9% of all cases ascertained in this age group were first diagnosed on or before the 5th day of life, the usual time of newborn discharge. Interestingly, this proportion was higher in whites, 157/162 = 96.7%, than in blacks, 58/67 = 86.6%, chi^2 = 7.1, P<.01. (The data are from J.D. Erickson, personal communication.) Of course these proportions would be lower if children who are not diagnosed until after age 1 year were included in the denominators, but data on this age group are unavailable. It appears unlikely however, that the proportions would drop below 70% to 80% even if these were included.

PHENOTYPIC SURVEILLANCE

The three surveillance methods described to this point — i) use of birth certificate reports and other routinely gathered data on the phenotype of Down's syndrome, ii) proportional analysis of laboratory data on interchange trisomic Down's syndrome and Patau's syndrome, and iii) use of data in birth defect registries or other birth defect reporting systems — all review and analyze existing records in a search for trends.

A fourth method which will provide data on other cytogenetic aberrations is phenotypic surveillance of livebirth infants for syndromes suggestive of chromosome aberrations, followed by cytogenetic investigation of suspicious cases. This requires a supervisory authority to ensure that all livebirths are examined by knowledgeable individuals and that appropriate referrals for cytogenetic evaluation are carried out. This may appear easy to implement, but one team of individuals is likely to be able to cover only a few hospitals and thus to screen no more than 10,000 livebirths a year. Moreover all livebirths must be screened, and this requires very close collaboration with nursing services and pathologists to ensure that even newborns who expire will receive systematic phenotypic evaluation and, if appropriate, cytogenetic study.

We estimate that, depending on how extensive cytogenetic information is needed, at most 1 to 2% of infants need be examined cytogenetically at a cost of 100 to 150 (1981) dollars each.Thus the laboratory cost of surveillance per 10,000 children would be about $10,000 to $30,000. If one restricted interest to suspect or certain occurrences of autosomal trisomic phenotypes and to the few other rare pathognomonic phenotypes associated with chromosome abnormality, e.g. cri du chat, only about 20 cases per 10,000 would need to be studied cytogenetically at cost of only 2,000 to 3,000 (1981) dollars. In many — but arguably, not all — of these instances cytogenetic analyses would be warranted on clinical grounds alone.

Results of cytogenetic abnormalities detected by phenotypic screening have been reported in the literature, although such surveys have usually been undertaken as part of an evaluation of the incidence of all birth defects. Reports limited to cytogenetic abnormalities have been cited by Mehes.(16) As may be inferred from the numbers reported screened in such surveys, which have usually taken several years to accumulate, it is difficult to generate adequate numbers for surveillance purposes.

Nevertheless, such systematic surveys are particularly attractive because they can serve a variety of purposes in addition to screening for cytogenetic abnormalities.

They can, for example, detect phenotypes which are associated with specific locus mutations (6) and birth defects which may reflect the teratogenic effects of toxic substances.

PRENATAL CYTOGENETIC DIAGNOSIS

The methods discussed to this point all focus upon cases observed in livebirths with a modal gestational length of 40 weeks. The proportion of fetuses affected by cytogenetic abnormalities is of course much greater at earlier stages. (See estimates in reference 12.) In jurisdictions where prenatal diagnosis of cytogenetic abnormalities is widely practiced, analysis of these data is pertinent for surveillance of rates of cytogenetic mutations. To avoid bias attention must be restricted to diagnoses in women with no known cytogenetic risk factor, aside from maternal age, and strict adjustment must be made for the maternal-age composition of the population to allow fair comparisons over time and between populations.

Data on rates of specific chromosome abnormalities diagnosed at amniocentesis, as reported to the U.S. Interregional Chromosome Register and the New York State Chromosome Registry, are given in tables 5 and 6. There is a marked drop in incidence of 47,+21, in cases reported to the Interregional Chromosome Register, in 1980 and 1979, compared to the two preceding years, but no such change for 47,+18. The observation is provocative; we do not know how to interpret it at present. No such change is seen in the pooled experience for the New York State Chromosome Registry for the same years.

Data derived from anmiocentesis studies are among the most useful for surveillance because cytogenetic diagnoses are done for clinical reasons in a defined group of women whose only cytogenetic risk factor is advanced maternal age. (Some younger women are included who request amniocentesis because of an increased fetal risk of a metabolic error or neural tube defect, and whose amniotic fluid is cytogenetically analyzed "incidentally" because the specimen is available readily for analysis.) However, the entire group in which such data are accumulated is, at least in the U.S.A., not representative of the general child-bearing population. It is likely to be of higher socioeconomic status than the rest of the population, and it clearly selectively excludes those whose religious and personal beliefs are opposed to abortion. Thus surveillance based solely on amniocentesis data may miss effects of agents concentrated in younger women or in the underrepresented socioeconomic and religious groups.

EMBRYONIC AND FETAL DEATHS

What further discoveries would advance the goals of surveillance? We have not discussed cytogenetic study of spontaneous embryonic and fetal deaths because systematic collection and study of specimens is likely to be too expensive to justify for surveillance purposes alone. However, if readily evident phenotypic signs of chromosome abnormalities were identified in spontaneously aborted embryos and fetuses, pathologists could assist surveillance by phenotypic screening and selective cytogenetic study of aberrant cases. Perhaps, for example, lymphoedema and/or cervical cystic hygroma would serve as such useful phenotypic indications of 45,X abortuses. Further developmental work on phenotypic markers in embryos and fetuses would be of great pertinence to surveillance.

For monitoring, as opposed to surveillance, there may well be an economic rationale for systematic cytogenetic study of all detectable spontaneous embryonic and fetal deaths, as well as livebirths. However, the costs of such studies, including the cost of specimen collection from the target population, are likely to be considerable. (See further discussions in references 14 and 18.)

Investigators of mammalian mutagenesis often suggest that the measurement of "dominant lethals" be applied to studies of mutations in human populations. For lower animals strictly controlled experiments may allow a plausible inference that embryonic and fetal deaths are produced by mutagenic effects of compounds which are administered *preconceptually* to the male parent. The only analogue in humans might be monitoring of the rate of all embryonic and fetal deaths where only the male parent has been exposed to a putative mutagen, as in some occupations, and where there is no possibility of maternal exposure. Otherwise embryonic and fetal deaths may result from nonmutagenic teratogens or other embryotoxic mechanisms. In practice, if paternal exposure continues into pregnancy, it is very difficult to exclude maternal exposure to the toxic substance via the father's work clothes, skin or even seminal fluid. Therefore, following the rate of embryonic and fetal deaths alone is not likely to be useful for mutation surveillance, whatever information may be derived on embryotoxicity.

ACKNOWLEDGMENTS

The United States Interregional Chromosome Register is supported by a contract from the National Institute of Child Health and Human Development to the University of Oregon.

REFERENCES

(1) Center for Disease Control. Congenital Malformation Surveillance Report January-December, 1979. (1980)
(2) Evans, J.A., Hunter, A.G.W., Hamerton, J.L. *J. Med. Genet. 15*, 43-47 (1978)
(3) Fleiss, J.L. Statistical Methods for Rates and Proportions. New York:John Wiley, 237-255 (1981)
(4) Fryers, T., Mackay, R.I. *Early Hum. Develop.3*, 29-41 (1979)
(5) Griffiths, A.J.F., Lowery, R.B., Renwick, S.H.G. *Environ. Health Persp. 31*, 9-11 (1979)
(6) Holmes, L.B., Vincent, S.E., Cook, C., Cote, K.R. *In*: Hook, E.B., Porter, I.H., eds. Population and Biological Aspects of Human Mutation. New York: Academic Press, 351-360 (1981)

(7) Hook, E.B. *In*: Morton, N.E., Chung, C.S., eds. Genetic Epidemiology. New York:Academic Press, 483-528 (1978)
(8) Hook, E.B. *New Eng. J. Med. 299,* 1036-1038 (1978)
(9) Hook, E.B. *Mutat. Res. 52,* 427-439 (1978)
(10) Hook, E.B. *In* de la Cruz, F., Gerald, P.S., eds. Trisomy 21 (Down syndrome): Research perspectives. Baltimore:University Park Press, 3-67 (1980)
(11) Hook, E.B., Cross, P.K. *J. Med. Genet. 18,* 29-30 (1981)
(12) Hook, E.B. *Lancet 2,* 169-172 (1981)
(12a) Hook, E.B., Albright, S.G. *Amer. J. Hum. Genet. 33,* 443-454 (1981)
(12b) Hook, E.B. *In:* Hook, E.B., Porter, I.H. eds. Population and Biological Aspects of Human Mutation. Academic Press:New York, 167-190 (1981)
(13) Hook, E.B. *In*: Bora, K.C., ed. Chemical Mutagenesis, Human Population Monitoring, and Genetic Risk Assessment. Amsterdam:Elsevier-North Holland, (in press)
(14) Jacobs, P.A. *In*: Morton, N.E., Chung, C.S., eds. Genetic Epidemiology. New York:Academic Press, 463-482 (1978)
(15) Lowry, R.B., Jones, D.C., Renwick, D.H.G., Trimble, B.K. *Teratology 14,* 29-34 (1976)
(16) Mehes, K. *Lancet 1,* 1262-1263 (1973)
(17) Sutherland, G.R., Clisby, S.R., Bloor, G., Carter, R.G. *Med. J. Austral. 2,* 58-61 (1979)
(18) Warburton, D., Stein, Z., Kline, J., Susser, M. *In*: Porter, I.H., Hook, E.B., eds. Human Embryonic and Fetal Death. New York:Academic Press. 261-287 (1980)

Table 1 — Trends in Rates of Down's Syndrome in Manitoba 1965-1974

Maternal age	Crude Rates			Standardized Rates		
	1965-1969	1970-1974	Change	1965-1969	1970-1974	Change
<35	$\frac{68}{79,744}$ = 0.85	$\frac{66}{81,944}$ = 0.81	-5%	0.84	0.85	+1%
≥35	$\frac{52}{10,647}$ = 4.88	$\frac{45}{5,988}$ = 7.52	+54%	4.46	7.38	+65%
All (including not stated for crude rates)	$\frac{121}{90,396}$ = 1.34	$\frac{112}{57,946}$ = 1.27	-5%	1.18	1.47	+25%

All rates are per 1,000 livebirths. Data are from Evans et al, 1978, and Evans, personal communication.

Table 2 — Trends in Rates of Down's Syndrome in Upstate New York 1965-1974 Reported on Birth Certificates

Maternal age	Crude Rates			Standardized Rates		
	1965-1969	1970-1974	Change	1965-1969	1970-1974	Change
<35	$\frac{226}{730,432}$ = 0.31	$\frac{191}{655,848}$ = 0.29	-6%	0.31	0.30	-3%
≥35	$\frac{186}{84,712}$ = 2.20	$\frac{126}{45,950}$ = 2.74	+25%	2.16	2.81	+30%
All (including not stated for crude rates)	$\frac{412}{815,387}$ = 0.51	$\frac{317}{702,055}$ = 0.45	-12%	0.49	0.54	+10%

Table 3 – Upstate New York: Temporal Trend in Standardized Rates per 1000 of Down's Syndrome Reported on Birth Certificates

Maternal Age	1965	1966	1967	1968	1969	1970	1971	1972	1973	1974	1975	1976	1977	1978
<35	0.24	0.29	0.30	0.33	0.41	0.31	0.29	0.30	0.37	0.22	0.31	0.33	0.28	0.27
≥35	2.08	1.70	2.12	2.52	2.55	2.72	2.34	2.37	3.68	3.04	3.40	2.45	2.16	2.27
											(3.47)	(2.72)	(2.63)	(2.54)
All	0.42	0.42	0.48	0.54	0.62	0.54	0.49	0.50	0.68	0.49	0.60	0.53	0.46	0.46
											(0.61)	(0.56)	(0.51)	(0.49)

*Reference population is Upstate New York white livebirths 1963-1974.
Rates in parentheses are those after adjustment for cases diagnosed at amniocentesis.

Table 4 – Percentage of Down's Syndrome Cases with Mutant Interchange Trisomy*

Maternal Age	Year of Birth												
	1968	1969	1970	1971	1972	1973	1974	1975	1976	1977	1978	1979	1980
<30	5.4-7.3	4.9-4.9	2.1-3.2	3.8-3.8	1.8-2.7	7.4-9.3	8.2-9.0	7.8-8.6	3.9-6.8	6.9-7.8	1.0-2.0	0.0-4.9	7.2-7.2
	(55)	(103)	(93)	(106)	(113)	(108)	(122)	(116)	(104)	(116)	(100)	(102)	(112)
≥30	3.2-3.2	1.0-1.0	0.9-0.9	0.8-0.8	3.3-3.3	2.1-2.1	2.0-2.0	1.3-1.3	0.0-0.0	2.2-3.3	3.7-3.7	2.3-4.7	3.6-3.7
	(63)	(98)	(109)	(122)	(92)	(95)	(99)	(150)	(118)	(91)	(108)	(86)	(83)
All	4.2-5.1	3.0-3.0	2.0-2.5	2.2-2.2	2.4-2.9	4.9-5.9	5.4-5.9	4.1-4.5	2.3-3.6	4.8-5.8	2.4-2.9	1.1-4.8	5.6-5.6
	(118)	(201)	(202)	(228)	(205)	(203)	(221)	(266)	(222)	(207)	(208)	(188)	(195)

*Numbers in parentheses are the total number of Down's syndrome cases to the New York State Chromosome Registry, by April 15, 1981, excluding known out-of-state residents. Approximately 5% to 10% of the 47,+21 cases did not have year of birth indicated; they were assigned here to year of report to the Registry. There is only one such interchange trisomic case, assigned here to 1971. Maternal age is not stated for about 15% of the 47,+21 cases and 6% of the interchange trisomies. They are distributed here according to the proportions of those with known maternal age. The range in percent of mutant interchange trisomy reflects instances in which one or both parents have not been investigated to determine if the translocation is present. In about 25% of such instances the case is inherited, in 75% it is mutant. For further analyses from somewhat different viewpoints, see references 12a and 12b.

Table 5 — U.S. Interregional Chromosome Register Data: Rate per 1000 of Cytogenetic Abnormalities in Women Aged 35 and Over Studied at Amniocentesis Because of Advanced Maternal Age, With No Other Risk

	1977-1978	1979	1980 (first half)
Trisomy 21			
Crude rate	$\frac{27}{1842} = 14.7$	$\frac{6}{1606} = 3.7$	$\frac{2}{1037} = 1.9$
Standardized rate	12.7 ± 2.5	3.7 ± 1.5	2.1 ± 1.5
Trisomy 18			
Crude rate	1.6	1.9	2.9
Standardized rate	1.5 ± 0.9	2.2 ± 1.3	3.5 ± 2.0
All other clinically significant (including 47,XXX)			
Crude rate	6.5	8.7	4.8
Standardized rate	6.8 ± 2.0	9.4 ± 2.5	5.0 ± 2.3

Table 6 — New York State Chromosome Registry Data: Rate per 1000 of Cytogenetic Abnormalities in Women Aged 35 and Over Studied at Amniocentesis Because of Advanced Maternal Age, With No Other Known Cytogenetic Risk

	1977	1978	1979	1980
Trisomy 21				
Crude rate	$\frac{20}{1788} = 11.2$	$\frac{26}{2577} = 10.1$	$\frac{33}{3830} = 8.6$	$\frac{52}{5354} = 9.7$
Standardized rate	9.8 ± 2.2	10.4 ± 2.0	9.9 ± 1.8	10.7 ± 1.5
Trisomy 18				
Crude rate	4.5	3.1	1.8	2.1
Standardized rate	3.9 ± 1.4	3.7 ± 1.3	2.1 ± 0.8	2.8 ± 0.9
All other clinically significant (including 47,XXX)				
Crude rate	8.4	7.0	6.0	6.7
Standardized rate	8.2 ± 2.1	6.7 ± 1.6	5.7 ± 1.2	7.1 ± 1.2

Cases are those reported by New York State laboratories only.

Towards Documenting Human Germinal Mutagens: Epidemiologic Aspects of Ecogenetics in Human Mutagenesis

John J. Mulvihill

Clinical Epidemiology Branch, National Cancer Institute, Bethesda, Maryland, U.S.A.

ABSTRACT To date, no human germinal mutagen has been directly documented. In the effort to do so, all human carcinogens, teratogens, and clastogens become suspect. Laboratory evidence of somatic cell mutations and of non-human mutagenesis is most convincing when confirmed by observations in human patients or populations. In practice, mutation epidemiology has been mostly limited to field studies of polymorphic variants in the offspring of survivors of the atomic bombs in Japan and to consideration of monitoring the frequency of birth defects and sentinel phenotypes. Perhaps alternate approaches based on the successes of cancer and birth defect epidemiology deserve exploration. Possibilities include: 1) the alert practitioner (case reports by observant clinicians of unusual exposures of parents of the first instance of a genetic disease in the family); 2) personal or familial clusters of genetic disease (e.g., multiple mutant diseases in an individual or a family, including cancer and birth defects); 3) retrospective case-control studies of environmental and genetic histories of parents whose offspring represent a new mutation; and 4) offspring of individuals with intense exposure to potential mutagens (e.g., occupational cohorts and long-time survivors of cancer). Any strategy to detect human mutagens should recognize the likelihood of an inborn variability of host susceptibility to mutagens.

ECOGENETICS IN MUTAGENESIS

Xeroderma pigmentosum and ataxia-telangiectasia, though rare in nature, are household words in mutation biology. Both conditions represent a genetic predisposition to a toxic reaction when the affected person is exposed to certain environmental agents in doses that cause no ill effects in normal human beings. The term for such gene-environment interactions is ecogenetics (25), as discussed in this volume and elsewhere by G.S. Omenn (34). By analogy with the better known term, pharmacogenetics (which means the study of variations in response to drugs), ecogenetics refers to variations in response to any environmental agent.

Besides xeroderma pigmentosum and ataxia-telangiectasia, genetic traits in human beings which predispose to cancer on exposure to environmental agents include hemochromatosis with dietary iron, the Turner syndrome with estrogen treatment, and the Purtilo X-linked lymphoproliferative syndrome, and nine others (26). The list is short, but establishes a principle that could apply to the origins of most human cancers; namely, cancer usually results from the interaction of a susceptible genotype with certain environmental agents. The same principle of ecogenetics is expected to apply to human mutagenesis. Specifically, some genotypes--some people--might well be more susceptible than others to mutation by environmental agents; and, some might be genetically resistant to environmental mutagens. How might an appreciation of ecogenetics help in identifying environmental mutagenesis in human beings?

TO SHOW A HUMAN GERMINAL MUTAGEN

The assumption must be made that the final measure of prime concern is human hereditary disease. By human, I mean that the real object of experimental mutagen testing is the human species. Hereditary indicates germinal mutations are more worrisome than somatic ones, which are surely of concern as possible steps toward cancer or as warnings of possible germ cell damage. Disease, because mutations that cause disease are worrisome, and neutral or helpful mutations are not.

Recently James V. Neel editorialized, "The evaluation of the genetic implications for humans of increasing exposures to ionizing radiation and complex chemicals presents one of the most difficult epidemiologic issues ever faced by biomedical science" (32). He called for further deliberations of possible methods to document human germinal mutagenesis; so, the remainder of this essay responds to that invitation to scientists worldwide. Discourses over the last decade may seem to exceed tolerable limits (5,7,8,29,31,33,40); yet, the questions remain urgent and largely refractory to most approaches.

To date, no human germinal mutagen has been directly documented. In practice, the search has been confined to the survivors of the atomic bombs exploded in Hiroshima and Nagasaki (37). By a number of parameters, there has been "no clearly statistically significant effects of parental exposures on the offspring characteristics. . .studied, but the various indicators of possible genetic damage all are in the direction expected if an effect was indeed produced" (37). Besides these studies in Japan, monitoring the frequency of birth defects and sentinel phenotypes has been urged (12,31,40), but has not yet proved useful. Clearly, other approaches deserve consideration.

CARCINOGENS, TERATOGENS, CLASTOGENS, MUTAGENS

The generalization can be made that agents that tend to cause cancer tend to cause birth defects, and, in one system or another, tend to break chromosomes and to be mutagenic (22).

The chemical, physical, and biological agents that are known to cause both cancer and birth defects in human beings (though not necessarily both in the same person) are tobacco, ionizing radiation, stilbestrol, alcohol, hydantoin, androgens, antimetabolites, and herpes virus. For example, cigarette smoking is the best known human carcinogen; the offspring of smoking women are, on average, smaller and lighter than children of non-smokers (1). Radiation of children and adults causes leukemia and other cancers, and, in the embryo, small head size and mental retardation (24). Three chemicals, stilbestrol, alcohol, and hydantoin, have been associated with cancer and birth defects in the same individual (22). Although R.W. Miller pointed out that the mechanisms of carcinogenesis and teratogenesis by the same agent may well be different (22), the fact remains that, when an agent is shown to be carcinogenic, it should be suspected of being teratogenic. The principle could be extended to include a suspicion of mutagenicity, when either carcinogenicity or teratogenicity is shown.

CANCER EPIDEMIOLOGY: LESSONS FOR MUTATION EPIDEMIOLOGY

By taking clues from human carcinogenesis and teratogenesis, some guess might be made about which of many agents that are mutagenic in non-human or in vitro tests should be most suspected as human mutagens in vivo. But how can an environmentally induced human germinal mutation be demonstrated? Just as human mutagens can be suspected from the lists of human carcinogens and teratogens, guidance for present research strategies can be gleaned from the successes of cancer and birth defects epidemiologists. A similar comparison of childhood cancer and birth defects epidemiology was made 11 years ago (20). Future approaches will surely depend on laboratory assays of human DNA or its products. Such assays are developing so fast that their usefulness in population studies can hardly be imagined.

According to R.W. Miller, in this volume and elsewhere (23), nearly all human carcinogens have been first detected and defined by alert practitioners who recognized unusual features in their patients' environmental or medical histories. He repeats the story of Dr. Miyanishi, an intern at Hiroshima University, who asked his 30-year-old patient, "Why, at such an early age, do you have lung cancer?" The patient said it was probably because he had worked at a mustard gas factory during the War, ten years earlier. The answer was correct, and another respiratory carcinogen was discovered by an alert practitioner (21,39).

A second method of cancer epidemiologists is to study clusters of cases. Color maps are available that show geographic variations in cancer death rates in the U.S. (17) and China (9), among other nations. The standard presentations give average distributions, for example, for lung cancer deaths in U.S. white males, 1950-1975 (Fig. 1A). This static distribution can be supplemented by a dynamic map to answer the question, "Where are rates changing the fastest?" (Fig. 1B). Finally, true time-space clusters can be identified for special scrutiny; for example, 15 U.S. counties had rates that were high on average and were increasing rapidly (Fig. 1C).

Besides geographically, cancer may cluster in individuals as multiple primary malignancies or in so-called cancer families (28).

Analytical approaches include the case-control study, which compares risk factors in persons with cancer and those without; for example, "Did persons with urinary bladder cancer use more saccharin than controls?" (13). Such a study starts

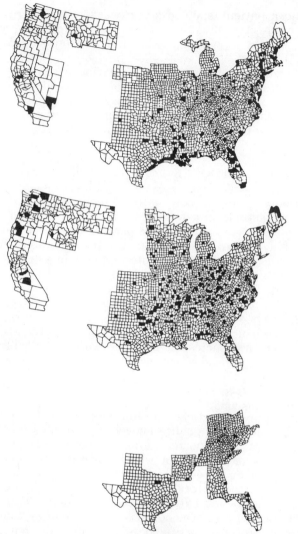

Fig. 1. Lung cancer deaths in white U.S. males, 1950-1975. Shaded areas show statistically significant high frequencies. A. Static distribution of standardized mortality ratios (where, on average, lung cancer rates were highest). B. Dynamic distribution (where rates increased the fastest, compared to total U.S. trends for the same tumor). C. Statistically significant time-space clusters. (Courtesy F.W. McKay, Clinical Epidemiology Branch, National Cancer Institute, Bethesda, Maryland).

with a cancer patient and looks backwards. In contrast, the prospective cohort study starts with a suspected exposure, like the atomic bombs, and asks, "Who in the exposed group got cancer?" (4).

The same four strategies for detecting carcinogens have been useful in detecting teratogens: alert practitioners, case clusters, case-control studies, and prospective cohort studies. Analogous ways for documenting a human mutagen deserve further consideration.

THE ALERT PRACTITIONER

Confronted with a new case of achondroplasia or bilateral retinoblastoma in a family, an alert practitioner might note some peculiar exposure of the parents of the new mutant offspring. [Of course, it is well known that sporadic (that is nonfamilial) cases, especially of autosomal dominant traits, are associated with advanced age of the father (10).] A thorough clinical inquiry for potential mutagen exposure would elicit the occupational, recreational, and medical exposures of both parents to x-rays and chemicals. In theory, the probing would include exposure of the mother's ovaries prenatally, when the oogonia are in first meiosis. In other words, the mutagenic history of the proband's maternal grandmother would be desirable.

Such clinical anecdotes of mutagen exposure must surely exist, but I was unable to ascertain them from the literature. Perhaps they should be solicited from clinical geneticists who may be reluctant to publish anecdotal experience. On the other hand, there are reported patients who could represent a cluster of mutational events for which there is no comment on parental exposures. A recent example is the report of a one-year-old infant with bilateral Wilms tumor, botryoid sarcoma, microcephaly, arhinencephaly, and cataracts (30). The authors wrote, "The multiplicity of neoplasms and congenital anomalies in this case suggests a genetic defect." They did not give a mutagen history, except to report the parents were healthy and the family history negative.

Another incompletely reported coincidence of mutations comprises three U.S. patients with malignant hyperthermia (an autosomal dominant trait) and Burkitt lymphoma, all reported within two years of each other (15). The chance of one such case in a year was calculated to be 10^{-12}. Another example is a newborn with Down and Klinefelter syndromes, who died with congenital leukemia at one month of age (36).

CLUSTERS OF GENETIC DISEASES

Such clinical observations illustrate the second strategy, the study of clusters of genetic diseases. The phenomenon of time-space clustering has guided recommendations for population-based surveillance of genetic and congenital diseases, especially the so-called sentinel phenotypes. (Of course, some clustering of genetic diseases is due to founder effect and inbreeding and is not helpful in revealing environmental mutagens.) Rather than discuss the usefulness of population surveillance, which is much debated (12,29), I shall focus on a special type of clustering, not emphasized elsewhere: the coincidence of several genetic diseases in a person or a family. Insights into the causes of cancer and birth defects seem to come fastest from the multiple occurrences of both growth disorders. Examples include the infant with multiple birth defects (38), the person with multiple primary tumors (28), and individuals or families with an excess of cancer, birth defects, or both (18,27,35). Within the last year, at least five families were reported, each with a thalassemia and another single mutant gene disorder (Fig. 2).

To the extent that cancer represents somatic mutations, families with an excess of cancer also illustrate clustering of

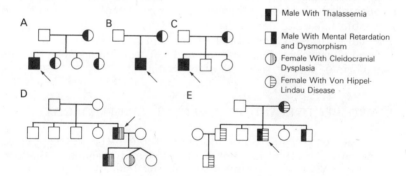

Fig 2. Familial clusters of mutant globin genes (thalassemias) with other genetic diseases. Arrows indicate probands. A-C. New alpha-chain mutation leading to hemoglobin H disease, mental retardation, and birth defects in three boys whose fathers' blood pictures failed to show the expected globin defect (41). D. Proband was a new double mutant for beta-thalassemia and cleidocranial dysplasia (2). E. Independent assortment of von Hippel-Lindau disease and beta-chain mutant (16).

mutational disease, especially in the presence of other genetic diseases. We are following a family in which three females in a sibship of 13 and their father had the autosomal dominant Saethre-Chotzen syndrome of craniosynostosis, facial asymmetry, and hypertelorism (18). Nasopharyngeal carcinoma developed in one dysmorphic girl, two brothers had Hodgkin disease, and another had embryonal cell carcinoma of the testis. Chromosomes were normal, including spontaneous levels of sister chromatid exchanges, as were serum titers of antibodies to Epstein-Barr virus. Minor immune dysfunction was seen in vitro; but, the explanation for the familial coincidence of cancer and birth defects might come sooner from mutation biology than from immunology.

David I. Hoar called such clustering of disease, "hard luck families," and asked whether they represent familial mutagen hypersensitivities, as seen in Drosophila as the mutator mutants, the meiotic mutants, and increased sensitivity to mutagens (11). Using fibroblast colony survival assays, he found dominantly transmitted hypersensitivity to three chemical mutagens in such families.

Similarly, in vitro radiosensitivity was found in members of a family in which four of six children died of acute myelogenous leukemia (3). Their mother had rectal carcinoma 14 years following—and probably as a result of—radiotherapy for cancer of the cervix. Fibroblasts from the mother and her two available leukemic daughters were moderately radiosensitive in the hands of N. Torben Bech-Hanson in M.C. Paterson's Chalk River Laboratory. The husband and two remaining normal sons had normal colony survival after radiation.

RETROSPECTIVE AND PROSPECTIVE EPIDEMIOLOGY

As for formal epidemiologic studies, the retrospective case-control method has not been tried and perhaps with good reason, in view of its high cost and the lack of an hypothesis. Nonetheless, the U.S. Centers for Disease Control has begun a $1.5 million study of 7000 children with birth defects and 3000 controls. If the concern were strictly mutation, design problems would be numerous. Well diagnosed cases of new dominant or X-linked mendelian traits would have to be assembled, preferably those that have proven to be germinal by the birth of affected offspring. Questions would be asked about environmental factors of the parents of the proband and even of the maternal grandmother. An ideal strategy immediately suggests the need for interviews of four generations in a family! Not an easy task, but one that might have to be considered if pressure to document a human germinal mutagen continues.

Finally, the prospective cohort design is the only one now in progress. The prototype is the famous study of new mutants among offspring of survivors of the atomic bombs (37). Another cohort under study, by A. Czeizel in Hungary, is the survivors of attempted suicide by drug ingestion, described elsewhere in this volume. Other types of study subjects are parents with hazardous occupational exposures and persons with cancer.

MUTATION IN CANCER PATIENTS

Perhaps cancer patients are the human cohort now exposed to the largest doses of potential mutagens, both physical and chemical, that is, radiotherapy and drugs. Of course, cancer treatment is often designed to interact specifically with DNA and to block its function. All pregnancies initiated by cancer patients are of research interest for diverse reasons. Pregnancies occurring before cancer serve as a control experience and as a measure of cancer heritability. Pregnancy during a woman's cancer therapy measures teratogenicity and, if treatment antedated conception, mutagenicity as well. Pregnancies after cancer treatment measure gonadal survival and potential germinal mutagenicity.

Experience to date is limited, but three recent reports from a cooperative clinical trial group (19) and single large cancer centers (6,14), included 155 pregnancies during or after chemotherapy in 109 women with cancer, and the wives of seven cancer patients (Table 1). Just over half the patients had Hodgkin disease. Abortion was elected in 16% of the pregnancies, an expected high frequency. Of the remaining gestations, 74% ended with normal liveborns, 8% with spontaneous abortion or stillborns, 8% in premature or small-for-dates babies, and 10% with some "abnormality." Whereas the Stanford group saw no birth defects (14), the two studies I collaborated in (6,19) reported as abnormalities very minor anomalies, such as hair patch, pelvic asymmetry, ear tags, sacral and pilonidal dimples, febrile seizure, and nevus. Most clinicians would deem these seven children normal, especially since they had isolated defects. Three major malformations—hydrocephaly, cleft lip and palate, and polydactyly with abruptio placentae and stillbirth—occurred in offspring of women on chemotherapy during pregnancy and more likely

Table 1

OUTCOMES OF PREGNANCIES
DURING OR AFTER CANCER CHEMOTHERAPY

Reference	Total Cases	Pregnancies	Elective abortions	Normal	Fetal Death	"Abnormal"
19-Hodgkin disease	37	54	7	27		8 (+6 LBW)*
19-other cancers	29	31	3	21		1 (+1 LBW)
6-several cell types	30	42	10	24		4
(-males only	7	12	2	6		3)
14-Hodgkin disease	20	28	5	21		0 (+3 LBW)
TOTAL	116	155	25	93		13 (+10 LBW)

*LBW = low birth weight

represent teratogenicity than mutagenicity. In my opinion, only two anomalies in the 130 pregnancies could represent mutational disease. Hydrocephaly occurred in a baby conceived conceived one year after the mother stopped chemotherapy for Hodgkin disease (19). The other was congenital hip dysplasia in a baby born six years after completion by the mother of chemotherapy for Ewing sarcoma (6). But neither disease is among the sentinel phenotypes or cytogenetic syndromes that are considered useful in monitoring populations for mutagens.

CONCLUSIONS

My title, "Epidemiologic aspects of ecogenetics in human mutation," is more a hope than a fact. There is every reason to think human hereditary disease can be caused by environmental agents, but no human germinal mutagen has been documented to date. Many agents, almost too many, can be suspected from in vitro and lower organism testing. Since germinal mutation in the intact human being is the subject of most concern, agents on the short lists of human carcinogens and teratogens should be most suspected and studied. Successful documentation of a human germinal mutagen is likely to depend on laboratory measures of DNA damage.

If epidemiologists are going to contribute to documenting human mutagens, the ways in which human carcinogens and teratogens were detected and defined may have lessons for mutation epidemiologists. Apparently neglected strategies include hypothesis-generating case reports by alert practitioners, especially concerning new mutants and clusters of genetic diseases in a patient or family. Case-control studies might be considered and, finally, prospective studies might include suicidal patients and cancer patients.

REFERENCES

(1) Abel, E.L. Human Biol. 52, 593-625 (1980)
(2) Alexander, W.N. and Ferguson, R.L. Oral Surg. 49, 413-418 (1980)
(3) Bech-Hansen, N.T., Sell, B.M., Mulvihill, J.J., and Paterson, M.C. Cancer Res. 41, 2046-2050 (1981)
(4) Beebe, G.W. Epidemiol. Rev. 1, 184-210 (1979)
(5) Berg, K., Ed. Genetic Damage in Man Caused by Environmental Agents. New York: Academic Press, 1979
(6) Blatt, J., Mulvihill, J.J., Ziegler, J.L., Young, R.C., and Poplack, D.G. Am. J. Med. 69, 828-832 (1980)
(7) Bloom, A.D., Ed. Guidelines for Studies of Human Populations Exposed to Mutagenic and Reproductive Hazards. White Plains: March of Dimes Birth Defects Foundation, 1981
(8) Brusick, D. Principles of Genetic Toxicology. New York: Plenum Press, 1980
(9) Editorial Committee Atlas of Cancer Mortality in the People's Republic of China. Beijing: China Map Press, 1981
(10) Jones, K.L., Smith, D.W., Harvey, M.A.S., Hall, B.D., and Quan, L. J Pediatr. 86, 84-88 (1975)
(11) Hoar, D.I. Can. J. Genet. Cytol. 21, 435-442 (1979)
(12) Hook, E.B. Environ Res. 25, 178-203 (1981)
(13) Hoover, R.N. and Strasser, P.H. Lancet 1, 837-840 (1980)
(14) Horning, S.J., Hoppe, R.T., Kaplan, H.S., and Rosenberg, S.A. N. Engl. J. Med. 304, 1377-1382 (1981)

(15) Lees, D.E., Gadde, P.L., and Macnamara, T.E. Anesth. Analg. (Cleve.) 59, 514-515 (1980)
(16) Martinez, R., Cabezudo, J., Vaquero, J., Areitio, E., and Barbolla, L. J. Neurol. Neurosurg. Psychiatry 44, 461 (1981)
(17) Mason, T.J., McKay, R.W., Hoover, R., Blot, W.J., and Fraumeni, J.F., Jr. Atlas of Cancer Mortality for U. S. Counties: 1950-1969. DHEW Publication No. (NIH) 75-780. Washington: Government Printing Office, 1975
(18) McKeen, E.A. and Mulvihill, J.J. Am. J. Hum. Genet. 29, 72A (1977)
(19) McKeen, E.A., Mulvihill, J.J., Rosner, F., and Zarrabi, M.H. Lancet 2, 590 (1979)
(20) Miller, R.W. In: Hook, E.B., Janerich, D.T., and Porter, I.H. Eds., Monitoring, Birth Defects and Environment: The Problem of Surveillance. New York: Academic Press, 1971, 97-111
(21) Miller, R.W. In: Anfinsen, C.B., Potter, M., and Schechter, A.M. Eds., Current Research in Oncology 1972. New York: Academic Press, 1973, 1-14
(22) Miller, R.W. Natl. Cancer Inst. Monogr. 52, 59-63 (1979)
(23) Miller, R.W. In: Hollaender, A. and deSerres, F.J. Eds., Chemical Mutagens 5, 101-126 (1978)
(24) Miller, R.W. and Mulvihill, J.J. Teratology 14, 355-358 (1976)
(25) Mulvihill, J.J. J. Natl. Cancer Inst. 57, 3-7 (1976)
(26) Mulvihill, J.J. In: Harris, C.C. Moderator, Ann. Intern. Med. 92, 809-825 (1980)
(27) Mulvihill, J.J., Gralnick, H.R., Whang-Peng, J., and Leventhal, B.G. Cancer 40, 3115-3122 (1977)
(28) Mulvihill, J.J. and McKeen, E.A. Cancer 40, 1867-1871 (1977)
(29) Murphy, E.A. Environ. Health Perspect. In press. (1981)
(30) Nakamura, Y., Nakashima, T., Nakashima, H., and Hashimoto, T. Cancer 48, 1012-1015 (1981)
(31) Neel, J.V. Perspect. Biol. Med. 14, 522-537 (1971)
(32) Neel, J.V. Science 213, 1205 (1981)
(33) Newcombe, H.B. In: Hollaender, A. Ed., Chemical Mutagens 3, 57-77 (1973)
(34) Omenn, G.S. and Motulsky, A.G. In: Cohen, B.H., Lilienfeld, A.M., and Huang, P.C. Eds. Genetic Issues in Public Health and Medicine. Springfield: Charles C Thomas, 1978, 83-111

(35) Parry, D.M., Safyer, A.W., and Mulvihill, J.J. <u>J. Med. Genet.</u> 15, 66-69 (1978)
(36) Prikhodchenko, A.A., Aksenovich, T.I., Iakovchenko, N.N., Chasovskikh, G.G., and Fediukova, T.V. <u>Vopr. Okhr. Materin. Det.</u> 25, 64-67 (1980)
(37) Schull, W.J., Otake, M, and Neel, J.V. <u>Science 213</u>, 1220-1227 (1981)
(38) Smith, D.W. <u>Recognizable Patterns of Human Malformation.</u> Second Edition. Philadelphia: W.B. Saunders, 1976
(39) Wada, W, Nishimoto Y., Miyanishi, M., Kambe, S, and Miller, R.W. <u>Lancet 1,</u> 1161-1163 (1968)
(40) Vogel, F. and Rohrborn, G. <u>Chemical Mutagenesis in Mammals and Man.</u> New York: Springer-Verlag, 1970
(41) Weatherall, D.J., Higgs, D.R., Bunch, C., Old, J.M., Hunt, D. M., Pressley, L., Clegg, J.B., Bethlenfalvay, N.C., Sjolin, S., Koler, R.D., Magenis, E., Francis, J.L., and Bebbington, D. <u>N. Engl. J. Med. 305,</u> 607-612 (1981)

Epidemiological Follow-Up Study on Mutagenic Effects in Self-Poisoning Persons

Andrew Czeizel

Department of Human Genetics, National Institute of Hygiene, Budapest, Hungary

ABSTRACT The pregnancy outcomes of self-poisoning persons /342 males and 960 females/ and their matched controls were studied. The different types of fetal deaths have not shown an obvious increase after self-poisoning. The rates of newborn with low birth weight were not higher. The birth prevalences and distribution of congenital abnormalities do not indicate any increase of Mendelian monogenic or chromosomal disorders.

INTRODUCTION

The mutagenesis epidemiology /2/ offers descriptive, analytic and "experimental" methods for the study of different end-points /molecular-biochemical, cytogenetical and phenotypic-clinical/ of mutations. That is evidences of the experimental type may be obtained by a critical assessment of unplanned experiments. The subject of the experimental epidemiological approach may be the populations at risk exposed to specific mutagens. The self-poisoning persons who in general attempt suicide by taking extremely large doses of drugs may represent an adequate population at risk for the study of mutagenicity of chemicals in man. The long term carcinogenic and mutagenic effects of these chemicals may be important in the subsequent lives of surviving suicide cases and probably in their offspring as well. These persons require particular attention because on the one hand: the majority of suicide attempters is found among the young of reproductive age. On the other hand: attempted suicides show a surprisingly high frequency: according to our previous calculation /1/ the rate of self-poisoning persons within the 15-34 year age-group is approximately 6%, i.e. 1 in 17 persons.

Since 1976 we have systematically studied persons who attempted suicide by chemicals. The short term effects of self-poisoning have been studied by cytogenetic, SCE, HGPRT, sperm abnormality and immunochemistry methods. Our preliminary data were published in 1980 /1/. Here I am summarizing the results of our epidemiological cohort study on the long term effects of self-poisoning. Planning this survey our defining specific question was: could extremely large doses of chemicals produce germinal mutations in semilethally intoxicated persons with epidemiologically detectable consequences in their reproduction?

MATERIAL AND METHODS

The study sample consisted of /i/ persons having attempted suicide /ii/ with extremely large doses of chemicals mainly barbiturates /at least with one day unconsciousness/, /iii/ below the age of 30 /presuming a later reproductive activity,/iv/ treated in 1960-67 /having time for the realisation of reproduction/, /v/ in the largest special department for intoxicated persons in Budapest: the Korányi Hospital. The persons of control sample /i/ who had undergone appendicitis, hernias and varicosity surgery, /ii/ in the surgical department of the same hospital, /iii/ in 1960-67 were matched for /iv/ sex, /v/ age and also /vi/ the district. A matching for /vii/ the socio-economic status was attempted, however, the available data did not allow us a reliable evaluation. Of course, our cases were evaluated as self-controls too, i.e. before and after the "event": self-poisoning or operation. The ascertainment of the stratified study and control samples involving nearly 7,500 and 7,400 cases, respectively, were obtained from the medical files of the Korányi Hospital.

Every second self-poisoning person having had matched control was visited at home for a personal interview by the help of printed questionnaires with purpose-designed data. Apart from congenital abnormalities of the offspring we had no possibility to check on medical records in an adequate size of samples. All male index cases have a matched control. However, owing to the female preponderance of suicide attempters /68% in Hungary, 1965-74/, the so-called "concordant pairs" were developed from female index cases and they have only one matched control person. Three interviewers did not know whether they met persons of the study or of the control groups. The data collection was repeated by other interviewers in 5% of the study and the control groups. The repeatibility rate of pregnancy outcomes was 90.0% and 80.6% in control and index cases.

The survey has raised a serious ethical problem. Every precaution was taken to protect the confidentiality of cases. Thus the index and control cases and their families were not informed about the true purpose of the survey and the previous attempted suicide of index cases. They were asked to take part voluntarily in an ongoing study of family planning in Budapest. Only 28.9% of self-poisoning males and 19.0% of self-poisoning females /$p < 0.01$/ replied positively to the question connected with their possible previous serious chemical intoxication treated in hospital.

RESULTS

Only 34.9% and 43.8% of ascertained index and control cases, respectively, could be evaluated. The main cause of loss is the "move off" to a new place whose address was unknown by neighbours /61.5%/. This rate is significantly higher both in males and females of the study group. The expected higher rate of death owing to the repeaters was not found on the basis of known data in the study group /o.5%/. Refused responses were higher in index than in control cases /3.1% vs. 1.o%/. In spite of these facts the selection bias concerning mutagenic consequences of self-poisoning does not seem to be considerable.

The analysis of marital status has shown 6.o% and 4.4% lower rates in the male and the female study subgroups, and it mainly may explain somewhat lower occurrences of pregnancies. On the basis of a slightly higher rate of induced abortions an increased birth control seems to be plausible in the study group. However, the occurrence of pregnancies was significantly higher after self-poisoning than after operation in both sexes.

The rates of sterility and delayed conceptions /$>$12 months/ were not higher in the study group.

The outcomes of pregnancies were studied concerning partly the distribution of fetal deaths /Table 1/ partly the conditions of livebirths /Table 2/. Spontaneous abortions were more frequent after self-poisoning in males than in matched controls. However, females have no similar differences. /The higher values after the "event" in general can be explained partly by the higher fetal death rates of advancing age-specified groups, partly by the lesser recall bias./ The rates of ectopic pregnancies did not show significant differences among groups and subgroups studied, however, in general they were higher than the registered national average. Neither did stillbirth rates show significant differences between the study and the control groups. The values of this variable also exceeded the national average. The rates of total fetal deaths were higher only in males, however, both before and after self-poisoning. Even the

Mutagenic Effect of Self-Poisoning   643

Table 1. Occurrence of fetal death before and after the events studied /"Accepted" pregnancies = total pregnancies - induced abortions/

Group	Sex	Accepted pregnancies occurred	Spontaneous abortion No / %	Ectopic pregnancies No / %	Stillbirth No / %+	Total No / %
Self-poisoning persons	Male	before: 93 after: 328	5 / 5.4+++ 31 / 9.5	1 / 1.1 5 / 1.5	2 / 2.3 11 / 3.8	8 / 8.6+++ 47 / 14.3+++
	Female	before: 386 after: 1,096	37 / 9.6+ 156 / 14.0+	8 / 2.1 3 / 2.1	5 / 1.5 21 / 2.2	50 / 13.0 180 / 16.4
Matched control cases	Male	before: 220 after: 282	8 / 3.6+++ 8 / 2.8+++	0 / 0.0 1 / 0.4	4 / 1.9 7 / 2.6	12 / 5.5+++ 16 / 5.7+++
	Female	before: 356 after: 456	34 / 9.5 57 / 12.5	4 / 1.1 7 / 1.5	6 / 1.9 11 / 2.8	44 / 12.4 75 / 16.5
Registered national average			~13.5	~0.2	~1.0	~15.0

+ per all births

+ p < 0.05
++ p < 0.01
+++ p < 0.001

inaccuracies of definitions of fetal deaths diminish the validity of these variables. Summing up: the different types of fetal deaths do not seem to have a real increase.

Table 2. Occurrence of livebirths with low birth weight and major congenital abnormalities before and after the events studied

Group	Sex	Livebirth occurred	Low birth weight No	%	Congenital abnormalities No	%
Self-poisoning persons	Male	before: 85	3	3.5	2	2.4
		after: 281	25	8.9	4	1.4
	Female	before: 336	19	5.7	4	1.2
		after: 916	96	1o.5	25	2.7
Matched control cases	Male	before: 2o8	13	6.3	2	1.o
		after: 266	19	7.1	5	1.9
	Female	before: 312	17	5.5	4	1.3
		after: 381	28	7.4	5	1.3
Registered national average				~11.o		~4.o

Within livebirths the per cent of newborn with low birth weight did not show a significant difference between index and control cases. Within the index cases only females' offspring born after self-poisoning had a higher rate. A particular accent was laid on congenital abnormalities: All reported children were checked on personally or on the basis of medical records. Misreported /12 cases, e.g. epilepsy, asthma/ and minor anomalies /15 cases, e.g. hydrocele, umbilical hernia/ were excluded. The birth prevalences and distribution of congenital abnormalities do not indicate any increase, including Mendelian monogenic or chromosomal disorders. The following 29 congenital abnormalities were found in the offspring of index cases after self-poisoning: cong. heart defect /ASD II: 2, VSD: 2, FT: 1/: 5; undescended testicle: 5; cong. dislocation of hip: 4; cong. scollosis: 2; encephalocele; cong. hydrocephalus; cleft palate; accessory auricle with deafness; cong. hypertrophic pyloric stenosis; intestinal atresia; anal atresia; renal dysplasia; cong. talipes equinovarus; hemivertebra; anodontia partial; Down syndrome; muscular dystrophy: in the son of a female index case.

The early death including mainly infant death and different kinds of diseases, e.g. tumours /1 osteosarcoma/ and autoimmun conditions also have no higher rates in index cases' offspring born after self-poisoning.

DISCUSSION

Our conclusions are limited owing to the restrictions of "non-captive" /2/ self-poisoning persons and to the well-known difficulties of the epidemiological approach, first of all the recall bias and retrospectively underreported values. Nevertheless, the negative results, namely that no higher rates were found in the phenotypic end-points of germinal mutations of self-poisoning persons are a reply to the specific question of our survey. Consequently /i/ data are available about the long term effect of mutagenic consequences of extremely large doses of chemicals in man. Because of recent "epidemic of self-poisoning" /1/ this information is of public health importance. /ii/ Furthermore the experimental epidemiological approach may give us a human model for the study of environmental chemical mutagens. Thus it may provide a fairly good comparison with the mutagenic effect of extremely large doses of ionizing radiation e.g. in the victims of the atomic bombs. Finally /iii/ these data confirm the efficacy of the Darwinian germinal selection.

Mankind has some unfavourable features: militarism, self-destruction, etc. However, among its favourable traits one of the most beautiful is that we can benefit from our own disasters, even from suicide.

Acknowledgement: I am grateful to I.Szentesi for the organizing of epidemiological study and to M.Szabuka, M.Zalatnay and T.Takács for the data collection of index and control cases and to Z.Marczis for the data processing.

/1/ Czeizel, A.: Human Population Monitoring and Genetic Risk Assessment /ed. Bora, K.C./ Elsevier/North Holland Biomedical Press B.V. 1981.
/2/ Miller, J.: Mutagenesis epidemiology /in press/

The Role of the Epidemiologist in Evaluating Human Genetic Risk

James R. Miller

Central Research Division, Takeda Chemical Industries, Ltd., Osaka, Japan California,

Epidemiologic studies in human populations are the ultimate means of determining genetic risks in man. Surveillance of a large population (e.g. newborns) or monitoring of individuals (and their families) exposed to a known or suspected mutagen are the two principal approaches to such studies. Each demands different epidemiologic strategies, usually case-control studies in surveillance and cohort studies in monitoring. The five primary elements of an epidemiologic study — characteristics of the population (including controls), description of the cause, the genetic characteristics to be used as endpoints, dose/response relations and time/response relations — must be clearly defined or determined. The endpoints may be somatic (structural chromosome aberrations and SCEs) or germinal (structural and numerical chromosome aberrations, sentinel phenotypes, biochemical variants, and embryonic and fetal deaths). The detection of the former necessitates follow up of the exposed individuals (and their offspring) to determine ill health consequences and evidence of germinal effects.

A multitude of non-human mutagen test systems now exist, and I imagine most people would hope that these would be adequate for evaluating human risk. The thought that the epidemiologist might have a role in risk evaluation is disturbing because it is an admission that the other systems are fallible and that, collectively, human beings are the ultimate test system. In fact, the admission, though disturbing, is realistic, and the sooner it is appreciated the sooner appropriate epidemiologic studies will be established. Such studies require careful design and time-consuming planning; this is frustrating for those who want quick, inexpensive answers to complex biological questions. Unfortunately, the purely scientific aspects of the problem are confounded by ethical, social, and economic issues, which frequently create difficulties and impede the implementation of sound epidemiologic investigations. In this presentation, I will discuss some requirements, limitations, and current deficiencies of epidemiologic studies for evaluating human genetic risk.

SURVEILLANCE AND MONITORING

There are two principal approaches to studying human groups. The first, continuous surveillance of a large population, may take several forms but can be divided roughly into routine newborn surveillance of the type now underway in several countries to detect changes in the birth prevalence of certain congenital defects (1), and surveillance of an entire population about to undergo a major change that entails a perceived risk (e.g. the introduction of a chemical plant or the opening of an uranium mine in a given region). The second approach involves monitoring of individuals (and their families) exposed to a known or suspected mutagen. This too can take many forms depending upon whether the exposure is of long or short duration and whether it is ubiquitous or confined to a large or small circumscribed population. Although some of these subgroups may be quite large (e.g. ubiquitous long exposure; tobacco smokers), others may be too small to obtain a meaningful sample size for detecting germinal mutations.

EPIDEMIOLOGIC STRATEGIES

Different surveillance or monitoring situations demand different epidemiologic strategies. During the continuous surveillance of a population not at any identifiable risk, if a base line mutation rate (determined for several endpoints by previous experience) is exceeded to some predetermined agreed-upon extent, then a case-control study would have to be initiated to detect the responsible mutagen. This involves matching the affected index cases with appropriate control infants, and searching for evidence of an exposure in the parents of the index infants not experienced by the parents of the control infants. Case-control studies are cheaper and generally easier to carry out than cohort studies, to be discussed presently, but the chances of uncovering "the" offending agent are not great.

The continuous surveillance of a large population at some perceived risk (usually low) and the monitoring of special "high risk" ad hoc groups requires a cohort approach. (The effect of smoking — the ubiquitous long-exposure class — could also be investigated by the case-control method; in this instance, the index cases would not be exposed individuals but children who exhibited specified genetic traits). In a cohort study, a specified outcome, i.e. evidence of mutation, is followed in "exposed" and "non-exposed" (control) groups (cohorts). The choice of controls in such studies is often difficult and controversial. For a large population undergoing continuous surveillance, another community free of exposure and with similar agreed upon demographic characteristics might serve as a control. In the case of ad hoc high risk groups, confounding variables pose problems for proper control selection. These are factors, other than the one under examination, that might influence the mutation rate directly or indirectly. Some might be obvious (e.g. smoking), but most will be subtle and difficult to quantitate.

Since cohort analysis is the strategy used most frequently in human mutation epidemiology, the remainder of the presentation will focus on it.

ELEMENTS OF AN EPIDEMIOLOGIC INVESTIGATION

There are five primary elements to an epidemiologic study: (1) the population — the characteristics of the individuals who will comprise the study group and appropriate controls, (2) the cause — the environmental exposure about which there is concern, (3) the result — the genetic characteristics that are to be used as endpoints, (4) dose/response relations — the dose of exposure required to produce a specific degree of effect, and (5) time/response relations — the lapse between the initiation of the cause and the expression of the endpoint (2).

The population. Certain aspects of populations have been mentioned already. It is essential that the individuals who are to be included in the study be clearly determined before it begins (e.g. all those living in a defined geographic area, or working in a specific industry, or receiving a particular medication, or associated with a specific life style). Since two generations, an exposed and its offspring, must be studied for germinal effects, it is essential that the characteristics of both be adequately defined.

The selection of proper controls for a given study presents a formidable challenge. The principal difficulty is created by variables that might confound the outcome. Such variables can be controlled by exclusion (e.g. individuals with a history of exposure to radiotherapy when somatic chromosome aberrations are used as an endpoint), by matching of cases and controls for some factor, and by proper statistical analyses for others.

The cause. The nature and perceived hazard of the exposure will vary considerably. Regardless of these, two facts must be clearly defined: whether the parental exposure involves mother, father, or both, and some estimate of the dose of exposure. The former may seem to be straightforward, but apparent paternal exposure in the work place may be confounded by maternal exposure if the agent under study is inadvertently carried home on clothing, or on the body. Estimate of exposure dose may be achieved in several more or less satisfactory ways: (1) the estimate may of necessity be crude and involve a simple statement of location in time and place with respect to the agent, and some comment on the possible route of exposure, (2) it may be based on a measurement of the environment (e.g. air or water levels of the agent), (3) it may be measured in body fluids, hair, teeth, etc., (4) it may be derived from a questionnaire of life style habits, or (5) it may be based on non-genetic pathologic evidence (e.g. liver damage, chloracne, abnormalities of sperm count or morphology).

The result. Very few of the many potential genetic endpoints are of practical value in monitoring. Because of this, certain characteristics — cancer, reproductive performance, birthweight, sperm abnormalities, and sex ratio — which are relatively easy to determine, are often used for the purpose. However, it must be kept in mind that these are also influenced by non-genetic factors and hence are not "clean" genetically; they are useful as indicators and may be used to supplement true genetic endpoints.

Evidence for a change in mutation rate can be sought at the somatic as well as the germinal level. However, since the ultimate concern is the ill-health consequences of germinal mutation and somatic events are only indicators of potential ill health, they must be followed-up to determine their real significance in this regard. There are two measures of somatic mutation — structural chromosome aberrations and sister chromatid exchanges (SCEs). A review of screening human populations for mutations induced by environmental pollutants using human lymphocytes has been published by Natarajan and Obe (3), and a statement on the epidemiologic aspects of studies of somatic chromosome breakage and sister chromatid exchange is being prepared for Committee 5 of ICPEMC by Hook. Significantly increased rates of occurrence of either structural aberrations or SCEs requires two types of long term follow up — on the exposed individuals to ascertain the ill health consequences of the somatic events, and on their offspring to seek evidence for germinal damage. Such studies, which are not easy to design and execute, are facilitated by record linkage technology. Unfortunately, few countries maintain vital and health records in a way that makes such follow up readily feasible. However, recent studies have demonstrated how cohorts of industrial workers can be effectively followed by the use of record linkage (4). This type of investigation requires accurate identification of individuals; the use of unique identifiers on all vital and health records is the most efficient way of achieving this. Since most countries lack such identifiers, it is essential that documents include adequate information to make linkage possible by other means; Smith (5) has provided a list of items that should be included on "starting" and "endpoint" records.

At present, there are no tests for determining point mutations in somatic cells, but several are under development and might be practical shortly.

Germinal mutations may be measured by changes in the rates of structural and numerical chromosome abnormalities, sentinel phenotypes, protein variants, and embryonic and fetal death. Hook has outlined in some detail various aspects of the use of chromosome aberrations for surveillance and monitoring (6,7,8).

Sentinel phenotypes comprise 13 autosomal dominant traits that can be unambiguously diagnosed as soon after birth as possible. Achondroplasia, aniridia, acrocephalosyndactyly (Apert syndrome), osteogenesis imperfecta, and retinablastoma are the most satisfactory traits for monitoring purposes, but it should be recognized that genetic heterogeneity may confound some diagnoses. Matsunaga (a) believes that retinoblastoma may be a valuable measure of both somatic and germinal mutations.

Protein variants can be studied by electrophoresis to detect charged amino acid substitution in functional proteins and by enzyme activity to detect major losses of protein function or quantity. Thirty-nine proteins are now being examined by the former method to survey a newborn population in Michigan (10) and to monitor children of atomic bomb survivors in Japan (11). Thirteen enzymes are being examined by the latter method in the Michigan study (12). Though none of the possible mutant variants will produce ill health, an increase in their frequency would suggest a general increase in mutation rate that could be expected to lead to ill health, hence, a population in which an increase in mutations affecting protein structure or function should be followed intensively for indications of an increase in genetic ill health.

Embryonic and fetal deaths may be used for monitoring (13). However, they are only of value as a precise indicator of genetic damage when cytogenetic studies are carried out on the abortuses (14).

Dose/response relations. Precise determination of exposure dose may be difficult but there are several ways in which estimates dose/response relations can be determined: (1) suggestive information may be available in the literature on human or animal studies, (2) the initial survey may be restricted to those individuals exposed to high levels (comparison would be with a zero exposure group): if the results were negative no further study would be required; if they were positive, the study would proceed with the use of more detailed estimates, (3) an exposed population may serve as its own control if endpoint rates can be determined before, during, and after exposure.

Time/response relations. The relation of the time of exposure to the suspected mutagen and the time of subsequently observing a given endpoint should be critically defined. Ideally, for germinal mutations, this would require specifying the stage of germ cell development at which exposure took place. Given the complexities surrounding the interpretation of related data in the mouse (15), it is clear that, in practice, the best that could be hoped for is determining relation of time of exposure to a specific reproductive event; even this may be difficult.

PROBLEMS

There are many problems associated with implementing effective studies; only two will be emphasized: the question of what to do in a catastrophe, and the issue of sample size.

In many situations, there is adequate time to design a proper study. However, the unexpected catastrophe presents a challenge because decisions are made rapidly under great pressure. Attention is focussed, understandly, on immediate acute ill-health effects in the exposed population; long term mutagenic, teratogenic, and carcinogenic effects are neglected or treated negligently. (An epidemiologic committee to study the Seveso accident began functioning one and a half years after the event (16)). The eventuality of a catastrophe can best be met by a series of reference centres throughout the world staffed with teams of experts prepared to move into disaster areas at short notice. Such teams would work from internationally agreed upon protocols; some relevant protocols are being developed by WHO consultation group. In order to reduce costs and assure maximum manpower efficiency the reference centres should have the expertise to conduct studies in cancer and malformation epidemiology as well as mutation epidemiology, and should be located in centres where active epidemiology is already taking place.

Sample size is a critical factor in the design and eventual evaluation of results. Although it may be tempting "to do everything possible" (perceived or real social responsibilities may exert pressure as well), the limitations of each genetic endpoint should be recognized. Germinal mutations are rare events and the sample sizes necessary for a given endpoint will depend on its baseline rate and the size of the increase in rate that might be willingly undetected.

CONCLUSIONS

No agent has been proven to be a germinal mutagen in man. This is a sobering fact, given the prodigous amount of effort devoted to studying the populations of Hiroshima and Nagasaki exposed to atomic bomb radiation ([17], [18]). Regardless, it seems prudent to remain vigilant in the face of genuine concern about the mutagenic potential or vast numbers of agents to which many human beings are exposed routinely. The existence of a battery of non-human mutagen test systems guarantees that some potentially deleterious compounds will be effectively screened out, but biological realities dictate that the human being is the ultimate test organism. The purely scientific aspects of mutation monitoring seem substantial: well established biochemical, cytogenetic, and clinical endpoints exist and more sophisticated ones are being developed. However, the formal design and execution of studies in mutation epidemiology are beset with major problems. The most difficult of these is the proper choice of controls, but the proper definition of an exposed population, the measurement of exposure, and the estimation of accurate dose/response and time/response relations are also critical. Compounding all of these are broader societal concerns about ethical and economic issues. As long as these concerns persist, implementing existing scientific knowledge will not be easy.

REFERENCES

1) Flynt,J.W., and Hay,S. Contrib.Epidemiol.Biostat.1,44-52 (1979). 2) MacMahon,B., and Pugh,T.F. Epidemiology: principles and methods,Boston:Little Brown, (1970). 3) Natarajan,A.T., and Obe,G. Ectoxicol. Environ.Safety 4,468-481 (1980). 4) Smith,M.E., and Newcombe,H.B. Can.J.Publ.Health (in press, 1981). 5) Smith,M.E., In:Bora,K.C.,ed. Chemical mutagenesis, human population monitoring, and genetic risk assessment. Amsterdam:Elsevier/North Holland (1981). 6) Hook,E.B. Environ. Res. 25 (in press,1981). 7) Hook,E.B. In:Bora, K.C.,ed.Chemical mutagenesis, human population monitoring, and genetic risk assessment. Amsterdam: Elsevier/North Holland (1981).8) Hook,E.B. These Proceedings. 9) Matsunaga,E. Mutation Res. (submitted for publication, 1981). 10) Neel,J.V.,Mohrenweisser,H.W., and Meisler,M.M.Proc.Natl.Acad.Sci.USA 77,6037-6041 (1980). 11) Neel,J.V.,Satoh,C.,Hamilton,H.B.,Otake, M.,Goriki,K.,Kageoka,T.,Fujita,M.,Neriishi,S., and Asakawa, J. Proc.Natl.Acad.Sci.USA 77,4221-4225 (1980). 12) Mohrenweisser, H.W. Proc.Natl.Acad.Sci.USA (in press,1981). 13) Kline,J.,Stein,Z.,Strobino,B.,Susser,M., and Warburton,D. Am. J.Epidemiol.106,345-350 (1977). 14) Warburton,D.,Stein,Z., Kline,J., and Susser,M. In:Hook,E.B., and Porter,I.A.,eds. Human embryonic and fetal death.New York:Academic Press (1980). 15)Preston,R.J. These Proceedings. 16)Bisanti,L.,Bonetti,F., Caramaschi,F.,Del Corno,G.,Favaretti,C.,Giambelluca,S.E., Marni,E.,Montesarchio,E.,Puccinelli,V.,Remotti,G.,Valpato,C., Zambrelli,E., and Fara,G.M.Acta Morphol.Acad.Sci.Hung. 28, 139-157 (1980). 17)Schull,W.J.,Otake,M. and Neel,J.V. Science 213, 1220-1227 (1981). 18) Schull, W. J. These Proceedings.

The Clinician's Role in Monitoring for Diseases from Chemicals in the Environment

Robert W. Miller

Clinical Epidemiology Branch, National Cancer Institute, Bethesda, Maryland, U.S.A.

ABSTRACT Generally, alert practitioners and sometimes non-medical observers have been the first to recognize environmental chemicals that cause birth defects or cancer in man. Epidemiologic studies and laboratory research then add to the evidence for a causal relationship. We need to increase the number of such observers and encourage the referral and investigation of patients or families with etiologically interesting disease patterns. The future for the alert clinician may lie more in defining peculiarities of host susceptibility than in identifying specific environmental mutagens, teratogens, or carcinogens.

INTRODUCTION

Virtually all known human teratogens and carcinogens were first recognized by alert practitioners (6). Registries and epidemiologic studies have been useful mainly for testing hypotheses, rather than generating them. Laboratory research has occasionally identified carcinogens subsequently found to affect man, but the observers usually were not aware of the laboratory studies, e.g., those concerning vinyl chloride and diethylstilbesterol. Alert clinicians have described important genetic diseases that predispose to cancer from radiation or Epstein-Barr virus, but genetic diseases that predispose to chemical carcinogens are not yet known.

EFFECTS ON GERM CELLS

In July 1975 a Taiwan-born practitioner in Hopewell, Virginia, noted severe neurologic disability in a patient and thought it might have been occupationally induced. By chance, the physician circumvented public health referral channels and sent a blood specimen directly to the Center for Disease Control, USPHS (7). The specimen was rich in chlordecone, from exposures at work. Other workers were similarly affected, some were sterile, and of 32 family members studied, 30 had chlordecone in their blood from the chemical tracked home on the workers' clothing (2).

In 1977 a group of workers in California, unable to have children, suspected sterility from the chemicals to which they were exposed at work. Their physicians were unimpressed, so the workers brought their semen specimens for examination to a clinical laboratory (7). In this way the sterility due to dibromochloropropane (DBCP) was discovered (15)--by workers in this instance, and not by their physicians.

Sterility is the only environmental effect on the germ cell that the practitioner has a chance to discover. All other measures of genetic damage require careful epidemiologic study; e.g., infertility, effects on the sex ratio, birth weight, other body measurements, stillbirth and neonatal death rates, F_1 mortality, congenital malformations, and abnormalities in biochemical measures.

TERATOGENESIS

Practicing physicians have discovered damage to the embryo from a variety of chemical agents; e.g., thalidomide, alcohol, phenylhydantoin, androgens, folic acid antagonists, tetracycline and warfarin sodium (6). None of these were known beforehand to be teratogenic in experimental animals. The malformations were distinctive or ordinarily rare. The increase in phocomelia from thalidomide was 50,000-500,000 times normal (13).

Epidemiologic and laboratory studies subsequently added to the evidence for a causal relationship. Surveillances for birth defects have been a poor means for detecting previously unknown teratogens, but have been valuable in testing hypotheses concerning suspected environmental carcinogens (5). A variety of mechanisms can lead to teratogenesis (16). The correlation between responses in animal experimentation and human teratogenesis has consequently been poor. Aspirin or cortisone produce cleft palate in rodents, but not in man. Thalidomide produces phocomelia in man but not in rodents (11). How unfortunate it would have been had use of these drugs been limited in man because of the effects found in the experimental laboratory. Exemptions of chemicals for human use because of differences in metabolism or mechanism from those of experimental animals may (9) or may not be possible. At a meeting in Martinique in 1976, F. Clarke Fraser stated that after 25 years of study, he still did not know why cortisone causes cleft palate in some species but not in man. Warkany has said, "(Animal studies) gave us important concepts about general principles of teratogenesis, but one cannot simply apply observations in mice, rats or monkeys to man." (14). It looks, then, as if human teratogens will continue to be discovered in the future as they have been in the past through alert clinical observations.

CARCINOGENESIS

Practicing physicians have been the first to discover the carcinogenicity of asbestos, the halo ethers, aromatic amines, chromium and nickel, among other chemicals (6). Usually individual cases or a cluster of cases is reported first. Asbestos is an example. A practitioner in South Africa observed a cluster of patients with mesothelioma. The only historical feature in common was that all had lived near open-pit asbestos mines as children (6). Previously, no one could have suspected that asbestos might be carcinogenic in man. Laboratory studies followed, and cancers were produced in rats (12). Epidemiologic studies, retrospective at first and prospective later, confirmed the hypothesis generated clinically.

The list of new human carcinogens is not growing quickly. Understanding of cancer biology, however, has moved ahead vigorously through clinician-recognized host susceptibility to cancer. In 1957 Boder and Sedgwick played a major role in separating ataxia-telangiectasia (AT) from a quagmire of ataxias and angiomatous disorders. Peterson et al. suggested that AT predisposes to lymphoma, on the basis of a handful of cases. Overreaction to conventional radiotherapy as treatment for the lymphomas caused an acute life-threatening reaction in two children with AT. Observation of a third such case in Birmingham, England, led to tests of the skin fibroblasts in culture for sensitivity to radiation as indicated by cell survival. In this way, new understanding of molecular biology came from a series of observations starting at the bedside (reviewed by Miller (4)).

In a similar fashion the chromosomal abnormalities now regularly found in acute myelogenous leukemia began with the clinical definition of rare disorders that have high risk of leukemia; i.e., Down's, Fanconi's and Bloom's syndromes (8,10,17). Another example: specific chromosomal deletions were found in patients with congenital malformations and retinoblastoma or Wilms' tumor. In one set of identical twins with the deletion and the identical malformation syndrome, only one of the twins developed Wilms' tumor, indicating a subtle additional influence, presumably environmental (3). These observations come from human disorders for which no animal model is yet known.

The future for alert clinical observations concerning chemical carcinogenesis seems now to lie more in studies of host susceptibility than in discovering new environmental carcinogens. Increased host susceptibility may be indicated by the excessive occurrence of certain cancers with specific preexistent disease or familial aggregation of disease, not necessarily neoplasia.

Unlike laboratory studies of teratogenesis, those for carcinogenesis show a high correlation with the results of short-term assays for mutagenesis. From such studies hypotheses have been generated concerning chemicals that may be human carcinogens; e.g., tris (2,3-dibromopropyl), used as a fire-retardant in children's sleepwear (1). It is unlikely that alert practitioners can play a role in testing this hypothesis; epidemiologic studies of occupational exposures would be more promising.

CHARACTERISTICS OF ALERT CLINICIANS

Alert practitioners are often young, Australian or

working in Seattle. Young physicians have not yet concentrated their full attention on diagnosis and therapy. Some are still curious about etiology. Australians, as they called attention to the hazards of thalidomide, tetracycline, methotrexate for psoriasis in pregnancy, and Reye's syndrome, seem to have been following the example set by Gregg in 1941 when he showed by a simple retrospective study that a cluster of infants with cataracts was due to rubella during pregnancy. Seattle has had, until recently, the leadership of David W. Smith and Judith G. Hall, with their exceptional capacity to recognize "new" syndromes and their causes.

The propensity to think etiologically may be found more often outside medicine than within it. Especially adept are politicians, business men, journalists, and detectives, who use this sort of thinking in their daily work; i.e., in political polls, market research, developing a news story and tracking from a crime to a criminal.

THE FUTURE

The future for this form of clinical etiology does not look bright unless we:

- Screen medical students for an aptitude to think etiologically and encourage them to develop this talent.
- Screen high school and college students in the same way, and encourage those who are exceptional to consider a career in medicine.
- Establish positions in medical schools and other medical centers for clinical etiologists.
- Advise practitioners to refer etiologically interesting cases to specialists for evaluation.
- Develop a means to honor practitioners who make outstanding observations.

REFERENCES

(1) Blum, A., and Ames, B.N. Science 195, 17, (1977)
(2) Cannon, S.B., Veazey, J.M., Jr., Jackson, R.S. Am. J. Epidemiol. 107, 529 (1978)

(3) Kolata, G.B. Science 207, 970 (1980)
(4) Miller, R.W. In: Harnden, D.G., and Bridges, B.A. (eds.) Ataxia-telangiectasia. John Wiley & Sons, Publ., London. In press.
(5) Miller, R.W., and Flynt, J.W., Jr. In: Brent, R.L., and Harris, M.I. (eds.) Prevention of Embryonic, Fetal and Perinatal Disease. DHEW Publ. No. (NIH) 76-853, Washington, D.C., (1976), pp. 287-293
(6) Miller, R.W. In: Hollaender, A., and deSerres, F.J. (eds.) Chemical Mutagens 5, 101-126 (1978), Plenum Publ. Corp.
(7) Miller, R.W. J.A.M.A. 245, 1548-1551 (1981)
(8) Miller, R.W. Cancer Res. 27 (Part 1), 2420 (1967)
(9) Nebert, D.W. Brit. Med. J. 283, 537-542 (1981)
(10) Rowley, J.D. New Eng. J. Med. 305, 164 (1981)
(11) Schardein, J.L. Drugs as Teratogens. CRC Press, Inc., Cleveland, Ohio (1976)
(12) Stanton, M.F., Layard, M., Tegeris, A., Miller, E., May, M., and Kent, E. J. Natl. Cancer Inst. 58, 587 (1977)
(13) Sullivan, F.M. Pediatrics 53 (Part II), 798-799 (1974)
(14) Warkany, J. Pediatrics 53 (Part II), 820 (1974)
(15) Whorton, D., Krauss, R.M., Marshall, S., and Milby, T.H. Lancet 2, 1259 (1977)
(16) Wilson, J.G., and Fraser, F.C. Handbook of Teratology V2, Plenum Press, New York (1977)
(17) Yunis, J.Y., Bloomfield, C.D., and Ensrud, K. New Eng. J. Med. 305, 135 (1981)

Risk Estimates and Assessment

Carcinogenic Potency

Bruce N. Ames, Lois S. Gold, Charles B. Sawyer, and William Havender

Department of Biochemistry, University of California, Berkeley, California and Biology and Medicine Division, Lawrence Berkeley Laboratory, Berkeley, California, U.S.A.

ABSTRACT Man is exposed to an extremely large number of carcinogens from a variety of sources: cooking food, natural chemicals in our diet, inhalation of burnt material, and man-made chemicals. To improve risk assessments, risk/benefit judgments, and regulatory policy, an index number of carcinogenic strength (or potency) for each chemical is desirable to help in setting priorities. Because quantitative information on the capacity of various chemicals to cause cancer in <u>man</u> is not available, one must turn to animal bioassays. We have been engaged for several years in creating a comprehensive data base which incorporates the animal bioassays reported in the world's literature which are suitable for determining a potency value. We describe briefly an index of carcinogenic potency, the TD_{50}, and our progress toward setting up a data base on the estimated potency values of several thousand cancer tests. The range of carcinogenic potency is over 10 million-fold. The carcinogenic potency scale can be used in calibrating short-term tests to show their limitations and strengths. We discuss the limitations of animal cancer tests as related to promoters, anti-carcinogens, and extrapolation to humans. Many of these problems are likely to be related to oxygen radicals and their role in mutation, promotion, aging, and cancer, and human defense systems against oxidative and radical damage.

CARCINOGENIC POTENCY DATA BASE

We have been able to work out the theory of an index of carcinogenic potency, the TD_{50} (or tumorigenic dose$_{50}$) (1). The TD_{50} is the daily dose which gives 50% of the test animals tumors by the end of the standard life-span for that species. Equivalently, it is the daily dose-rate required to decrease by half the probability of an animal remaining tumor-free throughout life. The lower the value of TD_{50}, the less the amount of that chemical required to induce tumors in half the animals; hence, the smaller the TD_{50}, the more potent it is. TD_{50} can be calculated for any tissue or tumor type or any combination of tissues and tumor types. Richard Peto (of Oxford University) and Malcolm Pike (of the University of Southern California) have developed the method we are using to estimate potency.

The TD_{50} has the merit of taking into account several factors which improve the estimate of carcinogenic potency and permit comparisons of diverse test results. Our calculation takes into account whatever spontaneous tumor incidence occurs in control animals. Where data are available about the time of death and tumor incidence of each animal, we calculate a lifetable TD_{50} using this information; because animals given high doses of chemicals frequently die early due to chemical toxicity, failure to account for this early mortality could lead to underestimates of the true potency. The TD_{50} has the additional merit that the dose-rate to be estimated (that which gives cancer to half of otherwise tumor-free animals) is usually not far from an actual dose used in an experiment which yields statistically positive results; therefore, only a small extrapolation from experimental observation is necessary.

In order to assure accurate comparisons of potencies among experiments with different protocols, species, and routes of chemical administration, we standardize the dose level for each experiment. The daily dose to an animal on a milligram per kilogram of body weight per day basis averaged over a lifetime is calculated for each dose group and is used to calculate TD_{50}.

We calculate a TD_{50} and its confidence limits for each site and tumor type which an author considers biologically or statistically significant; TD_{50}'s are also calculated for all tumor-bearing animals and for two mandatory sites in each species, whether or not the results are significant, so that comparisons can be made among bioassays regardless of results. All TD_{50}'s are calculated from data on the number of animals with a particular tumor rather than the total number of tumors.

In the case of a negative (i.e., statistically nonsignificant) bioassay where a TD_{50} cannot be calculated, we still calculate a lower confidence limit for the TD_{50} which shows the sensitivity of the bioassay. These sensitivities, which can differ enormously, depend on the dose levels used and on the experimental design. A negative test is described as excluding TD_{50}'s below a certain limit, rather than simply as "negative." Some experiments have such small numbers of test animals and use such low doses that they could not have detected any but the most potent carcinogens. The comparison of research designs and dose levels will sometimes make it possible to reconcile positive and negative results with the same compound: for example, if two such tests examined different regions of the dose range, they need not be contradictory.

We have designed a graphic display for presenting the results of our calculations in a readily comprehensible format. For each significant site in an experiment the plot provides descriptive information about sex, strain, species, site and tumor type, length of exposure, and length of the experiment. It also presents the TD_{50} value, its statistical significance and confidence limits, the shape of the dose-response relationship, the number of tumors and of animals at risk for each control and dose group, the administered dose levels in mg/kg/day, the author's opinion about tumorigenicity, and the reference. All of the information we use for the plot is stored in an easily accessible data base. We have completed our calculations for about 1200 experiments on 250 chemicals and are continuing to process the rest of the literature.

ANALYSIS

We are just starting on the analysis of the data base. We have found that carcinogenic potency values (TD_{50}'s) in rodents vary for different chemicals over a tremendous range--about 10 million-fold. This can be seen in Figure 1, which is a preliminary analysis of the results of the National Cancer Institute (NCI) bioassays of female mice. For compounds in Figure 1 the NCI concluded that the chemical was carcinogenic, and our own lifetable calculations indicate a statistically significant association between tumor incidence and chemical administration. For the most potent carcinogen in Figure 1, TCDD (2,3,7,8-tetrachlorodibenzo-p-dioxin), a daily dose per kilo in the hundreds of nanograms range would induce tumors in 50% of female mice. For the least potent carcinogen in Figure 1, 4-chloro-o-phenylene diamine, it would take a dose of more than one gram per kilo to have that effect. This great variability in the intrinsic carcinogenic power of different chemicals is an important aspect of developing a policy response to chemical hazards for humans. Assessments of chemical hazards to people need to combine estimates of the extent and levels of human exposure and estimates of carcinogenic potency (2,3). We are currently using our data base to secure information on many issues relevant to human risk assessment, such as the variability of TD_{50} between sexes, strains, and species, and the concordance between different tests on the same chemical (2).

EXTRAPOLATION FROM ANIMAL TESTS TO MAN: OXYGEN RADICALS

Extrapolation from animal tests to man is complicated by such difficulties as species variation in metabolism and the need to extrapolate to low doses. These issues have been discussed at length. We would like to consider here several other difficulties which relate to oxygen radicals.

Toxicity by oxygen radicals has been suggested as a major cause of cancer, heart disease, and aging (reviewed in 4 and 5). Oxygen radicals and other oxidants appear to be toxic in large part because they initiate the chain reaction of lipid peroxidation (rancidity). Lipid peroxidation generates various reactive species--such as

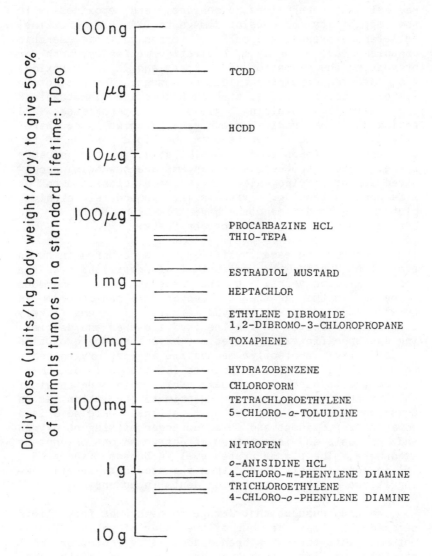

Range of carcinogenic potency in: Female Mice

Figure 1

radicals, hydroperoxides, aldehydes, and epoxides—with the capability of causing damage to DNA, RNA, proteins, cellular membranes, and cellular organization. Aerobic organisms have an array of protective mechanisms both for preventing the formation of oxidants and lipid peroxidation and for repairing oxidative damage. The protective systems include enzymes, such as superoxide dismutase and the selenium-containing glutathione peroxidase, and antioxidants and radical scavengers, such as α-tocopherol (vitamin E) and β-carotene in the lipid portion of the cell and glutathione and ascorbic acid in the aqueous phase. These protective mechanisms are now being recognized as anticarcinogenic and, in some cases, even as life-span extending. Vitamin E, ascorbate, selenium, glutathione, and β-carotene have all been shown to be anticarcinogenic in animal cancer tests.

A marked increase in life-span has occurred in human evolution during the descent from prosimians over the past 60 million years. At the same time an enormous decrease in the age-specific cancer rate has occurred in humans compared to short-lived mammals. It seems likely that a major factor in lengthening life-span and decreasing age-specific cancer rates may have been the evolution of effective protective mechanisms against oxygen radicals (reviewed in 5). We propose (5) that one of these protective systems is plasma uric acid, the level of which increased markedly during primate evolution as a consequence of a series of mutations. Uric acid is a powerful antioxidant and is a scavenger of singlet oxygen and radicals and is about as effective an antioxidant as ascorbate. The plasma urate level in humans (about 300 µM) is considerably higher than the ascorbate level, making it one of the major antioxidants in humans.

We also suspect that damage to membranes from lipid peroxidation may be one of the modes of action of promoters. This topic has been discussed by Professor Troll at this meeting. Promoters are clearly one of the complications in quantitative risk assessment extrapolation.

Thus, differences in antioxidant defense mechanisms and life-span complicate quantitative extrapolation from rodents to man. It is not so clear exactly how to extrapolate from a TD_{50} in mg/kg/day in a short-lived species such as a rat to a long-lived species such as man. In addition, both genetics and nutrition may contribute to human variability in these antioxidants and cause large differences in risk. For example, Hirayama (6) has shown considerable decrease in lung cancers among smokers who eat green and yellow vegetables regularly compared to those who do not. We believe that methods for measuring DNA damage in individual humans and other biochemical approaches to human epidemiology will be needed to help in human risk assessment for environmental mutagens and carcinogens. These methods will complement information from short-term tests, human cancer epidemiology, and quantitative analysis of animal cancer tests.

ACKNOWLEDGMENTS

This work was supported by DOE contract DE-AT03-76EV70156 to B.N.A., NIEHS Center Grant ES-01896, and NIEHS/DOE Interagency Agreement 222-Y01-AS-10066 through the Lawrence Berkeley Laboratory.

REFERENCES

(1) Ames, B. N., Hooper, K., Sawyer, C. B., Friedman, A. D., Peto, R., Havender, W., Gold, L. S., Haggin, T., Harris, R. H., and Rosenfeld, M. In: Banbury Report 5. Ethylene Dichloride: A Potential Health Risk? B. Ames, P. Infante, and R. Reitz, eds. (Cold Spring Harbor Laboratory, Cold Spring Harbor, New York), pp. 55-63 (1980).

(2) Hooper, K., Gold, L. S., and Ames, B. N. In: Banbury Report 5. Ethylene Dichloride: A Potential Health Risk? B. Ames, P. Infante, and R. Reitz, eds. (Cold Spring Harbor Laboratory, Cold Spring Harbor, New York), pp. 65-81 (1980).

(3) Gold, L. S. In: Banbury Report 5. Ethylene Dichloride: A Potential Health Risk? B. Ames, P. Infante, and R. Reitz, eds. (Cold Spring Harbor Laboratory, Cold Spring Harbor, New York), pp. 209-225 (1980).

(4) Ames, B. N., Hollstein, M. C., and Cathcart, C. In: Lipid Peroxide in Biology and Medicine. K. Yagi, ed. (Academic Press, New York), in press (1981).

(5) Ames, B. N., Cathcart, R., Schwiers, E., and Hochstein, P. Proc. Natl. Acad. Sci. USA 78, in press (1981).

(6) Hirayama, T. Nutrition and Cancer 1, 67-81 (1979).

The Alkylation of Macromolecules as a Basis for the Evaluation of Human Genetic Risk

Carl Johan Calleman

Department of Radiobiology, Stockholm University, Stockholm, Sweden

ABSTRACT Determination of the degree of reaction of nucleophiles in macromolecules provides a <u>unitary dosimetric basis</u> for the extrapolation of genetic effects of chemical mutagens from experimental organisms to man. Hemoglobin offers an easily available dose monitor in experimental animals and in man. For *in vitro* test systems other dose monitors are at hand. In order to facilitate risk estimation and in order to avoid erroneous conclusions in risk identification it is suggested that dosimetry of reactive compounds/metabolites should be prescribed in protocols for standard test systems.

INTRODUCTION

With the advent during the past decade of a number of new microbial and mammalian screening systems for mutagenic potency (1, 3) the number of compounds or mixtures of compounds demonstrated to possess mutagenic activity has greatly increased. This imposes the necessity of identifying those compounds which are actually operative as mutagens/cancer initiators in human populations and of determining their contributions to human genetic risk.
Research with these aims has been impeded by the absence of a unitary dosimetric basis for chemical mutagens (8, 9, 15) which could form a reliable basis for extrapolations from experimental data to risk in man.

The majority of ultimate mutagens/cancer initiators are, chemically speaking, electrophilic reagents, RX (16).

As such they react with nucleophilic groups, Y, of proteins and nucleic acids according to the equation

$$RX + Y \xrightarrow{k_y} RY + X \qquad (1)$$

where k_y denotes the second order rate constant of the reaction between RX and Y (17).

If dose, D, is defined as the time-integral of the concentration of free electrophilic reagent ($\int_t [RX]dt$) either *in vitro* or *in vivo* it can be shown (12) that

$$D = \frac{1}{k_y} \times \frac{[RY]}{[Y]} \qquad (2)$$

where Y is in principle any nucleophile in the system under study.

The degree of reaction of electrophilic reagents with nucleophiles in macromolecules may provide a dosimetric basis by means of which human exposures to chemical mutagens can be quantitatively assessed.

Suitable Macromolecules for In Vivo Dose Monitoring. The *in vivo* dose of free electrophilic reagent bears a more direct relationship to genetic risk than the dose in the outer environment, since it takes into account all the processes affecting the *in vivo* fate of a compound such as its rates of uptake, enzymatic activation and disappearance through deactivation, reaction with nucleophiles and excretion. The degree of modification of the DNA is directly proportional to the *in vivo* dose (eq. 2) .

The monitoring of doses of electrophiles by measurements of their reaction products with DNA has a great appeal because of its direct relationship to mutagenesis and presumably to cancer initiation. The immunological techniques already developed in some laboratories (21) for determinations of modified DNA-bases hold great promise for dose monitoring.

However, *in vivo* doses of electrophiles can also be determined by measurements of their reaction products with nucleophilic amino acids *e.g.* cysteines or histidines in a protein such as hemoglobin, which has a number of attractive features as an "internal dosimeter". These features include: a) Blood samples containing large amounts of hemoglobin can be drawn with relative ease from human subjects. b) The relatively long life-span of hemoglobin, about 18 weeks in man, is appropriate for dose monitoring, the degree of reaction giving an integrated dose over the entire life-span of the erythrocyte (24). Appropriate is here taken to mean appropriate for the estimation of average doses which are representative for a person subjected to chronic or intermittent exposures of electrophiles from *e.g.* dietal, industrial or endogenous sources. c) The reaction products of electrophiles with hemoglobin so far investigated have been stable during the lifespan of the erythrocyte. This is in contrast to certain reaction products with the DNA which are chemically and enzymatically excised. d) Doses in other organs than blood can be estimated from the erythrocyte dose. In the general case, but not always, they will equal the erythrocyte dose (11).

About thirty radiolabeled compounds, many of which require metabolic conversion to electrophilic intermediates, have been shown to bind covalently to hemoglobin in experimental animals (5). So far, no compound believed to derive its mutagenicity from its electrophilic reactivity has failed to give covalent reaction products with hemoglobin.

The Identification of Environmental Mutagens Actually Operative in Man. The totality of compounds in the environment to which humans are exposed, presents a bewildering array to the analytical chemist. Since the routes of metabolism into electrophilic agents do not easily lend themselves to prediction it is very difficult to identify the mutagens actually operative in man by a search for "premutagens". If, instead, the analysis is undertaken on the level of reaction products with hemoglobin, R-Hb, the compounds possessing electrophilic reactivity, possibly after metabolic conversion, can be sorted out and their identity inferred from the nature of the R-groups.

Bailey *et al.* (2) have determined the "background level" of methylations of cysteine residues in hemoglobin in thirteen different animal species, including man. The degree of methylation in the human subjects was about 16 nanomoles of methyl-cysteine per gram of hemoglobin. One methylating agent to which all humans are exposed is the ubiquitous dimethylnitrosamine (DMNA). Can this agent explain the methylations observed in hemoglobin? In West Germany (22) the weekly intake of DMNA has been estimated

to 7 μg. From the relationship between injected amount of
DMNA and degree of methylation of hemoglobin determined in
animal experiments at low doses (18, 20) it can be estimated that the human intake of DMNA would result in a
steady-state level (24) of about 0.6 picomoles of methyl-
-cysteine per gram of hemoglobin. Thus, unless the endogenous production of DMNA is significantly higher than its
intake from food stuffs, DMNA only explains a small fraction of the methylations observed. If these methylations
in hemoglobin are not produced by enzymatic, non-random
reactions, which has yet to be ascertained, they may
reflect a major contribution to human genetic risk from
methylating agents.

In Vivo Dose Monitoring of Specific Environmental Mutagens
in Man. As yet, the only electrophile for which *in vivo*
doses have been estimated in humans is the directly alkylating agent ethylene oxide (EO) (7). In a study of the
degrees of hydroxyethylation of histidine residues of
hemoglobin in a group of EO sterilizers whose outer exposure doses were relatively well-defined, it was estimated
that a weekly outer exposure dose, D_{exp}, of 1 ppmh at
medium physical activity yields a degree of alkylation of
about 10 picomoles of N^τ-hydroxyethylhistidine (g Hb)$^{-1}$
corresponding, according to eq. (2), to a weekly *in vivo*
dose of 0.1 μMh of EO (7). Studies in the mouse (12, 23)
and in the rat (19) have shown that the dose of EO in the
cell nuclei, as determined from the degree of alkylation
of guanine-N-7 in the DNA (eq. 2) of different tissues
(liver, spleen and testes), is approximately equal to the
erythrocyte dose as determined from the degree of alkylation of hemoglobin. In the absence of indications to the
contrary, ethylene oxide may therefore be assumed to be
uniformily distributed also in man.

Risk Estimation Based on Rad-Equivalents. In a number of
forward mutation systems the mutagenic effectivenesses of
monofunctional alkylating agents, including EO, have been
shown to be proportional to their rates of reaction with
nucleophilic groups with a relatively low nucleophilic
reactivity (8, 9, 10) presumably including the O^6-atom
of guanine in the DNA. Moreover, the dose of a given
alkylating agent producing the same frequency of mutations
as 1 rad of γ-radiation has been shown to be relatively
constant in forward mutation systems with organisms

belonging to different phyla (8, 9, 10). This dose leads to about one alkylation in 10^7 weakly nucleophilic groups (n=2 in the Swain-Scott scale (17)) in the DNA.

Due to the constancy of the rad-equivalent dose in different organisms it has been tentatively used in cancer risk estimation (7), under the assumption that cancers are initiated by forward mutations in somatic cells.

For EO the rad-equivalent dose is about 10 µM hr. Combining this dose with the estimated relationship between <u>in vivo</u> dose and exposure dose and the relatively well-established risk coefficient for radiation-induced leukemia (14, 25) we obtain:

$$0.1 \frac{\mu Mh}{ppmh} \times 0.1 \frac{rad}{\mu Mh} \times \frac{30 \cdot 10^{-6} \text{ leukemias}}{rad \cdot person} = 3 \cdot 10^{-7} \frac{\text{leukemias}}{ppmh \cdot person}$$

| $\frac{D_{in\ vivo}}{D_{exp}}$ | Rad-equivalent dose of EO | Risk coefficient for radiation-induced leukemia | Risk coefficient for EO-induced leukemia |

The validity of this risk coefficient has been indicated in an epidemiological study of a group of Swedish EO operators (13), although the small number of people renders this quantitative confirmation uncertain.

The use of the rad-equivalent as a unit may be a practical tool in risk estimation. Risk estimates in terms of rad-equivalents may be used for taking preventive measures on a well-defined scientific basis where collective risks from the sum of all the agents contributing to genetic harm in man have to be considered.

For EO a TLV of the order of 1 ppm appears reasonable in that it would involve the same cancer risk as is considered acceptable for radiological workers (7, 14).

The uncertainties involved in the suggested procedure for risk estimation require consideration: Firstly, neither the primary lesions induced by chemical mutagens and radiations nor the mechanisms of their repair are likely to be the same and it is not mechanistically understood why 1 rad of γ-radiation seems to produce the same mutation frequency in forward mutation systems as a degree of alkylation of about 10^{-7} of weakly nucleophilic groups in the DNA.

Secondly, the ratios between *e.g.* point mutations and chromosomal aberrations produced by different chemical mutagens and ionizing radiation vary considerably. Partly on this basis it has been suggested that rad-equivalents should be determined for each genetic end-point separately (4). This caution is certainly justified and it should be recognized that the application in cancer risk estimation of rad-equivalents from forward-mutation systems is only tentative.

Dose Monitoring in Mutagenicity and Carcinogenicity Tests. Many different test systems are currently in use or under development with the purpose of identifying chemical mutagens/cancer initiators in the environment potentially posing a genetic risk to man.

The correlations between mutagenicity in different test systems and between mutagenicity and carcinogenicity are often good, but far from absolute (3). Varying causes may explain the lack of qualitative or quantitative correlation between the response of a given genotoxic agent in different test systems: a) In the absence of dosimetric measurements the response cannot be related to dose. Doses may differ as a result of different metabolism of premutagens in the systems (differences in induction status of S-9 mixtures, plasma concentration in cell culture media, *in vivo* metabolism in mammals, etc.) b) Different sensitivities of the test systems. The absence of response of a compound is still too rarely considered in relation to the sensitivity of the particular test system (11). c) Different mechanisms of induction of the mutations scored by the system (forward versus back mutation, point mutation versus chromosomal abberation, repair-deficient versus wild type strains, etc.)

In *in vitro* mutagenicity test systems doses can be monitored either by the degree of reaction of the DNA or of a model nucleophile added to the medium such as N-acetyl histidine methylamide (6). Doses in whole animals in mutagenicity and carcinogenicity tests can be determined as discussed above from the alkylation of macromolecules. In a cancer test in rats exposed to ethylene oxide this has already been done (19). If doses are determined, then the sensitivity of a test system can be defined in terms of the lowest dose of *e.g.* a standard laboratory alkylating agent or of γ-radiation (11) giving a statistically significant response.

As a result of dose monitoring with concomitant estimates of the sensitivities of test systems, qualitative

correlations between systems are expected to be improved
and the risk of drawing erroneous conclusions from negative
results lessened. Moreover, the amount of information which
can be derived from genetic toxicity testing will be
greatly increased. Such information may prove to be of
value in providing data which may be extrapolated to risk
in man. For the above reasons it is suggested that dose
monitoring should be prescribed in protocols for standard
test procedures for genotoxic activity, in particular
cancer tests.

ACKNOWLEDGEMENTS

Thanks are due to the Swedish Work Environment Fund
and the Swedish Natural Science Research Council for
providing travel grants and financial support for this
research.

REFERENCES

(1) Ames, B.N., Durston, W.E., Yamasaki, E. and Lee, F.D. *Proc. Natl. Acad. Sci. USA 70*, 2281-2285 (1973).
(2) Bailey, E., Connors, T.A., Farmer, P.B., Gorf, S.M. and Richard, J. *Cancer Res. 41*, 2514-2517 (1981).
(3) *Banbury Report 2, Mammalian Cell Mutagenesis.* Eds. Hsie, A.W., O'Neill, J.P. and McElheny, V.K., Cold Spring Harbor Laboratory, (1979).
(4) Bridges, B.A. in *Radiobiological Equivalents of Chemical Pollutants*, 83-85, IAEA, Vienna (1980).
(5) Calleman, C.J. in *The Biological Monitoring of Exposure to Industrial Chemicals*. Hemisphere, Washington, D.C. In press.
(6) Calleman, C.J. and Poirier, V., to be published.
(7) Calleman, C.J., Ehrenberg, L., Jansson, B., Osterman-Golkar, S., Segerbäck, D., Svensson, K. and Wachtmeister, C.A. *J. Environ. Pathol. Toxicol. 2*, 427-442 (1978).
(8) Ehrenberg, L. in *Radiobiological Equivalents of Chemical Pollutants*, 11-22, IAEA, Vienna, 1980.
(9) Ehrenberg, L. *ibid* 23-36, IAEA, Vienna, 1980.
(10) Ehrenberg, L. and Hussain, S. *Mutat. Res. 86*, 1-113 (1981).
(11) Ehrenberg, L. and Osterman-Golkar, S. *Teratogenesis Carcinogenesis, and Mutagenesis 1*, 105-127 (1980).

(12) Ehrenberg, L., Hiesche, K.D., Osterman-Golkar, S., and Wennberg, I. *Mutat. Res. 24*, 83-103 (1974).
(13) Hogstedt, C., Malmqvist, N. and Wadman, B. *J. Am. Med. Ass. 241*, 1132-1133 (1979).
(14) ICRP, International Commission for Radiation Protection, Publication 26, Pergamon Press, Oxford (1977).
(15) Lee, W.R. in *Chemical Mutagens: Principles and Methods for Their Detection*, Vol 5. Eds. Hollaender, A. and de Serres, F.J. 177-202, Plenum Press, N.Y. 1978.
(16) Miller, E.C. and Miller, J. A. in *The Molecular Biology of Cancer*, Ed. Busch, H. 377-402, Academic Press, N.Y. 1974.
(17) Osterman-Golkar, S., *Studies on the reaction kinetics of biologically active electrophilic reagents as a basis for risk estimates*, Thesis, University of Stockholm, Stockholm, 1975.
(18) Osterman-Golkar, S., Ehrenberg, L., Segerbäck, D. and Hällström, I. Mutat. Res. 34, 1-10 (1976)
(19) Osterman-Golkar, S., et al. To be published.
(20) Pereira, M.A. and Chang, L., *Chem-Biol. Interactions 33*, 301-305 (1980).
(21) Poirier. M.C., Santella, R., Weinstein, I.B., Grunburger, D., and Yuspa, S.H., *Cancer Res. 40*, 412-416 (1980).
(22) Preussmann, R.. Eisenbrand, G. and Spiegelhalder, B. in *Environmental Carcinogenesis*. Eds. Emmelot, P. and Kriek, E. 51-71, Elsevier, Amsterdam 1979.
(23) Segerbäck, D. to be published.
(24) Segerbäck, D., Calleman, C.J., Ehrenberg, L., Löfroth, G. and Osterman-Golkar, S., *Mutat. Res. 49*, 71-82 (1978).
(25) UNSCEAR, United Nations Scientific Committee on the Effects of Atomic Radiation: Sources and Effects of Ionizing Radiation. United Nations, N.Y. 1977.

Radiation as a Mutagen/Carcinogen Model and a Regulatory Norm

Per Oftedal

Department of General Genetics, University of Oslo, Oslo, Norway

ABSTRACT Among mutagens and carcinogens, radiation is the most studied, most easily detected, and most extensively controlled agent. The philosophy behind the regulation reflects both insight in mechanisms, of action and in end points, and in the exposure patterns of workers and populations. - The present recommendations of the ICRP say, in principle, that the doses should be kept "as low as reasonably achievable". In detail, dose limits for various organs, exposure types, and work situations, are set to equalize end point risks. Risk estimates are based on detailed analysis of human and experimental material. The heterogeneity of this base is considerable. The general acceptance of relatively simple regulation encompassing such variation leads to hope for the development of similarly basic regulatory ground rules for chemicals.

Radiation protection has a history of some 50 years. One of the first statements was made in 1928 to the International Radiology Congress and concerned the use of the dosimetric unit, the r (10). Since then, what came to be the International Commission on Radiological Protection (ICRP) has constituted a generally accepted international authority. With the increased importance of radiation and of radiation protection, of relevance to a larger section of the community, the philosophy behind the recommendations has become more generally based, and the ICRP is now an important contributor in the fields of risk and protection philosophy in general.

Radiation has a number of physical aspects which makes it a useful model for the development of a protection philosophy. It is relatively simple to measure, and the physical endpoint of biological significance - the absorbed dose - can be estimated within rather narrow limits. On the other hand there is sufficient complexity - especially after the advent of radioactive nuclides - in terms of radiation quality, of dose rate, of different organ specificities, and of super historical time spans - to make the dosimetric situation a challenge and a test for many variables. Correspondingly, the biological sensitivities and end points are varied and complex and display a spectrum of effects in the same way that other genotoxic agents do. Finally, the socially significant features - which are obviously important in the regulatory context - are manifold and complex, ranging from the straightforward and uncomplicated controlled use of radiation in lifesaving therapy and diagnostics, through economically and socially important industrial activities, and into military and political areas of great complexity and wide implications.

The central concept of the ICRP recommendations remains the regulation of exposure of those persons who receive radiation as a consequence of their professional occupation (4).

The central philosophy of the ICRP as it stands today (5) contains a number of elements. Three of them are stressed in particular:
- that doses should never exceed the stated limits
- that any exposure should be justified
- that work should be organized so that the total risk should be minimal.

The first point means that the occupational whole body dose should never exceed 5 rem per year, with further limitations as to age and rate of accumulation, special conditions (e g pregnancy) and partial exposure.

The second point is an admonition that a risk-benefit approach should be used. If there is no benefit, no risk is acceptable.

And the third point implies that in a given situation it may be preferable to accept a certain calculated risk from radiation rather than a greater calculated risk due to other causes, in doing the same job without using radiation.

These aspects are capped and covered by a general principle termed "alara": <u>a</u>s <u>l</u>ow <u>a</u>s <u>r</u>easonable <u>a</u>chievable. The recommendations apply to occupational exposures of individuals. It is reckoned that these dose limits will prevent all non-stochastic damage, and reduce the frequency of stochastic type damage to acceptable levels.

One consequence of this philosophy is that no recommended population dose limit is given. The reasoning is that the foreseeable exposures are so small as to prevent all measurable effects. The exposures should instead be related to the benefits accrued through the relevant activities, and a total dose over the future (dose commitment) of one manrem per 1MW year of electric power produced has been proposed (7).

The quantitative implementation of the recommendations are left to the responsible - generally national - authorities, and obviously a body like ICRP has no jurisdictional authority. On the other hand, the underlying ICRP numbers are generally accepted, and the recommendations are used as regulatory norms in national radiation protection legislation and services all over the world.

The risk numbers presented by the ICRP are based on all available scientific evidence and experience, carefully scrutinized and evaluated in expert committees. The derivation of the actual risk estimate is made with reference to a normalized whole-body exposure in the occupational situation, considering radiation efficiencies, organ sensitivities, age (18 to 70) and - with regard to genetic effects - probability of procreation. The total risk for cancer comes to one in 100 per Sv, or 10^{-4} per rem.

The genetic risk number is 0.4 per 100 per Sv (5). From this normalized standard risk pattern, which is based on a wide range both of materials and data, one may work backward again to any exposure situation - a particular radionuclide, a given age distribution - and obtain a risk number pertaining to that defined situation.

The object of the exercise is to create the basis for a balanced risk estimate, which may enable one to give risk numbers for as many different situations as possible, total body irradiation, or differential exposure of the various organs, or age groups. Above all this type of exercise is useful when comparing different situations, not least by way of identifying what aspects of the situation are most important to obtain the most desireable result. These procedures have been accepted and are to a considerable extent applied (1).

The type of balance I have discussed so far is in principle concerned with the exposure situation. A corresponding analysis and balancing out procedure for the other side of the scale would pertain to the biological effects. This is a much more difficult problem, and the analysis has only just begun (6). In principle, one is trying to agree on a comparable scale for such different effects as immediate loss of life in an industrial accident, loss of life ten to twentyfive years hence after exposure to carcinogenic radiation, and a life - short or long - of reduced quality in the next generation or two due to infliction of genetic damage. Obviously, some terms can be relatively easily - possibly cynically - quantified, for instance by measuring the effect in terms of loss of working years, or by adding up the societal costs of therapy and institutional care for the spectrum of disabilities caused by accidents and exposures. It becomes much more difficult to quantify the anxiety which may well prevail before the cancer becomes manifest, or the disappointment and sorrow felt after having a subnormal child. But in some relations this kind of analysis and calculation may still be possible, and is of value in particular in relativistic situations.

If a comparison is made with identical assumptions applied to the alternatives offered, the result may be valid and useful even though there are flaws in the premises and techniques used.

This balance of risks vs benefits can be illustrated in many ways. A general feature of this type of presentation is that it seeks to illustrate in two dimensions effects and processes taking place in a multidimensional matrix, where the different elements have variable values - quantitatively but also qualitatively - and the presentation is deceptively far too simple. I have used another diagram before, to indicate the approximately diametrically opposite evaluation which can be argued for the same facts, in a scientific in contrast to a societal context (8).

I think it is important to realize that when moving from the pure, basic, presumably non-weighted science - the data - and into the realm of regulation and protection, politics - in the widest sense of the word - raises its ugly/beautiful head immediately, and changes - distorts/corrects - the picture. This leads to the need for a stratification of the process. It is imperative - especially as the procedures are set up - to recognize where one is at, and to identify the transfer of authority and the change of framework of reference as one moves from the pure scientific area and by various intermediates into the final level of implementation of the actual working rules.

A corresponding situation in research policy has been portrayed even more simply, and therefore more strikingly, as a triangle of tension between the mutually interacting elements of _data_, _theories_, and _values_ (3). Only when these elements are recognized, analyzed and brought together, is it possible to understand and evaluate the dynamics involved in this type of activity.

In radiation protection, the observational material is _relatively_ general, simple and limited. Local and ad hoc interpretations and arguments can therefore be mustered at short notice and as needs arise. There are on the books

many reviews and evaluations instituted by local
groups or governments, coming to varied conclusions aimed at the local needs, all going back
to the same series of observations, practically
all of which are already compiled and reviewed
in the UNSCEAR publications (11).

When looking at this type of reviews - and
when working with one, as I have done several
times - the principal and basic difference between
the international group reviews and the national/
local group reviews are quite obvious. One need
not point to the appearance and withdrawals of
some reports to realize that on the national
scene - any national scene - the combination of
strength of personality, scientific authority,
ecconomic control and political power may lead
to a greater risk for scewed presentations and
biased judgements, than in international fora
where the net is cast wider and multifaceted
views are more independently proposed and defended. The strength and sovereignty of international bodies should thus be cherished and protected,
although admittedly the edge of the argument may
be dulled by the needs of a broadly based
compromise.

I think the ICRP may fairly claim to have
found one workable approach, in restraining itself to recommendations founded both on analysis
of the data available, and on general principles
of broad applicability, leaving the detailed
legislation and regulation to the politically
responsible authorities. However, other solutions
are possible. - A contrast is provided by the
Codex Alimentarius organization, whose area of
reference is much more complicated - all kinds
of practices influencing the unintended presence
of all kinds of chemicals in foods - and area of
application more varied - international trade -
than that of radiation protection (2).

Codex Alimentarius issues limits for the
acceptable amount of given substances or conditions in given products, based on a combined
scientific and practical information, but developed through a system of political/administra-
tive interaction between the national authorities
and the international secretariat(s), giving a

working compromise with strong non-scientific overtones. But it retains nevertheless a predominant element of the initial purpose - to protect the public against undesireable effects of food contaminants.

The recommendations of the ICRP are simple and brief, and there are underlying principles which I think may be of value whichever form or purpose regulations may have.

The ethical basis for an acceptable radiation protection policy may be discussed in many different ways. I have found (9) that the most rational and satisfactory way to cut the cake is to look at it in three sections: With <u>honesty</u>, with <u>consistency</u>, and with <u>generosity</u>. These are rules that might be set to govern our lives and behaviour both in private and in public, and so it is perhaps neither a surprising statement nor a novel discovery.

When applied to radiation protection I have taken this to mean that <u>honesty</u> would entail conveyance of all information - including that of being ignorant - to all those concerned. <u>Consistency</u> means applying the same criteria and standards in all cases, to form a set of rules where everyone may know where they stand. And - or but - with <u>generosity</u>, which means nearly the opposite of consistency, but which opens the possibility for a balanced decision based on weighted arguments, even when they are incongrous.

If these same ground rules could be applied to the formulation and application of regulations in protection against chemicals, it would be a good thing.

REFERENCES
(1) Beninson, D. Justification and Optimization in Radiation Protection. *5th International Congress of the International Radiation Protection Association,* Jerusalem 1980.
(2) Codex Alimentarius. *Procedural Mannual FAO/WHO 4th ed,* 1975.
(3) Galtung, J. Methodology and Ideology. Eiler's, Copenhagen 1977.
(4) ICRP Publication 9. Recommendations. Pergamon 1966.

(5)   ICRP Publication 26. Recommendations. *Annals ICRP 1 no 3.* Pergamon 1977.
(6)   ICRP Publication 27. Problems involved in the Development of an Index of Harm. *Annals ICRP 1 no 4.* Pergamon 1977.
(7)   Lindell, B., and Löfveberg, S. Nuclear Power, Man and Safety (in Swedish). *Publica,* Stockholm 1972.
(8)   Oftedal, P. Extrapolation of mutagenic test results to man. In *Mutagenesis in Submammalian Systems.* Ed. G.E. Paget, MTP 1979.
(9)   Oftedal, P. Ethical norms in the use of radiation and nuclear power. *5th International Congress of the International Radiation Protection Association,* Jerusalem 1980.
(10) Recommendations of the International X-ray Unit Committee. *The Second International Congress of Radiology,* July 23-27, 1928. Norstedt & Sons, Stockholm 1929.
(11) UNSCEAR. Sources and Effects of Atomic Radiation. *United Nations,* N.Y. 1977 (and earlier).

Hiroshima and Nagasaki: Three and a Half Decades of Genetic Screening

William J. Schull,[1] James V. Neel,[2] Masanori Otake, Akio Awa, Chiyoko Satoh, and Howard B. Hamilton

Radiation Effects Research Foundation, Hiroshima, Japan

ABSTRACT Since mid-1946, the birth cohorts of Hiroshima and Nagasaki have been serving as the bases for a number of investigations of the potential genetic effects of the atomic bombings of these cities. Screening has involved clinical examinations, mortality surveillance and more recently, biochemical and cytological studies. This communication summarizes the findings to date, examines briefly the strengths and weaknesses of the various screening modalities which have been used, and suggests alternatives for the future.

INTRODUCTION

   Few, if any other human population has been scrutinized for evidence of mutation more closely than the children of the survivors of the atomic bombings of Hiroshima and Nagasaki. These efforts began over thirty-four years ago and continue (13,18,21,22). At one time or another many of the various surveillance strategies which have been suggested (see, e.g., 9), have been employed; these include searches for changes in frequency of certain population characteristics, e.g., the occurrence of major congenital defect and mortality in fixed cohorts; sentinel phenotypes;

[1] Present address: Center for Demographic and Population Genetics, University of Texas Health Science Center, Houston, Texas 77025, U.S.A.
[2] Department of Human Genetics, University of Michigan, Ann Arbor, Michigan 48109, U.S.A.

chromosomal abnormalities; and biochemical variants of a structural or kinetic nature. Diverse as these alternatives seem their aims are the same, namely, to estimate the probability of mutation per unit exposure to ionizing radiation and to ascertain the public health implications of an increase in the mutations measured. Unfortunately, as we shall subsequently describe, they serve these ends unequally well.

This presentation reviews the contribution each of these approaches has made to the assessment of the mutagenic effect of exposure to ionizing radiation in this population and on the basis of all evidence to date, estimates the "doubling dose" in man. It examines briefly the strengths and weaknesses of the various screening modalities which have been used, as they are revealed by these studies and suggests alternatives for the future.

POPULATION CHARACTERISTICS

The first steps toward a continuous surveillance of the children born in Hiroshima and Nagasaki subsequent to the atomic bombings were taken in 1946 and a full-scale program was initiated in 1948, primarily, as necessitated by the times and circumstances, morphological in nature. This surveillance utilized, as a case-finding mechanism the post-war rationing system which then obtained in Japan, a provision of which entitled pregnant women who registered their pregnancies after the fifth lunar month to access to supplementary rations. A variety of lines of evidence revealed that the vast majority of eligible women availed themselves of this opportunity (13). Through this means, it was possible to identify most (more than 95%) of all pregnancies in these cities which persisted for at least 20 weeks of gestation and upon termination of these pregnancies to examine the outcomes. These clinical observations were supported by an autopsy program which sought to study as many stillborn infants or infants dying in the first few days of life as possible, and a second examination of some 20 percent of surviving infants 8-10 months after their births.

The indicators of possible genetic effects which could be extracted from a program of examinations of newly born infants, such as that just described -- all, of course, confounded by a variety of extraneous factors -- were sex, birth weight, viability at birth, presence of gross malformation, occurrence of death during the neonatal period and physical development at age 8-10 months. A comprehensive analysis of the data which had accumulated on these metrics of possible damage through 1953 suggested that this extensive clinical program had reached its logical conclusion. Borderline findings with respect to the effect of radiation on the sex-ratio and survival of liveborn infants, however, prompted a continuation of the collection of data on these latter two variables. The study on sex-ratio was extended to embrace essentially all births occurring in Hiroshima and Nagasaki from May 1946 through December 1962. Eventually over 140,000 births were involved, but these data now seem less relevant, or at least less readily incorporated into an assessment of the mutagenic effects of ionizing radiation in man than then. The simple theory of sex-linked inheritance on which predictions of an effect of parental exposures on the sex ratio rested seems no longer tenable. Moreover, recent developments -- notably the recognition of X-chromosome inactivation, of the probable preferential inactivation of paternally derived X-chromosomes, and of the occurrence of chromosomal abnormalities that can obscure a simple anatomic assessment of sex -- make it difficult to predict the effects of parental exposure on the sex ratio. We have pursued these issues further elsewhere (21,22), and shall not do so again now. Suffice it here to state that we see no way to "correct" sex ratio data for the occurrence of the phenomena just described and therefore no way to incorporate them easily into a doubling dose estimate. We conclude only that these data do not contravene the dose estimates that emerge from the other indicators of genetic damage. The study of survival which ensued, the F_1 Mortality Study, still continues. It focuses on three age and sex-matched cohorts, namely, 1) all infants liveborn in the two cities between May 1946 and December 1958 one or both of whose parents were within 2000 meters of the hypocenter at the time of the bombing -- the so-called proximally exposed, 2) an age and sex-matched cohort randomly drawn from the remaining births in the two cities during this same period, for which one parent was exposed at 2500 meters or beyond (the distally exposed) and the other either similarly exposed or not exposed at all and 3) a second age and sex-matched cohort randomly drawn from the births in these cities in these years where neither parent was exposed (5). Recently, through perusal of the pertinent household registers (koseki), these cohorts of children have been enlarged to include births that have occurred in the period from January 1959 through December 1980.

Let us now examine briefly the findings of these two programs, that is, the early clinical studies and the continuing mortality surveillance. But first some words seem indicated about the estimates of dose which have been used in this analysis. The so-called T65 dosimetry, which has served as the basis for the dose-response analyses of data collected under the auspices of the Atomic Bomb Casualty Commission and its successor, the Radiation Effects Research Foundation for the last decade and a half, assigns to each survivor separate estimates of gamma and neutron exposures, and total kerma based upon his or her stated distance from the epicenter in Hiroshima or Nagasaki and the shielding which may have attenuated the "free-in-air" dose (8). The appropriateness of these estimates has recently been questioned (6,7). It is now maintained that the neutron exposures in Hiroshima and Nagasaki may have been only 15-23% those previously conjectured, and moreover, that the energies involved were much "softer", that is, less penetrating. These developments do not affect individual exposures in Nagasaki greatly, for it has always been believed that neutron doses in this city were small, but their effect in Hiroshima seems more substantial. A precise assessment does not exist as yet, however, for at the moment individual exposures under the newly proposed dosimetry can only be estimated for those individuals who were exposed in the open. Building and body shielding factors have not as yet been calculated. One can, of course, use the new "free-in-air" estimates with the old attenuation factors to obtain person-specific estimates. But this is only a temporary expedient, since it seems clear that the new attenuation factors will be different from those associated with the T65 dosimetry for, as we have previously stated, the neutrons were less energetic. Overall, it now appears that the reassessment will lead to a reduction in estimated total kerma exposures in Nagasaki but possibly an increase in exposures in Hiroshima. This could, in turn, alter the slopes of the regression relationships which have been estimated, and hence the "doubling dose" to be described shortly.

<u>Clinical Findings</u>: In 1948 through 1953, as a part of the clinical program we have described, 76,617 pregnancy terminations were studied of which 6,535 are unsuitable for analysis for a variety of reasons (see 6.4 in (13) for the

bases for their exclusions), mostly because the requisite data were incomplete or the pregnancy was unregistered. We have summarized the clinical findings on the 70,082 acceptable pregnancy terminations under the rubric "untoward pregnancy outcome." The latter includes pregnancies which terminated in a child with a major congenital defect, who was stillborn, or who died during the first week of life. Genetic theory predicts that these various outcomes should occur in proportion to the radiation dose of the parents because of the induction of mutations with deleterious effects.

The distribution of "untoward outcomes" by parental dose when the cities and sexes are combined, and the results of an analysis which regresses the frequencies of these events on parental exposures and a variety of concomitant variables such as parental ages and inbreeding, which are known to influence their occurence has been described *in extenso* elsewhere (21,22). Suffice it here to state that, based on an assumed, conservative RBE of neutrons of 5, these events increase in frequency with increasing parental dose although not statistically significantly so. The regression coefficient per 100 rems of gonadal exposure is 0.001824 (standard error = 0.003232).

If one assumes that during the interval covered by this study, characterized by an infant and childhood mortality rate of about 7 percent, approximately one in each 400 liveborn infants died before reaching maturity because of mutation (point or chromosomal) in the preceding generation, the zygotic doubling dose, based on the regression estimate above, becomes simply 0.0025/0.0000182 or 137 rems. Thus, the gametic doubling dose would be 69 rems. We believe the assumption of one in 400 is conservative and that values as high as one in 200 are defensible. Note if one of the latter alternatives is correct, then the doubling dose would be increased accordingly, or if the RBE is higher (see (1) for some evidence that this may be so), this would also be true.

Mortality: As originally constructed, the F_1 mortality cohort consisted of 52,621 live, single births where the exposures of both parents were known. To this group has recently been added 22,984 births in the years from January 1959 through December 1980 of which 11,196 are to parents of known exposure status. It should be noted that most of these latter parents were too young at the time of their exposures to have had children in the years from 1946 through 1958. Deaths within the original cohort and its extension are ascertained through examination of the household censuses (koseki) required by law. A copy of the death schedule is obtained on all individuals who have died since last their koseki was perused. Normally a five-year cycle intervenes between one scrutiny of the koseki and the next.

Among the 63,817 individuals where the exposures of both parents are known, there have been 3,786 deaths (3,552 in the original cohort and 234 in the extension). Our remarks here will be restricted to the original cohort, for two reasons. First, the bulk of the deaths which have been recorded involve individuals in this portion of the data, and second, individual doses have not as yet been assigned to most of the parents in the extension for although their distances from the hypocenter are known, their shielding has not as yet been evaluated. When last the deaths in the original cohort were analyzed (see 21), 3,231 deaths had occurred; the years at risk ran from May 1946 through December 1971. Some 321 additional deaths (about 10% more) have occurred in the years 1971-1980. The results of the analysis to which reference has been made (21) and one which includes the years 1972-1980 are shown in Table 1.

Table 1 Increments or decrements in the frequency of death in the F_1 mortality cohort per 100 rem of gonadal dose (T65 estimate), based on an assumed RBE of 5. The standard error of each regression coefficient is indicated in parentheses.

Variable	Years at risk of death	
	1946-1971[1]	1946-1980[2]
Concomitant sources of variation ignored		
Joint parental exposure	0.000927	0.000513
	(0.001140)	(0.001165)
Gametic doubling dose	135 rem	244 rem
Concomitant sources of variation considered		
Joint parental exposure	0.001130	0.000730
	(0.001900)	(0.001390)
Year of birth	-0.005880	-0.006207
	(0.000307)	(0.000316)
Gametic doubling dose	110 rem	171 rem

1 Data drawn from Neel, Kato and Schull (1974)
2 Current analysis

We present two alternative analyses, namely, one which does and one which does not take cognizance of concomitant variation, notably year of birth and parental ages which could influence the risk of death. Observe that the gametic doubling dose remains essentially unchanged with the addition of the additional eight years of data. Concomitant variation affects the estimates somewhat, but not greatly when one considers the errors inherent in the estimates.

Given the oft-argued relationship of mutation (somatic as well as germinal) to cancer, it is interesting to note that 72 of these 3,552 deaths were ascribed to malignancy. The most common of these latter causes of death was leukemia; indeed, 35 of the 72 deaths were ascribed to this malignancy. A moment's perusal of Table 2 suffices to establish that no clear trend in the occurrence of these causes of death with radiation exists as yet. It is important to note that there

Table 2 Deaths from cancer in the original F_1 cohort by parental exposure category, Hiroshima and Nagasaki, 1946-1980

Cancer	Parental Exposure				Total
	Neither exposed	Father only	Mother only	Both exposed	
Leukemia	11	7	13	4	35
Acute	9	5	8	2	24
Other	2	2	5	2	11
	(0.63)*	(1.18)	(0.71)	(0.36)	(0.67)
Lymphosarcoma	3	-	1	1	5
Stomach	3	-	4	2	9
Retinoblastoma	1	-	-	-	1
Wilms' tumor	1	-	-	1	2
Brain	3	-	1	-	4
Bone	-	1	1	1	3
Other	4	-	3	6	13
Total	26	8	23	15	72
Individuals at risk	17,226	5,592	18,224	11,199	52,621
	(1,51)*	(1.34)	(1.26)	(1.34)	(1.37)

*Deaths per 1000 individuals at risk

is no unequivocal increase with exposure in deaths attributed to leukemia; such an association has been alleged.

SENTINEL PHENOTYPES

It has frequently been argued that one potentially useful mutational surveillance strategy involves the search for changes in the rate of occurrence of isolated cases (within the family) of certain phenotypes, so-called sentinel ones, which have a high probability of being due to a dominant mutation. Among such phenotypes are aniridia, chondrodystrophy, epiloia, neurofibromatosis, retinoblastoma, and possibly neuroblastoma. Some of these are, or have been until recently, invariably fatal tumors, and most are presumably readily diagnosable. Here, we shall also include among surveys of sentinel phenotypes those clinical (as contrasted with cytogenetical) surveys which have sought to establish the existence of changes in the frequencies of phenotypes such as Down's, Klinefelter's and Turner's syndromes.

Insofar as the phenotypes earlier enumerated are concerned, the only ones which have been under constant scrutiny in Hiroshima and Nagasaki have been those associated with childhood malignancies, that is, leukemia, retinoblastoma, Wilms' tumor and the like. Tumor Registries which have existed in both cities since 1958 make possible the identification of these occurrences as well as malignancies among cancer prone diseases such as those associated with immunodeficiency (e.g., ataxia telangiectasia) or neurofibromatosis. In the years these Registries have existed we have encountered only one case of retinoblastoma and two cases of Wilms' tumor. Table 2 gives the exposures of the parents of these individuals. Patently there is no persuasive evidence of a relationship between their occurrence and parental radiation, but the numbers are abysmally small.

Three separate studies of Down's syndrome and the syndromes associated with sex chromosomal aneuploidy have been made. Schull and Neel (19) failed to find a relation between maternal radiation and Down's syndrome. Indeed, at face value, the frequency of the latter malformation decreased with maternal exposure. Eighteen individuals with Down's syndrome were recognized clinically among 25,843 children examined. In 12 cases neither parent was exposed (out of 12,995 individuals); in three cases the mother only was exposed (n = 7615); in no instance was the father only exposed (n = 1916); and in three cases both parents were exposed (n = 3317). Maximum exposure was less than six rad in any of the six cases where one or both parents were exposed. No significant differences existed between the cities in case occurrence (8 in 14,034 in Hiroshima; 10 in 11,809 in Nagasaki). Slavin, Kamada and Hamilton (23) studied some 92 individuals with Down's syndrome drawn largely from eight schools for mentally retarded children, five in Hiroshima Prefecture and three in Nagasaki. Twelve additional non-institutionalized cases were ascertained through five large hospitals in Hiroshima. Of these 92 cases, in nine instances one or both parents were in Hiroshima or Nagasaki at the time of the 1945 atomic bombs, and the child conceived subsequently. Seven of these cases were simple trisomy 21, one was a trisomy with a D/D translocation, and the final case involved mosaicism. Among these nine cases only two had exposures of greater than one rad -- one mother received 8 and another 16. These estimates are the "free-in-air" T65 doses. Although appropriate denominators for these cases are not clear, even the case numbers themselves are not compelling.

Finally, a buccal smear survey was conducted in April-May 1963 on 8192 students in Hiroshima who were scheduled for dental screening examinations at two junior and five senior high schools in Hiroshima (15). Subsequent studies including repeat buccal smear analysis, physical examination and chromosomal analysis were conducted on selected cases. Most of these students (all save 921) were born between April 1946 and March 1949. Of the smears obtained on these 8192 students, 7141 (4481 males, 2660 females) were deemed acceptable for analysis. No abnormalities were seen among the 2660 females, but three cases of Klinefelter's syndrome were seen among the 4481 males. None of the latter three were conceived by parents exposed to the atomic bombing of Hiroshima. It must be emphasized that the sample studied is not a representative one necessarily, and thus these data have limited value in the estimation of the dose-response relationship. But they do provide some indirect evidence of the difficulties inherent in the phenotypic approach. No less than two of these three males would not have been detected clinically because of their normal mental status and minor physical abnormalities, such as reduction in axillary hair or gynecomastia.

Patently, these studies do not lend themselves, singly or collectively to doubling dose estimation so sparse are the data.

CHROMOSOMAL ABNORMALITIES

Effort to assess radiation related chromosomal damage in the survivors and their offspring began as early as 1948. This undertaking was less than successful. Reliable information had to await the development in the late 1950s of the use of leukocyte cultures with phytohemagglutinin and the other technological steps now in vogue. Accordingly, on the basis of a pilot study conducted in 1967, an investigation of the children of exposed parents was initiated in 1968, the subjects being drawn initially from the cohorts established for the F_1 Mortality Study. Awa and his colleagues have described this study (2,3). Since the youngest children in the study are 13 and the oldest are now 34, the survey will not yield adequate data on the frequency of cytogenetic abnormalities associated with increased mortality rates, such as unbalanced autosomal structural rearrangements and autosomal trisomies. The data on sex chromosomal abnormalities and balanced autosomal structural rearrangements should, however, be relatively unbiased even now.

Awa has recently reported the frequency of sex-chromosome aneuploids in the children of distally exposed parents (beyond 2400 m) to be 13/5058 or 0.00257 (3). It should be noted that as a result of further study of the cytogenetic material, this figure differs by one case from the number given in (21). The frequency in the children of proximally exposed (within 2000 m), with an average conjoint parental gonadal dose of 87 rem at a neutron RBE of 5, is 16/5762 or 0.00278. These figures, though not statistically significantly different, can be taken at face value to generate zygotic and gametic doubling doses. We find the former to be 1071 rem and the latter 535.

BIOCHEMICAL VARIANTS

Electrophoretic techniques for the identification of abnormal protein molecules created a new potential approach to an assessment of the genetic effects of A-bomb exposure. This approach, like the cytogenetic one, is free of many of the ambiguities inherent in the study of population

characteristics or sentinel phenotypes, and has, as a
result been vigorously advocated (9,11,14). Accordingly,
after a pilot study that extended from 1972 through 1975,
a full-scale investigation employing this technique was
undertaken in Hiroshima and Nagasaki in 1976 (12,16).
The subjects are being drawn from the cohorts of children
born to the proximally and distally exposed parents identified for the mortality study previously described. The
same blood sample serves the needs of this biochemical
program as well as the cytogenetic one. Each child is
examined for rare electrophoretic variants of 28 proteins
of the blood plasma and erythrocyte, and since 1979, a
subset of the children is further examined for deficiency
variants of 10 of the erythrocytic enzymes.

A rare electrophoretic variant is defined in this
context as one with a phenotype frequency of less than 2%
in the population and an "enzyme deficiency" or "low activity" variant as one which results in an enzyme level
three standard deviations below the mean (or less than 66%
of normal activity). When either variant is encountered,
its occurrence is first verified and then blood samples
from both parents are examined for the presence of a
similar variant. If the variant is not also found in one
or both parents, and a discrepancy between putative and
biological parentage is improbable, it presumably represents a mutation. Some eleven different red cell antigenic
systems and the HLA phenotypes are used to explore discrepancies of the kind identified.

Satoh and her colleagues have recently estimated that
they have information on the equivalent of 419,666 locus
tests on children born to parents whose average conjoint
gonadal dose was approximately 59 rem (16). Two probable
mutations have been observed; one is a slow migrating
variant of glutamate pyruvate transaminase and the other
a slow migrating variant of phosphoglucomutase-2. No
probable mutations have been seen in the equivalent of
282,848 locus tests on children whose parents, one or both,
were distally exposed, that is, received less than 1 rad.

As yet the data on enzyme deficiencies or low activity
variants is too preliminary to provide much insight into
the mutational process. Indeed, even in the case of the
electrophoretic variants, their number remains too small
to provide a meaningful estimate of the doubling dose in
man. It warrants note, however, that these data do not
contravene the other estimates we have described.

THE DOUBLING DOSE

Elsewhere to obtain a combined doubling dose we have taken the simple average of the three individual doubling doses -- based on untoward pregnancy outcome, F_1 mortality and sex chromosome aneuploids -- as a value representative of the "true" doubling dose. This seemed permissible, for all three estimates were fairly similar despite differences in the sample sizes on which they were based. If we were to follow this practice here the gametic doubling dose would be 258 rem. But it is patent that this value is very heavily influenced by the data on the frequency of chromosomal abnormalities, data derived from the smallest of the several samples analyzed. A better estimate would seem to be the weighted average of the individual estimates where the weights are the inverses of the variances of the several estimates.

As rough approximations to these variances, we take

$$V(d_D) = d_D^2 \left[\frac{V(\alpha)}{\alpha^2} + \frac{V(\beta)}{\beta^2} - 2 \frac{Cov(\alpha,\beta)}{\alpha\beta} \right]$$

where d_D is the estimated doubling dose, α is the intercept of the dose response function fitted (or the frequency of the event of interest in the "standard" population), and β is the slope of the fitted dose response (here assumed to be linear). We find the three standard deviations of the estimates to be 93, 388, and 2416 for untoward pregnancy outcomes, F_1 mortality and sex chromosome aneuploids, respectively. Note the coefficient of variation is especially large in the case of sex chromosome aneuploids; the doubling dose lies well outside the dose interval covered by the observations. Be this as it may, the weighted average of the estimates is 139 rem with a standard deviation of approximately 157. We may assume the true variance to actually be somewhat larger since the doubling dose estimates for untoward pregnancy outcome and F_1 mortality are not wholly independent (neonatal deaths contribute to both estimates), nor is α known without error, as assumed in two instances.

DISCUSSION

Space does not permit a measured discussion of the strengths and weaknesses of the various strategies we have discussed. Patently, however, they vary. Measures of mortality and congenital malformation have immediate meaning to the public, whereas the occurrence of a new mutant form of glutamate pyruvate transaminase does not. But to the geneticist, events of the latter kind have a precision in the measurement of the frequency of mutation not to be found in changes in the frequencies of variables where the role of genetic factors are uncertain. Similarly, mortality surveillances, such as the one in Hiroshima and Nagasaki, are inexpensive, but the biochemical genetics program is not. Undoubtedly technological developments will, with time narrow these cost differences. The issue of relevance to the public will, of course, remain. Neel (14); see also Editorial, Science, 11 September 1981) has suggested biochemical and immunological measures which may better serve these disparate, seemingly irreconcilable needs, and has pleaded for an increasing effort to improve the data base on which estimates, such as those set forth here, are based. These ends warrant the endorsement of the entire scientific community.

REFERENCES

(1) Abrahamson, S. IAEA Symposium on Biological and Environmental Effects of Low Level Radiation. Vol I, 3-7 (1976)
(2) Awa, A. A., Bloom, A. D., Yoshida, M. C., Neriishi, S. and Archer, P. G. Nature 218, 367-8 (1968)
(3) Awa, A. A., Honda, T., Neriishi, S., Shimba, H., Amano, T. and Hamilton, H. B. Proc. 6th Int. Cong. Hum. Genet., Jerusalem, Israel (1981)
(4) Ishimaru, T., Ichimaru, M., Mikami, M. RERF Technical Report 11-81 (1981)
(5) Kato, H., Schull, W. J. and Neel, J. V. Am. J. Hum. Genet. 18, 330-3 (1966)
(6) Kerr, G. Fourth Symposium on Neutron Dosimetry, Munich-Neuherberg (1981)
(7) Loewe, W. and Mendelsohn, E. Technical Report D-80-14, Lawrence Livermore Laboratory (1980)
(8) Milton, R. C. and Shohoji, T. ABCC Technical Report 1-68 (1968)
(9) Neel, J. V. Perspectives Biol. Med. 13, 522-37 (1971)
(10) Neel, J. V., Kato, H. and Schull, W. J. Genetics 76, 311-26 (1974)

(11) Neel, J. V., Mohrenweiser, H., Satoh, C. and Hamilton, H. B.   Radiation Effects Research Foundation TR 4-77 (1977)
(12) Neel, J. V., Satoh, C., Hamilton, H. B., Otake, M., Goriki, K., Kageoka, T., Fujita, M., Neriishi, S. and Asakawa, J.   Proc. Nat. Acad. Sci. USA 77, 4221-4 (1980)
(13) Neel, J. V. and Schull, W. J.   NAS-NRC Publication No. 461 (1956)
(14) Neel, J. V., Tiffany, T. D. and Anderson, N. G.   In Hollander A. (Ed) Chemical Mutagens Vol 3, 105-40 (1973)
(15) Omori, Y., Morrow, L. B., Ishimaru, T., Johnson, K. G., Maeda, T., Shibukawa, T. and Takaki, K.   ABCC Technical Report 11-65 (1965)
(16) Satoh, C., Awa, A. A., Neel, J. V., Schull, W. J., Kato, H., Hamilton, H. B., Otake, M. and Goriki, K.   Proc. 6th Int. Cong. Hum. Genet., Jerusalem, Israel (1981)
(17) Schull, W. J. and Neel, J. V.   Science 128, 343-8 (1958)
(18) Schull, W. J. and Neel, J. V.   Am. J. Publ. Hlth. 49, 1621-9 (1959)
(19) Schull, W. J. and Neel, J. V.   Lancet i, 537-8 (1962)
(20) Schull, W. J., Neel, J. V. and Hashizume, A.   Am. J. Hum. Genet. 18, 328-338 (1966)
(21) Schull, W. J., Otake, M. and Neel, J. V.   Science 213, 1220-7 (1981)
(22) Schull, W. J., Otake, M. and Neel, J. V.   In: Hook, E. (Ed) Human Mutation: Biological and Population Aspects (1981)
(23) Slavin, R. E., Kamada, N. and Hamilton, H. B.   ABCC Technical Report 2-66 (1966)

A Model for Low Dose Risk Estimates

Carl Johan Calleman

Department of Radiobiology, Stockholm University, Stockholm, Sweden

ABSTRACT A model relating doses, D_i, and levels, L_i, defined on consecutive steps (i) from exposure to response or risk in man is proposed for the study of dose-response relationships. The rationale for the model is that the kinetics of the biological and chemical processes determining the transfer functions $D_{i+1}=f_i(D_i)$ between different steps can be subjected to direct study by measurements of the mutagen or its reaction products. If, for all transfer functions, $f_i'(D_i) > 0$ at $D_i=0$ then the response will not exhibit a threshold dose. The applicability of the model is exemplified by ethylene, an agent which has not been directly shown to be mutagenic, but the mutagenicity of which can be inferred from a combination of chemical and genetical evidence.

INTRODUCTION

The doses and concentrations of chemical mutagens/cancer initiators received by human populations are low in comparison to those yielding significant responses in most test systems for such agents. For comparison, the lowest radiation dose giving a significant response in standard cancer tests is of the order of 10^2 rad. In contrast, a yearly radiation dose of about 10 rad would be sufficient to explain the total cancer incidence in Sweden. Clearly, the sensitivity of test systems can be increased by increasing the size of tests, but for practical reasons, given the squared dependence of the resolving power on the size of

the test (7), this is only feasible to a certain limit. Low-
-dose risk estimates for mutations and cancers consequently
have to be based on extrapolations from dose-regions accessible to direct observation in test systems or in epidemiological studies (16), to the low-dose region of practical
interest.

The models upon which extrapolations to the low-dose
region are based are loaded with implications as to how
risks posed to humans by genotoxic agents should be handled.
If, on the one hand, the extrapolation implies the existence
of a threshold dose below which no response is induced, then
exposures below this threshold dose is to be considered
safe. On the other hand, if a threshold dose is not implied,
the particular agent has to be subjected to a risk-benefit
analysis by regulatory bodies and acceptable risks posed to
humans settled upon.

The introduction of the concept of an "apparent no-
-effect level" does not resolve the long-standing controversy of the shape of dose-response curves at low doses, rather
it creates an ambiguity as to whether the absence of a significant response at some low dose should be explained by
the necessarily limited sensitivity of the test system or by
biochemical or genetical processes creating a true threshold.

A Model for the Study of Dose-Response Relationships. A
suggested method in extrapolating data from animal experiments to man is to find the mathematical function which best
fits the experimental points and assuming this mathematical
function to adequately express risks to humans as a function
of dose, albeit with parametrical changes (21). Unless the
choice of mathematical function is based on the most reasonable biological processes determining the response, low-dose
risk assessment may, however, be seriously distorted. In
order to facilitate low-dose risk assessment based on chemical and biological theory, a model (Table 1) has been proposed (11) which interrelates doses and levels (D_i and
L_i) defined on different consecutive steps (i) from exposure level (L_1) to response or risk in human populations.
Doses defined at different steps are related to each other
by transfer functions, $D_{i+1}=f_i(D_i)$ which can be studied experimentally by determinations of doses and levels of
premutagens, A, ultimate mutagens, RX, or their reaction
products, RY. The mathematical functions interrelating doses
describe the kinetics of the processes which determine the
transfers between consecutive steps. The function most suitable to describe a dose-response curve will then be the pro-

Table 1. System of steps from exposure level/dose of genotoxic agents to biological response simplified from (9, 11).

Step*	Processes determining kinetics (examples)	Designation	Dimension (examples)
L_1, D_1	Exposure to premutagen A or ultimate mutagen RX	Exposure level Exposure dose	ppm ppm h
	$f_1(L_1)$ or $f_1(D_1)$ ↓ uptake		
L_2	Absorbed amount of A or RX	Pharmacological dose	mg (kg b.w.)$^{-1}$
	$f_2(L_2)$ ↓ activation, deactivation, transport		
D_3	RX in tissues	Tissue dose	mol (kg b.w.)$^{-1}$h
	$f_3(D_3)$ ↓ transport, diffusion		
D_4	RX in immediate environment of target molecules (often DNA)	Target dose	-"- ("DNA-dose")
	$f_4(D_4)$ ↓ alkylation, etc		
L_5	Chemically changed critical centers in target macromolecules	Molecular dose	mol RY(mol Y$^-$)$^{-1}$
	$f_5(L_5)$ ↓ repair, replication, promotion		
L_6	Response or risk in man		P=probability

*D denotes a dose in the strict sense with the dimension concentration x time; L is a level (concentration).

duct of the individual transfer functions between the step where dose was defined and the response.

Low Dose Risk Estimate of Ethylene. The idea underlying the model in Table 1 is that the only information available about the response at low doses is that about the kinetics of the processes determining this response. How such information can be utilized can be exemplified by the cases of two olefines, vinyl chloride and ethylene.

Some of the processes determining the transfer functions in Table 1 are saturable, meaning that they will, in the general case, follow Michaelis-Menten kinetics:

$$\frac{-dS}{dt} = V_{max} \times \frac{S}{K_m + S}$$

where S is the concentration of the substrate which may be A, RX or RY in the model. Examples of saturable processes are, metabolic activation and deactivation, error-free and error-prone DNA repair. Such processes are expected to give rise to a response deviating from linearity at high levels, L_i, of substrate. If the deviation will be positive or negative will depend on which processes are predominating in the particular case.

Maltonis' data for vinyl chloride-induced liver tumours (18) fit very well to a dose-response-curve following Michaelis-Menten kinetics with a negative deviation ($f'' < 0$). Presumably due to saturation of the metabolic conversion of vinyl chloride to chloroethylene oxide the fraction of animals developing tumours asymptotically approaches a maximal value of about 0.25 as the level, L_1, of vinyl chloride is increased.

Michaelis-Menten kinetics of the metabolic activation of the other olefine, ethylene, is probably the reason that it has never, to the best of our knowledge, been demonstrated to be positive in mutagenicity test systems. In a cancer test the compound failed to yield a significant response (4). Yet, ethylene was shown in 1977 (8) to be metabolically converted to ethylene oxide, an agent the mutagenicity and carcinogenicity of which must now be considered established (10, 12). It has later been shown by Segerbäck (22) that the degree of hydroxyethylation of hemoglobin, D_3 in blood, and of DNA in liver, kidney, and spleen, L_5, of the mouse depens linearly upon the concentration, L_1, in the air at low values of L_1. According to Andersen et al. (1), the uptake of ethylene in rats follows Michaelis-Menten kinetics with a K_m of about 220 ppm and a V_{max}=8.6 10^{-6}mol (kg b.w.)$^{-1}$h^{-1}. Thus, the maximal dose, D_3, of ethylene oxide attainable in animal experiments with ethylene will asymptotically approach that produced by chronic exposure to a level L_1, of about 6 ppm of ethylene oxide (19). From what is known about the long term study in rats of the carcinogenic potential of ethylene oxide at the Carnegie-Mellon Institute (12) this seems to be near the level giving a significant response. Thus the carcinogenic and mutagenic potential of ethylene may escape detection in normal scale tests for such effects. Ethylene, can however not be considered innocuous as the dose-response curve of its metabolite, ethylene oxide, has been shown to be linear in many mutagenicity test systems (10) including in human cells (20). In $E.\ coli$ Sd-4, Hussain (14) has shown the dose-response curve to be linear down to very low doses amounting to less than one hydroxyethylation of guanine-O^6 per bacterium. Combining the evidence from the kinetics of the processes determining the transfers from L_1 of ethylene to response it can be concluded that the risk associated with exposure to ethylene will depend linearly upon exposure dose at levels, L_1, below K_m. A quantitative estimate of the risk connected with human exposure to ethylene has to await the determination of $in\ vivo$ doses of ethylene oxide in persons occupationally exposed to well-defined concentrations of ethylene.

Threshold or no Threshold ?. An exhaustive scrutiny (11) of all the processes determining transfer functions between doses and levels from L_1 to L_5 in Table 1 is not within the scope of this presentation. However, for a discussion of the existence of threshold doses of chemical mutagens these processes may be subdivided into two categories: 1) Processes which follow, or approach, first order kinetics at low values of L_i. These processes are expected to give rise to linear transfer functions at low doses, $i.e.$ $D_{i+1} = k_i D_i$. Examples of such processes are uptake, diffusion, active and passive transport, metabolic activation and deactivation as well as thermal reactions. 2) Processes which follow first-order kinetics, the rates of which are changed above some level of substrate owing to the induction of enzymes, $e.g.$ of metabolic activation or DNA repair, resulting in higher or lower values of f_i' above the level where enzymes were induced (23).

Thus all known processes determining transfers from L_1 to L_5 in Table 1 are expected to give rise to linear transfer functions with slopes greater than zero at D=0.

In certain experiments relating L_5 to mutational response, such as exposure of cells from (healthy) humans to UV-light (17) a threshold dose or, at least a very low response at low doses, has been indicated. Presumably, the omni-presence of UV-light has necessitated the development of highly efficient DNA repair mechanisms against lesions induced by this agent.

There are three major reasons why it would be dangerous at present to conclude from this experiment, or from others with methylating agents (15), that thresholds exist for chemical mutagens in the general case. Firstly, thresholds have especially been demonstrated for methylating and ethylating agents and UV-light. For other agents thresholds are rarely indicated.

Secondly, the rate of repair of *e.g.* methylating agents have amply been demonstrated to vary between organs (13). The resulting mutagenic response will depend also on the rate of replication in different organs.

Thirdly, individuals vary in their capacities to repair premutagenic lesions in the DNA, an extreme example of which is provided by persons afflicted with Xeroderma pigmentosum. Thus, when considering human populations, mutations will presumably not exhibit no-effect levels even for methylating agents or UV-light.

Special Considerations for Carcinogenesis. In the case of radiation mutagenesis, where doses are defined at D_4 in Table 1, most researchers seem to agree that mutation frequencies depend linearly upon dose (3) at low doses. Moreover, at present the majority of scientists seem to agree that carcinogenesis is initiated by a mutation in a somatic cell. Yet, there is no general agreement about the shape of the dose-response curve of radiation(or of chemically) induced cancer (3). Part of the explanation to this controversy may be that the dose-response curves of initiation and the promotive step(s) also considered to be necessary for the development of tumours are rarely studied independently.

In the general case, the probability of developing a tumour, will be the product of its probabilities of initiation and promotion(s) (11). The induction of liver tumours seems to depend linearly upon radiation dose when promoted by carbon tetrachloride (6). Also in organs where a promotive situation is already at hand for unknown reasons, but evidenced by a high spontaneous frequency of tumours, tumour frequencies seem to depend linearly on dose (24). Assuming that the sensitivity of a population to promoting agents is a quality depending on a high number of variables it seems reasonable to describe it with a normal distribution. The probability of developing a tumour as a result of exposure to an agent which is capable of both initiation and promotion (a complete carcinogen) would then be best described by the product of the probability function (the integrated normal distribution) for promotion and a linear function for initiation (2). Given the very rapid approach to zero at low values of the probability function, complete carcinogens are expected to yield thresholded curves unless promotion is exerted by other agents already present. When considering cancer risks posed to human populations it has to be assumed that the probability of promotion is high (cancer is a common disease). The cancer incidence is thus expected to increase linearly for low doses of a single added cancer initiator (5).

ACKNOWLEDGEMENTS

Thanks are due to the Swedish Work Environment Fund and the Swedish Natural Science Research Council for providing travel grants and financial support for the research.

REFERENCES

(1) Andersen, M.E., Gargas, M.L., Jones, R.A. and Jenkins, L.J. Jr. *Toxicol. Appl. Pharmacol.* 54, 100-116 (1980).
(2) von Bahr, B., work in progress.
(3) Biological Effects of Ionizing Radiations, BEIR III, National Academy Press, Washington, D.C. (1980).
(4) Chemical Industry Institute of Toxicology (CIIT) Docket No. 12000, October 15, (1980).
(5) Crump, K.S., Hoel, D.G., Langley, C.H. and Peto, R. *Cancer Res.* 36, 2973-2979 (1976).
(6) Curtis, H.J. and Tilley, J. *Radiat. Res.* 50, 539-542 (1972).
(7) Ehrenberg, L. in *Handbook of Mutagenicity Test Procedures*. Eds. B. Kilbey et al. 419-459, Elsevier/North-Holland (1977).
(8) Ehrenberg, L., Osterman-Golkar, S., Segerbäck, D., Svensson, K. and Calleman, C.J. *Mutat. Res.* 45, 175-184 (1977).

(9) Ehrenberg, L. and Osterman-Golkar, S. *Teratogenesis, Carcinogenesis and Mutagenesis 1*, 105-127 (1980).
(10) Ehrenberg, L. and Hussain, S. *Mutat. Res. 86*, 1-113 (1981).
(11) Ehrenberg, L., Moustacchi, E. and Osterman-Golkar, S. Dosimetry of genotoxic agents and dose-response relationships of their effects. Working document for ICPEMC.
(12) Fishbein, G.W. Occupational Health & Safety Letter, December 8 (1980).
(13) Goth, R. and Rajewski, M.F. *Proc. Natl. Acad. Sci. USA, 71*, 639-643 (1974).
(14) Hussain, S. Personal communication.
(15) Jenssen, D. Low-Dose Mutagenicity in Mammalian Cells, Thesis, University of Stockholm, Stockholm (1981).
(16) Land, C.E., *Science 209*, 1197-1203 (1980).
(17) Maher, V.M., Dorney, D.J., Mendrala, A.L., Konze-Tomas, B. and McCormick, J.J. *Mutat. Res. 62*, 311-323 (1979).
(18) Maltoni, C., *Ambio 4*, 18-23 (1975).
(19) Osterman-Golkar, S. and Ehrenberg, L. Symposium on the "Metabolism and pharmacokinetics of Environmental Chemicals in Man", Sarasota, Florida, June 7-12, 1981, in press.
(20) Poirier, V. Personal communication.
(21) Proposed System for Food Safety Assessment *Food and Cosmetics Toxicology 61*, suppl. 2, (1978).
(22) Segerbäck, D., to be published.
(23) Segerbäck, D. and Ehrenberg, L. *Acta Pharmacol. Toxicol*, in press (1981).
(24) Staffa, J.A. and Mehlman, M.A. Quoted by Clayson, D.B. *Mutat. Res. 86*, 217-229 (1981).

Risk Estimations Based on Germ-Cell Mutations in Mice

U. H. Ehling

Institut für Genetik, Gesellschaft für Strahlen- und Umweltforschung, Neuherberg, Federal Republic of Germany

ABSTRACT The yields of gene mutations induced by chemical mutagens in mice depend on the spermatogenic stages, the sex and the different treatment conditions. The knowledge of these factors is the foundation for the assessment of the genetic hazard. A linear dose response relationship and the similarity between spontaneous and induced mutations are suppositions for the estimation of the genetic risk. For the extrapolation of the data from mice to men, the different metabolic capacities for activation and detoxification have to be taken into account. Two experimental approaches are available for the risk estimation. First, specific locus mutations can be used to compare the increase in total mutation frequency from a particular dose of a mutagen with the spontaneous level. The doubling dose for specific locus mutations will be compared with that for other genetic endpoints in mice. Second, dominant mutations in mice can be used to quantify the genetic risk of the first generation. Both approaches can be used to calculate the individual risk and the risk of the total population. Other possible extrapolation strategies will be discussed.

INTRODUCTION

The final goal for mutagenicity testing is the risk assessment of chemical mutagens. Procedures for the estimation of chemically-induced genetic damage in a population have to be developed. For the determination of the amount of human ill health arising from an increase in the mutation rate due to exposure to mutagenic agents, it is essential to recognize that in spite of extensive efforts, the attempts to detect radiation-induced genetic damage in man, so far, have proved inconclusive. Therefore, the UNSCEAR-Report (14) stated that there is no alternative for "using results obtained with experimental animals in estimating rates of induction in man. The limitations of such a procedure are obvious when it is realized that animal species differ from each other in their susceptibility to the induction of genetic changes by radiation and that there is no evidence indicating whether the genetic material of man is more or less sensitive to radiation than that of other animal species. The only mammal which has been studied in some detail with respect to radiation genetics is the mouse. Results of mouse experiments must therefore form the main basis for the assessment of genetic risk in man." This conclusion is likewise important for the attempt to estimate the risk of chemical mutagens to man. Using the mouse data for the risk estimation of chemical mutagens, it is necessary to adopt a principle enumerated in the BEIR-Report: "Use simple linear interpolation between the lowest reliable dose data... In order to get any kind of precision from experiments of manageable size, it is necessary to use dosages much higher than are expected for the human population. Some mathematical assumption is necessary and the linear model, if not always correct, is likely to err on the safe side" (1).

One way of expressing the genetic risk is the determination of the doubling dose. The doubling dose can be defined as the dose necessary to induce in a specific germ cell stage as many mutations as occur spontaneously. This approach is used for data obtained with the specific locus method in mice (2). The other approach, developed by us, is the direct method of estimating total phenotypic damage induced in a single generation. This method was originally applied to dominant mutations affecting the skeleton of mice. The most efficient technique now available to study the induction of dominant mutations in mice is the biomicroscopic examination of the eye with a slit lamp, which is an easy method for the detection of dominant cataract mutations. The advantage of this genetic endpoint is that hereditary cataracts observed in the mouse can be directly compared with the manifestation of cataracts in man (4).

In the paper the following topics will be discussed: 1st, the doubling dose approach based on the induction of

specific locus mutations; 2nd, the direct estimation of
the risk based on the induction of dominant cataract mutations in mice; and 3rd, other possible alternative extrapolation strategies.

DOUBLING DOSE APPROACH

The doubling dose approach is used to estimate the
genetic risk of procarbazine, cyclophosphamide and isoniazid. The data base for the estimation is the induction
of specific locus mutations in mice. The specific locus
method was described in detail in reviews by Ehling (2)
and Searle (12).

PROCARBAZINE

The drug is used in combination treatment of Hodgkin's
disease. The dose-response is linear for point estimates
of the mutation frequencies after treatment of A_S spermatogonia with 0, 200, 400, and 600 mg/kg (6). The doubling
dose for procarbazine based on the regression coefficient
is 110 mg/kg (Table 3). By comparing the therapeutic dose
of 215 mg/kg with the doubling dose of 110 mg/kg, one may
conclude that the procarbazine treatment of a patient with
Hodgkin's disease would induce 2 times as many mutations
as arise spontaneously, provided man and mouse are equally
sensitive. The calculation of the individual risk is of
importance for the patients treated with the drug. The
individual risk can serve as a guideline for counseling
a patient if an abortion is advisable. In addition, from
the individual risk in combination with the frequency of
the disease, one can calculate the population risk.
    Based on the assumption that all patients were treated
with procarbazine and that the reproduction of the patients
is normal the population risk of procarbazine was calculated. We estimated that 22 new mutations can be expected in
1 million births owing to the procarbazine treatment. This
calculation was made without the correction for dose fractionation and sex differences (6).
    Fractionation of the dose reduces the yield of mutations in male mice by approximately a factor of 3 (3).
A further reduction is due to the fact that oocytes are
less sensitive to the induction of mutations than spermatogonia (Table 1).

Table 1: Induction of specific locus mutations in female mice

Compound	Dose (mg/kg)	Interval between treatment and conception			
		Up to 7 weeks		More than 7 weeks	
		No. of offspring	No. of mutations at 7 loci	No. of offspring	No. of mutations at 7 loci
Procarbazine	400	8 075	0	20 094	2
	600	15 527	0	28 565	1
	total	23 602	0	48 659	3
Mitomycin C	2	2 847	1	4 956	0
	4	1 515	0	1 204	0
	total	4 362	1	6 160	0

Control rate 2.1 or 5.6×10^{-6} (10)

If spermatogonia and oocytes were equally sensitive to the induction of specific locus mutations by procarbazine, one would expect 23 mutants in a total of 72 261 offspring. The expected frequency of 23 is significantly different from the observed three mutations ($P<10^{-4}$). This result indicates that oocytes are less sensitive than spermatogonia for the induction of specific locus mutations.

If we correct our original calculation for the fractionation effect in spermatogonia and for the sex differences one can conclude that very likely not 22 new mutations can be expected due to the procarbazine treatment, but only 5 in 1 million births.

It is interesting to compare the induction of specific locus mutations with procarbazine in female mice with the results in radiation experiments (10). Irradiation of female mice with fission neutrons has shown that the interval between irradiation and conception has a striking effect on the mutation frequency observed in the offspring. In the first seven weeks after irradiation the mutation frequency was high. For conceptions occurring more than seven weeks after irradiation, the observed mutation frequency was near zero. This observation is in agreement with the mitomycin C experiment, but not with the results of the procarbazine treatment. The mutant in the mitomycin C experiment was born 22 days posttreatment, while in the procarbazine group the mutants were born 70, 74 and 243 days posttreatment. The highest frequency observed in the mitomycin C experiment is 8-23 times higher than the spontaneous mutation frequency, while in the procarbazine experiment the induced rate is 2-6 times higher than the control frequency. More data are needed to determine the factors which influence the chemically-induced mutation frequency in females.

Fractionation of the dose for spermatogonia and sex differences are two factors which influence the yield of mutations, another important factor is the differential spermatogenic response.

CYCLOPHOSPHAMIDE

The induction of specific locus mutations after i.p. injection of cyclophosphamide demonstrates the differential spermatogenic response (Table 2).

Table 2: Cyclophosphamide-induced specific locus mutations in male mice (dose: 120 mg/kg)

Mating intervals (days)	Average litter size	No. of offspring	No. of mutations at 7 loci	Mutations per locus per gamete $\times 10^5$	Doubling dose (mg/kg)
1- 7	4.4	1 627	3	26.3	3.8
8-14	4.8	1 794	3	23.9	4.2
15-21	4.9	1 744	1	8.2	13.0
22-42	7.1	2 678	1	5.3	21.3
≥ 43	7.4	12 573	1	1.1	320

The reduction of the average litter size indicates the induction of dominant lethal mutations. The induced specific locus mutation rate in spermatozoa is 85 times higher than the mutation frequency in spermatogonia. The mutation frequency in spermatogonia is not significantly different from the control frequency (P=0.51). Based on these data, we can calculate the doubling dose for spermatozoa to be 3.8 mg/kg and for spermatogonia to be 320 mg/kg. In dermatology, doses from 2-8 mg/kg/day for 6 days are used. It is very likely that in spermatozoa the doses are additive. A total of 12-48 mg/kg would increase the individual risk by a factor of 3-13, if man and mouse were equally sensitive. In other words, an offspring of a patient conceived immediately after the treatment has a 3-13 times higher chance of carrying a mutation than a child from an untreated father. However, after approximately three months this risk is drastically reduced. In addition, fractionation of the dose in spermatogonia may further reduce the genetic risk.

The statement that fractionation of the dose in spermatozoa is additive is based on the following observations: 1st, we observe a very close correlation between the induction of dominant lethals - indicated in Table 2 by the reduction in litter size - and specific locus mutations. 2nd, in a series of dominant lethal experiments with dose fractionation the results indicate no reduction in the mutation yield (5).

ISONIAZID (INH)

After i.p. injection of 2x125 mg/kg INH 24h apart a total of 11 specific locus mutations were observed in 84 728 offspring derived from treated spermatogonia. Out of 11 mutants 5 came from one cluster at the non-agouti locus. The mutation rate per locus per gamete is either 1.2 or 1.9×10^{-5} according to the mathematical treatment of the cluster (3). In another series (4x125 mg/kg INH), 3 mutations were observed in a total of 30 063 offspring (1.4×10^{-5} mutations/locus/gamete). The control frequency in these experiments was 11 mutants in 169 955 offspring. Isoniazid is used for the treatment of tuberculosis. In 1970 a total of 511 kg of INH was sold in the Federal Republic of Germany. According to this figure we can estimate the population dose for INH to be 4 mg/kg. If we treat the cluster as one event, we can calculate the doubling dose for spermatogonia on the basis of the dose-effect-curve. In this case the doubling dose is 625 mg/kg. Assuming equal sensitivity to mutation induction in males and females, one can calculate the population risk as the ratio between the population dose and the doubling dose times two, for the possible mutation induction in females, times the spontaneous mutation frequency. The critical point for the INH extrapolation is the assumption that men and mice are equally sensitive. This assumption for INH is questionable.

The extrapolation of data from one species to another can only be justified if we have proper knowledge about the metabolization of the compound. In the mouse INH is hydrolysed to the carcinogenic and mutagenic compound hydrazine, but in rats and hamsters this does not occur (11). However, I would like to emphasize that we have to encourage research in the field of comparative pharmacokinetics in order to improve risk estimation. The geneticists can improve the situation by comparing the induction of dominant lethal mutations or chromosome aberrations in two or more species.

DIRECT ESTIMATION OF THE GENETIC RISK FOR THE FIRST GENERATION

The direct estimation of the genetic risk for the first generation is possible on the basis of dominant cataract mutations in mice. To determine the frequency of dominant mutations affecting the lens, the F_1 offspring of treated and control mice are examined with a slit lamp at 4-6 weeks of age. The advantage of the dominant mutations is that hereditary cataracts observed in the mouse can be directly compared with the manifestation of cataracts in man (4, 8).

Ethylnitrosourea (ENU) was used to study simultaneously the induction of specific locus mutations and the induction of dominant cataracts. After i.p. injection of 160 mg/kg of ENU 21 specific locus mutations were observed in 5 667 offspring derived from exposed spermatogonia, in the experiment with 250 mg/kg of ENU 54 specific locus mutations and 10 dominant cataract mutations were observed in 9 352 offspring. The dose-effect-relationship for the induction of specific locus mutations by ENU is linear. In the latter experiment 5.4 times more specific locus mutations than dominant cataract mutations were observed. However, in man there are 20 well established dominant cataracts known. A similar number can be expected for the mouse. In the specific locus experiment we score only 7 loci. On a per locus basis the specific locus mutations are probably 16 times more sensitive to the induction of mutations by ENU than the cataract mutations. It is therefore desirable to estimate the induction of dominant mutations directly.

If one assumes that dominant cataract mutations are representative for all dominant mutations, then one can calculate, from the ratio of dominant cataract loci to the total number of dominant loci, the overall genetic damage in the first generation. If the mutation rate for the induction of dominant cataracts is representative for all dominant mutations, we have to multiply the mutation rate by 36.8. The multiplication factor is derived from McKusick's tabulation of "Mendelian Inheritance in Man" (9). He described 736 well established autosomal dominant traits in man. The number of well established dominant cataracts in man is 20. The ratio of 736/20 gives a multiplication factor of 36.8 for the total dominant genetic damage, assuming the same ratio exists for man and mouse. For the purpose of the calculation we will assume a population dose of 1 mg/kg of ENU. If man and mouse were equally sensitive to the induction of dominant mutations by ENU we could calculate the expected frequency of new dominant mutations: $4.3 \times 10^{-6} \times 36.8 \times 1 = 158$ dominant mutations of clinical importance that can be expected in the first generation for 1 million live births if the spermatogonia were treated with 1 mg/kg of ENU. For the total population damage we have to multiply the mutation rate of spermatogonia with the sensitivity factor for females.

This estimate can be checked independently by using the rate of induced dominant mutations affecting the skeleton of mice. I may add that the estimated frequencies of radiation-induced dominant mutations agreed well when based on either dominant cataracts or dominant skeletal mutations (4).

ALTERNATIVE EXTRAPOLATION STRATEGIES

In view of the experimental resources required for the proposed risk estimation, it seems also necessary to think of alternative possibilities for the risk calculation for chemical mutagens. One alternative is the comparison of doubling doses for gene mutations in mice with other genetic endpoints. The other alternative is the parallelogram approach suggested by Sobels.

DOUBLING DOSE OF CHEMICALS FOR DIFFERENT GENETIC ENDPOINTS

A comparison of doubling doses for different genetic endpoints in mice is given in Table 3.

Table 3: Doubling doses of chemicals in mg/kg for different genetic endpoints in mice

Compound	Germ cell stage (*)	Specific locus mutations	Heritable translocations (a)	Dominant lethals	Somatic mutations (b)	Sperm abnormality (c)
CP	pg	4	8	69**	N.T.	130
	g	320	-	-		
MMS	pg	2**	4	30**	22**	180
	g	17	-	-		
Procarbazine	pg	43**	43	213**	15**	120
	g	110**	-	-		
Mitomycin C	pg	-	4	2**	1**	1
	g	≤ 1**	-	-		

CP = Cyclophosphamide, * pg = postspermatogonia, g = spermatogonia, ** Calculation of the doubling dose is based on the regression coefficient; the other values are on point estimates. (a) I.-D. Adler; Teratogenesis, Carcinogenesis and Mutagenesis 1, 75-86, 1980 and W.M. Generoso et al., Mutation Res. 76, 191-215, 1980. (b) A. Neuhäuser-Klaus, Arch. Toxicol. (in press). (c) A.J. Wyrobek and W.R. Bruce, Chemical Mutagens 5, 257-285, 1978.

For cyclophosphamide and MMS the same mating intervals were used for the comparison of specific locus mutations and dominant lethals in postspermatogonia. For both compounds the calculated doubling doses differ by an order of magnitude for the induction of mutations in spermatozoa and spermatids. In general, however, I think that this comparison is encouraging. It may be worthwhile to extend the data base for this comparison and at the same time to define the calculation of the doubling dose.

For a systematic exploration of the comparison of doubling doses for different genetic endpoints I would like to reemphasize that one underlying assumption for the calculation of the doubling dose is a linear dose response relationship. It follows that several doses should be tested for one mutagen. Another assumption is the similarity between spontaneous and induced mutations. In addition, the following criteria should be taken into account: 1st, the calculation of the doubling dose should be based on the regression coefficient. 2nd, to reduce the fluctuation for the calculated doubling doses the use of a well established historical control is recommended. In postspermatogonial germ cell stages a well defined mating interval should be used for the establishment of a dose-effect curve. This mating interval should give a high mutation frequency, but at the same time it is mandatory that the frequency of fertile matings is similar in the control and the experimental group. The data which fulfill these criteria are marked in Table 3 with two stars.

PARALLELOGRAM METHOD

The parallelogram method was developed by Sobels (13). The idea is to find some suitable indicator of genetic damage in somatic cells which can be measured in both experimental animals and man. One possibility is to determine the alkylation per nucleotide in mammalian cells in vitro and in mouse germ cells. One can then determine the mutation frequency in mammalian cells in vitro and estimate the mutation frequency in mouse germ cells. This estimated frequency can be compared for model compounds with the observed frequency of gene mutations in the mouse. This approach can be combined with Ehrenberg's proposal that the alkylation of haemoglobin may be a suitable indicator of DNA damage (7). The interesting parallelogram approach will be tested in a cooperative study, which will be supported by the Environmental Research Programme of the European Communities.

CONCLUSION

It is possible to estimate the incidence of dominant and recessive mutations in humans induced by chemicals based on the extrapolation of data for the induction of gene mutations in the germ cells of the mouse. If we consider the number of substances traded worldwide and the number of unknown chemicals which will probably be found to be mutagenic and essential, and therefore where we have to perform a risk estimation, it is mandatory to develop alternative methods for the estimation of the genetic risk of man due to chemical mutagens. In addition to these tasks it is necessary to improve our model system for the estimation of chromosomal diseases and to develop approaches to estimate the complex constitutional and degenerative diseases. The last two problems must also be solved for the estimation of the radiation-induced genetic damage in man. We have laws to protect the human genome from radiation genetic damage. It is now imperative to develop guidelines for the controlled use of chemical mutagens similar to the regulations which already exist for radiation protection.

ACKNOWLEDGEMENT

The experimental work was carried out under contract No. 136-77-1 ENV D of the Environmental Research Programme of the European Communities.

REFERENCES

(1) BEIR-Report (Biological Effects of Ionizing Radiation) National Academy Press, Washington, D.C. (1980)
(2) Ehling, U.H. In: A. Hollaender and F.J. de Serres (eds.) Chemical Mutagens, Vol. 5, Plenum Publishing Corporation, New York, pp. 233-256 (1978)
(3) Ehling, U.H. *Arch. Toxicol. 46*, 123-138 (1980)
(4) Ehling, U.H. *Umschau 80*, 754-759 (1980)
(5) Ehling, U.H. 6. Tagung der Gesellschaft für Umwelt-Mutationsforschung, Darmstadt, 19.-20. Februar 1981
(6) Ehling, U.H., Neuhäuser, A. *Mutation Res. 59*, 245-256 (1979)
(7) Ehrenberg, L. In: V.K. McElheny and S. Abrahamson (eds.) Banbury Report 1, Cold Spring Harbor Laboratory, pp. 157-190 (1979)
(8) Kratochvilova, J., Ehling, U.H. *Mutation Res. 63*, 221-223 (1979)
(9) McKusick, V.A. Mendelian Inheritance in Man. The Johns Hopkins University Press, Baltimore (1978)
(10) Russell, W.L. *Proc. Natl. Acad. Sci. USA 74*, 3523-3527 (1977)
(11) Schöneich, J., Braun, R. Environmental Mutagens. Akademie-Verlag, Berlin, pp. 63-76 (1977)
(12) Searle, A.G. *Mutation Res. 31*, 277-290 (1975)
(13) Sobels, F.H. *Arch. Toxicol. 46*, 21-30 (1980)
(14) UNSCEAR-Report (United Nations Scientific Committee of the Effects of Atomic Radiation) United Nations, New York (1966)

The Mutational Specificity of Gamma-Rays Is Influenced by Dose: Implications for Thresholds and Risk Estimation

Barry W. Glickman

Laboratory of Molecular Genetics, National Institute of Environmental Health Sciences, Research Triangle Park, North Carolaina, U.S.A.

ABSTRACT Using the *lacI* system of *Escherichia coli*, the mutational specificity of γrays was determined over a broad range of doses and compared with the spectrum of spontaneously arising mutants. At the higher doses tested, 7.5 to 30 kR, both transition and transversion events were scored. The spectra produced by these doses superficially resembled that of spontaneously occurring mutations, although differences were detected at specific sites. However, after a dose of 5 kR, the mutants produced were largely, if not exclusively, the result of transition events. The mutational spectrum obtained at this dose differed from both the spontaneous spectrum as well as the γray spectrum recovered after higher doses. It is proposed that the differences in spectral quality reflect the induction of the error-prone "SOS" repair pathway or the saturation of a specific error-free repair pathway by increasing amounts of damage. The relevance of dose-dependent spectral variation is discussed with respect to the problem of risk estimation and threshold effects.

INTRODUCTION

The *lacI* system of *Escherichia coli* provides a method for determining mutagenic specificity at a large number of sites (2,3). We have previously made use of this method to compare the mutational spectrum of γrays with that found to occur spontaneously (4). One reason for these experiments was to examine the assumptions of qualitative equivalence inherent in the so-called "doubling-dose" method of risk estimation (e.g.

ref. 1). Our results suggested that there was an overall similarity between the spontaneous and γray-induced spectra. However, several sites behaved significantly differently and we questioned the value of the doubling-dose concept in the light of these qualitative differences. Moreover, kinetic studies in which the induction of the total lacI mutant population was compared to the sub-population of amber and ochre mutants, revealed that these two classes arose with different kinetics. The suggestion was made that this difference might reflect changes of mutational specificity related to the repair capacity of the cell and the induction of the "error-prone" repair process known as "SOS" repair.

In this study, we have undertaken an examination of the mutational specificity of γrays over a range of doses, and have found that mutational specificity does indeed appear to alter with dose.

MATERIAL AND METHODS

The rationale behind the system and the experimental strategy have been summarized previously (4). Unless otherwise indicated, all materials and methods were as described by Coulondre and Miller (2,3). For the spontaneous spectrum, 100 colonies were gridded from each independent culture, whereas for the induced spectra, 25 or 50 colonies were gridded from each culture, depending upon the dose. During the course of these experiments about 100,000 colonies were gridded for the spontaneous spectrum and 70,000 for the induced spectra. The kinetic data were obtained by direct plating onto phenylgalactose-supplemented plates. Up to a dose of 10kR, where survival is greater than 20%, there was direct proportionality between the number of cells plated and the mutant yield. At higher doses, reconstruction experiments provided the basis for correcting for lethality in calculating mutation frequencies.

RESULTS

Cellular survival and mutagenesis following exposure to γrays are summarized in Table 1. At the low doses (2.5 and 5 kR), SOS (as measured by the induction of prophage λ in a lysogenic derivative of this strain)

DOSE (kR)	SURVIVAL	FREQUENCY[a] OF lacI⁻ (TOTAL)	RELATIVE INCREASE	FRACTION[b] OF AMBER AND OCHRE	FREQUENCY[a] OF AMBER AND OCHRE	RELATIVE INCREASE
0	100%	600	1.0	108/7760 (1.4%)	8.4	1.0
2.5	86%	600	1.0	50/2100 (2.4%)	16	1.9
5.0	58%	800	1.3	164/4330 (3.6%)	29	3.4
7.5	46%	1000	1.7	64/1900 (3.9%)	39	4.6
10.0	33%	1200	2.0	315/6700 (4.7%)	56	6.7
20	16%	2300	3.8	177/2000 (8.9%)	205	24
30	5%	4000	6.7	216/2000 (10.8%)	433	51

[a] Frequency per 10^8 survivors
[b] Determined by supressibility with *supE*, *supB*, *supC* and *supF*.

Table 1. Survival and mutagenesis of NR 3835 following γ-irradiation. The mutants scored are lacI⁻ which were then later screened for amber and ochre mutants as described in the text.

is only partially induced, while at higher doses (7.5 kR or greater), the maximum level of prophage induction is achieved (data not shown). As can be seen in Table 1, at low doses the factor of induction of amber and ochre mutants is greater than the factor for the total $lacI^-$ population. At doses too low (2.5- 5kR) to increase the frequency of the total $lacI^-$ population significantly, the frequency of amber and ochre mutants is increased substantially. The doubling dose for the population of nonsense mutations is about 3kR compared to a dose of about 10 kR required in the case of the total $lacI^-$ population. These results are comparable to what we observed previously (4). In order to properly compare the mutational spectra obtained at low levels of mutagenesis, it was first necessary to establish a spontaneous spectrum containing a sufficient number of isolates to allow a valid comparison. Because of their greater ease in analysis, we restricted substantial parts of these experiments to the study of amber mutants. The spontaneous spectrum (Table 2) shows the distribution of 446 independently isolated spontaneous amber mutants. These data are summarized on the basis of the single base change required to generate each mutant in Table 3. Fully three-quarters of the spontaneous mutations are transitions, although this fraction drops when the G:C→A:T transition hotspots occuring at 5-methylcytosine sites (5) are removed from consideration.

Following a dose of 5 kR, the frequency of amber and ochre mutants is increased three-fold (Table 1). About 90% of the mutants analysed arose by transition (Table 3). In fact, the transversion events present in the 5-kR spectrum can virtually all be accounted for on the basis of the expected contribution of spontaneously arising transversion mutants.

At doses greater than 5 kR, the proportion of transversions increases appreciably. This can be seen after 7.5 kR even though the sample size is rather small. The distributions of mutants following doses of 10 and 30 kR are virtually identical (Table 2) and transversion events account for roughly 40% of the amber mutants (Table 3).

DISCUSSION

The mutational spectrum following exposure to 5 kR

	0kR	5kR	10kR	30kR
G:C→A:T				
Am 5	6	11	5	4
6	99	25	3	0
9	12	5	10	10
15	71	9	6	8
16	21	9	2	5
19	22	9	3	5
21	13	6	3	5
23	15	1	3	5
24	21	7	11	10
26	12	4	4	4
31	1	4	4	8
33	9	10	3	5
34	29	11	2	1
35	2	2	3	1
G:C→T:A				
Am 2	2	1	3	2
7	10	0	2	4
10	4	0	2	4
12	9	0	1	2
13	6	1	5	5
17	4	1	0	1
20	7	0	2	5
25	7	3	9	2
27	25	1	2	5
28	2	0	3	5
A:T→T:A				
Am 11	5	1	0	1
18	22	2	2	3
32	0	0	2	1
36	2	0	1	3
A:T→C:G				
Am 3	6	0	5	8
4	7	1	2	2
14	2	2	1	1
22	6	0	2	1
30	1	0	0	1
G:C→C:G				
Am 1	4	1	4	3
8	0	0	2	1
29	0	0	0	0

Table 2. Distribution of *lacI* amber mutants on the basis of the base substitution required to generate the mutant. The spectra include both spontaneously arising and γ-ray induced mutants.

DOSE : 0kR	SITES AVAILABLE	SITES FOUND	NUMBER	% OF TOTAL
G:C→A:T	14	14	333	75%
-hotspots	11	11	134	(54%)
G:C→T:A	10	10	76	17%
A:T→T:A	4	4	11	2.5%
A:T→C:G	5	5	22	5%
G:C→C:G	3	1	4	1%
TOTAL			446	

DOSE : 5kR				
G:C→A:T	14	14	113	89%
-hotspots	11	11	68	(83%)
G:C→T:A	10	5	7	5%
A:T→T:A	4	2	3	3%
A:T→C:G	5	2	3	3%
G:C→C:G	3	1	1	1%
TOTAL			127	

DOSE : 7.5kR				
G:C→A:T	14	6	19	63%
-hotspots	11	3	7	(39%)
G:C→T:A	10	5	5	17%
A:T→T:A	4	1	1	3%
A:T→C:A	5	1	5	17%
G:C→C:G	3	0	0	0%
TOTAL			30	

DOSE : 10kR				
G:C→A:T	14	14	62	54%
-hotspots	11	11	51	(50%)
G:C→T:A	10	9	29	25%
A:T→T:A	4	3	5	4%
A:T→C:G	5	4	12	11%
G:C→C:G	3	2	6	5%
TOTAL			114	

DOSE : 30kR				
G:C→A:T	14	13	75	56%
-hotspots	11	10	66	(53%)
G:C→T:A	10	10	35	26%
A:T→T:A	4	4	8	6%
A:T→C:G	5	4	13	10%
G:C→C:G	3	2	4	3%
TOTAL			135	

Table 3. Distribution of spontaneous and γ-ray induced amber mutants on the basis of the base substitution required to generate the mutant. The hotspots are amber sites 6, 15 and 34 and where the percentages given exclude the hotspots, the contribution of these sites to the total has been removed.

largely represents transitional events and is considerably different from the spontaneous spectrum. Most of these transitions did not occur at spontaneous hotspots, confirming that γrays do not cause the deamination of cytosine (4). The 5-kR spectrum also differs substantially from spectra produced at higher levels of γ-irradiation (Table 3). At higher doses the fraction of transversion events increases to greater than 40% of the recovered single base substitutions. These results clearly demonstrate an alteration in mutational specificity relative to the dose of ionizing radiation given.

In the present study, a dose range was chosen which would allow an examination of the effect of the induction of SOS repair. THe spectrum produced at 5 kR represents the spectrum of mutations produced when SOS is poorly induced (although we cannot exclude the possibility that mutation only occured in those few cells where SOS was "on"). This spectrum, containing primarily transition events, differs from the induced spectra where SOS was fully induced. Current models of SOS repair suggest that the induction of SOS results in a reduction of DNA replication fidelity allowing the bypass of "blocking lesions" (6-8). Such a model may well account for our observation of an alteration in spectra as SOS is induced. Alternatively, a repair system may exist which is capable of removing a premutational lesion which potentiates transversions. To date, however, neither such a lesion nor repair system have been identified.

One of the methods currently employed in quantifying genetic hazards to man from exposure to ionizing radiation involves the use of the so called "doubling dose" (e.g. ref.1). Extrapolation based upon the doubling-dose concept requires the assumption of qualitative equivalence at the molecular level between the induced and spontaneous genetic changes. Though there is no *a priori* reason to expect the kinds of changes to be recovered in treated and untreated samples to be similar, genetic systems having sufficient resolution to test this hypothesis are only now being developed. In our original study of γray-induced mutational specificity (4), while there were exceptions, we reported some similarity between the induced (30 kR) and spontaneous spectra. The present study has revealed, however, that at lower γray doses, mutational specificity can be dramatically different from the spontaneous spectrum. While the philosophy of the doubling-dose concept was not to estimate genetic risk in *Escherichia coli*, results from a genetic system with the potential determine mutational specificity demonstrate differences at the molecular level between induced and spontaneous mutagenesis and serious doubts about the applicability of this concept raise.

The low yield of $lacI^-$ mutants and the resultant statistical uncertainties preclude the determination of a useful estimate of the specific-locus rate at 5 kR. A particular advantage of the $lacI$ system is that, by being approachable by genetic analysis at the molecular level, a degree of sensitivity can be obtained comparable to reversion analysis but without depending upon a single site. For this reason, it has proven possible to detect significant increases in mutation frequencies at doses where the overall frequency of $lacI$ mutants has not been significantly increased. A base substitution rate of 9×10^{-13} per nucleotide pair per rad per gene copy can be calculated. The elevated sensitivity available in the $lacI$ system by screening for nonsense mutants raises the question of real vs. imaginary thresholds and demonstrates the importance of the genetic system utilized.

REFERENCES

(1) BEIR Report, National Academy of Sciences-National Research Council, Washington (1972).
(2) Coulondre C. and Miller J. *J. Molec. Biol. 117*, 525-576 (1977).
(3) Coulondre C. and Miller J. *J. Molec. Biol. 117*, 577-606 (1977).
(4) Glickman, B.W., Rietveld, K. and Aaron, C.S. *Mutation Res. 69*, 1-12 (1980).
(5) Miller, J. Coulondre, C. and Farabaugh, P.J. *Nature 274* 275-280.
(6) Radman, M. In "Molecular and Environmental Aspects of Mutagenesis" (eds. Prakash, L. et al.) Thomas, Springfield, Ill., 128-146 (1974).
(7) Radman, M. In "Molecular Mechanisms for Repair of DNA" (eds. Hanawalt, P.C. and Setlow, R.B.) Plenum Press, New York, 335-344 (1975).
(8) Witkin, E.M. *Bacteriol. Rev. 40*, 869-907 (1976).

Apparent Threshold and Its Significance in the Assessment of Risks Due to Chemical Mutagens*

Yataro Tazima

National Institute of Genetics, Mishima, Japan

ABSTRACT Unlike ionizing radiation, several intervening processes exist for chemical mutagens from their intrusion into animal body until they get to the target cells. Hence, it may not be unreasonable to assume the presence of apparent threshold at very low dose range administered in vivo. In order to elucidate the validity of this assumption following data were collected: (a) point mutations of mammalian cells in culture, (b) specific loci mutations in the silkworm oocytes, and (c) point mutations of transplacentally treated cells of hamster embryos. The presence of threshold was demonstrated or at least reasonably assumed for most chemical mutagens in vivo experiment of the silkworm. Similar evidences have also been revealed with use of Sylian hamster embryos. The importance and significance of the study of apparent threshold is discusssed.

INTRODUCTION

Whether or not there is a threshold in the dose-mutation frequency relationship is an important problem in the assessment of risks due to chemical mutagens. The same is true for ionizing radiations. But, basic difference between the two is that ionizing radiations, having strong penetrability, can easily reach target tissues and/or cells in the body, whereas chemical mutagens are liable to be

*This work was supported by Grant in Aid from the Ministry of Education, Science and Culture, Japan

weakened before they reach target cells, due to several hampering mechanisms; such as absorption by organic substances, inhibition of penetration by tissue membranes and metabolic decomposition, *etc*. Such intervening mechanisms may operate independently of the administered dose, but manifest their effects invalidating the mutagenicity of a given chemical at lower dose range. Consequently, it is highly likely that there may exist a no effect dose range, within which a substantial yield of mutation could not be recognized.

Threshold in such sense has been called apparent threshold, on which little attention has been paid so far. Instead, another sense of threshold, called true threshold, has been the center of interests of many workers who were interested in repair mechanisms. In this paper, I shall demonstrate the presence of threshold of both senses and stress the importance of the research on the apparent threshold in the assessment of risks due to chemical substances.

DEFINITION

When we discuss the dose-effect relationship for chemical mutagens it is necessary to distinguish clearly between the effect per administered dose and the effect per dose actually incorporated into the target cells.

Threshold is termed as the dose limit below which no significant increase in biological effects could be recognized. It has been classified into two categories; an apparent threshold and a true threshold. The former is observed when the effects are plotted against administered doses and the latter plotted against actually incorporated doses. However, such term should be revised. Because, the "true" is the antonym of "false," but "apparent threshold" is not "false threshold." I, therefore, propose to use "genuine threshold" instead of "true threshold."

ANALYSIS OF DATA

In order to make clear the problem of threshold in chemical mutagenesis, we organized a research group and carried out detailed analysis of dose-mutation frequency relationship, each member using his most experienced test system. The systems used varied from point mutations of mammalian cells in culture to specific loci mutations in the silkworm, including eight different systems. However, as to the point mutations in cultured cells, enough data have not been obtained yet to discuss the absence or presense of any thresholds. Hence, relevant data have been collected from the literature and used for discussion. Experimental data, in which chromosome aberrations were used as criteria, were discarded. Because, most chromosome aberrations are known to exhibit dose-response curves of exponential type which causes the difficulty in identifying the threshold that may exist in the one hit response curve at low dose range.

Thus, selected were three kinds of experimental data; (a) point mutations with mammalian cells in culture, (b) specific loci mutations in the silkworm oocytes, and (c) point mutations of hamster embryo cells that were treated transplacentally.

1) *In vitro* treatment of mammalian cells in culture

The results of *in vitro* experiments obtained by our group failed to conclude the presence of threshold, all exhibiting almost linear curve. Employing 8-AG resistant mutation system of human diploid fibroblasts Kuroda (8) examined the dose-mutation relationship for sterigmatocystine, Trp-P-1 and Trp-P-2. Okada and Nakamura (9) carried out similar experiments with mouse L5178Y cells with regard to the mutagenicity of EMS, STC and Trp-P-1, using MTX resistance and 6-TG resistance as markers. The dose-frequency curve thus obtained were linear showing a slope of 1 when plotted on log-log scale.

Recently, two extensive mutation test data have been published independently by Clive et al.(3) and Amacher et al.(1), both using L5178Y cells. As markers for mutation detection, Clive et al. used loci of thymidine kinase (TK) and HGPRT, and Amacher et al. used TK locus only. They examined the dose-response relation over wide range of administered doses for various compounds.

Of 43 compounds investigated by Clive et al., 5 compounds (2AAF, B(α)p, ethylene dibromide, methyl iodide and AF-2) were suspicious about the presence of threshold. Mutation frequencies obtained at lower dose range were either lower than those of control group or not significantly different from the control. The most remarkable was the case of 2AAF (2a3 and 2c in Table 2 of reference (3)). The compound represented at low doses lower mutation frequencies than the control in both cases. Except for those 5 compounds, Clive's data did not show up any such type of response. Amacher et al. examined the dose-mutation frequency relationship for 30 compounds. Four of these (anthracene, auramine O, caffeine and pyrene) gave the relation that indicates the presence of no effect range. Induced mutation frequencies at lower doses fluctuated around the spontaneous mutation level (anthracene in Table 3 in reference (1)). But majority of compounds did not show threshold.

The other examples indicating the presence of threshold were those reported by Duncan et al.(4) for 7-BrMeBA (7-bromomethylbenz(α)anthracene) and 7-BrMe-12-MeBA (7-bromomethyl-12-methylbenz(α)anthracene). The survival curve comprised three components and the first component did not show up any increase in mutation frequency up to a certain chemical dose.

In summary, when mutagenic response was examined with cultured mammalian cells, chemical substances do not exhibit, in most cases, the presence of no effect dose range, but a few substances have so far been found to show the presence of no effect dose range. Those cases could be mentioned as examples of genuine threshold.

2) Specific loci mutations of silkworm germ cells

In case of germ cell treatment, several modifying processes could intervene from the intrusion of chemical substance into the animal body until they get to the target cells. Accordingly, it is highly likely in such cases that we may observe the threshold in response curves.

In order to test the validity of this supposition, the specific loci system with silkworm oocytes can be conveniently used. In this system the test substance is injected, at mid-pupal stage, into the body cavity of the female, which conceives 400-500 developing eggs per one individual. Injected substance will then be incorporated into the ooplasm and react with DNA of late prophase nuclei. The mutations thus produced can easily be detected as pink or red (mosaic) eggs among dark colored wild type ones.

By employing this system, Murakami (10,11) investigated the dose-mutation frequency pattern by administering DES at variable doses. He obtained a sigmoid shaped dose-effect curve at or higher doses than $2.5 \cdot 10^{-2}$ µg per mother pupa. But, treatment at lower doses produced no significant increase in mutation frequencies at all. Using the similar test system, he examined the dose-response pattern of several chemical substances. He tested, as weak mutagens, chlorodiazepoxide·HCl and diazepam, both are known minor tranquilizers, and, as strong mutagens, sterigmatocystine and aflatoxin B_1. Strong mutagenic pyrolysates, Trp-P-1 and Trp-P-2 did not respond to this system. Three compounds except sterigmatocystin, exhibited sigmoid type or nearly two component curves (Fig. 1). Fig. 2 depicts the dose-mutation frequency curve obtained for chlorodiazepoxide.HCl. The curve represented typical sigmoid shape and treatment with doses lower than 100 μg per pupa gave mutation frequencies almost similar to or not significantly larger than those obtained for the non-treated control.

As indicated in Fig. 1, strong mutagens such as AFB_1 also represented two component curve, indicating a slope of approximately of one on log-log scale at higher doses, but the slope turned to far more gentle slope at lower doses. However, it should be noticed that linear relation on log-log scale plot is represented by the function $y=x^a$, where x is the dose and y is mutation frequency, and constant a determines the slope. Conveniently, when $a=1$, it indicates one hit curve and $a=2$ two hit curve. When a is smaller than 1, the function implies that production of mutation is less efficient than one hit. Since a takes 1 at higher doses, but smaller than 1 at lower dose range, the curve clearly indicates that mutagenic potency of the chemical mutagens was hampered at low dose range. Even when existence of threshold could not be proved, we may assume theoretically the presence of no effect dose range, if we get a smaller than 1. Similar reaction pattern was observed by von Borstel in yeast after AF-2 treatment (2). He called this phenomenon partial-target mutagenesis and assumed several possibilities.

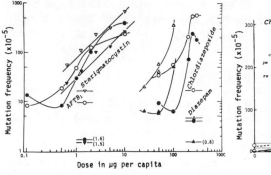
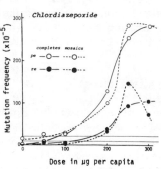

Fig.1. Dose-mut. freq. curves for AFB_1, STC, Cl-diazepoxide and Diazepum on log-log scale
(Murakami et al.,1979)

Fig.2. Dose-mut. freq. curves for Cl-diazepoxide on ordinary scale
(Murakami et al.,1979)

The results obtained in the silkworm were in good accord with our expectation that threshold may appear in such a test system where mutation may occur after several intervening processes. Further the results suggested that the existence of no effect dose range may be rather common when we deal with chemical mutations induced in the germ cells. The observed threshold in those cases could be classified into apparent threshold.

3) Transplacental treatment of hamster embryonic cells

Another data to be used for the discussion of the threshold are those of transplacental treatment of embryonic cells of Sylian hamster. Test substances were injected intraperitoneally into a pregnant mother, so as to treat the embryos for a definite time interval. The embryos thus treated were excised and chopped up into fine pieces. Chopped pieces were digested with trypsin and cells were transferred into culture medium in flask. Employing those cells Inui investigated the mutation response pattern in relation to the dose of chemicals maternally administered (5,6,7). Mutations for 8-AZ resistance and uovain-resistance were used as markers. Since the system permits the intervention of several modulating processes of injected chemicals, we may expect the presence of threshold, also in this experiment, at low dose range. Inui examined dose mutation frequency relation for several mutagens, e.g. X-rays, AF-2, sodium nitrite, aminopyrine, Trp-P-1, Trp-P-2 and DMN. Linear relations were observed after the treatment of X-rays and two strong mutagens, Trp-P-1 and Trp-P-2. In contrast, AP, $NaNO_2$ and DMN showed exponential type of dose response curves and mutation frequencies of lower dose treatment group fell into the error fluctuation range of the non-irradiated control.

DISCUSSION

Studies with silkworm germ cells suggested that the existence of threshold in dose-mutation response curves of germ cells might be rather common phenomenon for most chemical mutagens, except for some strong ones. The same idea could be applied even to the case of man. Experiments on hamster embryos with transplacental administration of chemicals appeared to support this generalization.

Whether or not threshold is present, or how large is the observed dose range of no significant effect may vary depending on the activities of intervening processes in the test animal; *i.e.* metabolic activities, inhibition of penetration, scavenging activity and repairability of lesions, *etc*. Some of these mechanisms may operate independently of the administered dose but the effects will manifest themselves only at lower dose range showing no or negligibly small effects.

Furthermore, it is almost certain that there may exist dose-rate effect or concentration effect in the process of chemical mutagenesis. This is very important when we estimate chemical risks to man. Conversely, when we detect the mutagenicity of a chemical substance and estimate the width of no effect dose range, we may be able to reveal, by using this phenomenon as a clue, unknown mechanism(s) that act antagonistically to mutation induction. In these context, the author would like to stress the importance and significance of the study of apparent threshold.

REFERENCES

1) Amacher, E. D. et al. *Mutation Res. 64*, 391-406 (1979)
2) von Borstel, R. C. *IAEA-SM-202/511*, 361-368 (1976)
3) Clive, D. et al. *Mutation Res. 59*, 61-108 (1979)
4) Duncan, M. E. et al. *Mutation Res. 21*, 107-118 (1978)
5) Inui, N. et al. *Mutation Res. 41*, 351-360 (1979)
6) Inui, N. et al. *Proc. Japan Acad. 55B*, 286-289 (1979)
7) Inui, N. et al. unpublished data
8) Kuroda, Y. *Res. Rept. to Ministry of Education, Science and Culture B34-R20-2*, 10-14 (1979), *B51-R20-25*, 13-16 (1980), *B109-R20-6*, 12-14 (1981)
9) Okada, S. et al. *Ibid. B34-R20-2*, 21-25 (1979), *B-51-R20-25*, 25-28 (1980), *B109-R20-6*, 17-19 (1981)
10) Murakami, A. *Ibid. B34-R20-2*, 34-37 (1979), *B51-R20-25*, 41-44 (1980)
11) Murakami, A. *Mutagens and Toxicology 4*, in press

Thresholds and Negative Slopes in Dose-Mutation Curves

R. C. von Borstel

Basel Institute for Immunology, Grenzacherstr, Switzerland, and Department of Genetics, University of Alberta, Edmonton, Alberta, Canada

ABSTRACT   Three characteristics of dose-mutation curves are considered: (1) There is a threshold for induced mutations. It is the background noise created by the spontaneous mutation rate. (2) Partial-target mutagenesis reflects either different or changing sensitivities in the cells being exposed. (3) The negative slope found at high doses in dose-mutation curves can be attributed to the heterogeneity of a cell population with respect to mutagen sensitivity or to differential saturation by the mutagen of a mutagenic DNA repair system.

INTRODUCTION

Ever since Marcovich [14] plotted the induction of temperate bacteriophage by X-irradiation, it has become conventional to plot the mutation frequency against the dose on log-log plots where the kinetics of mutation induction can be seen by inspection. Such curves have a slope of one or two, indicating that either one or two events are required in order to obtain a mutation. When the slope is three or more, the usual interpretation is that three or more events are required. For instance, the incidence of colon cancer in human beings, when plotted against age, has a slope of six, and it has been suggested that six events are required to induce this type of carcinoma [6]. Similarly, one can make a case for seven events to induce stomach cancer and fourteen to induce cancer of the prostate [20].

THE APPARENT THRESHOLD

The spontaneous mutation rate itself sets a threshold (1, 5) which Tazima (22) calls the apparent threshold. The spontaneous mutation rate is the background noise and induced mutations are not recognizable as such until they rise significantly above this level.

The variability of the spontaneous mutation rates from population to population tells us that they are under intrinsic control, therefore heritable and subject to natural selection (2). Undoubtedly, repair mechanisms have evolved to process lesions in DNA caused by both intrinsic and extrinsic sources. Even with DNA repair systems being present, induced mutations cannot be completely eliminated. The presence of the DNA repair systems simply requires that higher doses of a mutagen be applied before the mutations can be observed above the apparent threshold. The balance of mutagenic (error-prone) and nonmutagenic (error-free) repair processes sets the spontaneous mutation rate.

PARTIAL-TARGET KINETICS

Another, more obscure kind of threshold is occasionally seen - the partial target. By definition, partial-target kinetics is observed when the dose-response for an event (e.g. mutation, chromosomal aberrations, or cell transformation) exhibits a slope of less than one on a log-log plot of events/surviving cell against the dose. Partial-target kinetics have been observed in different cell types after exposure to chemical mutagens or radiation (3). It is useful to present some mechanisms.

(1) A population that is a mixture of mutagen-sensitive and mutagen-resistant cells will occasionally exhibit partial-target kinetics over the transition part of the curve, i.e., between the part where the slope is that of the sensitive cells and the final part of the slope is that of the resistant cells (3). This presumes that the cells sensitive to mutation induction are also sensitive to killing.

(2) If there is an inducible nonmutagenic DNA repair pathway, as the dose increases, more and more lesions in the DNA are channelled into the newly induced nonmutagenic repair route, thus flattening the mutation response curve.

An inducible error-free repair system has been described for bacteria (19) and it is likely that one exists for mammalian cells as well (15, 17). The double-strand-break repair pathway in yeast acts through chromosome recombination and this nonmutagenic pathway is inducible also (8).

(3) The mutagenic repair route is saturated with lesions to be repaired before the error-free routes become saturated. The lesions are then channelled into the nonmutagenic repair systems, and the induced mutations would tend to reach a plateau.

(4) There may be a progressive inhibition of a mutagenic repair pathway by a mutagen.

Glickman (9) pointed out that one type of mutation induced in E. coli increases with dose disproportionately in frequency when compared with other types of mutations. If the kinetics of the total mutation frequency exhibits a slope of one on a log-log plot, then if one type increases at the expense of a second type, then the second type of mutation must follow partial-target kinetics. Such a phenomenon could come about from a mechanisms listed above.

Calleman (7) has suggested that an activation system could be saturated with a premutagen, and the mutations or cancers would reach a plateau. This is an example of the formal equivalence of activation processes and DNA repair processes (4). It has been proposed that chemical reactions might take place which would increase the presence of a mutagen at something less than that proportionate to the quantity of the premutagen (3) and Calleman's suggestion is one way that the dosage could be altered.

NEGATIVE SLOPES IN DOSE-MUTATION CURVES

When mutations are induced in cells, the initial increase often reaches a plateau which is followed by a decline in mutation frequency (13). A satisfactory explanation for this phenomenon was provided by Pittman (18). He demonstrated that when the cell population is heterogeneous with respect to sensitivity to X-irradiation, then the mutations are induced first in the sensitive group of cells. This group is killed because they are more sensitive to killing, causing a decline in the frequency of mutants, and a secondary rise in mutant frequency is caused by mutations being induced in the resistant population. A similar phenomenon was observed by Zimmermann and Schwaier (23; cf. 21) with N-nitroso-N-methylacetamid. Equations for handling such data have been provided by Oftedal (16) and von Borstel (1).

Ultraviolet radiation and most chemical mutagens do not act differentially on various stages of the cell cycle. Yet in those experiments which have used high doses, the mutational increase is very often followed by a decline. In their development of mutational yield equations for ultraviolet irradiation experiments on yeast, Haynes and Eckardt (10, 11, 12) have termed differential probabilities for clone formation by mutant and nonmutant "delta effects." It is not clear how a mutated cell would know that it was mutated and thereby survive less well than nonmutant cells (or survive better). They carefully used a formal approach and thereby avoided the problem.

A possible mechanism should be considered to explain the negative slopes in mutation induction curves in homogeneous cell populations: at higher doses of a mutagen there may be a differential saturation of a mutagenic repair process. Thus lesions formerly channelled into the mutagenic repair system would be handled almost completely by nonmutagenic repair systems. Such a mechanism could account for an effect which is occasionally observed. At higher doses the induced mutation frequency actually decreases to a level below the level of spontaneous mutations. Thus at higher doses, the spontaneous lesions themselves are channelled into nonmutagenic repair systems.

From the above it is obvious that all the biological mechanisms used to explain partial targets will also explain negative slopes of survival curves.

Acknowledgements: I am grateful to C.M. Steinberg for his critical reading of the manuscript.

(1) von Borstel, R.C. in *The Biological Basis of Radiation Therapy* (E.E. Schwartz, ed.) pp. 60-125, Lippincott Co., Philadelphia (1966)
(2) von Borstel, R.C. *Japanese Jour. Genetics 44* (Suppl.) 102-105 (1969)
(3) von Borstel, R.C. in *Biological and Environmental Effects of Low-level Radiation*, Vol. I, pp. 361-368, International Atomic Energy Agency, Vienna (1976)

(4) von Borstel, R.C. and Hastings, P.J. in *DNA Repair and Mutagenesis in Eukaryotes* (W.M. Generoso, M.D. Shelby, and F.J. de Serres, eds.) pp. 159-167, Plenum Press, New York (1980)
(5) von Borstel, R.C., Moustacchi, E., and Latarjet, R. in *Late Biological Effects of Ionizing Radiation*, Vol. II, pp. 277-290, International Atomic Energy Agency, Vienna (1978)
(6) Cairns, J. *Cancer: Science and Society*, W.H. Freeman and Company, San Francisco (1978)
(7) Calleman, C.J. *Proc. Third Intern. Conf. Environ. Mutagens, Tokyo and Misima*
(8) Fabre, F. and Roman, H. *Proc. Nat. Acad. Sci. U.S. 74*, 1667-1671 (1977)
(9) Glickman, B.W. *Proc. Third Intern. Conf. Environ. Mutagens, Tokyo and Misima*
(10) Haynes, R.H. and Eckardt, F. in *Radiation Research, Proc. Sixth Intern. Cong. Radiation Res.* (S. Okada, M. Imamura, T. Terasima, and H. Yamaguchi, eds.) pp. 454-461, Japanese Assoc. Radiation Res., Tokyo (1979a)
(11) Haynes, R.H. and Eckardt, F. *Can. Jour. Genet. Cytol. 21*, 277-302 (1979b)
(12) Haynes, R.H., Eckardt, F., and Kunz, B.A. *Proc. Third Intern. Conf. Environ. Mutagens, Tokyo and Misima*
(13) Hollaender, A. and Emmons, C.W. *Cold Spring Harbor Symp. Quant. Biol. 9*, 179-186 (1941)
(14) Marcovich, H. *Nature 174*, 796 (1954)
(15) Montesano, R., Brésil, H., and Margison, G.P. *Cancer Res. 39*, 1798-1802 (1979)
(16) Oftedal, P. *Mutation Res. 1*, 63-76 (1964)
(17) Pegg, A.E. *Biochem. Biophys. Res. Commun. 84*, 166-173 (1978)
(18) Pittman, D.D., *Jour. Bacteriol. 71*, 500-501 (1956)
(19) Samson, L. and Cairns, J. *Nature 267*, 281-283 (1977)
(20) Scherer, E. and Emmelot, P. in *Carcinogenesis: Identification and Mechanisms of Action* (A.C. Griffin and C.R. Shaw, eds.) pp. 337-364, Raven Press, N.Y.(1978)
(21) Schwaier, R. *Z. Vererbungsl. 97*, 55-67 (1965)
(22) Tazima, Y. *Proc. Third Intern. Conf. Environ. Mutagens, Tokyo and Misima*
(23) Zimmermann, F.K. and Schwaier, R. *Z. Vererbungsl. 94*, 261-268 (1963)

International Cooperation and Problems in Developing Countries

Problems of Environmental Mutagens and Carcinogens in Developing Countries: Introduction to Panel Discussion

Per Oftedal

Institute of General Genetics, University of Oslo, Oslo, Norway

The title of the panel discussion we are about to begin is "Problems of Environmental Mutagens and Carcinogens in Developing Countries". The reasons for this gathering are quite obvious, and training courses and encouragement of research and control activities in mutagen identification and containment has been high on the priority list of the International Association as long as it has existed.

The reasons behind this interest are many, but let me mention a few:
a) Mutagen research has been during the last few years a lively sector of the biological sciences, with implications for our understanding both of developmental and evolutionary questions. Therefore it is important that younger nations should become familiar with this scientific area.
b) Many of the developing countries have special conditions in their climate, environment, fauna and flora, where natural and artificial mutagens may be in need of identification and control.
c) The impact of chemical techiques in agriculture and pest control may have a much more extreme and serious aim and effect in the virginal situations of developing countries.
d) And finally, developing countries are less prepared to exercise the quality control and the environmental monitoring which has been built up in industrial countries over periods of development.

In summary, the capacity for environmental monitoring and control of mutagens/carcinogens are part and parcel of the creation of an independent and autonomous national or regional strucure. This is important in relation to the demands that the integration of new element into the modern world society carry with them.

The role of the IAEMS has been to a large extent expressed as organizer of training courses in test techniques. Any mention of this activity would be incomplete without the name of Dr. Alexander Hollaender being included. Over the years he has continued to initiate and organize and help to finance dozens of courses, lectures, visits by experts, in all corners of the world. He has excerted a great influence through his efficient and never-tiring capacity for knowing just what to do when, in order to start the ball rolling, and to keep it going untill the task has been completed. I think he must have friends and find gratitude in practically every country of the world. At one time or another, he has been in contact, tested the possibilities, and if at all feasible he has got something organized.

I know our first three speakers - Dr. de Nava from Mexico, Dr. Lim-Sylianco from The Philippine, and Dr. Smutkupt from Thailand - have all had much of their stimulus and encouragement from Dr. Hollaender. I know also that Dr. Ramel - who will give an overview of the activities as seen from the more advanced modern research institute - has had much scientific inspiration and benefit from his contacts with Dr. Hollaender.

Dr. Higginson - who is speaking on behalf of IARC has had less contact with Dr. Hollaender - and he has missed something good. That probably goes to some extent for the representative of UNEP - Dr. Leaca from Bankok as well, but then Dr. Hollaender has been an environmental Protection Programme all by himself.

Evaluation of Short-Term Tests for Mutagenicity in the International Program for Chemical Safety

Frederick J. de Serres

National Institute of Environmental Health Sciences, Research Triangle Park, North Carolaina, U.S.A.

The International Program for Chemical Safety (IPCS) is sponsored by the World Health Organization, the United Nations Environment Program and the International Labor Organization. The first meeting of the Program Advisory Committee was held in April 1980 and a tentative work plan was developed. The objectives of IPCS are as follows:

1. To evaluate the effects of chemicals on human health and the environment,
2. To develop guidelines on exposure limits of chemicals in air, food, water and the working environment,
3. To develop methodology for toxicity testing, epidemiological and clinical studies, and risk assessment,
4. To coordinate laboratory testing and epidemiological studies when an international approach is appropriate,
5. To promote research on dose-response relations and on mechanisms of biological action of chemicals, and
6. To develop information for coping with chemical accidents, to promote technical cooperation on control of toxic substances in Member States, and to promote training and development of manpower.

The IPCS became operational in June 1980 when a separate Central Unit was established at WHO headquarters in Geneva, operating under WHO procedures. One of the outputs of IPCS in the areas of mutagenesis and carcinogenesis will be the evaluation of specific effects of chemicals as well as the development of guidelines on methodology for measurement of the effects. The first meeting of the IPCS Working Group on the Application of Short-term Tests to Predict Mutagenic and Carcinogenic Potential was held in Geneva, April 30 - May 1, 1981. The meeting of the Working Group was convened by the IPCS Central Unit to discuss and make plans for projects which address the IPCS general goal of developing and promoting appropriate methodology for identifying genotoxic agents, and determining the predictive value of short-term tests for chemical carcinogens. A project submitted by the National Institute of Enviromnental Health Sciences (NIEHS) calls for an international collaborative testing study to identify in vitro assay systems which reliably detect genotoxic chemicals difficult to detect in bacterial mutation assays.

Project I, Part 1 - Short-term In Vitro Tests

An extensive international collaborative research study called the International Program for the Evaluation of Short-term Tests for Carcinogenicity (IPESTTC) was carried out between 1977 and 1979. A report of the results obtained in this program has been published[1] in which tests on 42 coded samples by some 60 scientists using 30 different assay systems are described. Three broad conclusions from the study were (1) bacterial mutation assays are good predictors of chemical carcinogenicity, (2) in vitro eukaryotic assays provide valuable complementary test results and (3) in vivo assays are not good candidates as first screens for mutagenic activity which may be indicative of carcinogenic potential but may be more valuable in resolving inconclusive results from in vitro tests. Additional data from IPESTTC indicate that in vitro tests for transformation, chromosome damage and gene mutations may serve to detect genotoxic agents which are not detected by bacterial mutation assays.

OBJECTIVE   The objective of the proposed IPCS study is to identify in a battery testing approach, one or more in vitro eukaryotic tests which complement bacterial mutation assays by reliably detecting chemical carcino-

---

[1] Evaluation of Short-term Tests for Carcinogens: Report of the International Collaborative Program. F. J. de

gens which are difficult or impossible to detect with
bacterial assays. Although it is not necessary that the
complementary assays detect the same carcinogens, it is
important that they have appropriately high sensitivities
(ability to show a positive response with known carcinogens) and high specificities (ability to show a negative
response with known noncarcinogens). It is further
recognized that such complementary assays cannot be
selected strictly on the basis of the proposed study.
It is anticipated that other sources of data (e.g.
IPESTTC and the US Environmental Protection Agency's
Gene-Tox Program) will be utilized.

TEST CHEMICALS

Two groups of chemicals were chosen for testing.
The first group consists of 8 carcinogens that are
known to be difficult to detect in short-term in vitro
mutagenicity assays, particularly bacterial assays. The
second group of chemicals that was selected have been
judged noncarcinogenic following what is currently
considered to be adequate testing. The 8 carcinogens
are 4-dimethylaminobenzine, o-toluidine, hexamethyl-
phosphoramide, safrole, 3-aminotriazole, urethane,
formaldehyde and acrylonitrile. The 3 chemicals with
negative evidence of carcinogenicity are benzoin,
caprolactam and isopropyl N(3-chlorophenyl) carbamate.
All of these test chemicals are available commercially
and it is proposed that a single batch of each chemical
be purchased. After ascertaining chemical purity of each
batch to determine the presence of any impurities, appropriate quantities of test chemicals will be sent to all
participating laboratories.

The advantages and disadvantages of requiring that
the short-term test be conducted with investigators not
knowing test chemical identities were discussed at
length by the Working Group. Despite the potential for
bias in data collection and interpretation, it was the
general feeling of the Working Group that problems with
shipping, laboratory safety and testing justified
identifying test chemicals for investigators.

ASSAY SYSTEMS

The Working Group felt that the selection of assays

Serres and J. Ashby (EDS), Elsevier North-Holland, New
York, NY, 1981.

should strongly emphasize, but not be limited to, eukaryotic in vitro assays since the objective of the project is to iedentify one or more short-term assays to complement the bacterial mutation systems in identifying chemical carcinogens. Furthermore, since much of the groundwork for the proposed study has been laid down in the IPESSTC, and in order to keep the study to a manageable size, it was decided that the specific assay system would be selected from those (1) which were conducted in the IPESTTC and (2) with additional endpoints which have a demonstrated or anticipated likelihood of detecting genotoxic activity of the chemicals under test. The assay systems proposed are listed in Table 1.

TABLE 1

INTERNATIONAL PROGRAM ON CHEMICAL SAFETY
ASSAYS PROPOSED FOR IN VITRO STUDY

Project I - Part 1

Transfromation
- BHK-21
- C3H 10T1/2
- BALB/c 3T3
- Syrian Hamster Embryo
- Viral Enhancement

Yeast
- Nondisjunction
- Gene conversion/mitotic recombination
- Forward and reverse mutation

Drosophula
- Sex-linked Recessive Lethals
- Chromosome Translocations

Mammalian Cells in Culture
- Gene Mutations
- Unscheduled DNA Synthesis
- Chromosome Aberrations
- Sister-chromatid Exchange

PROTOCOLS

All investigators will be required to submit a protocol prior to testing. Advantage will be taken of recent testing experience in the IPESTTC as well as published protocols such as those resulting from the US

EPA Gene-Tox Program. Initial protocols will be submitted to all prospective participants for review and revision prior to final protocol recommendation.

Project I, Part 2 - Short-term In Vivo Tests

The organization and planning of a second study to evaluate short-term in vivo tests for genotoxic agents was discussed by the Working Group. The purpose of the study would be to determine the responses of short-term in vivo genotoxicity tests to two pairs of chemical analogs that show large potency differences in in vitro assays, plus an additional 2 chemicals. The chemicals recommended for study are shown in Table 2. The assay systems proposed for the in vivo study are given in Table 3.

TABLE 2
INTERNATIONAL PROGRAM ON CHEMICAL SAFETY
OBJECTIVE

Project I - Part 2
Determine the responses of short-term in vivo genotoxicity tests to pairs of chemicals that show large potency differences in in vitro assays.

Chemical	IPESTTC Results	
	In Vitro	In Vivo
Benzo(a)pyrene	+++	+
Pyrene	+	-
2-Acetylaminofluorene	+++	+
4-Acetylaminofluorene	+	-
Hexamethylphosphoramide	-	+
O-Toluidine	+	-

ORGANIZATION AND FUNDING

Coordination of Project I will be divided between the Manager of IPCS at WHO (Professor M. Mercier) in Geneva and the Chairman of the Working Group (Dr. F. J. de Serres) at NIEHS in the Research Triangle Park, NC. The IPCS Central Unit in Geneva will be responsible for overall management, whereas the Chairman of the Working Group will assume responsibility for coordination among participating scientists.

TABLE 3
INTERNATIONAL PROGRAM ON CHEMICAL SAFETY
PROPOSED ASSAYS IN VIVO STUDY

Project I - Part 2
    Drosophila
        Sex-linked Recessive Lethals
    Mouse
        Spot-test
        Chromosome Aberrations
        Micronucleus Test
        Dominant Lethal Test
        Sister-chromatid Exchange
        Sperm Morphology
        Unscheduled DNA Synthesis in Germ Cells
        Alveolar Macrophage
    Hamster
        Pouch Granuloma
    Rat
        Lung Ademona

Financial support for individual investigators will be through the lead institutions in the different Member States. The IPCS Central Unit will be primarily responsible for the purchase and destribution of test chemicals and for supporting required travel and meetings.

Problems of Environmental Mutagens and Carcinogens in Developing Countries: Synthesis of the Problems

C. Ramel

Wallenberg Laboratory, University of Stockholm, Sweden

Although it is almost four decades since it became clear that chemicals can give rise to mutations, it has been a slow process to demonstrate the implications of that finding with respect to human health. It is not until the last decade that the magnitude of the problems really has been fully anticipated. Thanks to the development of new and sensitive screening methods it has become clear that man is exposed to a wide variety of mutagens and carcinogens, which presumably are responsible for at least the vast majority of cancer in humans. The recognition of which chemicals - man made or natural - that are of particular importance in that respect is a general problem as relevant for industrial as non-industrial countries.

Even if the basic practical problems in chemical mutagenesis and carcinogenesis no doubt are universal, there are certain aspects and problems of particular relevance for developing countries. I will try to give a survey of such aspects.

It seems to me appropriate first of all to emphasize that many problems in this area in the developing countries are, to a great extent, a question of scientific communication. Scientific front line discussions take place in international journals or conferences, which formally are open to any country. However, in reality we all know that comparatively few countries have the economic and scientific capacity ot contribute to and fully exploit such channels of communication. The result of course is that new scientific information diffuse very slowly or not at all to many non-industrialized countries. It is easy to foresee that such a lack of communication may have fatal consequences in the present context when one is discussing chemical mutagens and carcinogens in the human environment. It is obvious that industrialized countries in this case have a responsibility beyond ordinary and more or less passive information service. The international courses and workshops that Dr. Alexander Hollaender has initiated under the auspieces of IAEMS, UNEP and other international organizations, therefore fulfil a fundamental function in this respect.

The art and magnitude of the problems in the developing countries is often difficult to grasp for several reasons. In many countries epidemiological investigations are next to impossible to perform and only the most dramatic effects at the population level can be traced. Monitoring of the environment for mutagenic and carcinogenic chemicals require expertize and equipment, which simply does not exist in developing cuntries and the control of chemicals used for various purposes is highly fragmentary. The danger of chemicals is often not known at all and the use of such chemicals consequently not questioned or even recorded.

The moral responsibility for the exposure of developing countries to dangerous chemicals very often falls on industrial countries responsible for the manufacturing and the export of chemicals. There are numerous examples of ruthlessness and thoughtlessness from industrialized countries. Thus the export of methyl mercury dressed wheat to Iraque 1971 was a grave mistake, the consequence of which anybody with some knowledge of the effects of methyl mercury could have suspected. The situation with other pesticides is not much better. Agricultural pests are of course a fundamental problem particularly in those developing countries, where the crop return may mean life or death. Chemical companies sometimes take advantage of this situation in a rather disgusting fasion and enforcing an undiscriminated use of pesticides beyond any objective reason. This problem was subjected to a vivid and highly interesting discussion during the workshop in Manila last year. Particularly Dr. de Nava gave a well-informed and frightening insight in the situation in Latin America concerning the use of chemical pesticides. According to her chemical companies even manage to sell pesticides against pests which do not exist in that part of the world.

During the work shop in Manila participants from different Asiatic countries witnessed the difficulties to control the activities of the chemical industries in general. The legal regulation of the developing countries usually lag behind the industrial countries in this respect. In the industrial countries the chemical industries are surrounded by more and more restrictions. It is then tempting to move out the whole production to developing countries with less rigorous pollution control and occupational health regulations, rather than to alter the production in accordance to domestic regulatory requirements. In the developing countries on the other hand the establishment of new industries imply a possibility to improve a hopeless economy and severe unemployment situation - two vital problems in the third world. Under such circumstances it is difficult to maintain strict health and pollution control.

The variation in cancer incidences between different countries and populations point to the fact that the vast majority of human cancer depends on environmental factors. The data however indicate that man made chemicals are not the principle source of cancer induction, but rather food and living conditions. A factor of particular importance in tropical countries is mycotoxins and especially aflatoxin from moulded nuts and seeds. The high incidence of liver cancer in some African and Asian countries apparently can be traced back to the exposure to mycotoxins from bad storage conditions of the crop. The storage facilities in its turn is influenced by economical, political and other factors. Perhaps the increased liver cancer reported by some scientists in Vietnam after the war may depend on increased aflatoxin exposure rather than the exposure to the defoliation chemical "Agent Orange" with 2,4,5,-T and tetrachlorodioxin.

During the last few years a wide variety of naturally occuring mutagens and carcinogens have been found in ordinary food items. The health hazard involved with such chemicals are difficult to evaluate at present for several reasons. The exposure to them presumably vary greatly between different populations, which magnify the problems considerably.

No doubt the situation concerning the health hazard from both synthetic and natural genotoxic chemicals in developing countries calls for some action. However no action will occur without the appropriate background knowledge among responsible authorities in the countries themselves. It is difficult for the developing countries to collect this information in isolation by themselves but this requires a collaboration with industrialized countries. As I emphasized before the international work shops and laboratory courses, which have been organized in developing countries during the last few years are the most important means of disseminating basic information for further practical use and regulatory actions. The way such work shops and training courses are organized is of course of great importance for their efficiency and this has to be under constant consideration when organizing these activities. I will discuss some points in this field from my own experience.

It is in the first place necessary to establish what can realistically be achieved in developing countries, considering the limited resources available. The possibilities for most of these countries to perform biological tests are very limited and restricted to such tests which do not require sophisticated equipment. The primary source of information concerning genotoxic risks of chemicals inevitably will therefore not come from local mutagenicity screening but from literature data on chemicals found in the environment. The purpose of laboratory traning and demonstrations is, in my opinion, not in the first place to encourage a large scale screening of chemicals, but rather to get the participants acquainted with various test systems and how to critically interpret test results. This is of importance in order to understand and evaluate literature data. The training courses therefore have to cover a wide range of test systems, even if these test systems never actually are put up by the participants.

It is however clear that there is a need of screening of chemicals also by developing countries themselves. This need does not in the first place concern individual chemicals, for which literature data already exists. Instead it seems to me more important to apply the screening to complex environmental samples - from industries, air and water pollution and food items. Such a screening will make it possible to identify hazardous environments for further investigations and control.

The most useful test system for this screening is Ames test with Salmonella or other bacterial tests. The procedure has been facilitated by the fact that rat S9 mix is now available commercially.

Although bacterial tests no doubt are the most practical ones for a prescreening of environmental chemicals it has to be emphasized that they are not sufficient for a final identification of genotoxic risks. Other test systems have to be applied for a verification. For chromosomal aberrations the most practical and sensitive in vivo test system in my opinion is the micronucleus test on erythrocytes in mice. The scoring of micronuclei can be done with a minimum of training and it is easy to obtain sufficient material. Other widely used test systems such as mammalian cell culture and Drosophila require more experience but can of course be applied by laboratories with appropriate experience and equipment.

The selection of test systems and the amount of screening to be done will inevitably be guided by available resources. It is however essential to stress that unless the screening is done in a proper way it may do more harm than benefit. This applies for instance to the size of the experiments. The literature is swarmed with data which is of very little value because of totally unsufficient material. In practice the most obvious statistical problems pertain to megative data. Even the most potent mutagen may be negative in a small experiment and therefore one always has to be aware of the risk for false negative data because of too limited material, that is, the β- error or type B error in statistical terms. A negative result cannot be evaluated without knowing the confidence limits or which maximum effects the results have excluded. In the large international journals one is aware of these aspects, but in more local publications serious statistical mistakes of this nature may escape discovery before publication.

The work shops in developing countries have usually had a dual purpose. Firstly the aim has been to give a basic theoretical and practical knowledge in the field of genetic toxicology. Secondly they have functioned as scientific symposia. This coupling of educational and scientific presentations have, according to my experience, been valuable, but it has put some demand on basic knowledge particularly in genetics and molecular biology among the participants. This has to be taken into consideration in selecting participants and also for the actual organization of the courses. Firstly reasonable requirement for participants seems to be a knowledge equivalent to a master's degree. Secondly and more critical the participants should have possibilities to prepare for the course at an early stage. It is therefore desirable that the work shops are organized sufficiently in advance for the participants to make appropriate completion of their knowledge. This would be greatly facilitated if written course material could be put to their disposition far ahead of the work shop. Considering the rapid increase of such work shops and courses the time may have come to publish a book specifically designed for these courses for developing countries.

I would finally like to emphasize that the collaboration between industrialized and developing countries in the field of genetic toxicology should not stop after a work shop. As important but perhaps more difficult and less glamorous is the following up of the work shop in the different countries - that is to provide them with further help and advice. A mechanism for that side of the activities is still largely lacking, but it certainly deserves some discussion.

Southeast Asian Conference on the Detection and Regulation of Environmental Mutagens, Carcinogens and Teratogens

Clara Y. Lim-Sylianco

Department of Chemistry, University of the Philippines, Quezon City, The Philippines

There is an emerging awareness on environmental mutagens, carcinogens and teratogens in the Philippines. Growing awareness was triggered by the Southeast Asian Conference on the Detection and Regulation of Environmental Mutagens, Carcinogens And Teratogens which was held at the University of the Philippines in October 1980 which was participated by Japan, United States of America, Switzerland, Sweden, Mexico, Korea, China, Malaysia, Thailand, Singapore and Indonesia. The problem is how to sustain this awareness and dissipate this to reach a greater portion of populace in the Philippines including those involved in regulatory mechanisms.

In the Philippines, awareness on environmental mutagens, carcinogens and teratogens is only emerging. Growing awareness was triggered by the Southeast Asian Conference on the Detection And Regulation of Environmental Mutagens, Carcinogens And Teratogens. This was held at the Asian Institute of Tourism and the Natural Science Research Center, University of the Philippines, Diliman, Quezon City, October 19-31, 1980.

The conference was realized because of very inspiring and very effective help and guidance of an advisory council composed of:

    Dr. Alexander Hollaender
    Dr. Yataro Tazima
    Dr. Walderico M. Generoso

Dr. Tsueno Kada

Financial assistance was given by the following institutions:
1. United States Agency for International Development/ Man and the Biosphere Programme
2. United States National Cancer Institute
3. Japan Society for the Promotion of Science
4. United Nations Environmental Programme
5. Sandoz A.G., Basel, Switzerland
6. International Union of Pharmacology, Toxicology Section
7. University of the Philippines
8. National Research Council of the Philippines
9. National Environmental Protection Council of the Philippines
10. Philippine Environmental Mutagen Society

Lectures and laboratory workshops on methodology were handled by faculty participants composed of the best minds from Japan, United States of America, Switzerland and Sweden.

There were eight faculty participants from Japan:
Dr. Tsueno Kada
Dr. Yukiaki Kuroda
Dr. Taijiro Matsushima
Dr. Masaaki Moriya
Dr. Sayaka Nakai
Dr. Masao Sasaki
Dr. Yasuhiko Shirazu
Dr. Yataro Tazima

There were four faculty participants from the United States of America:
Dr. Jeanne N. Manson
Dr. Walderico M. Generoso
Dr. Michael J. Prival
Dr. James Selkirk

Four faculty participants came from Europe:
Dr. Peter Maier   Switzerland
Dr. Bernhard E. Matter   Switzerland
Dr. Claes Ramel   Sweden
Dr. Friedrich E. Wurgler Switzerland

The number of participants in lecture sessions and those trained in the laboratory workshops are shown:

	Lecture	Laboratory Trainees
Philippines	150	14
Japan	8	-
United States of America	4	-
Peoples Republic of China	2	2
Indonesia	1	1
Korea	1	1
Malaysia	1	1
Mexico	1	-
Singapore	1	1
Thailand	1	1
Sweden	1	-
Switzerland	3	-

The lectures given by the faculty participants were on the following topics:
1. Toxicology In Relation To Human Health
2. Basic Principles of Mutagenesis, Carcinogenesis and Teratogenesis
3. Reviews on Methodology
4. Relationship Between Mutagenesis, Carcinogenesis and Teratogenesis
5. Mutagens, Carcinogens and Teratogens In The Environment
6. Comutagenesis and Cocarcinogenesis
7. Antimutagens
8. Naturally Occuring Mutagens
9. Toxicological and Risk Evaluations
10. Regulatory Mechanisms
11. Regulatory Guidelines

The laboratory workshops conducted by the faculty offerred training for two weeks on the following methods:
1. Rec Assay
2. Ames Test
3. Host-Mediated Assay
4. Micronucleus Test
5. Tests Using Drosophila
6. Cytogenetic and Somatic Cell Genetic Tests
7. Dominant Lethal and Heritable Translocation Tests
8. Tests Using Yeast

Most of those who attended the conference from the Philippines were teachers from colleges from all over the country. Three regulatory agencies were represented; Food and Drug Administration, Fertilizer and Pesticide Authority and the Ministry of Health.

Although growing awareness on environmental mutagens, carcinogens and teratogens has spread to several classrooms all over the country, reaching out to those involved in regulatory mechanisms is still a problem.

The Philippines is only starting to industrialize. Setting up of industries is being encouraged to help improve national economy. Already, several of these new industries are releasing mutagens, carcinogens and teratogens into the air and waterways. Workers are exposed to a number of health hazards continously. No regulations are imposed regarding occupational hazards. Because of ailing financial situations, several industries cannot put up water treatment plants or even simple anti-pollution devices. Although fines are imposed on industries polluting the environment, other means can be used to circumvent regulations and control.

There are regulatory agencies headed by a chosen few who are not well equipped with expertise to be able to assess experimental data on mutagens, carcinogens and teratogens. There are no research arms in these regulatory agencies.

There are attempts to screen pesticides to be marketed and train individuals for application of pesticides. Inspite of this, there are pesticides found in our commercial centers that are banned in other countries. There are those who are untrained who are applying pesticides in banana plantations, rice fields, vegetable and fruit concerns.

There is a regulatory agency that depends on information released from other countries on banned drugs and food additives. Information is not usually updated enough to catch drugs banned elsewhere and used in the country.

Only a handful few are trained to do research on mutagens, carcinogens and teratogens. They are usually connected with educational institutions and are saddled more for teaching than for research.

The Philippine Environmental Mutagen Society for the last three years has been doing its part in spreading information and encouraging more to go into research on mutagens, carcinogens and teratogens. It has gained some progress. The Southeast Asian Conference strengthened its position. Recently, the National Science Development Board and the National Research Council of the Philippines gave financial support to some members of the Society to do research on the mutagenicity potential of Philippine Medicinal Plants. This has been designated as a priority program.

Before the Southeast Asian Conference, there was only one group doing mutagenicity research. After the conference, two more groups emerged.

So much has to be done. The Southeast Asian Conference and the Philippine Environmental Mutagen Society have only touched superficially those involved in regulatory mechanisms.

A Report of the Southeast Asian Workshop on the Methods for Detecting Environmental Mutagens and Carcinogens, February 4 - 17, 1979

Sumin Smutkupt

Department of Applied Radiation and Isotopes, Faculty of Science, Kasetsart University, Bangkok, Thailand

ABSTRACT: Thai people are involuntarily exposed to hundred of imported chemicals, especially pesticides every year. With strong assistance and encouragement from Dr. Hollaender and Dr. Tazima, the Workshop was organized. The main objectives were (1) to acquaint junior research scientists with the principles and methods currently used for the detection of chemical mutagens, and (2) to disseminate the ideas and knowledge of the mutagenic effects of environmental chemicals that may cause genetic hazards to man. Among 120 participants present for the opening, 16 were registered full time. The Workshop was successfully carried out and the objectives were met. Details of the workshop activities were summarized in a report. After the Workshop, the Thai participants formed the Environmental Mutagen Research Group. The members meet approximately every two to three months. Three Thai EMS Newsletters were already published. As a result of our regular meetings, a project proposal on mutagenicity screening of pesticide residues and mycotoxins on agricultural products, food and feed stuff consisting of 17 subprojects was drafted. However, this proposed project needs to be reviewed and should be coordinated and established as a national project. It is hoped that after this panel discussion, a more intensive regional cooperation can be expected, which may lead this matter to setting up a foundation of the Southeast Asian Environmental Mutagen Society as Dr. Tazima and his colleagues have long cherished in mind.

Approximately 127,000 tons (one hundred and twenty seven thousand tons) of pesticides were imported into the country within a ten-year period of 1971 - 1980.

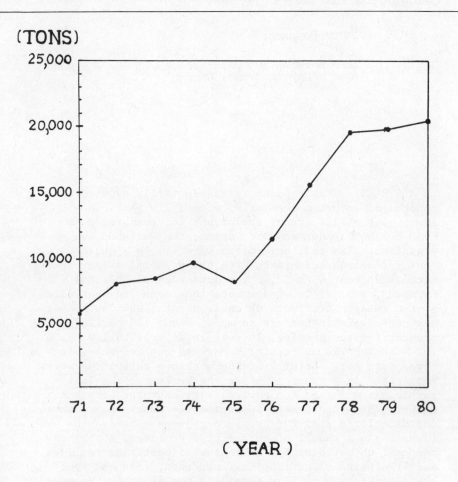

Fig. 1 Importation of pesticides into Thailand 1971 - 1980

Figure 1 shows the sharp annual increase of pesticides imported between 1975 and 1978.

Table 1. Amount and origin of certain pesticides imported into Thailand in 1980 (1).

Name	Year 1976-1980 tons	Year 1980 alone tons	Origin
Aldrin	189.58	20.02	Holland
BHC	2,060.74	775.32	Switzerland, Romania, Malaysia, UK
Chlordane	110.84	23.06	U.S.A.
DDT	8,086.03	1,007.02	U.S.A., China, Romania
DDVP	582.52	91.03	Spain, Japan, Holland, West Germany
Dieldrin	268.35	37.77	Holland
Endrin	478.99	23.21	Holland, U.S.A.
Heptachlor	134.32	36.28	U.S.A.
Toxaphene	4,150.66	860.19	West Germany, Switzerland, U.S.A., UK, East Germany, Nigeria, Netherland
2,4-D	6,624.53	305.36	Hungary, U.S.A., Spain, England
2,4,5-T	830.89	613.46	Canada
Total	23,517.72	3,792.99	14 countries

Table 1 shows the amount and origin of certain pesticides imported in 1980 alone, some of them are known as carcinogenic and all of them were banned in highly developed countries.

The residues of BHC, DDT, Dieldrin and Endrin are given as examples in Table 2. They were found in samples of vegetables, seeds of field crops, rice, meat and fish (2). Besides agricultural commodities, the residues of them were found in soils and water.

Table 2. Certain insecticide residues found in agricultural commodities in Thailand in 1979.

Sample	No. of samples analyzed	BHC		DDT		Dieldrin		Endrin	
		Residue found in sample (%)	Average residue (ppm)	Residue found in sample (%)	Average residue (ppm)	Residue found in sample (%)	Average residue (ppm)	Residue found in sample (%)	Average residue (ppm)
Vegetables	123	28.46	0.003	100	0.014	95.93	0.002	98.37	0.002
Peanut	24	66.67	0.011	100	0.156	75	0.097	75	0.019
Soybean	15	100	0.325	100	0.427	80	0.026	80	0.068
Sesame	37	62.16	0.044	100	0.256	75.67	0.015	51.35	0.012
Mungbean	75	65.33	0.012	100	0.140	93.33	0.114	100	0.147
Rice and Flour	82	30.48	0.004	68.29	0.021	36.58	0.026	17.07	0.002
Meat	157	–	–	77.70	0.192	22.93	0.029	5.09	0.002
Fish	143	–	–	70.62	0.149	11.18	0.011	6.29	0.039

With strong assistance and encouragement from Dr. Hollaender and Dr. Tazima, the Southeast Asian Workshop on the Methods for Detecting Environmental Mutagens and Carcinogens was organized during February 4 to 17, 1979 at Kasetsart University, Bangkok, Thailand.

The main objectives were :

(1) to acquaint junior research scientists with the principles and methods currently used for the detection of chemical mutagens and carcinogens,
(2) to disseminate the ideas and knowledge of the mutagenic effects of environmental chemicals that may cause genetic hazards to man.

At the opening ceremony there were about 120 participants. Approximately 50 - 70 of them attended the morning lectures during the course of the workshop. Sixteen participants were registered full time. They attended lectures in the mornings, laboratories, demonstrations, and discussions in the afternoons. Eleven international scientists and six national staff members served as the Faculty of the workshop. A twelve-day Workshop was successfully carried out and its detailed activities were summarized in a report.

After the Workshop our junior scientists who had recognized the problems of chemical mutagens in the human environment formed the Environmental Mutagen Research group to carry out the following functions :

(1) to detect environmental mutagens and carcinogens in the country.
(2) to exchange scientific information and materials, and
(3) to disseminate research findings to the public. There are 40 members at present.

The members meet approximately every two to three months. Three Thai EMS Newsletters were already published. As a result of our regular meetings, a project proposal on mutagenicity screening of pesticide residues and mycotoxins on agricultural products, food and feed stuff consisting of 17 sub-projects was drafted. However, this proposed project needs to be reviewed and should be coordinated and established as a national project.

In Thailand, we have laws and regulations concerning poisonous substances, such as the Poisonous Article Acts 1967, a Notification of the Ministry of Interior on Safety of Work in Connection with Environment (chemicals) 1977 and others.

The work on toxic substances is involving at least 21 divisions of 14 departments in 5 ministries. They are ministries of Agriculture and Cooperatives, Industry, Public Health, Interior, and Science, Technology and Energy. Besides ministrial organizations, there are 13 universities under the Office of Universities Affairs having work related to toxic substances.

In August this year, a National Committee on toxic substances comprising of 14 members, four of which were members of our EMS group, was appointed by the Government and it is hoped that activities concerning toxic substances of various agencies in the country can be coordinated through this Committee.

It is hoped also that after this panel discussion, a more intensive regional cooperation can be expected, and may lead this matter to setting up a foundation of the Southeast Asian Environmental Mutagen Society as Dr. Tazima and his colleagues have long cherished in mind. Thank you!

REFERENCES

1. Agricultural Regulartory Division, Imported Poisonous Articles 1980, Department of Agriculture, Ministry of Agriculture and Cooperatives, Bangkok, Thailand (1980)

2. Pesticides Research Branch, Pesticide Residues in Agricultural Commodities 1979, Department of Agriculture, Ministry of Agriculture and Cooperatives, Bangkok, Thailand (1979)

Environmental Mutagenesis and Carcinogenesis in Latin America

Cristina Cortinas de Nava

Instituto de Investigaciones Biomédicas, Universidad Nacional Autónoma de México, México, D.F., México

At present, the problems associated with environmental mutagens and carcinogens in Latin America appear less urgent than those related to the availability of food, potable water and energetics, or those raised by demographic growth unemployment and transmissible diseases. However, increasing industrialization, the trend to relocate high risk industries in underdeveloped countries, the uncontrolled use of pesticides and the planned development of uranium mining make necessary the evaluation of the accompanying toxicological risks in order to protect the ecosystem and human health.

When in 1975, Dr. Alexander Hollaender suggested both to Dr. Rogero Meneghini of Brazil and to me that we hold for the first time in our respectives countries a Workshop on Environmental Mutagenesis, we considered it a challenge due to the small number of groups which at that time were interested in this field. Inspite of the fact that we thought that it would be more opportune to hold one workshop for all Latin America, we accepted Dr. Hollaender's proposition and organized two workshops, one in Brazil and one in Mexico which were held in June and December 1977 respectively.

After these first two workshops, others were held in Argentina, Chile and Colombia. An important consequence of the workshops has been the creation of the Latin American Association of Environmental Mutagenesis

Carcinogenesis and Teratogenesis, during the First Latin American Symposium on this topic held in Mexico in June 1980.

A Course in Genetic Toxicology and a Workshop on Epidemiological Methods to Evaluate the Effects of Environmental Pollution on Health took place this summer in Mexico and others on Environmental Mutagens and Carcinogens are planned in Brazil in November 1981, in Peru, Chile and Bogota in 1982 and in Cuba in 1984.

These courses had been possible thanks to the support of: The International Agency for Development/Man and the Biosphere through Dr. Alexander Hollaender, the United Nations Environmental Program, the Panamerican Health Organization and the International Environmental Mutagens Society.

Since the first meetings in 1977, it has been considered indispensible to place the development in this field of research within the context of the problem confronted by each of the Latin American countries. From the beginning it was evident that, even when other urgent needs had to be attended in Latin America, serious problems of environmental pollution did exist that could not be avoided and that were growing due to the growth of industrialization. However, no single strategy to study environmental problems could be designed for all of Latin America since there is no one process of development for all these countries. Therefore, at each workshop, has been necessary to analyse regional problems as well as local resources.

Some Latin American countries have, in the past, achieved great advances in their development which have lead to high levels of education and of scientific, technological, and medical training. The influence of such advances is reflected in the morbidity, mortality, and demographic patterns of these countries, making these patterns similar to those of industrialized countries. Other developing Latin American Countries, are confronted with serious problems, which include a high degree of illiteracy, and accelerated demographic in--

crease, and an elevated rate of morbidity and mortality due to malnutrition and contagious diseases. Among the latter countries is Mexico, where a stage of rapid industrialization has produced problems proper to highly developed nations and which are associated with an increase in degenerative and genetic diseases.

Cuba inspite of its being a developing country, has overcome illiteracy, the high demographic increase, and contagious diseases, due to enormous advances in social medicine such that now the two primary causes of death are cardiovascular diseases and cancer. However, the majority of Latin American countries still have high unemployment and lack the necessities such as potable water, food, and energy. At the same time, some of these countries also face grave problems of pollution such as that caused by the massive use of pesticides and herbicides which are used not only in the agriculture to protect crops, but also to combat the insect vectors of transmissible diseases (Malaria, Chagas) and to destroy weeds from which narcotics are extracted. As a mater of fact pesticides, prohibited in other countries because they are carcinogens, are used throughout Latin America and, in the majority of these countries, there are no regulations concerning their use or, if such regulations do exist, they are not enforced for lack of supervision, ingnorance or corruption.

The great need to industrialize, to produce energy, and to fight unemployement affect our environmental and population health. Brazil is presently the major producer of asbestos in Latin America and just as other countries of the region is disposed to accept the asbestos industry which has been prohibited in those nations which now pretend to export all their polluting industries to third world nations.

Cuba actively exploits its nickel mines and will become in the near future an important exporter of this metal. Mexico will soon develop uranium mining, will establish industries to process it and will build nuclear reactors to meet the energy demands of the country. Like Ecuador and Venezuela, Mexico is one of the principal producers of petroleum and its petrochemical industries have now been extended throughout the country.

The risk associated with exposure to pesticides, asbestos, nickel, uranium and petroleum derivatives are well known and it is well to ask if steps have been taken in our countries to minimize them. It is also well to ask if the international industrial companies which have installations in Latin America meet the regulations required in their home countries to reduce environmental pollution. It is important to know, in turn, if the standards of exposures to chemical agents which are permitted in the idustrialized countries, are applied to our countries and if epidemiological surveillance, not only of the workers but also of the general population, are carried out.

In general, the answer to each of these questions is no and the reasons for this are based fundamentally on; the type of sociopolitical systems in power in Latin America, on the powerful economic interests of the international industrial companies, and on the lack of information and awareness of the community in general. What is even more serious is that, should these countries at this very moment, wish to effect a substantial change in this situation the necessary means - personnel, equipment, technology would not be available.

In view of the above, it is evident that whoever is insterested in studying environmental problems in Latin America must have an integrated vision of the problems that exist in each country or region. Due to the limited human, material, and technological resources, an improved coordination of effort to increase the efficiency and the relevance of the research is indispensible. The development of multidisciplinary projects is also of major importance. It is necessary to direct attention to the most urgent problems, this implies an establishment of priorities. Since it is known, that some of these problems may be solved or minimized based on present information, the first step is to gather the information, analyse it, and translate it into recommendations which can be applied to each situation.

We are aware that we must not respond emotionally and that it is not desirable to stop the development of our countries by closing all the industries that represent a risk. But at the same time, we feel that it is our responsability to bring to light those elements which exist at present to prevent or halt the above mentioned risks. Monitoring of the environment, identification of new agents which represent a risk, epidemiological surveillance, and development of technology appropiate to solving the problems which do exist, all present themselves as necessities. At the same time, it is urgent that we greatly increase the number of individuals capable of addressing themselves to such aspects of the problem, by diffusion of information and by expanded training programs. To achieve these ends, it is also important that active communication and collaboration between the different sectors -researchers, educators, businessmen, authorities and the community in general be maintained.

The Latin American Association over which I preside, decided from its beginning to include Carcinogenesis and Teratogenesis in its title to draw attention to the wide scope of its interest in addition to that of environmental mutagenesis. Now, we know that we should also integrate our efforts with all those groups from Latin America which are interested in solving environmental and health problems. We are actively working toward that goal.

Author Index

A

Alekperov, Urkhan ... 361
Ames, Bruce N. ... 663
Andersson, H.C. ... 483
Archer, Michael C. ... 557
Auerbach, Charlotte ... 37
Awa, Akio ... 687

B

Barclay, B. J. ... 121
Bartsch, Helmut ... 577
Baumen, Clifford F. ... 147
Beland, F. A. ... 385
Beranek, D. T. ... 385
Bochkov, Nikolay P. ... 423
Bridges, B. A. ... 47
Bruce, W. Robert ... 557

C

Cabral, J. R. P. ... 231
Calleman, Carl Johan ... 671, 701
Carmella, Steven ... 545
Chouroulinkov, Ivan ... 249
Cooper, Sheldon F. ... 147
Correa, P. ... 565
Cox, Roger ... 157
Cross, Philip K. ... 613
Czeizel, Andrew ... 639

D

De Flora, Silvio ... 527
de Nava, Cristina Cortinas ... 771
de Serres, Frederick J. ... 747
Delehanty, J. ... 81
Dooley, K. L. ... 385
Dusenbery, Ruth L. ... 147

E

Eckardt, F. ... 137
Edmiston, S ... 105
Ehling, U. H. ... 709
Elledge, Stephen J. ... 113
Evans, F. E. ... 385
Evans, H. J. ... 589

F

Falck, Kai ... 323
Fishbein, Lawrence ... 307, 371
Fox, J. G. ... 565
Fujii, Taro ... 399
Furuya, Keizo ... 545

G

Glickman, Barry W. ... 721
Ginsburg, H. ... 105
Göggelmann, W. ... 137
Gold, Lois S. ... 663
Goldberg, Mark T. ... 557

H

Hamilton, Howard B. ... 687
Hansson, K. ... 483
Hartley-Asp, B. ... 483
Havender, William R. ... 663
Hayatsu, Hikoya ... 521
Haynes, R. H. ... 121, 137
Hecht, Stephen S. ... 545
Heflich, R. H. ... 385
Hervers, F. ... 511
Hollaender, Alexander ... 21, 101
Hook, Ernest B. ... 613

I

Inoue, Tatsuo ... 331
Ishii, Kenji ... 505
Ishiwata, Hajimu ... 571

K

Kada, Tsuneo ... 355
Kadlubar, F. F. ... 385
Kamataki, Tetsuya ... 505
Kato, Ryuichi ... 505
Kihlman, B. A. ... 483
Kittrel, C. ... 431
Kouri, E. ... 209
Kraemer, Kenneth H. ... 603
Kunz, B. A. ... 121, 137

L

Latt, S. A. 431
LaVoie, Edmond J. 545
Lawrence, Christopher. 129
Lichtman, Marilyn R. 113
Lee, Myung M. 603
Lim-Sylianco, Clara Y. 759
Likhachev, A. J. 231
Little, J. W. 105
Little, J. G. 121
Loprieno, Nicola 259

M

MacLeod, Michael C. 497
Mansfield, Betty K. 497
Markham, B. 105
Miller, James R. 647
Miller, Robert W. 655
Moriya, Masaaki 331
Mount, D. W. 105
Mulvihill, John J. 625

N

Neel, James V. 687
Newberne, P. M. 565
Nichols, Warren W. 443
Nikbakht, Anne 497
Nomura, Taisei 223

O

Oftedal, Per 679, 745
Ohshima, Hiroshi. 577
Ohta, Toshihiro 331
Otake, Masanori 687

P

Paika, I. J. 431
Palitti, F. 483
Perry, Karen L. 113
Plewa, Michael J. 411
Ponomarkov, V. 231
Powrie, W. D. 347
Preston, R.Julian. 463
Pool, Beatrice Luise 295

R

Ramel, C. 753
Ray, Verne A. 197
Razzouk, Chehab 511
Roberfroid, Marcel, B. 511
Russell, W. L. 59

S

Sahar, E. 431
Sasaki, Masao S. 475
Satoh, Chiyoko 687
Sawyer, Charles B. 663
Schechtman, Leonard M. 209
Schmähl, Dietrich. 295
Schreck, R. R. 431
Schull, William J. 687
Scotto, Joseph 603
Searle, A. G. 169
Selkirk, James K. 497
Shahin, Majdi M. 379
Shirasu, Yasuhiko. 331
Smith, P. Dennis. 147
Smutkupt, Sumin 765
Sobels F. H. 81, 179
Sorsa, Marja 323
Stich, H. F. 347
Sugimura, Takashi. 3

T

Tanimura, Akio 571
Tannenbaum, S. R. 565
Tazima, Yataro 91, 729
Teramoto, Shoji. 331
Tezuka, Hideo 331
Tomatis, L. 231
Troll, Walter 217
Truhaut, René 241

U

Upton, Arthur C. 71

V

Vainio, Harri 323
Viehe, H. G. 511
Vogel, Ekkehart W. 183
von Borstel, R. C. 737

W

Walker, Graham C. 113
Weisburger, John H. 283
Williams, Gary M. 283
Wu, C.H. 347

Y

Yamazoe, Yasushi 505
Yanisch, C. 105
Yosida, Tosihide H. 455

Z

Zeiger, Errol 337
Zetterberg, Gösta 315